RENEWALS 458-4574
DATE DUE

WITHDRAWN
UTSA Libraries

WORKS ISSUED BY
THE HAKLUYT SOCIETY

Series Editors
W. F. Ryan
Gloria Clifton
Joyce Lorimer

FOUR TRAVEL JOURNALS

The Americas, Antarctica and Africa, 1775–1874

THIRD SERIES
NO. 18
(Issued for 2006)

THE HAKLUYT SOCIETY
Council and Officers 2006–2007

PRESIDENT
Professor Roy Bridges

VICE-PRESIDENTS
Dr John Bockstoce
Captain M. K. Barritt RN
Dr Nigel Rigby
Dr Sarah Tyacke CB
Professor Glyndwr Williams

COUNCIL MEMBERS
Dr Jack Benson (co-opted)
Captain R. J. Campbell OBE RN (2004)
Dr Daniel Carey (2007)
Ms Susannah Fisher (2005)
Francis Herbert (2006)
Bruce Hunter (co-opted)
Dr James Kelly (2003)
Jonathan King (2003)
Lionel Knight (2006)
Professor Robin Law FBA (2004)
Anthony Payne (2005)
Royal Geographical Society
  (Dr John H. Hemming CMG)
Dr Joan-Pau Rubiees (2005)
Dr Ann Savours Shirley (2007)
Professor William Sherman (2006)
Dr John Smedley (co-opted)
Professor D. W. H. Walton (2007)
Professor T. Youngs (2004)

CUSTODIAN TRUSTEES
Professor W. E. Butler
Dr John H. Hemming CMG
Dr Sarah Tyacke CB
Professor Glyndwr Williams

HONORARY TREASURER
David Darbyshire FCA

HONORARY SECRETARY AND SERIES EDITOR
Professor W. F. Ryan FBA

HONORARY ASSISTANT SERIES EDITORS
Dr Gloria Clifton   Professor Joyce Lorimer

HONORARY EDITOR ON-LINE PUBLICATIONS
Raymond John Howgego

HONORARY ARCHIVIST
Mrs Margaret Makepeace

ADMINISTRATOR
Richard Bateman
(to whom queries and application for membership may be made)
Telephone: 0044 (0)1428 641 850   E-mail: office@hakluyt.com   Fax: 0044 (0)1428 641 933

Postal address only:
The Hakluyt Society, c/o Map Library, The British Library, 96 Euston Road,
London NW1 2DB, UK

http://www.hakluyt.com

© The Hakluyt Society, 2007

Registered Charity No. 313168   VAT No. GB 233 4481 77

INTERNATIONAL REPRESENTATIVES OF THE HAKLUYT SOCIETY

Australia	Dr Martin Wood, Curator of Maps, National Library of Australia, Canberra, ACT 2601
Canada	Professor Joyce Lorimer, Department of History, Wilfred Laurier University, Waterloo, Ontario, N2L 3C5
France	Contre-amiral François Bellec, 1 place Henri Barbusse, F92300 Levallois, France
Germany	Thomas Tack, Ziegelbergstr. 21, D-63739 Aschaffenburg
Japan	Dr Derek Massarella, Faculty of Economics, Chuo University, Higashinakano 742–1, Hachioji-shi, Tokyo 192–03
Netherlands	Dr Leo M. Akveld, Hammerfestraat 48, 3067 DC Rotterdam
New Zealand	John C. Robson, Map Librarian, University of Waikato Library, Private Bag 3105, Hamilton
Portugal	Dr Manuel Ramos, Av. Elias Garcia 187, 3Dt, 1050 Lisbon
Russia	Professor Alexei V. Postnikov, Institute of the History of Science and Technology, Russian Academy of Sciences, 1/5 Staropanskii per., Moscow 103012
Spain	Carlos Novi, Calle de Vera 26, E-17255 Begur (Girona), Spain *and* 39 Hazelmere Road, Petts Wood, Kent BR5 1PA, England, UK
USA	Edward L. Widmer, The John Carter Brown Library, P.O. Box 1894, Providence, Rhode Island 02912

DONATIONS

The Hakluyt Society gratefully acknowledges donors in 2006–2007
including donors to the American Friends of the Hakluyt Society

The inclusion here of a record of recent financial contributions to the work of the Society is an innovation which we trust members will approve. The need for such contributions has grown inexorably in recent years and Council has thought it right that the generosity of individuals and institutions in this respect should be acknowledged, not just in the relatively ephemeral Annual Report or Newsletter but in the more permanent form of the pages of our published volumes. Certainly, the Society is extremely grateful for the contributions and bequests it has received. These help to make possible the endeavour which Richard Hakluyt himself inspired, and which the Society has continued for 161 years – the endeavour to record, to understand and to interpret the means by which, for better or worse, different regions and different peoples of the world have become connected with one another.

Mr Edward Alsip
Mr Robert C. Baron
Mr Herbert K. Beals
Dr Sanford H. Bederman
Mr Matthew S. Blum
Dr John Bockstoce
Mr Bruce P. Bogert
Mr James Breckenridge
†Mr Brian Bridges
Mr Charles Elk
Dr Norman Fiering
Mr Joseph H. Fitzgerald
Mr Richard H. Float
Mr & Mrs Albert & Mary Fullerton
Mr Martin L. Green
Mr Manuel Guerra
Mr Todd Hanson
Dr William L. Harris
Mr Warren Heckrotte
Mr Paul Herrup
John Carter Brown Library

Mr Stephen A. Kanter
Mr John Levin
Mr John H. Libcke
Mr Caedmon A. Liburd
Mr Stephen F. Lintner
Dr & Mrs Ross D. E. MacPhee
Mr Kenneth MacPherson
Mr William McKinstry
Mr Glen McLaughlin
Mr & Mrs Robert M. Norris
Dr James C. Orcutt
Mr R. David Parsons
Dr Norman C. Peeler
Mr Brian R. Pinto
Dr Hugh Raffles
Mr William S. Reese
Mr & Mrs Curtis & Joan Roy
Professor David Harris Sacks
Mr Robert F. Scholl
Mr Werner Schuele
Mr & Mrs Harold & Michelle Schwab

Mr Neil M. Silverman
Mr Kenneth J. Siple
Mr Stephen A. Skold
Mr David H. Stam
Mr Elmer Templeton
Mr Stuart Thro

Professor Andrew Walls (per the Scottish Institute of Missionary Studies)
Mr F. David Westcott
Mr Peter H. Wood
Mr Edgar L. Weber

The President and Council would like in particular to express their gratitude for a substantial bequest from the estate of the late Mr Raymond Jagger, a member of the Society for over thirty years. As with all donations and bequests to the Society, this will go directly to defraying the costs of publication.

Grants made to assist the publication of specific volumes are acknowledged in those volumes and in the Annual Report.

FOUR TRAVEL JOURNALS

THE AMERICAS, ANTARCTICA AND AFRICA, 1775–1874

Edited by
Herbert K. Beals
R. J. Campbell
Ann Savours and Anita McConnell
Roy Bridges

Published by
Ashgate
for
THE HAKLUYT SOCIETY
LONDON
2007

© The Hakluyt Society 2007

All rights reserved. No part of this publication may be reproduced, stored in a retrieval system, or transmitted in any form or by any means, electronic, mechanical, photocopying, recording or otherwise without the prior permission of the publisher.

Published for The Hakluyt Society by

Ashgate Publishing Limited
Gower House
Croft Road
Aldershot
Hants GU11 3HR
England

Ashgate Publishing Company
Suite 420
101 Cherry Street
Burlington
VT 05401-4405
USA

Ashgate website: http://www.ashgate.com

British Library Cataloguing in Publication Data
Four travel journals: the Americas, Antarctica and Africa, 1775–1874. – (Hakluyt Society. Third series; no. 18)
1. Bodega y Cuadra, Juan de la, 1743–1794 – Diaries 2. Stokes, Pringle – Diaries 3. Hay, Joseph Henry – Diaries 4. Wainwright, Jacob – Diaries 5. Explorers – Diaries 6. Discoveries in geography – History – Sources 7. California – Discovery and exploration – Spanish 8. South America – Discovery and exploration – British 9. Antarctica – Discovery and exploration – British 10. Africa, East – Discovery and exploration – German
I. Beals, H. K. (Herbert Kyle), 1934– II. Hakluyt Society
910.9'22

Library of Congress Cataloging-in-Publication Data
Four travel journals: the Americas, Antarctica, and Africa, 1775–1874 /
edited by Herbert K. Beals [et al.].
p. cm. – (Hakluyt Society, third series; 18)
Includes index.
ISBN 978-0-904180-90-9 (alk. paper)
1. Voyages and travels. 2. America – Discovery and exploration. 3. Antarctica – Discovery and exploration. 4. Africa – Discovery and exploration. I. Beals, H. K. (Herbert Kyle), 1934–
G465.F69 2007
910.9'034–dc22
2007001666

ISBN 978-0-904180-90-9
ISSN 0 904180 90 5

Typeset by IML Typographers, Birkenhead, Merseyside
Printed and bound in Great Britain by
the University Press, Cambridge

CONTENTS OF VOLUME

PART 1

The 1775 Journal of Juan Francisco de la Bodega y Quadra

Translated from the Spanish and edited by Herbert K. Beals   1

PART 2

The Journal of HMS *Beagle* in the Strait of Magellan by Pringle Stokes, Commander RN 1827

Edited by R. J. Campbell   141

PART 3

Journal kept by Midshipman Joseph Henry Kay during the voyage of HMS *Chanticleer*, 1828–1831

Edited by Ann Savours and Anita McConnell   253

PART 4

'A Dangerous and Toilsome Journey'. Jacob Wainwright's Diary of the Transportation of Dr Livingstone's Body to the Coast, 4 May 1873–18 February 1874

Translated from the German and edited by Roy Bridges   329

FOREWORD BY THE PRESIDENT OF THE HAKLUYT SOCIETY

When the Council of the Hakluyt Society decided in 2001 to publish what was initially labelled as a 'Miscellany' volume, it was prompted by the argument that there existed a number of texts featuring interesting and important voyages and travels which were too short to justify volumes to themselves but which might usefully be brought together in a composite publication. Previous volumes issued by the Society have sometimes contained a variety of texts but in all such cases there has been a connection between those texts in terms of theme, period or traveller. In its original conception, all that linked the three texts chosen in 2001 was that they concerned maritime activities. The decision in 2004 to include in the volume a fourth element featuring a land journey certainly removed even that much similarity although naval activity featured in the life of the writer concerned. The reception by readers of a volume containing four relatively short texts introduced by five editors will certainly be awaited with interest by the Council.

I am confident, however, that each of the components of this volume will prove of interest. The material in our recent Malaspina volumes alerted many of us to the importance of the work of Juan Francisco de la Bodega y Quadra on the Pacific coast of North America – a region on which the editor, Herbert Beals, is an acknowledged expert. Captain Campbell's excellent South Shetlands volume of 2000 means that his work is already familiar to members. Here he turns his attention to a hitherto little-known voyage in a famous ship, HMS *Beagle*. Our Vice-president from 2002–2007, Dr Ann Savours Shirley, is well known for her vast acquaintance with Polar exploits while her co-editor, Dr Anita McConnell, another former Council member, has a formidable knowledge of navigation and navigational instruments. Their subject, Midshipman Kay, may be among the youngest travel writers featured by the Society. Whether he was as young as the writer of the fourth text, Jacob Wainwright, is difficult to establish since the latter's true age is unknown. Whether or not he is our youngest featured writer, he is certainly the first black African. Rescued from slavery by the Royal Navy and much affected by and affecting the naval officer Cameron, Wainwright can be argued to have maritime connections, but his contribution is to record a remarkable journey from the centre of Africa to the coast.

What may not be immediately apparent to readers is that a volume with four texts and four introductions quadruples the work of the Society's Series Editor involved. I remark this because I wish to take this opportunity to say how much the Society owes to Professor Michael Brennan, who dealt with this volume in its earlier stages, and to Professor Will Ryan, who has overseen the later stages of preparing the materials for actual publication. This work and all the editorial activity in connection with other projects is unpaid and ensures that our volumes maintain the highest standards of scholarship as well as conforming to the Society's own conventions and formats.

Roy Bridges
June 2007

PART 1

THE 1775 JOURNAL OF JUAN FRANCISCO DE LA BODEGA Y QUADRA

Translated from the Spanish and edited by
HERBERT K. BEALS

CONTENTS

List of Illustrations and Maps	5
Preface	6
Abbreviations	8
Editorial Comments	9
Sources of the Text	9
Variant Spellings and Translations	10
Equivalents and Glossary	12
Eighteenth-Century Spanish Naval Ranks and British Equivalents	12
Weights and Measures	12
Spanish Terms which have been Translated into English	12
Spanish Terms which have not been Translated	13
Eighteenth-Century Spanish Compass Directions and English Equivalents	13
Spanish–English Glossary of Geographical Terms	14
INTRODUCTION	15
Prologue	15
1. The Spanish Lake	16
2. The Manila–Acapulco Trade and Navigation	27
3. English Challenge and Spanish Response	31
4. The 'Sacred Expedition'	41

5. The Bucareli Expeditions	54
6. Bodega y Quadra: *Marinero Limeño*	73

THE EDITED TEXT OF JUAN FRANCISCO DE LA BODEGA Y QUADRA'S JOURNAL — 85

Description of this [Socorro] Island	90
Opinion	92
Description [of Trinidad Natives]	96
Description of the Harbour	98
Departure from Puerto de la Trinidad	99
Departure from the place called Rada de Bucareli	108
Opinion	109
Departure from Puerto de los Remedios	114
Departure from the Entrada and Puerto de Bucareli	119
Departure from this [Bodega] Harbour	124
Departure from Puerto de Monterey	125

APPENDIXES

1. Extract from Francisco Antonio Mourelle's 'Navegación … 1775' for 14 July	127
2. The Schooner *Sonora*	131
Bibliography	133
Index	385

LIST OF ILLUSTRATIONS AND MAPS

Figure 1: Oil portrait by Julio García Condoy of Juan Francisco de la Bodega y Quadra Mollinedo (1744–94). Reproduced by permission of the Museo Naval, Madrid. Bodega is posed before the backdrop of Nootka Sound on the west coast of Vancouver Island. The scene depicts a twin-masted *goleta* 'schooner' with a native canoe nearby. At the time of his death, Bodega held the rank of *capitán de navío* equivalent to a post captain in the eighteenth-century British Royal Navy. 73

Figure 2: Anonymous oil portrait of Francisco Antonio Mourelle de la Rúa (1750–1820). Reproduced by permission of the Museo Naval, Madrid. Mourelle was second in command under Bodega aboard the schooner *Sonora* on her 1775 voyage. He was among the first European mariners to reconnoitre the entire NW coast of America up to 62°N latitude. He is credited with discovery of the Vava'o Group of the Tonga Islands. Upon his death he held the rank of *jefe de escuadra* equivalent to the rank of rear admiral in the eighteenth-century British Royal Navy. 127

Map 1: Voyage of the *Sonora*, 1775: From San Blas, Mexico, to Sea Lion Cove, Alaska and return. 87

Map 2: Olympic Coast, Washington: Destruction Island to Cape Elizabeth. 102

Map 3: Cape Elizabeth Bight: Cape Elizabeth to Point Grenville. 105

Map 4: Kruzof Island, Alaska: Mount Edgecumbe to Sea Lion Cove. 111

Map 5: Bodega's Map of Sea Lion Cove, Puerto de los Remedios. 113

Map 6: Bucareli Bay and Dixon Entrance. 116

Map 7: Bodega's Map of Bucareli Bay, Entrada o Puerto de Bucareli. 118

Map 8: Bodega's Map of Bodega and Tomales Bays, Puerto del Capitán Bodega. 124

PREFACE

At a time when the Hakluyt Society has recently issued an English-language edition of Alejandro Malaspina's Journal it seems fitting to issue a translation of an account of another Spanish expedition sent to explore the north-west coast of America sixteen years earlier in 1775. Compared to the Malaspina expedition, it was small scale and singularly ill-equipped for the challenge at hand. But it was a pioneering effort, in which great adversity was overcome and significant contributions made towards unlocking the mysteries of what was among the last temperate coastlines to be explored by European mariners. Two other explorers, Aleksei Chirikov (1741) and Juan Pérez (1774), had made earlier fully documented visits to the north-west coast. Another even earlier voyage, that of Sir Francis Drake (1577–80), probably reached and examined various parts of the north-west coast – but exactly which parts continues to be the subject of much controversy.

The voyage with which this work is concerned was neither a scientific expedition per se nor a rival in size or expense of later expeditions such as those of Captain James Cook (1776–80), Jean-François de Galaup de la Pérouse (1786–88), Alejandro Malaspina (1789–94), or George Vancouver (1791–95). It was under the command of Senior Lieutenant Bruno de Hezeta aboard the flagship, a frigate named *Santiago*, and escorted by a small schooner named *Sonora* captained by Junior Lieutenant Juan Francisco de la Bodega y Quadra. They had been ordered north in March 1775 by the viceroy of New Spain, Antonio María Bucareli, to complete what Pérez had set out to do the previous year – with mixed if not disappointing results. As with other eighteenth-century exploring voyages, expectations sometimes outpaced what could realistically be achieved in the course of a single voyage. Pérez had been instructed by Bucareli to reach 60°N latitude, search for evidence of Russian or other foreign activity, and go ashore wherever possible to lay formal claim to the coast in the name of the Spanish Crown. And as time and circumstance allowed, he was also to make an assessment of the region's natural resources and report on the characteristics of its indigenous population. It was a tall order for a single vessel not particularly well suited for exploration and lacking the assistance of a smaller-sized consort. Pérez returned south after reaching 54°40′N latitude; there would be no landings; no formal claims; only a general reconnaissance.

If the Pérez expedition of 1774 was sent on a 'mission impossible', prospects of the Hezeta-Bodega voyage of 1775 were scarcely much better – especially considering that the latitude it was expected to reach was increased from 60°N to 65°N. However, a change in leadership and the addition of an escort vessel made the task somewhat less daunting. This follow-up voyage had its disappointments too: it fell short of reaching 65°N latitude, forced to turn back two or three minutes south of 58°N – still an improvement over Pérez. As captain of the escort schooner *Sonora*, Juan Francisco de la Bodega y Quadra was determined to do everything in his power to meet as many of Bucareli's unrealistic expectations as possible. The 1775 expedition succeeded in going ashore in four locations – one in northern California, another in Washington State, and two in south-east Alaska – where formal possession-taking rituals were performed. Moreover, Hezeta and the *Santiago* succeeded in locating the mouth of the Columbia River. Maps of each of these localities were produced, together with an overall chart of the coast from Monterey Bay, California, north to Kruzov Island in south-east Alaska. It is something of a miracle that a majority of the officers and men of these voyages survived at all.

Up and down the chain of command, from viceroy to planners and participants in these Spanish voyages, all would come to realize that the coast above the forty-eighth parallel was a formidable, complex maze of intricate off-shore islands and channels, which would take years – and the mariners of several nations – to eventually accomplish a reasonably accurate cartographic depiction. The 1775 Journal of Bodega y Quadra is, as here presented in English translation, an eyewitness account of one of the more important early steps in that direction.

As I reflect back over the years spent understanding the events of Bodega y Quadra's involvement in the exploration of the North-west and Alaskan coasts, the names of many persons come to mind. They have assisted me in various ways, large and small. In the event that I may have overlooked someone of importance, I offer my apologies. I am especially indebted to my Hakluyt Society editors who have contributed so much to the publication of this work. In any event, the persons listed here are a good sampling of those who have been of help to me. (Canada) Greg Foster, Robin Inglis, John Kendrick and Freeman Tovell. (Quinault Nation) Francis Mason and Oliver Mason. (Spain) Salvador Bernabeu Albert, Dolores Higueras Rodríguez, Luisa Martín-Merás, Antonio Menchaca and Mercedes Palau. (United Kingdom) Michael Brennan, Andrew Cook, Andrew David, Robin Law and Will Ryan. (United States) Donald Cutter, Elizabeth Crownhart-Vaughan, Iris Engstrand, David Kroenke and Thomas Vaughan.

ABBREVIATIONS

ABC	Archives of British Columbia, Victoria, BC, Canada
AGI	Archivo General de Indias, Seville, Spain
AGN	Archivo General de la Nación, Mexico City, Mexico
AGS	Archivo General de Simancas, Valladolid, Spain
AHA	American Historical Association
BCHQ	*British Columbia Historical Quarterly*
BL	British Library, London
BLib	Bancroft Library, Berkeley, Calif.
CHM	*California History Magazine*
CHQ	*California Historical Quarterly*
CHR	*Catholic Historical Review*
GB	*Geographic Bulletin*
GM	*Geographical Magazine*
JIE	*Junto de Iconografía Español*
JPH	*Journal of Pacific History*
JW	*Journal of the West*
MC	*Map Collector*
MN	Museo Naval, Madrid, Spain
ND	*Northwest Discovery*
OHQ	*Oregon Historical Quarterly*
OHS	Oregon Historical Society, Portland, Ore.
PHR	*Pacific Historical Review*
PNQ	*Pacific Northwest Quarterly*
SHQ	*Southwestern Historical Quarterly*
TIJHD	*Terrae Incognitae, Journal for the History of Discoveries*
USNA	US National Archives
USNIP	*US Naval Institute Proceedings*
UWPA	*University of Washington Publications in Anthropology*
VISO	Archivo-Museo D. Alvaro de Bozán Viso del Marques, Ciudad Real, Spain
WHQ	*Washington Historical Quarterly*
WestHQ	*Western Historical Quarterly*

EDITORIAL COMMENTS

Sources of the Text

The original holographic text of Bodega's 1775 Journal or 'Diario' is in the Archivo General de la Nación, Mexico City.[1] Comparing it with other writings known to be in Bodega's own hand, this document was almost certainly written by him and not a copyist.[2] Another nearly identical text of the 1775 Journal, differently titled and written in a hand other than Bodega's, is in the Museo Naval, Madrid. The Spanish text of this document, referred to here in abbreviated form as 'Navegación',[3] was first published in *Colección de diarios y relaciones para la historia de los viajes y descubrimientos* (Madrid, 1943).[4] Subsequently, in an effort to correct printing errors and lines omitted in the transcription from the manuscript to its publication in the *Colección*, the original Spanish text was republished in a work titled *Juan Francisco de la Bodega y Quadra: El descubrimiento del fin del mundo* (Madrid, 1990), edited by Salvador Bernabéu Albert.[5] These are the main sources drawn upon in preparing the English translation of Bodega's 1775 Journal presented in this work.

Several other versions of Bodega's account of the *Sonora*'s 1775 voyage exist, which were written at various times after the expedition's return to the naval base at San Blas, Mexico. One example, 'Primer Viaje hasta la altura de 58° en un goleta de 18 codos de quilla … 1775',[6] is the Spanish text published in *Anuario de la Dirección de Hidrografía*

---

[1] The full Spanish title is: 'Año de 1775. Diario de la exploración hecha de las Costas septentrianales de California hecha D. Juan Francisco de la Bodega y Quadra en la goleta Sonora'. Or in English translation: 'Year of 1775. Journal of the Exploration made on the Northern Coasts of California undertaken by D. Juan Francisco de la Bodega y Quadra in the Schooner *Sonora*'. Archivo General de la Nación, Hist. 324, Mexico City.

[2] This is evident in comparing the handwriting of the original 1775 'Diario' MS with Bodega's self-compiled record of his 'Commands and Assignments …' written and signed by him at Havana, Cuba, 8 January 1784. See Bodega's personal file, Archivo-Museo Don Alvaro de Bozán, Viso del Marques, Ciudad Real.

[3] The full Spanish title is: 'Navegación hecha por Dn. Juan Francisco de la Bodega y Quadra teniente de fragata de la Real Armada y comandante de la goleta *Sonora*; a los Descubrimentos de los mares y costa septentrional de California. Año de 1775'. Or in English translation: 'Sea voyage made by Dn. Juan Francisco de la Bodega y Quadra, junior lieutenant of the Royal Navy and commander of the schooner *Sonora*; for discoveries on the seas and coast of northern California. Year of 1775.' Museo Naval, MS 622, Madrid. Copies of this MS also exist in the Archivo General de Indias, Seville, and in the Archivo Historico Nacional, Madrid. These manuscripts are accompanied by three large-scale maps of Puerto de Bucareli (Bucareli Bay), Puerto de los Remedios (Sea Lion Cove), and Puerto del Capitán Bodega (Bodega Bay).

[4] *Collection of Journals and Narratives for the History of Voyages and Discoveries*, 2, Madrid, 1943, pp. 102–33

[5] *Juan Francisco de la Bodega y Quadra: Discovery at the Ends of the Earth*, Madrid, 1990, pp. 55–110. In addition to Bodega's 1775 'Diario' this publication also includes the Spanish texts of his journals for the 1779 and 1792 expeditions to the north-west coast.

[6] 'First Voyage to Latitude 58° in a Schooner of 18 *codos* of Keel Length … 1775.'

(Madrid, 1865).[1] This version gives much the same account as the original 'Diario' and 'Navegación', but the text has been edited and in some cases rewritten with a view to its publication. An unpublished English translation of the 'Primer Viaje' made by G. F. Barwick, 1912, is in the Archives of British Columbia, Victoria, BC, Canada. Another example, in the Museo Naval, Madrid, identified here by the shortened title 'Comento',[2] includes accounts of both the 1775 and 1779 voyages brought together in an elaborate leather binding. These and other such accounts have not figured significantly in the translation of Bodega's 1775 Journal in this work.

Variant Spellings and Translations

In 1775, when Bodega y Quadra wrote his Journal, the spelling of Castilian Spanish had not been standardized. For example, the letters *B* and *V* were sometimes used interchangeably, as were also the letters *C* and *Q*, with the result that Bodega y Quadra is sometimes written Vodega y Cuadra. Wherever this occurs in the Journal's Spanish text, the letters *V* and *C* have been silently emended in the English translation to *B* and *Q*, respectively. In Canada and Great Britain, Bodega y Quadra has generally been known by his mother's surname: Quadra. But Spanish, Mexican and American historians favour Bodega y Quadra or simply Bodega.[3] The latter approach has come to prevail and is followed in this work. Another minor spelling issue has arisen over the surname of the commander of the 1775 expedition, Bruno de Hezeta. In his own signature he employed a *z*, but in the mid-nineteenth century a spelling reform replaced *z*'s with *c*'s, changing his surname to read Heceta. In recent times, however, scholars have come to prefer his original spelling which now generally prevails. The surname of Bodega's second officer aboard the *Sonora* has also been the source of some confusion. Although his surname has appeared in manuscripts and print variously as Maurelle, Maurell, or Morelle, the preferred spelling is now generally agreed to be Mourelle – the spelling adhered to in this work.[4]

Two other variant forms of names need explanation. The full name of the viceroy of New Spain during Bodega's second tour of duty as commandant of the Department of San Blas (1789–93) was Juan Vicente de Güemes Pacheco de Padilla, Conde de Revillagigedo (viceroy 1789–94). Be it noted that the last name in his title appears as a single word. In other mostly earlier instances, however, this name is divided into two words: Revilla Gigedo. The origin of this difference seems traceable to a time three

---

[1] *Annual Report of the Hydrographic Office*, 3, Madrid, 1865, pp. 279–94; this publication also includes a Spanish text of Bodega's account of the 1779 expedition, pp. 295–331.

[2] 'Comento de la Navegación y descubrimientos hecho en dos Viajes de Orden de Su Majestad en la Costa Septentrional de California desde la Latitud de 21 grados 30 minutos en que se halla el Departamento de San Blas, por Dn. Juan Francisco de la Bodega y Quadra, del Orden de Santiago y Capitán de Navío de Real Armada'. Or in English translation: 'Commentary on the Navigation and Discoveries made in two Voyages by Order of His Majesty on the Northern Coast of California, from Latitude 21 degrees, 30 minutes, in which the Department of San Blas is located, by Dn. Juan Francisco de la Bodega y Quadra, of the Order of Santiago, and Post Captain of the Royal Navy'. Museo Naval, MS 618, Madrid.

[3] Although there are exceptions, Spanish surnames usually consist of the father's name preceding the mother's. If a person prefers a single surname it normally is the father's.

[4] See Landin, *Mourelle de la Rua*, p. 8, n. 2.

decades earlier when the viceroy of New Spain was named Francisco de Güemes y Horcasitas, Conde de Revilla Gigedo (viceroy 1746–55), with the final name in his title divided into two words. This is obviously not the same viceroy who served during Bodega's time, and for reasons not entirely clear the single-word version came to prevail in spelling the latter viceroy's name – as adhered to in this work.[1] Another issue concerns the name of Lima's port city. In a Spanish-language context the name is *el Callao*, but in an English context the definite article *el* is usually omitted, as is the case in this work.

In the course of translating Bodega's eighteenth-century Spanish into modern English several issues were encountered as to what might be the best or most appropriate rendering. Some Spanish words have no clear counterpart in English; as for example, when the Spanish naval rank of *alférez* is compared to eighteenth-century British naval ranks one finds nothing exactly comparable. However, modern Spanish–English or English–Spanish dictionaries equate *alférez* with the term 'ensign', a rank somewhere between *guardiamarina* 'midshipman' and *teniente de fragata* 'junior lieutenant'. The translation of Bodega's 1775 Journal in this work employs 'ensign' as the closest English word to *alférez*. Another Spanish term, *piloto*, is often translated in other works as 'pilot' or italicized and not translated. Yet it seems well within the meaning of what in eighteenth-century British naval usage was called 'master' or 'sailing master', namely: the officer charged with the day-to-day operation of the ship and in particular the vital matter of navigation. A ship's captain could also be the master, especially on smaller vessels. In any event, in this work *piloto* has been translated simply as 'master'. In addition, other Spanish naval designations are related in some way to this term, such as *secundo piloto*, 'second master', *pilotín*, 'master's mate', and *piloto práctico*, 'coast or harbour pilot'.[2]

---

[1] See Bannon, *Spanish Borderlands Frontier*, pp. 175, 202.
[2] O'Scanlan, *Diccionario Marítimo Español*, p. 420 (editor's trans.).

EQUIVALENTS AND GLOSSARY

Eighteenth-Century Spanish Naval Ranks and British Equivalents

Capitán General	Admiral
Teniente General	Vice Admiral
Jefe de Escuadra	Rear Admiral
Brigadier de Real Armada	Commodore
Capitán de navío	Post Captain
Capitán de fragata	Commander
Teniente de navío	Senior Lieutenant
Teniente de fragata	Junior Lieutenant
Alférez de navío	[Senior Ensign]*
Alférez de fragata	[Junior Ensign]*
Guardiamarina	[Midshipman]*
Capellán	Chaplain
Cartógrafo	Cartographer
Piloto	Master
Secundo piloto	Second master
Pilotín	Master's mate
Piloto práctico	Coast or harbour pilot
Contramaestre	Boatswain
Secundo contramaestre	Boatswain's mate

* No exact equivalent

Weights and Measures

Spanish terms which have a direct or close English equivalent have been translated into English; terms for which there does not appear to be a direct or close equivalent have not been translated.

Spanish Terms which have been Translated into English

braza	5.48 feet = one fathom
cable	one-tenth of a nautical mile = one cable
legua	three nautical miles = one league
linea	one-twelfth of a *pulgada* = one line (an obsolete English term for one-twelfth of an inch)

milla náutica	the internationally accepted unit of distance at sea, one-sixtieth of a degree of latitude = one nautical mile (6,076 feet)
nudo	the internationally accepted unit of speed at sea of one nautical mile an hour = one knot
pie	0.91 feet = one foot
pulgada	0.91 of an inch = one inch
vara	33 inches = one yard

Spanish Terms which have not been Translated

arroba	approximately 25 lb
codo	18 to 22 inches (cubit)
toesa	approximately 6.4 feet (the French *toise*)
quintal	approximately 102 lb

Eighteenth-Century Spanish Compass Directions and English Equivalents

Spanish	*English*	*Spanish*	*English*
N¼NE	N by E	S¼SW	S by W
NE¼N	NE by N	SW¼S	SW by S
NE¼E	NE by E	SW¼W	SW by W
E¼NE	E by N	W¼SW	W by S
E¼SE	E by S	W¼NW	W by N
SE¼E	SE by E	NW¼W	NW by W
SE¼S	SE by S	NW¼N	NW by N
S¼SE	S by E	N¼NW	N by W

Spanish compass bearings and wind directions were based on a compass rose divided into eight *vientos* 'winds' (N, NE, E, SE, S, SW, W, NW), which were further divided into eight *medios vientos* 'half winds' (NNE, ENE, ESE, SSE, SSW, WSW, WNW, NNW); which in turn were divided into 16 *cuartas* 'quarter winds' (N¼NE, NE¼N, NE¼E, et cetera, as cited above). This divided the compass rose into 32 points of 11¼ degrees each. Another system, which was coming into use in Bodega's time, divided the compass rose into four numbered quadrants: First or NE; Second or SE; Third or SW; and Fourth or NW. Each quadrant was in turn divided into 90 degrees, with N and S at zero degrees.

Spanish–English Glossary of Geographical Terms

Spanish word	English meaning	Spanish word	English meaning
alto/a	height, upper	*estero*	creek, inlet
archipiélago	archipelago	*farallón(es)*	needle-like rock(s)
arena	sand	*golfo*	gulf
arrecife	reef	*isla*	island
arroyo	stream	*islote*	small island, islet
bahía	bay	*lengua*	tongue, language
bajo/a	shoal, low	*monte*	mountain
banco	bank	*nuevo/a*	new
boca	mouth	*piedra*	rock, stone
cabo	cape	*playa*	beach
camino	road	*península*	peninsula
canal	channel	*pueblo*	town, village
casa	house	*puerto*	port, harbour
castillo	castle	*punta*	point
cerro	hill, peak	*río*	river
cerrillo, cerrito	small hill, hillock	*roca*	rock
cordillera	mountain range	*sierra*	mountain range
ciudad	city	*tierra*	land
ensenada	cove, inlet, small bay	*torre*	tower
estrecho	strait	*volcán*	volcano

INTRODUCTION

Prologue

Among the rocky capes and headlands on the Olympic coast of Washington State, there is one that guards a peculiar and sombre secret. It was once called by its Spanish discoverers Punta de los Mártires (Point of the Martyrs), suggestive of what happened there over two centuries ago. All along this rugged coast, fog can be pervasive, enveloping the sea and land alike in a dreary, grey pall. And so it was on the morning of 14 July 1775, when seven Spanish seamen met their deaths attempting to go ashore in the lee of a point that is today called Cape Elizabeth, near the mouth of the Quinault River. The tragic fate of these men at the hands of the indigenous Quinault people, and the same unhappy fate suffered by six or seven of their own later in the incident heighten the drama of this first encounter of two greatly different cultures. But most such early meetings of Europeans and north-west coast Indians would prove rather more amicable, even if only guardedly so, and the reasons for the Quinaults' hostility in this particular case remain to this day difficult to explain.

Let no one think that the sailors who died that morning had any illusions about the dangers they faced on the open ocean and among the treacherous inshore reefs along this section of coast. Their officers combined with hard experience had schooled the crew well in the requirements of survival amidst the ocean's impersonal forces. But ironically, they were overwhelmed on terra firma by what can be the most dangerous and unpredictable factor in all creation: one's fellow human beings. Aroused to hostility by some real or imagined slight, or frightened by a perceived though unintended threat, or for reasons now imponderable so many years after the fact, they can commit acts that would otherwise be unthinkable.

For the victims of this small-scale clash, their hardships ended with sudden finality, while for their surviving shipmates, the torment and testing had only begun. A gruelling, protracted struggle with the very waters on which they sailed lay ahead, and there would come a time when they may have envied the fate of the seven who died martyrs' deaths on that remote and ill-starred beach. But when the surviving officers and men eventually returned to their home port at San Blas, Mexico, they had achieved what surely ranks among the more remarkable feats of navigational skill and daring in the history of North American coastal exploration. Their exploits not only served to redraw charts of the north-west coast, but the determination they exhibited in overcoming adversity remains to this day a compelling example of courage and resourcefulness – or as their captain, Juan Francisco de la Bodega y Quadra, expressed it, 'a glory for posterity'.[1]

---

[1] See below, p. 100.

By a curious turn of events, a copy of an account of this voyage kept by Francisco Antonio Mourelle, Bodega's second in command, was somehow obtained and spirited to London, where it was translated into English and published in 1781. This was not an ordinary occurrence, because most European nations then engaged in maritime exploration were not eager to have detailed records of their various expeditions given unauthorized public exposure. Whatever the reasons for the lapse of Spanish security in this instance, the Englishman who translated and published the document, Daines Barrington, jurist, antiquarian and publisher, could not help but admire the steadfast courage of the purloined journal's author. He observed that Mourelle 'seems to have been a most diligent navigator; whilst, to his honour, he always advises the proceeding to as high a northern latitude as possible, though some of his brother officers almost despair'.[1]

Because of the early translation and publication of Mourelle's journal, the 1775 voyage of the *Sonora*, a small, cramped, twin-masted *goleta* 'schooner' has become better known among English-speaking audiences than most if not all of the various Spanish voyages to the north-west coast of America in the eighteenth century. The *Sonora* was in fact only the escort vessel for a larger flagship, the frigate *Santiago*, captained by Bruno de Hezeta, the expedition's overall commander. Whatever the *Sonora*'s achievements were, the early publicity they received distorted historical understanding of the expedition. The eminent scientist and world traveller Alexander von Humboldt, for example, credited Bodega and the *Sonora* with locating the mouth of the Columbia River, when in fact that honour belonged to Hezeta and the expedition's flagship.[2] Another problem with Barrington's translation of Mourelle's journal concerns its errors in detail and excessive reliance on paraphrasing. So, while readers may think they have a reliable narrative of the *Sonora*'s sea venture rendered into English, the fact is otherwise. It also seems that the availability of Barrington's translation, whatever its errors and ambiguities, has helped to impede efforts to translate and publish the full text of Bodega's account of the schooner's voyage.

The leadership, courage and navigational skills of these two resourceful Spanish naval officers are such that both of their journals deserve to be available to a much wider readership than has hitherto been the case. The translation and annotation included in this work are offered in the hope that the glory Bodega, Mourelle and the *Sonora*'s crew earned and bequeathed can be better appreciated by an English-speaking posterity.

1. The Spanish Lake

What had brought these mariners to a part of the globe so remote from their Iberian homeland? It began with an agreement reached by Spain and Portugal in the small Castilian town of Tordesillas, scarcely two years after Columbus's famous voyage of discovery in 1492. This small yet historic place is located some 140 km north-west of Spain's present-day capital of Madrid and within twenty-eight km of Valladolid, a

---

[1] Mourelle, *Voyage of the* Sonora, p. 8.
[2] See Humboldt, *Political Essay on the Kingdom of New Spain*, 2, p. 365; and Beals, *For Honor and Country*, p. 41.

former Castilian capital. It witnessed the signing of a treaty that essentially divided the world into two mutually exclusive hemispheres of influence for the two Iberian maritime powers, complete with the blessings of Pope Alexander VI. By this agreement, the Spanish Crown was given sway over newly discovered lands westward of a meridian 370 leagues (about 1,100 nautical miles) west of the Cape Verde Islands. As for Portugal, its part of the bargain lay eastward of the demarcation line, with undiscovered lands there to come under Portuguese control. The signatories to this treaty, including its papal sponsor, affixed their names to the document hoping the two countries could thereby avoid wasteful conflict in their rush to establish overseas colonies in a world of exciting and rapidly expanding geographic discoveries.

Samuel Eliot Morison observed that the Treaty of Tordesillas 'inaugurated a period of friendly and intimate relations between Spain and Portugal, which made secrecy unnecessary to protect Portuguese interests in the Atlantic'.[1] Although it was not recognized at the time, the treaty had a beneficial potential for Spanish interests in an ocean that had yet to be discovered by Europeans. The Tordesillas agreement also held another surprise when it was learned that part of the South American mainland lay eastward of the line, placing it in the Portuguese zone. These lands, which were discovered in the spring of 1500 by the Portuguese mariner Pedro Alvares Cabral, came to be called Brazil.[2]

From a Spanish perspective, however, the exploit of an obscure Extremaduran named Vasco Núñez de Balboa, thirteen years later on the Isthmus of Panama proved to be of the highest importance. On 25 September 1513, this *conquestador* – or 'popular desperado' as one historian has styled him[3] – sighted a vast and shimmering oceanic expanse from the summit of a Panamanian mountain overlooking the Isthmus' west coast. Within four days, Balboa and his army were on the shores of this water body, which seemed to stretch away endlessly to the south, and for which reason it was dubbed Mar del Sur or the South Sea. Without knowing the ocean's true immensity, Balboa, his sword in hand, waded boldly into its warm, tropical surf, claiming it and all the shores it washed upon for his Castilian sovereigns.[4] Glory for Balboa, however, would prove fleeting. Within a year of the discovery that made him famous, the Spanish Crown, in whose service he had tried to labour but without formal authority, saw fit to depose him as *de facto* governor of the settlement he had helped found, Santa María de la Antigua de Darien. His replacement, a court favourite named Pedro Arias de Avila (also spelled Pedrárias Dávila), soon had the discoverer of the South Sea arrested, following which he was tried and convicted of treason. His execution in 1517 was by public beheading.

Once it was learned with certainty that another ocean lay beyond the lands Columbus and his successors had found, efforts were sparked to find a waterway through them to reach the South Sea. The Spanish Crown had a particularly strong

---

[1] Morison, *Portuguese Voyages to America in the Fifteenth Century*, p. 90.
[2] See Morison, *European Discovery*, pp. 210–29.
[3] See Parry, *Spanish Seaborne Empire*, p. 52
[4] At the time of Balboa's discovery, Castile's Isabella had been dead for nine years, nominally replaced by her daughter, Juana. But her father, Ferdinand of Aragon, was effectively head of the Spanish state, because the new queen of Castile proved to be mentally incompetent, earning her the sobriquet 'La Loca' ('the Mad'). See Chapman, *History of Spain*, pp. 208–9; Morison, *European Discovery*, pp. 200–205; Beaglehole, *Exploration*, pp. 8–9, 15; and Parry, *Spanish Seaborne Empire*, pp. 52, 60.

interest in this because, in addition to the claim Balboa had staked for Spain, this new ocean would seem to fall entirely west of the demarcation line, placing it squarely on the Hispanic side. What remained unknown was the new ocean's breadth, and whether the much-coveted Moluccas (also known as the East Indies or Spice Islands) fell in Spain's zone, if the Atlantic demarcation line were extended to the opposite or Pacific side of the globe.

By one of history's supreme ironies, the man destined to uncover the first indications of the true expanse of Balboa's South Sea was a Portuguese defector to Spain. In his Lusitanian homeland, he was called Fernão de Magalhães; Spaniards call him Hernando (or Fernando) de Magallanes; and in English, his name becomes Ferdinand Magellan. Whatever form his name may take, he is clearly the most important maritime discoverer after Columbus, and only the moon landings in our own time can be considered to have overshadowed his singular achievement.

As with Columbus, Magellan held certain erroneous geographic opinions that helped him to convince the Spanish Crown – not to mention himself – of the efficacy of another westward-sailing expedition to reach the Moluccan islands. Magellan, although less so than Columbus, underestimated the earth's circumference.[1] He was also wrong in believing that those always-beckoning, spice-rich isles lay in the Spanish zone established by the Tordesillas demarcation line, as extended from the Atlantic to the Pacific hemisphere.

In another article of his geographic faith, however, Magellan was spectacularly correct. He was convinced that somewhere south or south-west of the Brazilian lands discovered by Cabral, a strait existed offering passage through the continental land mass, or that the continent itself ended in much the same fashion as Africa at the Cape of Good Hope. Magellan established the existence of the passage, which has borne his name ever since, but he died without learning that he was correct about the latter assumption as well.

The expedition's departure from Seville in August 1519 was a sizeable spectacle, with banners and pennants snapping in the breeze from the masts of Magellan's five vessels. 'Discharging many pieces of artillery', writes Italian gentleman-volunteer Antonio Pigafetta, 'the ships held their fore staysails to the wind, and descended the river Betis, at present called Gadalcavir [Guadalquivir]'.[2] They slipped past the city's famed Torre del Oro (Golden Tower) on their way down the river's course to Sanlúcar on the Atlantic shore.

Once on the open ocean, the expedition steered south-west past the Canary and Cape Verde Islands, then on to Sierra Leone on the Africa coast. From there, Magellan

---

[1] According to Morison, *Admiral of the Ocean Sea*, p. 68, Columbus believed that the distance between the Canary Islands and Cipangu (Japan) was some 2,400 nautical miles, while the actual distance is no less than 10,600 nautical miles. This misconception, according to Columbus's second son and first biographer, Ferdinand, arose primarily from a faulty reading of computations by Muslim astronomer and geographer Alfragon (al-Farghāni) of the late 9th and early 10th centuries. See Keen, ed., *Life of the Admiral Christopher Columbus*, p. 16. There is less certainty about Magellan's beliefs concerning the earth's circumference. Being aware of the Atlantic's breadth as revealed by Columbus and other explorers, Magellan realized his voyage would require traversing more degrees of longitude than any previous westward-sailing European had achieved. Nevertheless, he had little idea of how vast an undertaking he faced to reach the Spice Islands. See Morison, *European Discovery*, p. 336; and Nowell, *Magellan's Voyage*, pp. 36–7.

[2] Ibid. p. 89.

crossed less than a thousand nautical miles to reach the South American mainland, a passage almost minuscule compared to the immense crossing that would later confront him. Before he was able to break out of the Atlantic, a serious mutiny had to be quelled, and he would suffer the loss of two ships. One, the smallest, a caravel named *Santiago*, ran aground on the Patagonian coast and had to be abandoned; the other, a larger vessel called *San Antonio*, was lost to a mutinous crew, which deserted the expedition and returned to Spain.

Despite his various troubles, Magellan saw his three remaining vessels, the flagship *Trinidad*, the *Concepción* and *Victoria*, successfully negotiate the labyrinthine passage separating Patagonia on the north from Tierra del Fuego on the south.[1] As the expedition neared the western terminus of the newly discovered Estrecho de Magallanes (Strait of Magellan), a boat was sent ahead to determine if the waters near a cape visible in the distance did indeed open on the great ocean Balboa had called the Mar del Sur.

The men returned within three days, notes Pigafetta and reported that they had seen the cape and the open sea. The captain-general wept for joy, and called that cape 'Cape Dezeado', 'for we had been desiring it for a long time'.[2]

On 28 November 1520, the expedition's three remaining ships left the strait astern, heading north, then north-west, into an ocean that Pigafetta began to call the 'Mar Pacifico'. As he later wrote: 'In truth, it is very pacific, for during that time [crossing it] we did not suffer any storm'.[3] And so, these Europeans – mostly from the Iberian Peninsula, and under the banners of Castilian Spain – entered upon the South Pacific Ocean, little knowing how vast an open expanse of blue water confronted them before they would again set foot on the green good earth of land.

After veering away from the South American coast, Magellan's flotilla sailed toward the setting sun, on a course trending west-north-west, then north-west, and eventually due west. In three month's time, only two small uninhabited atolls were encountered, neither of which offered much relief for the expedition's progressively ailing crews.[4] By chance, Magellan had set his course on a route extending for nearly nine thousand nautical miles before he would come upon any significant islands. For this entire span, they somehow endured without fresh food, except for fish or birds they may have caught.

'We ate biscuit', says Pigafetta, 'which was no longer biscuit, but powder of biscuits swarming with worms, for they had eaten the good ... We drank yellow water that had

---

[1] The name Tierra del Fuego 'Land of Fire' apparently arose from circumstances described in an account of Magellan's voyage attributed to one Maximilian of Transylvania. After noting that they had seen no humans on shore, he writes: 'But one night a great number of fires were seen, mostly on their left hand, from which they guessed that they had been seen by the natives of the region'. Ibid, p. 291.

[2] Ibid. p. 117. The cape Magellan was so pleased to reach is now called Cape Pilar. The name Dezeado (or Deseado, as it is spelled today) has migrated south-west to another cape nearby. Both capes are at the north-west extreme of Desolation Island. See Morison, *European Discovery*, pp. 386, 392.

[3] Nowell, *Magellan's Voyage*, p. 123.

[4] These two atolls have been identified by Morison, *European Discovery*, pp. 410–12, as present-day Puka Puka in the Tuamotu archipelago, south of the Marquesas Islands, and Caroline Atoll, a little less than 700 nautical miles to the west. Collectively, the two islands were called by the Spaniards Islas Infortunadas 'Miserable Islands'. Individually, the island he identifies as Puka Puka, Magellan called San Pablo; he dubbed the second one Isla de los Tiburones (Island of Sharks) because of the numerous sharks that swarmed thereabouts.

been putrid for many days. We also ate some ox hides that covered the top of the mainyards ... We ate sawdust from boards. Rats were sold for one-half ducado apiece.'[1]

Not surprisingly, many crew members began experiencing the ominous symptoms of scurvy: swollen, painful gums and aching extremities. Pigafetta informs us that nineteen of the twenty-five to thirty men who were beset by the illness died from it, although he himself escaped the affliction.[2]

Countering their miseries, Magellan and his men had the good fortune to be favoured by strong south-east – and later, above the equator, north-east – trade winds, so that their ships were sent scudding westward, free of storms, at what must have been a most respectable pace. On 6 March 1521, the expedition came into view of what appeared to be three substantial islands, not mere atolls. They were most likely Guam and its smaller neighbour to the north, Rota, the latter of which can appear to be two islands at a distance.[3] Turning south-westward, Magellan sought an anchorage on the west or lee side of the larger island, where he hoped to replenish his water supply and obtain fresh food.

The expedition's stay at Guam, however, did not prove to be as long or beneficial as Magellan had hoped. Before anyone could be sent ashore, the islanders swarmed about and even boarded the ships. 'They pilfered', says Pigafetta, 'whatever they could lay their hands on ... The men were about to strike the sails so that we could go ashore, but the natives very deftly stole from us the small boat that was fastened to the poop of the flagship'.[4]

Angered by this, the captain-general led a party of forty men ashore, who quickly burned forty or fifty native dwellings, along with a number of their lateen-rigged seacraft, called proas. The ship's boat was recovered, while seven natives paid with their lives for its theft. Magellan at first dubbed the place Islas de Velas Latinas (Isles of Lateen Sails), but in the face of native thievery he changed the name to Islas de Ladrones (Islands of Thieves).[5]

When the expedition's men went ashore at Guam, they no doubt hoped to replenish their much depleted reserves of drinking water and food. But the shore party seems to have been more punitive than logistical, although some water and fresh food must have been obtained to supplement the meagre rations on which the crews had barely subsisted for so many weeks.[6] Whatever the case, the expedition quickly resumed its westward course, and less than a week out of Guam, in mid-March 1521, forest-clad

---

[1] Nowell, *Magellan's Voyage*, p. 123.

[2] According to Morison, *European Discovery*, p. 413, Magellan had a 'supply of quince preserve', and that it 'doubtless came in useful as an anti-scorbutic for the afterguard'.

[3] Magellan's landfall in the Mariana Islands has generated controversy over its location. Morison, ibid, p. 415, had no doubt that Guam and nearby Rota were the islands sighted by Magellan, and that the former island was where his anchorage and subsequent landing occurred. Islands in another group of the Marianas – Saipan, Tinian and Aguijan – have also been considered possible candidates for the honour. An article by Rogers and Ballendorf, 'Magellan's Landfall in the Mariana Islands', in *JPH*, October 1989, pp. 193–208, argues persuasively in favour of Morison's Guam-Rota theory.

[4] Nowell, *Magellan's Voyage*, p. 129.

[5] See Morison, *European Discovery*, p. 415.

[6] Pigafetta's narrative does not explicitly say if food and water were obtained ashore on Guam. The account attributed to Maximilian of Transylvania describes taking on water at an island called Acaca, but it is not clear if this was Guam or an island in the Philippines. See Nowell, *Magellan's Voyage*, p. 293.

mountains hove into view on the western horizon, indicating they were approaching a sizeable landmass, offering prospects of relief from their extraordinary ordeal. They had reached the eastern edge of a sprawling and complex archipelago upon which other Spaniards would later bestow the name Las Islas Filipinas, (The Philippine Islands), to honour the Infante (Crown Prince) of Spain, destined to become King Philip II. As their European discoverer, Magellan named these islands San Lazaro after the biblical Lazarus, although the name would prove ephemeral.[1] His claim to the islands on behalf of the Spanish Crown, however, was an act of supreme and enduring importance for his adopted Castilian homeland and its growing overseas empire. Together with his epochal crossing of the widest expanse of ocean on the globe, Magellan's discovery of the Philippines lent substance to Spanish claims to the entire Pacific Basin.

Magellan had endured so many rigorous hardships and attained such remarkable navigational achievements that the dismal fate awaiting him in these islands seems both incongruous and improbable. After his officers and crews had regained their health, he allowed himself, eagerly but unwisely, to be drawn into a dispute among the local rulers. Masterful as he clearly was in his command at sea, Magellan, as events would prove, lacked the judgement and tactical skills essential to the conduct of amphibious warfare ashore.[2]

The end came at an island named Mactan, paralleling the east coast of the much larger island of Cebú. Magellan had cemented friendly relations with a newly-Christianized sultan named Humaban at a settlement on Cebú Island's east coast. And across the waters on nearby Mactan, another local leader, Zula, an ally of Humaban, was similarly well disposed towards the Spaniards. But a third local chief, also on Mactan Island, named Lapu Lapu, was a bitter rival of Zula. The situation is described in Pigafetta's account:

> [Zula] requested the captain to send him only one boatload of men on the next night, so that they might help him and fight against the other chief [Lapu Lapu] . The captain-general decided to go thither with three boatloads. We begged him repeatedly not to go, but he, like a good shepherd, refused to abandon his flock. At midnight, sixty men of us set out armed with corselets and helmets, together with the Christian king [Humaban] , the prince [Zula], some of the chief men, and twenty or thirty *balanquais* [native canoes].[3]

It was in the early morning hours of 27 April 1521, when Magellan commenced his ill-advised, star-crossed amphibious assault on the north-east shore of Mactan Island. The enemy, drawn up in three formations of some five hundred men apiece, greatly outnumbered the Spaniards and their native allies – with accounts differing at to exact numbers. The disparity in numbers was magnified because the men occupying the twenty or thirty *balanquais* played no part in the battle except to witness it from a

---

[1] Ibid., p. 135.
[2] Morison, *European Discovery*, p. 428, opines that 'Magellan broke every rule of amphibious warfare, as we learned them in World War II; ... he did not attempt surprise, he made no provision for gunfire support, he chose an unsuitable beach full of natural obstacles ... he timed the assault at low water when the rocks were most prominent, and he failed to coordinate his attack with native allies'.
[3] Nowell, *Magellan's Voyage*, pp. 168–9.

distance. With little more than sixty men at his disposal, Magellan had virtually no chance to prevail, even with firearms and crossbows; it is a wonder indeed that any of this small corps – which included Pigafetta – survived the fight at all.[1]

Magellan's death left his surviving officers and men in stunned silence. It was as if Cortés or Pizarro had been struck down on the eve of their respective conquests of Mexico and Peru. Why Magellan elected to gamble away his life, having just completed the first recorded transit of the Pacific Ocean and being poised to achieve the globe's first circumnavigation, is almost inexplicable. Pigafetta's remarks, which liken the captain-general's motives to those of a shepherd who refuses to abandon his flock, have a hollow ring. Whatever the reasons, the expedition would now be forced to find its way home without its remarkable and much-lamented leader.

The story of the expedition's subsequent actions and eventual return to Spain needs no recounting here in detail. Of the three ships that had reached the Philippines only one, the *Victoria*, under command of Juan Sebastián de Elcano (also spelled Juan Sebastián del Cano), actually accomplished the globe-circling feat. Attrition had so diminished the expedition's numbers that the *Concepción*, least seaworthy of the flotilla's three remaining vessels, was abandoned and sunk because there were not enough crewmen to man three ships. As for Magellan's original flagship, the *Trinidad*, it would suffer an equally dismal fate in an unsuccessful attempt to reach Panama by retracing the expedition's route back across the Pacific. Under command of Gonzalo Gómez de Espinosa, laden with a valuable cargo of East Indian spices, the *Trinidad* reached the Islas de Ladrones (Marianas). But adverse easterly winds forced Espinosa to steer northward, possibly as high as the forty-second parallel, where stormy conditions compelled the ship to return south-westerly to the Moluccas. There, in desperation, Espinosa surrendered to a Portuguese naval force that had been scouring Moluccan waters for Magellan's expedition. At Ternate, the *Trinidad* was relieved of its cargo of spices, together with its papers and log-books, and stripped, says Morison, 'of sails, lines and all valuable gear'.[2] Later, while the ship was riding at anchor, it was destroyed in a storm. Only four survivors, of whom one was the ship's captain, lived to see their Iberian homeland again.[3]

Touting the success of Magellan's expedition is not as easy as one might expect, because in some ways it was anything but successful. Its captain-general lay dead in the Philippines; less than a third of the expedition's officers and men survived.[4] Only one of its ships actually accomplished the earth's circumnavigation; three of the five vessels that had originally left Seville were now on the ocean bottom; and a fourth, seized by mutinous officers and men, returned home empty-handed and without glory, having never set eyes on Pacific waters. Moreover, by uncovering the enormous distances and many hazards to be overcome in reaching the Moluccas by sailing west, Magellan's heroic strivings and regrettable death served mainly to demonstrate the route's commercial impracticability.

---

[1] Pigafetta's graphic account of the battle and Magellan's death is given in Nowell, *Magellan's Voyage*, pp. 169–72. See also, Skelton's translation of the Beinecke MS in *Magellan's Voyage, A Narrative Account*, 1: 87–9.

[2] Morison, *European Discovery*, p. 455.

[3] For more detail concerning events after Magellan's death, see Nowell, *Magellan's Voyage*, pp. 173–270; Morison, *European Discovery*, pp. 438–73; and Skelton, *Magellan's Voyage*, 1, pp. 89–148.

[4] There is no agreement on the exact number of the expedition's losses. See Morison, *European Discovery*, p. 469.

As if these unwelcome results were not discouraging enough, the expedition failed to determine conclusively if the Moluccas were located on the Spanish side of the Tordesillas demarcation line extended to East Asia.[1] This was mainly because neither Magellan nor any other mariner of his time had a reliable means to fix their position east or west (longitude) in relation to a known location (prime meridian). The expedition's surviving officers had only the roughest estimates as to the distance they had sailed crossing the Pacific.[2] And in learning that a Portuguese naval squadron was operating in the Moluccas, the Spanish Crown was forced to consider the unhappy prospect that these spice-rich islands might indeed lie on the Portuguese side of the line, or even worse, were already under Lisbon's control or influence.

Despite so many reasons to lament its outcome, Magellan's voyage has long occupied a celebrated place of honour in the annals of maritime exploration and discovery.[3] If any doubt remained as to the earth's spherical shape, the arrival in Spain of Elcano and the *Victoria* ended it for all time. The voyage opened a vast oceanic region hitherto unknown to Europeans, which greatly exceeded in extent the most grandiose speculations of Europe's cosmographers. It offered the Spanish Crown an opening wedge in assessing the dimensions and potentials of the territories to which it believed the Tordesillas agreement had entitled it. And finally, by discovering the Philippines, something new to the European perspective, Magellan provided his adopted nation a basis to claim that important archipelago in the face of uncertainties as to which side of the demarcation line it actually fell.

The experiences of Magellan and his men in their epic crossing of the world's widest ocean were precursor of what subsequent explorers in the Pacific could expect. With sailing distances of magnitudes many fold greater than any previously experienced by mariners on other oceans, the design, provisioning, hygiene, and navigation of ships undertaking such lengthy voyages would have to be improved. There was also a warning in Magellan's death and the expedition's high casualty rate that commanders should exercise great caution before committing themselves or their men to hostilities ashore involving indigenous populations. One is reminded of the lapse of judgement that cost the life of Captain James Cook while visiting the Hawaiian Islands in 1779.[4]

No sooner had the *Victoria* made her way up the Guadalquiver River and cast anchor near the quay of Seville, than plans were underway for a follow-up expedition to the Pacific. Although command of this Secunda Armada de Moluca (1525) was nominally

---

[1] Although both the Moluccas and the Philippines are known now to fall clearly within the Portuguese zone, Spanish maps in the 17th century persisted in showing them on the Spanish side. See, for example, a map entitled 'Descripción de las Indias del Poniente', attributed to Antonio de Herrera, dated 1601, in Wagner, *Spanish Voyages*, between pp. 346–7.

[2] Beaglehole, *Exploration*, p. 38, says that 'one of Magellan's pilots was in error in his calculations at the Philippines by no less than 52°55″'.

[3] Morison, *European Discovery*, pp. 319–20, is almost effusive in praise of Magellan and his expedition's achievements, and similar sentiments are readily voiced by other evaluators. Baker, *History of Geographical Discovery*, p. 107, says, 'As a sailor, a geographer, an explorer, Magellan was a great man, greater perhaps than either Columbus or De Gama ...'. One of Magellan's principal biographers, Charles McKew Parr, in *So Noble a Captain*, pp. 376–7, observes that although the voyage 'was ostensibly an imperialistic episode in the political struggle between the royal houses of Spain and Portugal [Magellan was] not only a master navigator and scientific explorer, but that he was a man of noble character, great daring, and knightly ideas'.

[4] See Beaglehole, ed., *Journals of Captain Cook*, pp. 533–8.

in the hands of a landsman and soldier named Francisco García Jofre de Loaysa, it would be effectively under control of the more experienced and seasoned mariner, Elcano, who was named as the expedition's second-in-command and chief master. But there would be precious little glory for the leaders of this second expedition, only watery graves in the mid-Pacific. They, along with many of their crewmen, perished of scurvy or other illnesses long before the expedition raised Guam or the Philippines.[1] Some other way would have to be found if Spain was to capitalize on Magellan's bold venture into the Great Ocean.

Among the crewmen sailing with the Loaysa-Elcano expedition was the chief master's cabin boy, a certain Andrés de Urdaneta. This lad, then a mere seventeen years old, was destined to play an important role in mastering the formidable obstacles to transpacific navigation by ships reliant on wind for power. Nearly forty years elapsed before his opportunity to contribute would arise. After surviving the misfortunes of the Loaysa-Elcano expedition, Urdaneta remained in the Far East for some eight years. During this time, he seems to have acquired the navigational skills needed to be a master, as well as considerable knowledge of sailing directions and other information useful in navigating the waters of the western Pacific.[2] This included knowledge of native languages spoken in the region, and there is some possibility that he learned about Chinese merchant vessels regularly engaged in trade with the native inhabitants of the Philippines.[3]

In 1552, after a varied career that included official posts in New Spain and Central America, Urdaneta entered at the age of forty-four the Augustinian order as a friar.[4] Meanwhile, hope faded that trade between Spain and the Moluccas or the Philippines could be conducted over the long, arduous route Magellan and Elcano had pioneered. For one thing, Spanish vessels attempting to return home by circling the globe, as Elcano had done, were certain to court trouble with the Portuguese. Moreover, as Espinosa's failed attempt to cross the Pacific eastward had shown, prospects of finding a way home in this fashion were equally problematic. Although it would offer only partial amelioration of these difficulties, Spanish officials began to think of launching future expeditions to reach the Moluccas or the Philippines from New Spain to avoid the protracted and hazardous voyage around South America.

The first such attempt was made in 1526 by Alvaro de Saavedra Cerón, under sponsorship of his famous kinsman Hernando Cortés. Although the expedition reached the Moluccas, its ships and men suffered more or less the same fate as those of Magellan and Loaysa, never again to see either New or Old Spain.[5] Another venture was mounted in 1542 from a harbour called Navidad on the Mexican west coast. It was under the joint auspices of the newly installed viceroy of New Spain, Antonio de Mendoza, and Pedro de Alvarado, the conqueror of Guatemala. The expedition actually consisted of two components. One, under command of Ruy López de Villalobos, set out in October 1542 to reach the Philippines, sailing before the north-easterly trade winds. The other flotilla, commanded by Juan Rodríguez Cabrillo, set

---

[1] For a full account of the Loaysa-Elcano expedition see Morison, *European Discovery*, pp. 475–92.
[2] Ibid, p. 492.
[3] See Schurz, *Manila Galleon*, p. 69.
[4] See Morison, *European Discovery*, pp. 492–3.
[5] Ibid, pp. 488–91.

sail from Navidad in June 1542, some three months prior to Villalobos's departure, with orders to follow the North American coast north or north-westward as far as possible.[1]

History credits Cabrillo as the European discoverer of what is today the state of California.[2] As with Magellan, Cabrillo did not survive the voyage, and it fell to his chief master, Bartolomé Ferrelo, to see that the expedition went as far north as circumstances would allow. And he was successful in returning the expedition to its home port at Navidad. The surviving officers believed that they had reached latitudes as high as the forty-fourth parallel, but subsequent evaluators – Henry Raup Wagner, in particular – have been sceptical of such claims.[3] If the planners of this voyage thought that Cabrillo had a chance of linking up with Villalobos in the Philippines, the realities of the North Pacific's currents, winds and geography would soon enough dash any such fanciful expectations. But whatever failures and frustrations had beset the expedition, Cabrillo and Ferrelo were pathfinders who opened the way to high latitudes on the west coast of North America.

Returning to Villalobos's venture to reach the Philippines, circumstances were at first favourable. The expedition consisted of six ships and some 370 men. According to Morison, it discovered 'dozens of islands that Magellan and Loaysa had missed'.[4] In January 1543, Villalobos found the archipelago where Magellan had met his unhappy fate, only to run afoul of the Portuguese later in the Moluccas. A mere handful of the expedition's men survived to see their Iberian homeland again. Villalobos was not among them.[5] Despite so many rebuffs, and still without a proven means of returning to New Spain from the Orient, the Spanish Crown remained committed to finding a means by which the Great Ocean could be spanned for commercial purposes. When Philip II ascended the throne in 1556, he brought with him a determination, as Morison puts it, 'to nail down the Philippine Islands for Spain'.[6] There was still no way to prove conclusively on which side of the Tordesillas line the Philippines – or for that matter, the Moluccas – actually lay. And so, measures to occupy the islands Magellan had discovered, and which now bore the new Spanish king's name, were pressed with vigour.

It was in this new effort that the Spanish Crown turned to Fray Andrés de Urdaneta. His exploits and the knowledge he had acquired while navigating in the western Pacific, three decades earlier, had not escaped notice. Although his status as an Augustinian friar prevented him from assuming command of the expedition, a special post was created enabling him to accompany the fleet as an advisory master.

Fray Andrés de Urdaneta, according to Antonio de Morga, a high official in the Spanish colonial administration of the Philippines, 'left the court for New Spain and,

---

[1] See Kelsey, 'Finding the Way Home', pp. 155–6; and Cummins, ed., *Sucesos de las Islas Filipinas*, p. 54. An account of Cabrillo's discovery of Alta California is in Kelsey, *Juan Rodrigues Cabrillo*, pp. 143–63. According to Kelsey, ibid, p.115, Mendoza twice offered Urdaneta command of the expedition to the Moluccas, and both times he refused. 'No one knows why', says Kelsey, 'though some think he was even then toying with plans to enter a monastery.'

[2] Ibid., pp. 143–63; and Morison, *European Discovery*, pp. 628–31.

[3] See Wagner, *Spanish Voyages*, p. 77.

[4] Morison, *European Discovery*, p. 493.

[5] Ibid.

[6] Ibid.

since he was experienced and such a fine cosmographer, he offered to go with the Armada and discover a return route'.[1]

The expedition Urdaneta joined was under the command of Miguel López de Legaspi, an official of Basque origin stationed in Mexico City. The flotilla, comprising five ships with more than four hundred officers and men, set out in November, 1564, from Navidad on the Mexican west coast.[2] Neither its officers nor Urdaneta knew with certainty what their destination was to be until the Crown's sealed orders were opened when the expedition was one hundred leagues at sea. While it was probably no secret that the Philippines were uppermost among the Spanish Crown's interests, the Moluccas or even the South Pacific island known today as New Guinea were also possibilities. Urdaneta had in fact recommended this huge island on the grounds that he believed it to be the only one of the three that was actually on the Spanish side of the Tordesillas line – a belief that was in fact essentially correct.[3] But Urdaneta was not aboard to make strategic decisions; his task was to help find a practical way to sail eastward from the Orient to reach New Spain. In this objective, he would not disappoint his sovereign's hopes.

Once the expedition reached the Philippines, which was in mid-February, 1565, López de Legaspi set about to explore the archipelago and establish a fortified outpost on Cebú Island. Then, on 1 June 1565, one of the expedition's ships, the *San Pedro*, departed under the nominal command of López de Legaspi's seventeen-year-old grandson, Philip de Salcedo, to seek a way across the Pacific that would enable her and subsequent Spanish vessels to return to New Spain. Urdaneta was there to offer navigational advice to the ship's regular masters, and if we may judge by the successful outcome of the voyage it must have been sound.

'The actual work of pilotage', according to William Lytle Schurz, an eminent scholar of the Manila galleon trade, 'was now in the hands of Esteban Rodríguez and Rodrigo de la Isla Espinosa, but they followed a theoretical course which Urdaneta had mapped out five years before and consulted him in regard to all the larger details of the voyage'.[4]

Some interpreters of this discovery, which ranks among the most important contributions made by Iberian mariners to the European exploration of the Pacific, are disinclined to allow Urdaneta all or even most of the credit for its accomplishment. Henry Raup Wagner, an authority on Spanish exploration of the North American west coast, took exception to the belief that Urdaneta played a key role in discovering the return route. Wagner contended that the knowledge necessary to conduct the voyage was already well known, and that all Urdaneta 'may have contributed to the solution of the problem was a better understanding of the best season in which to sail'.[5]

There is also evidence that the *San Pedro* (with Urdaneta aboard) may not have been the first ship to discover the secret of reaching New Spain from the Philippines. One of the vessels that departed with the López de Legaspi expedition from Navidad was a

---

[1] Cummins ed., *Sucesos de las Islas Filipinas*, p. 54.
[2] Discussion of the López de Legaspi expedition is based on the following sources: Kelsey, 'Finding the Way Home', pp. 145–64; Morison, *European Discovery*, pp. 492–5; Schurz, *Manila Galleon*, pp. 22–3, 219–20; Wagner, *Spanish Voyages*, pp. 94–120; and 'Urdaneta and the Return Route', pp. 313–16.
[3] Kelsey, 'Finding the Way Home', p. 160.
[54] Schurz, *Manila Galleon*, p. 220.
[5] Wagner, *Spanish Voyages*, p. 110.

small tender of forty tons named *San Lucas*, under the command of Alonso de Arellano, with Lope Martín as master. Sometime before the expedition reached the Philippines this ship drew ahead of the fleet and became separated from it. The *San Lucas* eventually turned up on the Mexican coast at Navidad in early August 1565, two months before the *San Pedro* reached there.[1] Arellano and Martín claimed they had departed to the north-east from the Philippines, sailed off Japan to a point possibly as high as the forty-third parallel, where they turned eastward towards North America on more or less the same route Urdaneta advised the masters of the *San Pedro* to steer.[2]

It made little difference which ship first completed the *torno viaje* 'return voyage', as Spaniards refer to the achievement, for in either case the mystery was solved and the stage set for Spanish commerce to rule the Pacific's waves henceforth for over two centuries. Spanish masters on the Manila–Acapulco run were required to learn their craft well, especially on the eastward leg of the journey, where difficulties and dangers lurked in the vast, cold, blue-green waters of the treacherous North Pacific. Many of Spain's finest masters served on the Manila galleons at one time or another in their careers, including Francisco Antonio Mourelle, the second officer and master aboard the *Sonora* in 1775.

And so, efforts begun in 1513 with Balboa's chance discovery, and carried forward in 1520 with Magellan's gallant but ill-starred dash across the Pacific's vast breadth, came full circle with Urdaneta and the others who participated in the *torno viaje* of 1565. The prospect of the Great Ocean being one immense Spanish Lake, across which Spanish galleons could profitably ply, seemed beyond challenge as events of the last quarter of the seventeenth century began to unfold.

2. The Manila–Acapulco Trade and Navigation

When Magellan's expedition first reached the Philippines in March 1521, Antonio Pigaffeta's account of the voyage notes that the captain-general received some unusual gifts from one of the local island leaders. Not the least of these were 'three porcelain jars covered with leaves and full of raw rice'.[3] Later, on 29 July, when the expedition had moved south-west to Brunei, on the island of Borneo, Pigaffeta records details of trading with the natives there:

> Their porcelain is a sort of exceedingly white earth which is left for fifty years under the earth before it is worked, for otherwise it would not be fine … The money made by the Moros in those regions is of bronze pierced in the middle in order that it may be strung. On only one side of it are four characters, which are letters of the great king of China. We call the money *picis*. They gave us six porcelain dishes for one *cathils* [a measure of weight] … of quicksilver; one hundred *picis* for one book of writing paper; one small porcelain vase

---

[1] The *San Pablo* reached Navidad on 1 October 1565, but instead of putting in there, her captain, Felipe de Salcedo, ordered the ship to proceed to the nearby and far superior harbour at Acapulco, in which she anchored on 8 October. See Wagner, *Spanish Voyages*, p. 117.

[2] An account of the Arellano-Martín voyage is in Wagner, *Spanish Voyages*, pp. 110–12. See also Nowell, 'Arellano versus Urdaneda', pp. 111–20.

[3] Nowell, *Magellan's Voyage*, p. 138.

for one hundred and sixty *cathils* of bronze; one porcelain vase for three knives; one *bahar* [about 200 *cathils*] of wax for 160 *cathils* of bronze.[1]

Pigaffeta's understanding of these materials was not always reliable. The Ming porcelains and coins were not, as he implies, made by the Moros, but were instead produced in mainland China and traded out to the islands. Nevertheless, his remarks bear witness to some of the commodities that in time would be flowing in great quantities from China to Manila, and thence eastward across the North Pacific and eventually to the Mexican port of Acapulco. Writing in 1609, some seventy-six years later, Antonio de Morga, lieutenant-governor of the Spanish colony at Manila, describes the Manila-Acapulco trade in his time:

> Merchants and traders form the majority of the residents in the islands, on account of the opportunities provided by the abundance of merchandise (not counting the local goods) brought to the islands from China, Japan, Maluco, Malaca, Siam, Cambodia, Borneo and other parts. The merchants invest in this business, exporting annually on the galleons that go to New Spain (and also, at the present time, to Japan) where raw silk brings big profits. The proceeds from this trading is brought back to Manila and up to now the profits have been quite simply splendid.[2]

Morga's account includes an exhaustive list of commodities, exotic and commonplace, exported from Manila, among which were 'china-ware jars' and 'cakes of white and yellow wax'.[3] He likewise discusses how the ships engaged in this trade followed more or less the same route pioneered by Urdaneta to reach their destination on the Mexican west coast.

To begin their eastward passage, the galleons departed the port of Cavite (just south of Manila) in June or July, accompanied by the onset of monsoon squalls. To the bafflement of some observers, instead of heading directly north along the west coast of Luzon, the ships picked their way south-easterly through narrow and often dangerous insular waters. They reached the ocean at an opening variously known as the Embocadero, Capul Strait, or by its modern name, San Bernardino Strait – the most feared and risky passage of all.[4] Once they reached the open ocean, the galleons set their helms to steer easterly towards the Mariana (Ladrones) Islands, then northerly past the Volcano and Bonin islands to the waters off the east coast of Japan. There, in Morga's words, they faced 'a mighty sea of open water where the galleon may run free in any weather ... [crossing] for many leagues under whatever winds are met, as high as a latitude of forty-two degrees north, bound for the coast of New Spain'.[5]

There are no lonelier stretches of ocean than the North Pacific between the thirtieth and fiftieth parallels. It is an immense watery expanse uninterrupted by islands, and

---

[1] Ibid, pp. 192–3.
[2] Cummins, *Sucesos de las Islas Filipinas*, p. 304.
[3] Ibid., p. 309.
[4] Why they persisted in following such a dangerous and indirect way to exit the Philippines may simply be a case of mariners' conservatism: known dangers seem easier to cope with than those unknown. Schurz, *Manila Galleon*, p. 224, says that 'many proposals were, in fact, made to change the route to this more open and direct way'. The route along Luzon's west coast, however, had its own problems in that ships departing Cavite after the end of June faced the possibility of being pinned against the coast by a typhoon. It was not until 1777, two years after the Hezeta-Bodega expedition, and after numerous years of dispute among the galleon masters, that the departure route was officially changed to follow Luzon's west coast: see Schurz, ibid, p. 226.
[5] Cummins, *Sucesos de las Islas Filipinas*, p. 321.

even today it is not heavily trafficked by ships. In December 1941, Japan's Pearl Harbour strike force successfully eluded detection by steaming eastward, under radio silence, through this unfrequented zone along the forty-third parallel before turning south-easterly to descend on the Hawaiian Islands. When the Manila galleons sailed these same waters in the sixteenth, seventeenth and eighteenth centuries the area was no less devoid of islands as now. And the only ships venturing there were the Manila galleons themselves, an occasional derelict Japanese fishing vessel or coastal trader driven there involuntarily, and a handful of other European vessels exploring its edges or seeking to prey on the galleons.

While Morga places the galleons' eastward passage at latitudes as high as the forty-second parallel (the present-day Oregon-California line), sailing directions a century later indicate that the masters of these ships came to prefer a somewhat lower latitude for their North Pacific crossings. José González Cabrera Bueno, author of a navigational manual for galleon masters published at Manila in 1734, advised them, after having positioned themselves north-east of the Philippines and eastward of Japan in about 31°N latitude, 'to set your course steering east north-east until placing yourself in 36 or 37 degrees of latitude ... and from said position you will set your course steering east'.[1] Although winds were more likely to be favourable in higher latitudes, the masters seemed to have learned since Morga's time that cold temperatures and storminess were more prevalent above the fortieth parallel. It has also been suggested that some mariners may have believed prolonged stays in high latitudes to be an aggravating factor in scurvy.[2]

Whatever latitude any particular Manila galleon sailed across the North Pacific, her officers, sailors and passengers were always elated when they first sighted the *señas* 'signs' indicating the North American continent was not far over the eastern horizon. It meant that prospects of surviving their gruelling ocean-crossing ordeal would soon be looking up. The *señas* are described in Cabrera Bueno's words as 'various bulbs [*porras*] or heads like onions in Europe, their green or reddish brown stems being from three to four fathoms long; the water is turbid, similar to mud'.[3] Morga's description is more detailed and encompassing:

> Three hundred leagues before reaching land, certain signs are observed, as, for instance, *aguas malas*, which are round, purple and as big as a man's hand with a crest in the middle like a lateen sail, called *caravelas*. These continue until the vessel is one hundred leagues from land when certain fish are seen with half their body in the shape of a dog [seals or sea lions]. These frolic with one another near to the ship. After these 'little dogs' there come some 'sticks' [*porras*], which are very long, hollow, yellow grass-stalks with a ball at the end, floating on the water. About thirty leagues from the coast are many huge clumps of grass called 'rafts' [*balsas*] which are carried out to sea by the great rivers along the coast.[4]

---

[1] Cabrera Bueno, *Navegación*, p. 294 (editor's trans.); See also Dahlgren, *Were the Hawaiian Islands Visited by Spaniards*, p. 135.

[2] See Cook, *Flood Tide of Empire*, p. 31.

[3] Cabrera Bueno, *Navegación*, p. 294 (editor's trans.). The bulbous-headed objects described (called *porras* in Spanish) are known today as 'bull kelp' (*Nercocystis luetkean*). See Ricketts and Calvin, *Between Pacific Tides*, p. 415.

[4] Cummins, *Sucesos de las Islas Filipinas*, p. 322. *Aguas males* are jellyfish-like creatures called *Velella velella*. See Ricketts and Calvin, *Between Pacific Tides*, pp. 226–7. The 'sticks' (as they are called in this translation), refer to *porras* or bull kelp.

As the galleons neared the North American coastline, those on high-latitude courses – above the fortieth parallel – would often come in sight of a prominent cape. It acquired the name Cape Mendocino, probably to honour the first viceroy of New Spain, Antonio de Mendoza, and it became famous as a key landmark for the galleon masters. Curiously, what is today called Cape Mendocino, located at 40°26′N, is nowhere close to the latitude of the cape the galleon sailing masters knew by that name. Morga gives its latitude as 42°30′N and Cabrera Bueno as 42°07′N, both locations considerably closer to modern Cape Blanco at latitude 42°58′N.[1] If in fact the galleon masters did see present-day Cape Blanco, it would make some of them among the earliest Europeans to sight the coast of what is today Oregon. The final leg of their lengthy trans-Pacific journey took the Manila ships south-eastward, more or less paralleling the North American coast. Those on the more northerly track, and which came into visual contact with the coast near Cape Mendocino (or Blanco, as the case might be), would follow the coastline to Cape San Lucas, at the southernmost point of Baja California. Other vessels on more southerly tracks usually turned south-easterly immediately upon encountering the *señas* remaining out of sight of land until Cape San Lucas hove into view.

We are informed by an account of George Anson's expedition, sent in the 1740s to intercept Manila galleons, that when the galleons neared Cape San Lucas, their captains were under orders 'to fall in with the land ... northward of the [cape], where inhabitants are directed on sight of the vessel, to make the proper signals with fire'.[2] Thereupon, the galleon would send a launch ashore with mail from Manila destined for the Jesuit missionaries on the Baja California peninsula. The shore party would then return with 'refreshments ... and likewise intelligence whether or no there are any enemies on the coast'.[3] If the coast towards Acapulco was clear of foreign marauders, the ship would sail with all possible speed to the haven so long anticipated by crew and passengers alike.

When the galleon entered the splendid harbour at Acapulco, it was to the intense relief of everyone aboard and many anxious persons ashore. At a minimum, the voyage required four months at sea, and sometimes twice that number were needed.[4] A large part of the voyage took place entirely out of sight of land, under arduous and debilitating conditions. Surviving crew members and passengers unaccustomed to such hardships must have uttered more than a few prayers, vowing never to leave terra firma again. But when the cargoes were later unloaded to be sold at great profit in the markets of New and Old Spain, the hardships and trials of the voyage would dim somewhat in memory.

---

[1] See Cummins, *Sucesos de las Islas Filipinas*, p. 322, and Cabrera Bueno, *Navegación*, p. 275. Elsewhere in the latter work (p. 302) the latitude of Cape Mendocino is given as '41 and a half degrees', together with another unnamed cape located at 42°, leaving the impression that Cabrera Bueno was uncertain about their respective locations.

[2] Purves, ed., *'A Voyage Round the World'*, p. 103. Although Anson missed an opportunity to waylay the 1741 galleon off Acapulco or Cape San Lucas, he was later successful in capturing a *patache* (a twin-masted tender), which had been pressed into service on the Manila–Acapulco run. The ship was called the *Nuestra Señora de Cabadonga*, and Anson captured her in Philippine waters in June 1743, off the San Bernardino Strait on her way to Manila from Acapulco. It was from the documents captured in this incident that Anson learned details of the Manila–Acapulco navigation.

[3] Ibid., p. 103.

[4] See Schurz, *Manila Galleon*, pp. 263–5.

Compared to the eastward voyage, galleons seeking to cross the Pacific in the opposite direction faced a much less formidable task. West-bound ships, loaded mostly with silver extracted from Mexican, Peruvian or Bolivian mines in the form of bullion or coins, normally left Acapulco in February or at the latest by mid-March. In uncomplicated fashion, they sailed from Acapulco south-westerly, dropping down to somewhere between the tenth and fourteenth parallels to pick up the exceedingly dependable north-east trade winds. The masters merely kept the ship's bow on a bearing due west, running down the latitude they had selected until reaching the Mariana (Ladrones) Islands, where they put in temporarily at Guam. On the final leg of the voyage, they sailed for San Bernardino Strait (Embocadero), after which they threaded their way through Philippine inter-island waters to reach Cavite, the port city of nearby Manila.[1]

Had the only hazards faced by the west-bound Manila galleons been natural, there would have been little reason to worry about the crews, passengers and valuable cargoes they carried. But the ease of sailing west enjoyed by the galleons could also benefit vessels of other nations bent on their capture. At first, these treasure ships plied South Sea waters unmolested by foreign corsairs, but in 1578 a determined Englishman would usher in a new and troubling chapter in the history of Spain's transpacific trade.

3. English Challenge and Spanish Response

His name was Francis Drake – later to be Sir Francis – and he gained access to the Pacific Ocean in almost the same fashion Magellan had, fifty-eight years earlier in 1520. There are parallels between these two famous voyages of circumnavigation, but they cease when Drake and his ship, the *Golden Hinde* (formerly the *Pelican*), sailed beyond the Strait of Magellan. The depredations Drake inflicted on Spanish settlements and maritime commerce along the west coast of the Americas were without parallel. They met with little armed resistance. The Spanish Crown, believing in the validity of its exclusive claim to the Pacific and counting on the difficulties confronting ships attempting to enter the Pacific from the Atlantic, had neither fortified these west coast harbours nor stationed naval forces there of any consequence.

Students of Drake's voyage of circumnavigation have debated its many facets endlessly. Was the expedition sent out, as Henry Wagner believed, primarily to forge a place for England in the spice trade?[2] Or was it, as Zelia Nuttall has suggested, to found an agricultural colony on the north-west coast of America?[3] Was it to search for the Pacific entrance of the north-west passage? Or was it simply, as Peter Gerhard has suggested, 'to acquire loot at the expense of Spain'?[4] The expedition's purposes may have been any or all of these, or more. But whatever the case, Drake displayed a singular talent for the last mentioned of these several possible objectives. In addition to raiding coastal towns from Valparaiso, Chile, northward to Guatulco in Mexico, Drake

---

[1] Ibid.
[2] See Wagner, *Drakes's Voyage*, pp. 26–7.
[3] See Nuttal, ed., *New Light on Drake*, pp. xxxvi–xxxviii.
[4] Gerhard, *Pirates on the West Coast of New Spain*, p. 62.

successfully captured several Spanish vessels engaged in coastal trade. One of them, the *Nuestra Señora de la Concepción*, (alias *Cacafuego*), was taken en route from Peru to Panama in March 1579. She was laden with no less than twenty-six tons of silver bullion, thirteen chests of pieces-of-eight coins, eighty pounds of gold, not to mention 'a quantitie of jewels and precious stones, ... two very faire gilt silver drinking bowles, and the like trifles'.[1] The *Golden Hinde*'s cargo holds bulged with these and other spoils, while the *Concepción* was allowed to hasten away 'somewhat lighter than before to Panama'.[2]

Recognizing the great risk of being intercepted should he attempt to return home by way of the Strait of Magellan, Drake had recourse to two other options. He could sail north to search for the legendary and elusive north-west passage in hopes that he could thereby reach the North Atlantic and British home waters. Or he could cross the Pacific as Magellan had, hoping to elude capture by Spanish or Portuguese naval forces in the Philippines or Moluccas, then cross the Indian Ocean to reach the Atlantic and eventually home waters by doubling the Cape of Good Hope. Circumstances would dictate that he consider the north-west passage option first.

It was already too late to initiate a westward passage across the Pacific that would avoid the Philippine typhoon season. Drake seems to have learned of this from two captured Manila galleon masters, Alonzo Sánchez Colchero and Martín Aguirre.[3] There were other reasons to delay embarking on such a lengthy voyage, not the least of which was the need to careen and repair the *Golden Hinde*. They would also have to take on wood and water, and rest her crew for the ordeal that lay ahead. And so, on 16 April 1579 (OS), Drake departed the obscure Mexican port of Guatulco, south-east of Acapulco. He would first direct his attention to search for the north-west passage, but when that appears to have failed he sought an anchorage far enough north to be secure from Spanish pursuers, where he could prepare for crossing the Pacific when conditions would be more favourable.

The surviving records of Drake's sojourn on the North American west coast are sketchy and rife with ambiguities.[4] The difficulties mainly concern his search for the north-west passage and the location of the harbour in which his flagship and crew were prepared for crossing the Pacific. Before leaving Guatulco, Drake released his Iberian captives so they would have no direct knowledge of his whereabouts northward on the coast. Accompanied by a captured Spanish vessel, known only by the name of her owner, Panamanian merchant Rodrigo Tello, and manned by an English crew, Drake sailed the *Golden Hinde* some five to six hundred leagues or 1,500 to 1,800 nautical miles westward to gain a northing. He thereby avoided beating windward up the coast against the prevailing summer northwesters – quite likely on the advice of his captive

---

[1] Penzer, ed., *World Encompassed*, p. 46. See also Wagner, *Drake's Voyage*, p. 270.

[2] Penzer, ed., *World Encompassed*, p. 47.

[3] It was mid-April by the time Drake and the *Golden Hinde* were ready to depart Guatulco, more than a month and a half past the official deadline – the end of March – set for galleons departing Acapulco for the Philippines. See Hanna, *Lost Harbor*, p. 143.

[4] Upon Drake's return to England, the original logs, journals and charts of his voyage of circumnavigation were delivered to Queen Elizabeth I, never to be seen publicly again. See Wagner, *Drake's Voyage*, p. 229. Thus, much of what is known about the voyage must be assembled from sometimes conflicting secondary accounts and maps that are open to varying interpretations.

Iberian masters, or by consulting their charts and sailing directions. From this point on, however, many uncertainties plague our understanding of where Drake sailed.

The earliest published account of Drake's circumnavigation, Richard Hakluyt's 'The Famous Voyage' narrative (1589), states that on

> The 5 day of June 1579 [OS] being in 42. degrees [changed to 43° in the edition of 1600] toward the pole Arctike, we found the aire so colde, that our men being grievously pinched with the same, complained of the extremities thereof, and the further we went, the more the colde increased upon us. Whereupon we thought it best for that time to seek the land, and did so finding it not mountainous, but low plaine land, till we came within 38 degrees toward the line … . into a faire and good Baye, with a good winde to enter the same.[1]

In this account there is no indication that the *Golden Hinde* sailed any significant distance north of the forty-third parallel; nor any mention that Drake temporarily put into an anchorage anywhere near the forty-eighth parallel.

A somewhat different story is told in *The World Encompassed*, first published in 1628 and compiled by Drake's namesake nephew mostly from the notes of Francis Fletcher, the expedition's chaplain. It states that by 3 June 1579 (OS) they had sailed north

> till we came into 42 deg. of North latitude, where in the night following we found such alteration of heate, into extreme and nipping cold, that our men in general did grievously complaine thereof, … in sayling but 2 deg. farther to the Northward [to lat. 44°N] … it seemed a question to many amongst us, whether their hands should feed their mouthes, or rather keepe themselves within their coverts from the pinching cold that did benumme them.[2]

By this account, they evidently had reached the forty-fourth parallel, and on 5 June (OS) contrary winds and unusually inclement weather forced them to seek shelter on the coast. What they found was 'a bad bay, the best roade we could for the present meet with'.[3] It proved an inhospitable anchorage, however, and they soon departed, heading south to find a better refuge. The location of this 'bad bay', according to *The World Encompassed*, was at or near latitude 48°N (a location close to Cape Alava on the northern Washington coast), but there is no mention of it whatever in 'The Famous Voyage' narrative. Both accounts agree that a 'faire and good Baye' (in 'The Famous Voyage') or 'a convenient and fit harborough' (in *The World Encompassed*) was found in the vicinity of latitude 38°N, where Drake and his men remained for over a month preparing to cross the vastness of the Pacific Ocean.

There is a third early account of Drake's voyage that tells yet another version of his presence on the north-west coast. Because its authorship is uncertain, it has come to be known as 'The Anonymous Narrative'.[4] According to this account, after Drake's departure from Guatulco, he sailed northward to latitude 48°N, where the coastline

---

[1] Hakluyt, *Principal Navigations*, VIII, pp. 62–3. See also Hanna, *Lost Harbor*, pp. 351, 357.

[2] Hakluyt, *Principal Navigations*, VIII, p. 63.

[3] Penzer, *World Encompassed*, pp. 48–9.

[4] The MS of 'The Anonymous Narrative' is in the British Library's collection (Harley 280, fols 81–90). It covers only the portion of the voyage from the time Drake entered the Pacific Ocean until his return to England. The document went unpublished until 1854, when the Hakluyt Society included it as an appendix in its 1854 edition of *The World Encompassed by Francis Drake*, ed. W. S. W. Vaux. Wagner also included the 'Anonymous' text in his *Drake's Voyage*, pp. 264–85.

seemed to stretch away endlessly to the north and north-west. It led Drake, so the account says, to abandon the search for a north-west passage, whereupon he turned south hugging the coast until he came upon a suitable harbour at latitude 44°N. There, somewhere on the northern Oregon coast, this version places the location at which the *Golden Hinde* and her crew were made ready to cross the Pacific.

Much time and effort have been expended inquiring as to which account is more reliable.[1] Was 'The Famous Voyage' narrative, in the course of its editorial condensation, stripped of contents that mentioned reaching latitude 48°N and the 'bad bay' found thereabouts? If it was excised, was it done so to cloak something in secrecy? Or was it not there to begin with? On the other hand, is *The World Encompassed* a less redacted account, or merely an exaggerated one, in which its author has expanded upon the truth? As for 'The Anonymous Narrative', long unpublished and held to be unreliable, is it now to be regarded as more dependable than the other accounts? These are some of the questions that have plagued efforts to gain an accurate understanding of Drake's presence on the coastline of the land he called Nova Albion. The currently prevailing interpretation of the *Golden Hinde*'s sojourn on the north-west coast, is something of a compromise between these narratives (as well as several other secondary documentary and cartographic sources) runs more or less as follows.[2]

Voyaging out of Guatulco, the *Hinde* accompanied by Tello's captured vessel, reached a location off the Oregon coast between latitude 43° and 44°N. Foul weather and bone-chilling cold quickly forced them to abandon the search for a north-west passage. They closed with the coast, where they encountered a headland, possibly Cape Arago (lat. 43°20.5′N at its light), on the south side of which they found temporary shelter in a cove – the 'bad bay' – now named South Cove. Departing this exposed and unsafe anchorage, the vessels followed the coast south until sighting Point Reyes (lat. 37°59.7′N at its light), a landmark jutting prominently seaward from the general trend of the coastline. Within the shelter of this point, they found a remarkably spacious and inviting bay, with white cliffs that reminded them of the chalk cliffs of Dover. The ships located and entered a well-protected *estero* 'inlet' within the bay, where Drake and his men remained from 17 June to 23 July 1579 (OS). The bay and inlet, both of which have been given Drake's name, are located some twenty-three nautical miles north-west of the elusive and often fog-bound entrance to present-day San Francisco Bay. The evidence pointing to Drakes Bay as the place where the *Golden Hinde* and her crew were readied to cross the Pacific is plausible though largely circumstantial.

There are other interpretations of the evidence which assign more credence to *The World Encompassed* and 'The Anonymous Narrative' – particularly the assertion that Drake reached the forty-eighth parallel before turning back.[3] But none has gained

---

[1] See Hanna, *Lost Harbor*, pp. 85–102.

[2] As set forth in Aker and Von der Porten, *Discovering Portus Albionis*, pp. 7–30, and Morison, *European Discovery*, pp. 663–89.

[3] A British Columbia land surveyor, R. P. Bishop, in 'Drake's Course in the North Pacific', p. 159, remarks that the 'latitude of 48 degrees was generally accepted until the middle of the 19th century, when it became involved in the Oregon boundary question'. In the boundary negotiations, American representatives sought to minimize the northward latitude Drake may have reached in order to undermine British claims to the Oregon Country. See Greenhow, *Memoir, Historical and Political*, p. 37. Based on an analysis of a meteorological phenomenon variously known as the North Pacific High or Fleurieu's Whirlpool, Bishop

sufficiently wide acceptance to displace the Cape Arago-Drakes Bay thesis. It appears that Tello's captured Spanish vessel was abandoned wherever Drake stopped to prepare for crossing the Pacific. In the unlikely event the remains of this vessel should ever be found, speculation as to the whereabouts of Drake's anchorage would be muted though probably not silenced.[1]

When the *Golden Hinde* departed the North American continent, whatever latitude she had reached or wherever she was careened and repaired, there was nothing uncertain about the challenge Drake's actions posed for exclusive Spanish sovereignty in the Pacific. His buccaneering exploits were spectacular. And when he vanished into the Pacific's vastness only to reappear in November 1579, on the other side of the ocean at Ternate in the Moluccas, Spanish colonial officials along the entire Pacific seaboard, from the Strait of Magellan to Baja California, had good reason for alarm. The only thing Drake had failed to do was capture a Manila galleon – and that surely would be the ambition of the next enterprising English venture on Pacific waters.

The realization that such an attack might be in the offing did not produce an immediate response in Madrid or Spain's vicegeral capitals at Mexico City (New Spain), Bogota (New Granada), or Lima (Peru). But in 1585, after Archbishop Pedro Moya de Contreras became viceroy of New Spain, he wrote the Spanish king, Philip II, that he had ordered, on his own authority, the construction of two frigates, which were to sail northward from New Spain to explore the Alta California coast. The reasons for such an endeavour, the viceroy explained, were threefold: firstly, to search out suitable ports or harbours in which the east-bound Manila galleons might seek refuge in an emergency; secondly, to see if there was any truth to the rumoured existence of a north-west passage; and lastly to determine if New Mexico could be reached by sea.[2]

---

concluded that Drake's northernmost landfall was most likely in the vicinity of latitude 48°N. See Bishop, 'Drake's Course in the North Pacific', pp. 163–4. More recently, Kenneth Holmes, an Oregon historian, in 'Francis Drake's Course in the North Pacific,1579', pp. 5–35, similarly argues that wind and ocean currents, as well as the documentary evidence, favour Drake's northernmost landfall at the 48th parallel. Another even more recent and far more radical challenge to the Cape Arago-Drakes Bay thesis has been advanced by Samuel Bawlf, in *Sir Francis Drake's Secret Voyage* (2001) and *The Secret Voyage of Sir Francis Drake*; and Robert Ward in 'Drake and the Oregon Coast'; and 'Lost Harbour Found'. Bawlf and Ward theorize that Drake not only reached the 48th parallel, but that he explored a considerable distance farther north. They differ somewhat as to details of Drake's supposed northward probe, but they agree that Drake believed he had discovered the Pacific entrance of the north-west passage. Unable to pursue its exploration, they believe Drake returned south to an anchorage on the Oregon coast at present-day Whale Cove (lat. 44°45′N), where he prepared for the Pacific crossing. Bawlf and Ward further speculate that Hakluyt's 'The Famous Voyage' narrative was a deliberate fabrication put out as a cover story to cloak in secrecy Drake's supposed discoveries concerning the north-west passage, until a follow-up expedition could be mounted to take advantage of them.

[1] In his biography of Sir Francis Drake, Harry Kelsey suggests that Drake's anchorage may have been among southern California's Channel Islands. He bases this contention in part on a report in 1602 that Sebastian Vizcaíno sighted the wreckage of a vessel cast ashore on Catalina Island. This and other circumstantial evidence is adduced to suggest that Drake's anchorage could have been far south of the sites hitherto considered. See Kelsey, *Sir Francis Drake*, p. 182.

[2] See Wagner, *Spanish Voyages*, pp. 108–9, 132. Viceroy Moya's letter to the king, dated 22 January 1585, seems to have been inspired by a certain Fray Andrés de Aguirre, an Augustinian monk, who had sailed with Urdaneta on the Legaspi expedition of 1564–5. A letter by Aguirre to Viceroy Moya, undated, explaining the need for exploration of Alta California's coast is in Cutter and Griffin, eds, *California Coast*, pp. 10–17.

The ships were neither built nor the expedition carried out. Instead, the viceroy became convinced that it would be better to assign the task to ships sailing out of the Philippines on the northern track taken by the east-bound Manila galleons. The first Spanish attempt to undertake such a reconnaissance of the Alta California coast was made in 1587 by one Pedro de Unamuno.[1] For the voyage, he employed a 50-ton frigate named *Nuestra Señora de Esperanza*. She was a Manila galleon in name only, because she was far too small to carry anything but a fraction of a galleon's normal cargo. As for her accomplishments exploring the coast, except for a brief and unproductive stay at what is believed to be present-day California's Morro Bay (lat. 35°20′N), the *Esperanza* slipped into the harbour at Acapulco with precious little new information useful for navigation along the coast of Alta California – and absolutely nothing about the coast farther north.

Unamuno narrowly missed an encounter with English corsairs off the Baja California coast near Cape San Lucas. His frigate may have escaped notice by virtue of her small size, or if sighted not considered sufficiently valuable to attack. The corsairs were after larger prey, namely, the 700-ton Manila galleon *Santa Ana*, captained by Tomás de Alzola, and due to arrive in Acapulco sometime in November. The English squadron lurking off Cape San Lucas consisted of two vessels, the 120-ton *Desire* and the 60-ton *Content*, under the overall command of Thomas Cavendish. On 4 November 1587 (OS) – 14 November by Spanish reckoning – the lightly armed *Santa Ana* fell victim to these corsairs after a fight lasting five or six hours. One of the English participants, Francis Pretty, wrote:

> Their Captaine [Alzola] still like a valiant man, with his company stood very stoutly ... not yeelding as yet: Our General [Cavendish] encouraging his men a fresh with the whole noise of trumpets gave them the third encounter with our great ordinance.[2]

And so, punctuated by trumpet fanfares, the battle raged until Captain Alzola was forced to surrender. It took Cavendish and his men a week to sort through and take possession of some two hundred tons of the *Santa Ana*'s immense cargo of Asian commodities and merchandise. Pretty says that the galleon's captain and master

> certified the Generall [Cavendish] what goods they had within board, to wit, an hundreth and 22 thousand pezos of golde: and the rest of the riches was in silkes, sattens, damasks, with muske & divers other merchandize, and great store of al maner of victuals with the choise of many conserves of all sortes for to eate, and of sundry sorts of very good wines.[3]

Cavendish had the captured galleon towed into San Lucas Bay, where he allowed 190 survivors of the *Santa Ana*'s crew and passengers – among whom were a number of women – to go ashore with provisions, sails, planks, arms, and ammunition sufficient to sustain themselves and provide for their defence against the local native population.[4] A

---

[1] See Wagner, *Spanish Voyages*, pp. 139–53, for an account of Unamuno's voyage.
[2] Hakluyt, *Principal Navigations*, VIII, pp. 236.
[3] Ibid, pp. 236–7.
[4] According to Francis Pretty, Cavendish kept aboard the *Desire* two Japanese boys, Christopher (age 20), and Cosmus (age 17); and three youths born in Manila: Alphonso (age 15), Anthony de Dasi (age 13), and a nine-year-old whose name he failed to record. We also know from the same account that a Portuguese master named Nicholas Roderigo, and a Spanish master named Thomas de Ersola were also compelled to remain aboard. Except for the nine-year-old, they evidently were needed for various linguistic or navigational skills or knowledge they possessed. See Hakluyt, *Principal Navigations*, VIII, pp. 237–8.

near mutiny broke out aboard the *Content* over division of the spoils. And to conclude the episode, on 19 November (OS), Cavendish ordered the galleon set afire, and all watched as she burned to the waterline, consuming whatever cargo had not been carried off. This calamitous event dealt a severe blow to the merchants of New Spain. It deserved immediate and redoubled Spanish efforts to strengthen coastal defences and press ahead with exploration on the Alta California coast. But Madrid had greater concerns in the Atlantic, and the high cost of such exploring expeditions in the Pacific required that they be delayed. Except for Unamuno's largely unproductive voyage, the earliest indication that serious attention was being given the threat of corsairs in the Pacific concerns a voyage allegedly undertaken in 1592, some five years after the disastrous loss of the *Santa Ana*. A certain Greek master, Apostolos Valerianos, better known as Juan de Fuca, supposedly participated in two northward exploring expeditions secretly authorized by Luis de Velasco, Marqués de Salinas, who was viceroy of New Spain from 1590 to 1595.

Fuca's story begins with an assertion that he had been aboard the *Santa Ana* when she was captured by Cavendish, and that he, Juan de Fuca, had been robbed of 'sixtie thousand Duckets, of his owne goods'.[1] According to Michael Lok the Elder, an English merchant residing in Venice and the only known source of information about Juan de Fuca, the first of the two exploring voyages was aborted before its mission could be accomplished. But Fuca's second alleged voyage, undertaken in 1592, reportedly discovered 'a broad Inlet of Sea, between 47. & 48. degrees of Latitude: [which] hee entred thereinto, sayling therein more than twentie dayes'.[2] Thinking he had found the Strait of Anian, Fuca returned to Acapulco hoping that the viceroy in Mexico City would reward him handsomely for this discovery – or failing that, later by the king in Madrid. But these hopes all proved illusory, for which reason he sought to interest Lok in promoting an English expedition to relocate the strait. If there is any truth in these claims as reported by Michael Lok the Elder, Fuca's two voyages are among the best kept secrets in maritime history, as no archival or other evidence has ever been found to substantiate them.[3]

There was, however, a Spanish exploring voyage in 1595 sufficiently well documented to be considered authentic.[4] As with Unamuno's earlier voyage, this one also originated in the Philippines, with plans to reach the Alta California coast by following the standard eastward route of the Manila galleons across the North Pacific. The ship selected for the voyage was the *San Agustín*, a privately owned vessel of under 200-tons burden. Its owner, Pedro Sarmiento, a sea captain of some prominence in Manila, would be allowed to ship a modest cargo in compensation for the Crown's use of his

---

[1] Cook, *Flood Tide of Empire*, p. 539. Fuca's account was first published in London in 1625, under the title 'A Note made by me Michael Lok the elder, touching the Strait of Sea, commonly called Fretum Anian, in the South Sea, through the North-West passage of Meta incognita', in Purchas, *Hakluytus Postumus or Purchas His Pilgrimes*, 3, pp. 849–52. This work was reprinted by the Hakluyt Society in *Works*, extra ser. 14, Glasgow, 1906, pp. 415–21. Other reprints of this material are in Greenhow, *History of Oregon and California*, pp. 407–11, and Cook, *Flood Tide of Empire*, pp. 539–43.

[2] Ibid, p. 540.

[3] See Mathes, 'Apocryphal Tales of the Island of California', pp. 52–9, for a discussion of the lack of documentary evidence concerning Fuca's voyage and other such apocryphal expeditions.

[4] Velasco to Philip II, 6 April 1594, outlines the expedition's purpose. See Cutter, *California Coast*, pp. 28–9. An account of the voyage is also in Wagner, *Spanish Voyages*, pp. 154–67.

vessel for exploration. A Manila galleon master named Sebastián Rodriguez Cermeño – the senior master aboard the *Santa Ana* when it was captured on that fateful November day in 1587 – would be in command.[1]

The voyage went reasonably well across the North Pacific, but early in November 1595, when the *San Agustín* made landfall on the North American coast in the vicinity of the forty-second parallel, things began to go awry. In his diligence to carry out his orders, Cermeño insisted on getting close in to the shoreline – something his officers and men did not greet with enthusiasm. An attempt was made to anchor in a cove sheltered in the lee of a high, rounded promontory (probably 116 m [380 ft] high Trinidad Head, lat. 41°03′N). But it proved too rock-infested for Cermeño, and he decided to stand off.

The *San Agustín* was then steered south hugging the coast as close as possible. On 6 November 1595, about noon, 'a point of high land was discovered which revealed a great *ensenada*'.[2] It was apparently the same bay into which Drake is thought to have sailed the *Golden Hinde* sixteen years earlier. Unaware of this, or of any previous names – indigenous or otherwise – that may have been given the bay, its Spanish visitors named it 'La Baya de San Francisco'.[3] Although it seems to have proven a secure haven for his English predecessor, this bay would not be a place of happy memories for Cermeño, because as the *San Agustín* rode at anchor she was wrecked by a late-November storm blowing in from the south.

The ship's modest yet valuable cargo of Asian merchandise, not to mention most of the food supply, was lost or scattered along the beach. Survivors of the accident could count themselves lucky that a launch, carried aboard the *San Agustín* for surveying purposes, was not destroyed. It enabled them to save themselves in a remarkable open-boat voyage down the coast to Acapulco, but little else was accomplished. This failure convinced the new viceroy in Mexico City, Gaspar de Zúñiga, Conde de Monterey, that combining private enterprise with exploration was not such a good idea. And in a letter to the Spanish king, Philip II, the viceroy observed that: 'To all practical men it seemed that in making this exploration the better method would have been for the ship to sail from here [New Spain] and along the coast'.[4] Although the viceroy's opinion seemed eminently sensible, any mariner familiar with the difficulties of beating windward up the coast would have been less sanguine about it. The disappointing outcomes, however, of both the Unamuno and Cermeño voyages made trying a different approach imperative.

The task of exploring Alta California's coast would fall next to a veteran soldier, erstwhile Manila merchant and participant in pearling ventures in the Gulf of California. His name was Sebastián Vizcaíno, and he seems to have hailed from Extremadura, the same impoverished province from which many *conquistadores* such as

---

[1] Details and documents of Cermeño's 1595 expedition to explore the California coast are in Cutter, *California Coast*, pp. 28–39; Mathes, *Vizcaíno and Spanish Expansion*, pp. 44–50; and Wagner, *Spanish Voyages*, pp. 154–67. For more about Cermeño's involvement in the *Santa Ana* disaster, see Mathes, *Vizcaíno and Spanish Expansion*, pp. 20–22.

[2] Wagner, *Spanish Voyages*, p. 158.

[3] Ibid. Drake does not seem to have given a name to the bay in which he found shelter. On modern maps the name 'San Francisco' has migrated some 25 nautical miles to the south-east, while the bay believed to have been discovered but left unnamed by Drake now bears his name.

[4] Monterey to Philip II, 26 November 1597, in Cutter and Griffin, eds, *California Coast*, p. 99.

Balboa and Cortés had come.¹ There was little to dispute the soundness of Viceroy Zúñiga's choice for the job. Vizcaíno's long and honourable service to the Spanish Crown had earned him a solid reputation for trustworthiness and diligence. He was also well acquainted with the Baja California peninsula from his pearling ventures there. If the viceroy had any reason for concern, it was that Vizcaíno might be tempted to neglect his exploring duties to harvest pearls.

The expedition to Alta California was to be exclusively devoted to discovering, exploring and mapping harbours in which Manila galleons could seek refuge in an emergency.² Viceroy Zúñiga was adamant that no private pearling excursions were to be made in the Gulf of California. Anyone caught doing so could expect to be put to death. This iron-fisted policy was tempered somewhat when the viceroy allowed that such pearling rights might be granted after the exploring expedition had been satisfactorily completed. Vizcaíno had under his command a 200-ton flagship named *San Diego*, and two escort vessels, the *Santo Tomás*, an ex-Peruvian galleon, and a frigate called *Tres Reyes*. The expedition embarked from Acapulco on 5 May 1602, heading north-westward more or less on the same track Cabrillo had followed sixty years before – apparently without knowledge of what that earlier expedition had learned. Under orders to stay close to the coastline, Vizcaíno was forced to beat windward against the prevailing coastal northwesters. It would exact a heavy toll in time and the well-being of his men. But they pressed on, exploring and charting most of the islands and potential harbours of any importance from Cape San Lucas, on the southern extremity of Baja California, northward to a bay on the Alta California coast at the thirty-seventh parallel. This bay they named in honour of Viceroy Zúñiga, who was the Conde de Monterey. In all likelihood it was the same bay that continues to bear the count's name today. For Vizcaíno and his men, however, it was mid-December and the winter storm season was closing in on them. With sick and dying crewmen, and provisions running low, Vizcaíno decided to send one of his escorts, the *Santo Tomás*, back to New Spain to seek 'new succor in the way of provisions, men and ammunition, and speedy dispatch of these'.³ It proved an empty hope, as there would be no speedy dispatch of anything.

Despite the worsening winter weather, Vizcaíno decided to continue north with his two remaining ships to search for the bay in which the *San Agustín* had been lost, hoping to recover the remains of its cargo. Francisco de Bolaños, the chief master aboard Vizcaíno's flagship *San Diego*, had been a member of the *San Agustín*'s crew when it was wrecked in 1595, so he could help identify the location. On 6 January 1603, Bolaños confirmed they were in sight of the same point behind which Cermeño's ship had taken refuge. The *San Diego* anchored briefly in a cove beside this point, which

---

¹ His surname, Vizcaíno, suggests that he was from the province of Vizcaya, on the temperate and rainy Cantabrian coast of northern Spain, and therefore of Basque ancestry. Wagner writes: 'His name indicates, of course, that he was a Basque or of a Basque family', *Spanish Voyages*, p. 170. But Mathes argues persuasively that he was more likely a native of Spain's arid, western region of Extremadura – a contention based on Vizcaíno's military service in Portugal, which borders Extremadura, and his familiarity with plants native to arid regions. See Mathes, *Vizcaíno and Spanish Expansion*, p. 29, n. 10.

² Details and documents of Vizcaíno's 1602–3 expedition to explore the Alta California coast are in: Bolton, *Spanish Exploration in the Southwest*, pp. 43–103; Cutter, *California Coast*, pp. 46–117; Mathes, *Vizcaíno and Spanish Expansion*, pp. 54–107; and Wagner, *Spanish Voyages*, pp. 168–285.

³ Cutter, *California Coast*, p. 109.

they named Punto de los Reyes, because it was the feast day of the Epiphany (*día de los Reyes*). But Vizcaíno abandoned the search for the *San Agustín*'s wreckage out of concern for his escort vessel *Tres Reyes*, which had earlier become separated from the flagship.

Both ships would sail north on separate tracks. On 13 January 1603, the *San Diego* was off a cape at latitude 40°30′N, which Vizcaíno considered to be the Cape Mendocino familiar as a landmark to Manila galleon masters. A powerful winter storm with heavy winds out of the south-east propelled them farther north, possibly as high as the forty-second parallel within sight of the southern Oregon coast. There they took note of another cape, which they named Cabo Blanco de San Sebastián.[1] It was 19 January, and by this time the crew was so exhausted and sick that they could scarcely work the ship. Vizcaíno decided to return with the first wind out of the north-west and run before it southward to New Spain.

Meanwhile aboard the *Tres Reyes*, commanded by Martín de Aguilar with Antonio Flores as master, the same bad weather had exacted a similar toll. They may have reached the forty-third parallel, where they reported that the

> land makes a cape or point, which was named 'Cabo Blanco', and here the coast begins to trend to the north-east. Close to it a very copious and deep river was discovered … On attempting to enter it, the force of the current did not permit it.[2]

They named it Río de Santa Inéz, although later it came to be known as Río de Martín de Aguilar. The cape they described could well be present-day Cape Blanco (lat. 42°50.2′N at its Light), but the 'very copious and deep river' nearby has proven difficult to identify.[3] Whatever the case, Aguilar and Flores decided the crew of the *Tres Reyes* had suffered enough, whereupon they set a course for New Spain, running before a brisk north-west wind.

Sailing independently, the *San Diego* and *Tres Reyes* were both able to reach home waters, although Aguilar and Flores did not survive the return voyage. The expedition's third ship, the *Santo Tomás*, had also returned safely, but the reinforcements it had been sent to obtain were never forthcoming. Despite the ordeal they all had endured beating windward up the coast, Vizcaíno and his men were remarkably successful in charting principal harbours and anchorages of Alta California. Only the now-famous and often fog-bound body of water known today as San Francisco Bay eluded them.

---

[1] The cape was described by Fray Antonio de la Ascención, one of the ship's chaplains, as being 'of white earth close to some high sierras covered with snow'. San Sebastián was included in the name, says Ascención, because its discovery was on the 'eve of the glorious martyr, San Sebastián'. See Wagner, *Spanish Voyages*, p. 253. The cape is probably today's Cape Sebastian (lat. 42°19.5′N.).

[2] From Ascención's account in Wagner, *Spanish Voyages*, p. 255.

[3] There are four rivers that empty into the ocean in the vicinity of Cape Blanco. They include to the north the Coquille (18 nautical miles away) and the Sixes (less than one nautical mile away); and to the south: the Elk (2.5 nautical miles away) and the Rogue (26 nautical miles away). The Sixes and Elk rivers are small streams, which renders them unlikely candidates for the river reported by Aguilar. The Coquille and Rogue rivers are of moderate size, and therefore come closer to fitting the description 'very copious and deep', especially in mid-winter shortly after a large storm had hit the coast with copious precipitation. But they are 'close to' Cape Blanco in a relative sense only. There is also a remote possibility that Flores Lake, which is located some 4.5 nautical miles north of Cape Blanco, could have been mistaken for a river opening.

Vizcaíno became convinced that Monterey Bay was ideally situated to serve as a way station for Manila galleons, a view with which Viceroy Zúñiga readily agreed. Preparations were begun to mount an expedition to establish a settlement at Monterey, with Vizcaíno in charge of the venture. But in March 1604, a new viceroy, Juan de Mendoza y Luna, Marqués de Montesclaro, took office. Lacking his predecessor's zeal for colonizing the Alta California coast, he believed instead that the most effective and least expensive policy would be to let the region remain a vast unexplored wilderness, inhabited only by its indigenous population.

> For my part [Mendoza wrote to the king] I think that once it was explored and settled it would be common ground for friends and enemies where, besides finding shelter for their ships, they would find Spaniards with whom to treat for food, clothing and goods.[1]

It meant the abrupt cancellation of plans for Vizcaíno's return to Monterey, and there would be no further Spanish voyaging to Alta California or beyond until 1769. The Manila galleons would have to fend for themselves as best they could without a way station at Monterey. As for the native inhabitants ashore, it ensured that their indigenous ways of life would continue essentially uninterrupted by European contact for another century and a half.

4. The 'Sacred Expedition'

Spain's policy of benign neglect toward the coastline stretching north of Baja California would end almost as abruptly as it had begun. In 1741, two mariners in the service of the Russian Crown, Vitus Bering and Aleksei Chirikov, succeeded in reaching North America by sailing eastward from Kamchatka.[2] It opened the way for Russian *promyshlenniki* 'fur hunters' to begin exploiting the fur resources of the Aleutian Islands. Despite efforts to keep such activities cloaked in secrecy, word inevitably got out. Officials in Madrid and Mexico City grew increasingly uneasy about the presence of Russian explorers and fur hunters in the Aleutian Islands and on the nearby Alaskan mainland. It was viewed as a challenge to Spain's sovereign claims in the Pacific potentially as great or greater than depredations on Spanish shipping by English or Dutch privateers.

In 1768, another government-financed Russian exploring expedition, under command of two naval officers, Commander Captain Peter Krenitsyn and Lieutenant Michael Levashev, was launched and began charting the Aleutian Islands.[3]

---

[1] Wagner, *Spanish Voyages*, p. 277.

[2] Accounts of the Bering-Chirikov expedition are in Golder, *Bering's Voyages*, 1, pp. 25–348; Fisher, *Bering's Voyages*, pp. 120–51; and Müller, *Bering's Voyages*, pp. 78–128.

[3] Accounts of the Krenitsyn-Levashev expedition are in Coxe, *Account of the Russian Discoveries between Asia and America*, pp. 251–66; Masterson and Brower, 'Bering's Successors', pp. 112–20; and Dmytryshyn, Crownhart-Vaughan, and Vaughan, eds, *Russian Penetration*, pp. 245–52. Unlike the Bering-Chirikov expedition of 1741, the Krenitsyn-Levashev voyage had no foreign scientists or scholars aboard, presumably to minimize security breaches. But curiously, the Russian Tsarina, Catherine II (the Great), allowed a Scottish historian, William Robertson of Edinburgh, access to study Krenitsyn's journal and map in connection with a historical study on which Robertson was working and would eventually publish under the title *The History of America* (1777). Robertson had contacts in Spain for the portion of his work dealing with Spanish America, which conceivably could have been a source of information for Spanish officials concerning Russian activities in the North Pacific. See Makarova, *Russians on the Pacific*, pp. 8–9.

Considerable effort was made by the Russian Crown to keep the activities of this expedition secret, with somewhat better success than before. But eventually it would also be of no avail. It could have been coincidental, but 1768 was the same year in which the Spanish Crown decided to revive Vizcaíno's nearly forgotten proposal to establish an outpost at Monterey Bay.[113] This new enterprise, however, would exceed anything Vizcaíno may have envisioned. Instead of merely a stopover for Manila galleons at Monterey Bay, it would seek to establish *misiones* 'missions', *presidios* 'military posts', and *pueblos* 'settlements', not only at Monterey, but up and down the coast between the sites of present-day San Diego and San Francisco.

In addition to bringing Alta California under the control of the Spanish Crown, the expedition was intended to undertake the evangelization of its native population. This much-professed religious objective seems to have given rise to the project being unofficially dubbed 'The Sacred Expedition'.[114] Whatever the case, its less-advertised secular political aims were as much or more concerned with heading off the perceived Russian threat in Alaska than turning aboriginal Alta Californians into Roman Catholic Christians. Overall leadership for this undertaking fell to José de Gálvez, an energetic, intelligent, impatient, sometimes eccentric, career bureaucrat, who had been dispatched to New Spain in August 1765 in the capacity of *visitador general* 'inspector general'. As the personal representative of the Spanish king, Charles III, Gálvez was empowered to recommend a variety of administrative and economic reforms; his authority in some instances even exceeded that of the viceroy of New Spain himself – who was then Carlos Francisco de Croix, Marqués de Croix. In a word, Gálvez had the Crown's backing to do whatever was necessary to Hispanicize Alta California.[115]

Among Gálvez's many responsibilities was the task of overseeing the expulsion of all Jesuit missionaries from New Spain on the royal order of Charles III in 1767. The Society of Jesus, having been accused of sedition, was to be expelled from all Spanish territories at home and abroad.[116] Whatever truth there may have been in the allegations of sedition, the Jesuits had experienced only mixed success in their efforts to

---

[1] At least one earlier attempt to implement Vizcaíno's proposal is known, but it never materialized. In the late 17th and early 18th centuries the Jesuit missionary and explorer Eusebio Francisco Kino advocated advancing the Spanish frontier northward overland into Alta California as far as Monterey Bay. Kino's discovery that 'California no es isla, sino penisla' (California is no island, but a peninsula) led to his proposal for an overland expedition to initiate occupation of the region. His motivation was primarily religious, with little if any consideration given to political aims of the Spanish Crown – which may help to explain why the expedition failed to take place. See Bannon, *Spanish Borderlands Frontier*, pp. 65–9; Bolton, *Rim of Christendom*, pp. 474–8, and (ed.) *Kino's Historical Memoir*, pp. 351–4; Burrus, *Kino Reports*, p. 97; Chapman, *Spanish California*, pp. 18–23; and Weber, *Spanish Frontier*, p. 239.

[2] The name 'La Santa Expedición' 'The Sacred Expedition' has its origin, according to Theodore Treutlein, in a letter written by Inspector-General Gálves to Frey Francisco Palóu, on 9 January 1769. The phrase apparently was only used that one instance and was never intended as an official designation. Nevertheless, the name has gained wide currency among historians and writers as a convenient way to reference the expedition. See Treutlein, *San Francisco Bay*, p. 8; and Weber, *Spanish Frontier*, pp. 243, 449, n. 28.

[3] Concerning Gálvez's role in the effort to initiate the colonization of Alta California, see Bannon, *Spanish Borderlands Frontier*, pp.153–4; Chapman, *Spanish California*, pp. 68–71; and Weber, *Spanish Frontier*, pp. 237–9.

[4] On the expulsion of the Jesuits from Spanish territory, see Bannon, *Spanish Borderlands Frontier*, p. 153; Chapman, *Spanish California*, pp. 69, 72, 105; and Weber, *Spanish Frontier*, pp. 241–2.

bring the natives of the arid and barren Baja California peninsula into Christendom. In place of the Jesuits, Franciscans from the Apostolic College of San Fernando in Mexico City were hastily rushed into the breach created by the expulsion. Among them, Fray Junípero Serra would prove to be a key figure in the expedition to occupy Alta California.

Gálvez's grand strategy was to establish settlements in Alta California, which were envisioned eventually to become self-sustaining and serve as bases to support maritime expeditions dispatched into the far north to blunt the spread of Russian activities – and those of other nations as well, should they arise. But before such strategic aims could be achieved a naval base and shipyard would have to be established somewhere on the west coast of New Spain, north of and apart from the busy commercial entrepôt at Acapulco. The site selected by Gálvez was an obscure village named San Blas on the Pacific shore in the present-day Mexican state of Nayarit. Surrounded by insect-ridden mangrove swamps, cursed with a less than healthful climate, poorly suited for farming or ranching, and with a shallow, silt-prone harbour, it did not seem a felicitous choice. It did, however, possess one clear advantage in that its location offered the most direct route for ships sailing to or from Alta California or the outer coast of Baja California. As for climate, its most baneful conditions – enervating, muggy summers and autumns – did not persist year-round, and the winter and spring weather could seem by comparison almost salubrious. Whatever its advantages or disadvantages, the Naval Department of San Blas would serve, during the last three decades of the eighteenth century, as the principal facility supporting Spanish efforts to colonize Alta California and conduct exploration along the Pacific littoral as far north as Alaska.[1]

The plans to gain a Spanish foothold in Alta California would involve a two-pronged advance: one by land and the other by sea. The land party was itself divided into two groups. The first, under command of Mexican-born Captain Fernando de Rivera y Moncada, consisted of twenty-five *soldados de cuera* 'leather-jacketed soldiers';[2] accompanied by José de Cañizares, a promising young naval officer who was charged with keeping the captain's diary and making astronomical observations; and by Franciscan missionary, Fray Juan Crespí, a Majorcan native, who served as chaplain and diarist. There were also three muleteers to manage the pack train. And according to Crespi there were 'some forty California Indians, new Christians, from the last [northern] missions, for the labour of opening roads and other things that might come up'.[3]

---

[1] The reasons Gálvez chose to locate the naval base at San Blas have long been debated. It may have simply been, as one observer has argued, that the 'answer lies in the initial ignorance and lack of investigation of the site on the part of Gálvez and his subsequent refusal to admit the mistake once construction had started'. See Inskeep, 'San Blas, Nayarit', p. 139.

[2] These leather-clad soldiers were originally cavalry troops, who were, together with their mounts, burdened with well over 100 lb of various weapons, such as lances, swords, firearms, as well as protective leather shields and jackets offering protection against Indian arrows. See Weber, *Spanish Frontier*, pp. 217, 219.

[3] Bolton, ed., *Fray Juan Crespi*, p. 62. Eyewitness accounts of Rivera's advance party do not always agree on exact numbers. It appears from Cañizares's account that 15 of Rivera's 25 troops remained at Velicatá to assist in building a new Franciscan mission there. Crespi's account makes no mention of this. As for the California Indian participants, Cañizares places their number at 42; Crespi at 40. In any event, an uncertain number of the natives are known to have deserted, died or returned to their home missions in the course of travel. For this reason, an accurate count was difficult if not impossible then and remains so today. For Cañizares's account, see Thickens and Mollins, 'Putting a Lid on California', 31 Jun. 1952, pp. 109–24; 31 Sept. 1952, pp. 261–70; 31 Dec. 1952, pp. 343–54. For Crespi's account see Bolton, ed., *Fray Juan Crespi*, pp. 59–273.

This advance party set out on 24 March 1769 from Velicatá, on the northern frontier of Baja California. Its task was to blaze a route to San Diego Bay across an arid and often inhospitable landscape, little of which had ever been trod by Europeans – and from there eventually to Monterey Bay. The second contingent was led by the expedition's commander-in-chief, Catalonia-born Captain Don Gaspar de Portolá, military governor of the Californias. This group assembled at Mission Santa María, northernmost of the former Jesuit missions in Baja California. In addition to Captain Portolá, this party, escorted by ten soldiers, included the Majorcan friar, Junípero Serra, *padre presidente* 'father president' of the Baja California missions now in the care of the Franciscan order. On 11 May 1769 they began their northward trek first to Velicatá and thence followed the route pioneered by Rivera's advance party to reach San Diego Bay.[1]

As for the expedition's naval contingent, Gálvez's plans called for the overland parties to be supported initially by two ships – later backed up by a third. Freshly constructed at a shipyard on the banks of the Río Santiago near San Blas, the first two included the packet boats *San Carlos*, alias *El Toisón de Oro* (The Golden Fleece) and her sister ship *San Antonio* alias *El Príncipe* (The Prince).[2] Under command of Master First Class Vicente Vila, a native of Andalucia, the flagship *San Carlos* was first to depart San Blas on 26 September 1768. She was bound for La Paz, a well-sheltered bay on the inner coast of Baja California, about ninety nautical miles directly north of Cape San Lucas, the southernmost point of the peninsula. At La Paz, Gálvez had positioned in advance supplies, equipment, and additional personnel, to be taken aboard the *San Carlos*. She would then sail around Cape San Lucas and up the outer coast of Baja California to rendezvous with the overland parties at San Diego. Her departure from San Blas, however, was considerably delayed, which placed her in the Sea of Cortez at the peak season for hurricanes (late summer and autumn).

Arriving sometime in December 1768, the *San Carlos* took much longer than usual to sail from San Blas to La Paz, a distance of some 280 nautical miles.[3] And much to Gálvez's considerable chagrin, she was not only late but her tackle and rigging were in shambles; she was leaking and much in need of various repairs – suggesting a possible encounter with a hurricane. In any event, the inspector-general had foreseen such a contingency and was ready with a repair crew to board the *San Carlos* and restore her seaworthiness. After a fortnight to accomplish this, the *San Carlos* was ready to depart La Paz on 9 January 1769. On that occasion, after religious observances, Gálvez delivered a speech exhorting the assembled crew, among other things, 'to observe

---

[1] See Smith and Teggert, eds, *Diary of Gaspar de Portolá*, pp. 38–59.

[2] They were twin-masted vessels each with displacements of 195.2 tons each, resembling what is called in English terminology a snow. The snow was characterized by a 'snowmast' immediately aft of the mainmast from which a gaffsail could be rigged. *San Carlos*'s length on the main deck was 24.1 m (79.2 ft); *San Antonio*'s was somewhat less at 21.4 m (70.3 ft). Keel length of *San Carlos*, 19.7 m (64.6 ft); *San Antonio*'s 19.5 m (64 ft). In other respects the ships' dimensions were the same: width abeam 7 m (23.1 ft); width of hold, 4.6 m (15.1 ft); depth of hold, 3.2 m (10.4 ft). See Kenyon, 'Naval Construction', pp. 114–22.

[3] In Bolton, ed., *Historical Memoirs*, 2, p. 12, Francisco Palóu says the *San Carlos* 'arrived at La Paz in the early part of December'. A second source, Fireman, *Spanish Royal Corps of Engineers*, p. 98, places her arrival in mid-December 1768. A third source, Juan Manuel de Viniegra, quoted by Thurman in *San Blas*, p. 69, places the ship's arrival at La Paz on 25 December 1768. As Viniegra's reliability has been questioned, the best that can be said about the *San Carlos*'s arrival at La Paz is that it happened sometime in December. See, Weber, *Spanish Frontier*, p. 449, n. 39.

peace and harmony among themselves and obedience and respect to their superiors'.[1] It suggests, in addition to adverse sailing conditions, some form of dissension may have arisen on the voyage from San Blas to La Paz. Whatever the case, the inspector-general, sailing aboard another packet boat named *Concepción*, accompanied Vila and the *San Carlos* from La Paz as far as Cape San Lucas. From there, on 11 January 1769, the *San Carlos* embarked upon the open ocean.

Meanwhile, the *San Antonio*, captained by Master First Class Juan Pérez, a Majorcan native and veteran Manila gallleon master, was being readied to depart San Blas. It was almost November 1768, however, before Pérez was able to get underway to rendezvous with Vila and the *San Carlos* at La Paz. After leaving San Blas, Pérez and the *San Antonio* seem to have encountered some degree of adverse weather, but nothing as serious as Vila and the *San Carlos* had faced. Although the *San Antonio* received a lesser share of weather-inflicted damage, it was sufficient to prevent her from rounding the Baja peninsula's easternmost point and forced Pérez to abandon his effort to reach La Paz, seeking instead shelter at San Bernabé, an anchorage near Cape San Lucas. It was in the same vicinity Gálvez had bid farewell to Vila and the *San Carlos* only a few days earlier.

According to Franciscan historian Fray Francisco Palóu, Gálvez learned of Pérez's intentions to head for the San Barnabé anchorage, so when the *San Antonio* hove into view, the inspector-general was there ready to offer assistance.[2] His repair crew was well practised and ready to go into action immediately, 'It was', says Palóu, 'necessary to unload [the *San Antonio*] and examine the seams, through which water was leaking'.[3] By mid-February, the ship was careened, repaired, reloaded, and ready for the voyage northward to San Diego. After observing religious rites, the *San Antonio*'s officers and men were treated to the same exhortations delivered earlier by Gálvez to the *San Carlos*'s crew.[4] Both packet boats were now at sea committed to reach San Diego Bay and after that, the bay Vizcaíno had named in honour of the Conde de Monterrey.

Although the flagship *San Carlos* departed Cape San Lucas fully a month before the *San Antonio*'s departure from there, the latter was first to cast anchor at San Diego. In fact it took Vila and the *San Carlos* no less than 110 days to travel what Pérez and the *San Antonio* covered in only fifty-four. As the two ships were essentially identical in design and age, it is difficult to attribute this disparity solely to flaws in one of the ships. Other more likely possibilities include: inept seamanship, poorly constructed casks, lax water rationing or just plain bad luck. Vila's journal of the voyage paints a picture of persistent water shortages brought on by casks rendered leaky in rough seas. Vila explains: 'the strong wind and swollen seas from the north and NNE tossed the [*San Carlos*], making her roll and pitch violently and causing great loss in the contents of the

---

[1] Bolton, ed., *Historical Memoirs*, 2, p. 13. Weber, *Spanish Frontier*, p. 243, speculates that 'sabotage by mariners reluctant to sail' may have been a problem on the voyage from San Blas to La Paz. Gálvez's 'wise and tender speech', as Palóu characterized it, could be seen as a response to some unspecified insubordination.

[2] Bolton, ed., *Historical Memoirs*, 2, p. 14.

[3] Ibid., 2, p. 15. This was the *San Carlos*'s maiden voyage, and according to Kenyon, in 'Naval Construction', p. 120: 'She had left San Blas in poor condition, without needed repairs or even an inspection due to illness in the shipyard'.

[4] Bolton, ed., *Historical Memoirs*, pp. 16–17.

casks, owing to constant leakage'.[1] Because of this, he felt compelled to find a source of fresh water to replenish his dwindling supply – as well as to make certain repairs and collect firewood. Accordingly, on 17 February, within sight of an island they knew as Guadalupe, at latitude 29°N and about 160 nautical miles offshore, Vila ordered the *San Carlos*'s helmsman to alter course to the north-east in search of the mainland, in hopes of finding a suitable watering place.

They first came in sight of the mainland on the afternoon of 19 February. It was partly obscured in fog, making a close approach dangerous. After an unsuccessful search for several days, on 7 March they came upon an island. Vila calls it Isla de Cerros, although it is almost certainly the same one modern charts label Isla de Cedros (Cedros Island).[2] A launch, with several armed soldiers under command of Lieutenant Pedro Fages, was sent ashore to search for fresh water. By nightfall, they reported back with news the island did indeed have water, some of which they bottled and brought back for the ship's officers to test. 'The officers tasted it as did the surgeon, Don Pedro Prat', says Vila, 'and, although they found it somewhat brackish, they thought that, in our present circumstances, it was not to be disregarded'.[3] And so the collection of water commenced in earnest.

For eight days, 13 to 21 March, Vila's watering crews laboured tirelessly to fill no less than 159 water casks and transport them to the ship from various locations on the east shore of the island.[4] With her water supply replenished, the *San Carlos* would have to regain her northward track, which would quickly prove to be fraught with considerable difficulty. Not until 8 April did she come in sight of Guadalupe, the island where Vila's original decision to seek a watering place had been made. It meant that twenty-seven days had been devoted to restoring the ship's water supply. An additional eleven days, 9 to 20 April, passed before the *San Carlos* started once more to gain latitude. In all, thirty-eight days had been spent on this quest before Vila and the expedition's flagship were back heading in the direction of their immediate objective of San Diego.

On 19 April, the day before favourable winds began to blow, Vila recorded the death of Fernando Alvarez, boatswain's second mate and coxswain of the launch.[5] The cause of his demise is unclear, but it was a harbinger of things to come. Five days later, on 24 April, another death, that of Manuel Reyes, a master, was recorded, accompanied by mention of general sickness among the crew members.[6] Their ill health could have been symptomatic of that ancient scourge of the sea, scurvy. But the timing of Alvarez's and Reyes's deaths fairly soon after the detour at Cedros Island, suggests it may have had something to do with the water the crew was now drinking. In any event, whatever its cause, the growing incapacitation of the *San Carlos*'s crew did not bode well for reaching San Diego in a timely fashion.

The timing of the *San Carlo*'s arrival at San Diego was a matter of no small importance. Both the flagship and her consort were under orders requiring the first ship to reach San Diego to remain there twenty days, awaiting the arrival of the other

---

[1] Rose, ed., *Diary of Vicente Vila*, p. 19.
[2] Ibid., p. 43. In Spanish, *cerros* means hills or peaks; *cedros* means cedars.
[3] Ibid.
[4] Ibid., pp. 49–61.
[5] Ibid., p. 79.
[6] Ibid., p. 83.

ship or members of the overland expedition. If, at the end of twenty days, neither had appeared, the first ship to arrive was to set sail in search of Monterey Bay.[1] As Vila had no knowledge of exactly when the *San Antonio* departed Cape San Lucas, he could only conjecture as to when Pérez might arrive at San Diego – except that, the *San Carlos*, having the advantage of embarking from San Lucas a month earlier than the *San Antonio*, the former would seem most likely to be the first to arrive.

At two o'clock in the afternoon of 25 April, Vila reported sighting an island in latitude 33°15′N.[2] It was probably San Nicolas, a small outlying island known to be at about that same latitude. It was welcome news as it meant that the *San Carlos* was nearing the coast at a point where Vila could soon begin searching for the harbour at San Diego.

The next morning, Vila remarked:

> I was between four islands and the mainland; the country high and mountainous with several high ridges extending NW–SE, all of them covered with snow, like the Sierras Nevadas of Granada.[3]

These four islands were well known to Spanish navigators as part of the Channel Islands, an off-shore archipelago defining the Santa Barbara and San Pedro channels. They had been first explored by Cabrillo in 1542; and again by Vizcaíno in 1602. Manila galleon masters were also well acquainted with these islands and channels, which offered a degree of protection as a coastal shipping lane for the galleons threading their way south to Acapulco. Three of the islands, San Miguel, Santa Rosa, and Santa Cruz, stretch along the thirty-fourth parallel, where they form the seaward side of the Santa Barbara Channel. The fourth island, Santa Catalina, some sixty-five nautical miles south-east of the Santa Barbara Channel, marks the San Pedro Channel.

As the *San Carlos* ranged south-easterly along the coastline and through the San Pedro Channel, she passed the future site of the city of Los Angeles. Vila tried to stop and anchor in San Pedro Bay, where the present-day city of Long Beach is located. This attempt to stop appears to have been made to deal with another water supply shortage, but it was aborted by an unfavourable change in the wind. Vila thereupon decided to forgo any further attempts to go ashore and instead simply pressed on to reach San Diego as soon as possible.[4] On the morning of 28 April, the coast was engulfed in fog. Then, about 9 a.m., the fog began to lift, and Vila discerned a group of islets or rocks eight leagues to the south, which he recognized to be *Los Coronados*. 'They are', he noted, 'the best and surest marks for making the port of San Diego which is situated about five and a half or six leagues due north of these islands'.[5] Late in the afternoon of 30 April, the *San Carlos* reached the entrance to San Diego Bay, where she ploughed through a field of kelp so thick it almost brought the ship to a complete

---

[1] Bolton, ed., *Historical Memoirs*, 2, p. 18.

[2] Rose, ed., *Diary of Vicente Vila*, p. 83.

[3] Concerning the snow-covered mountains, Vila wrote: 'Following the notes of the sea-pilot Cabrera Bueno, I decided that they might be the ridges which the Philippine [sailors] call [Sierra] de Santa Lucia'. Loc. cit. It seems more likely, however, that they were either the Santa Ynez or Santa Monica mountains, north-west of present-day Los Angeles.

[4] After Vila decided against further attempts to stop, he wrote: 'I made sail to look for the port of San Diego, where we confidently expected to find the relief that we needed, especially the sick, as our water supply was already running short'. Ibid., p. 87.

[5] Ibid., p. 89.

halt. But Vila and his ailing crew managed somehow to push through and complete their entrance into the safe haven of this superb harbour. And there, awaiting them, was Juan Pérez and the *San Antonio*.[1]

Upon sighting his consort, Vila immediately broke out his colours, to which the *San Antonio* responded in kind – in addition to which, says Vila, she 'fired one gun to call in her launch which was ashore'.[2] On the following day, 1 May, Pérez and Vila met to assess their situation. The *San Antonio* had arrived at San Diego on 11 April, and in the ensuing nineteen days, during which Pérez had waited, he saw neither Vila's flagship nor any sign of the overland parties. The two ships were on the verge of missing each other as the twenty-day limit was about to expire. According to Vila, 'Pérez had conjectured that I had probably been the first to reach the port of San Diego, and then had gone on to Monterey'.[3] Without knowing the actual fate of Vila and the *San Carlos*, Pérez had reluctantly begun to prepare the *San Antonio* to depart for Monterey Bay, which he would have been obliged by orders to do. But it would not have been something easy to do, as roughly half of his crew was afflicted by scurvy. Pérez himself, says Vila, 'was also in poor health'.[4]

Despite the prevalence of scurvy among Pérez's crew, the *San Antonio*'s voyage to San Diego had evidently been less taxing than that of the flagship. Pérez had not found it necessary to stop in mid-voyage to restore his water supply. Instead, he had sailed directly to the Channel Islands, where he stopped at Santa Cruz Island to take on water, after which he followed the coast south-easterly without delay to San Diego.[5] The illnesses afflicting the flagship's crew clearly were more serious than among Pérez's men. According to Palóu:

> The cause of the gravity of the sickness of the flagship's crew was thought to be the water which they had to take on at the island of Cedros. It was so bad nothing could be cooked in it; the meat came out tougher than before it was put on the fire, and the same happened with the miniestra [?]; and as they drank the same water, for lack of any other, those who were already ill became worse and the plague seized upon the rest.[6]

Vila and Pérez quickly agreed that a supply of fresh water had to be located as soon as possible and shelter constructed ashore for the mounting numbers of sick crewmen and soldiers. The plan to sail on to Monterey Bay had to be postponed, and the officers instead began to contemplate the possibility of sending one of the packet boats back to San Blas for reinforcements and additional supplies. They found fresh water in relative abundance at the mouth of a river,[7] and the shelters ashore – described by Vila as built of 'brushwood and earth', and by Palóu as 'tents [made] of the sails for an infirmary'[1] – were presently completed. But no decision could be made about

---

[1] Ibid., p. 91.
[2] Ibid.
[3] Ibid., p. 95. Pérez's conjecture that the *San Carlos* had arrived first and, after waiting 20 days, had gone on to Monterey Bay was perfectly reasonable considering the *San Carlos* had left Cape San Lucas a full month ahead of the *San Antonio*.
[4] Ibid., p. 93.
[5] See Bolton, ed., *Historical Memoirs*, 2, pp. 16–18, for a brief account of the *San Antonio*'s voyage.
[6] Ibid., 2, p. 21. The exact meaning of 'miniestra' is uncertain, although it apparently refers to food of some sort.
[7] Probably the San Diego River, which then flowed into the bay from the north.

returning one of the ships to San Blas until the arrival of the expedition's commander-in-chief, Captain Gaspar de Portolá. Vila was also aware of plans for a third packet boat, the *San José*, to set sail with additional back-up.[2] In the event this relief ship arrived, says Vila,

> we flattered ourselves that we might still continue the voyage to Monterey, and that we might dispatch one of the vessels to San Blas ... to seek suitable reinforcements.[3]

It had been raining since sunrise on the morning of 14 May 1769. Then, at about 10 a.m. when the rain let up, members of Captain Rivera's overland party bestirred themselves to resume their march towards the bay where they soon expected to join their maritime colleagues.[4] José de Cañizares, Rivera's diarist, upon coming into view of San Diego Bay, would write: 'After travelling about a league we had the great joy of seeing in the distance the two packet boats, the *San Carlos* and *El Príncipe* [alias *San Antonio*], anchored in the port'.[5] But Cañizares's initial joy soon turned to concern as he became aware of the widespread sickness among the unfortunate sailors and soldiers who had survived their respective sea voyages only to fall victim to disease. Just two seaman of the *San Carlos*'s crew were well enough to perform their regular duties. And aboard the *San Antonio*, about half of the crew was down with scurvy. When many of the sick – mostly aboard the *San Carlos* – began to die, the survivors were forced to expend precious effort in the burial ashore of their shipmates.[6]

Members of Rivera's overland party were, by contrast, in comparatively good health – perhaps fatigued and somewhat malnourished – but otherwise with few if any apparent cases of scurvy. Galvéz and his advisors, in planning the expedition, may have thought its seagoing contingent would be there to assist the road-weary overland arrivals, when in fact the circumstances were very nearly the exact opposite. The packet boats were capable of transporting much heavier cargoes than overland mule trains, but at a considerable price in human suffering. To further compound the problem, the third packet boat, the *San José*, concerning which Vila held out hope for

---

[1] Rose, ed., *Diary of Vicente Vila*, p. 101; and Bolton, ed., *Historical Memoirs*, 2, p. 20.
[2] This packet boat, *San José* alias *El Descubridor* 'The Discoverer', was the first ship to be constructed at the shipyard within San Blas proper. Completed in 1768, she was somewhat smaller than both the *San Carlos* and the *San Antonio*. The *San José*'s keel length was 15.5 m (50.9 ft), compared to the *San Carlos*'s 19.7 m (64.6 ft) and the *San Antonio*'s 19.5 m (64 ft). As for width abeam, the *San José* measured 4.6 m (15.1 ft) compared to 7.0 m (23.1 ft) for both the *San Carlos* and the *San Antonio*. And the *San José*'s tonnage was 180 tons compared to the *San Carlos* and *San Antonio*'s 193 tons each. See Kenyon, 'Naval Construction', pp. 31, 122–3.
[3] Rose, ed., *Diary of Vicente Vila*, pp. 95, 97.
[4] See Thickens and Mollins, 'Putting a Lid on California', pp. 352–3.
[5] Ibid., p. 353.
[6] While the *San Carlos* was on the sea voyage from Cape San Lucas to San Diego Bay, Vila recorded by names and dates the deaths of two of his crewmen. They were: Fernando Alverez, boatswain 2nd mate (died 18 April) and Manuel Reyes, master (died 24 April). Both men were buried at sea. Once the *San Carlos* reached San Diego, Vila recorded by names and dates the deaths of three additional crewmen: Fernández de Medina, Philippine seaman (died 5 May); Manuel Sanchez, cabin boy (died 10 May); and Matheo Francisco, Philippine seaman (died 10 May). These three presumably were buried ashore at San Diego. There were many other deaths and burials not mentioned by name or date in Vila's account, possibly because their large numbers and his own weakened condition prevented him from keeping count. See Rose, ed., *Diary of Vicente Vila*, pp. 79, 83, 101, 103, 105.

assistance, failed to reach San Diego and was in fact lost at sea along with all hands and her entire cargo.[1]

It was 1 July 1769 when the main overland party, with Captain Portolá and Fray Junípero Serra, finally reached San Diego. The expedition was now reassembled, but under circumstances far from ideal. Portolá's entry in his diary for that day depicts the dismal conditions that had befallen the expedition:

> This day, we arrived at the camp at San Diego, where we found Captain Fernando de Rivera and his men and, in the port, [we found] anchored the *San Carlos* and the packet *El Príncipe* [or *San Antonio*]. The greater number of the land-volunteers were sick; of the naval force there remained only a few sailors, and in particular on the *San Carlos* nearly all the men had died. This predicament left them undecided [what to do].[2]

Despite the many difficulties the expedition faced, Portolá was determined to reach Monterey Bay by whatever means could be devised. For the present it would not be by sea. The *San Carlos*'s crewmen were either dead or too ill to work the ship. This, together with the failure of the relief ship *San José* to appear, led Portolá and Vila to assemble the remaining able-bodied seaman of the *San Antonio* to sail her back to San Blas to secure provisions and personnel to replace the expedition's losses. The voyage commenced on 9 July 1769, with Pérez in command. Although there would be nine more burials at sea, the *San Antonio* reached San Blas after a voyage of only twenty days, following which Pérez successfully obtained the needed reinforcements and supplies. On his return voyage, his orders were to bypass San Diego and sail directly to Monterey Bay, in the mistaken belief that the overland contingent would by then be established there and the *San José* would have brought relief for the expedition's members remaining at San Diego. In the course of his return north, however, Pérez learned from natives in the Channel Islands that the Spaniards ashore had reached the Monterey vicinity only to return hastily to San Diego. Thus informed, Pérez altered course and sailed directly for San Diego, where, on 24 March 1770, he went ashore to find the surviving expedition members in desperate circumstances. His commendable seamanship and willingness to risk sailing for San Diego, contrary to his orders, probably saved the expedition from ending in a disastrous failure.[3]

Portolá's overland bid to reach Monterey Bay faced a daunting march of over 700 km up the Alta California coast. He assembled a corps including, in addition to himself: Captain Fernando de Rivera, with a sergeant and eight soldiers; Lieutenant Pedro Fages with six of his Catalonian volunteers; an engineer, Miguel Costansó; and two Franciscan missionaries Fray Juan Crespi and Fray Francisco Gómez. The party was also accompanied by muleteers to manage the expedition's mule train.[4]

---

[1] See Bolton, ed., *Historical Memoirs*, 2, pp. 36–9, for a discussion as to what was known about the *San José*'s ill-fated voyage in 1769.

[2] Smith and Teggart, eds, 'Diary of Gaspar de Portolá', p. 19.

[3] See Bolton, *Historical Memoirs*, 2, pp. 105–8 and 275–7. When Pérez returned to San Diego in March 1770, he received a hero's welcome from the starving and beleaguered soldiers, sailors, missionaries, and other colonists.

[4] Portolá's diary, in his entry for 14 July, explicitly mentions Fages, his six volunteers and Costansó. See Smith and Teggart, eds, *Diary of Gaspar de Portolá*, p.19. Other participants, such as Rivera, a sergeant and eight soldiers under his command, are mentioned in Portolá's entry for 7 November 1769. See ibid. p. 39. As for the chaplains, Crespi and Gómez, their participation is fully documented in Crespi's diary. See Bolton, ed., *Historical Memoirs*, 2, pp. 109–260; or Bolton, ed., *Fray Juan Crespi*, pp. 122–273.

Spanish Borderlands historian Herbert Bolton has described Portolá's march north from San Diego as 'the California Anabasis'.[1] Whatever the aptness of this comparison, the expedition earned a memorable place in California's colonial history. It also bore witness to many unusual if not unique features of aboriginal California not readily discernable from offshore. Most of the trek followed well-established native trails, paths or roads. Over the entire length of the march, Portolá counted at least forty native villages ranging in size from the smallest of about eight inhabitants upward to some with a population as large as an estimated 800.[2] In most cases, the Spaniards were greeted in friendly fashion by the natives. When the explorers crossed the basin where present-day Los Angeles and its suburbs are located, they learned first-hand of the region's seismic infamy. At noon on 28 July 1769, a powerful earthquake shook the ground, which, according to Portolá 'lasted about half an Ave María and, about ten minutes later, it repeated though not so violently'.[3] More earthquakes continued to rumble. And in the same vicinity the expedition came upon swamps of tar or bitumen, which subsequently have gained fame as the La Brea Tar Pits.[4] As they continued their trek along the Santa Barbara Channel, the explorers came among the remarkable Chumash people, taking special note of their unusual planked seagoing canoes.[5]

As the expedition pushed northward to find Monterey Bay, Portolá and his officers found themselves in something of a quandary. When they reached the latitude at which the bay was supposed to be located (36°38′N), what they found failed to meet the high expectations they had envisioned of the bay and its harbour. When Vizcaíno discovered and named Monterey Bay in December 1602, he described it glowingly:

> [We] found ourselves to be in the best port that could be desired, for besides being sheltered from all the winds, it has many pines for masts and yards, and live oaks and white oaks, and water in great quantity, all near the shore.[6]

It was an exaggeration of the harbour's desirable qualities – especially the assertion that it offered protection from all wind directions – which led Portolá and his officers to conclude that the real Monterey Bay must lie farther north. In thus continuing their search, they would discover quite by accident what is today known as San Francisco Bay.

It was 31 October 1769, when Portolá and his party scrambled to the summit of 'a very high mountain' – apparently somewhere on Montara Mountain – from which they discerned in the distance to the north-west 'some small islands, a point of land,

---

[1] Bolton, *Outpost of Empire*, p. 21.
[2] Portolá's population estimates total 7,173 individuals, living in some 39 villages. Although it is difficult to assess the accuracy of these estimates, the aboriginal population of Alta California clearly was substantial.
[3] Smith and Teggart, eds, *Diary of Gaspar de Portolá*, pp. 21, 23.
[4] The *patanos de brea* 'swamps of bitumen', as Portolá called them – or *chapapote* according to the natives – are doubtless the La Brea Tar Pits. See ibid., pp. 22, 23, 'We debated', says Portolá, 'whether this substance, which flows from underneath the earth, could occasion so many earthquakes'. Ibid. p. 23.
[5] 'These canoes', wrote Portolá, 'though narrow, are eight yards in length, well made, and constructed of boards'. Ibid., p. 25. The Chumash planked canoes were unique and limited geographically to the Santa Barbara Channel area, nearby San Luis Obispo Bay, and the islands of Santa Catalina and San Clemente. They allowed the Chumash to travel between the various Channel islands and communicate with the mainland. The use of planks in construction of these canoes required large amounts of asphaltic caulking material, which likely came from the La Brea Tar Pits. See Heizer, *The Plank Canoes of the Santa Barbara Region, California*, pp. 220–21.
[6] Bolton, *Spanish Exploration in the Southwest*, p. 91.

and a bay'.[1] Portolá believed they were looking at the 'neighborhood of the Port of San Francisco as described in Cabrera Bueno's *Navegación*'.[2] In this surmise Portolá was undoubtedly correct. The 'small islands' correspond to today's Farallon Islands, the 'point of land' to Point Reyes, and 'the bay' to what is now called Drakes Bay. Shortly thereafter, on 5 November, they discovered

> a great arm of the sea, sixteen or twenty leagues [in extent], which the pioneers [a scouting party of eight soldiers led by Sergeant Ortega] said formed a sheltered port with two islands in the middle.[3]

This 'great arm of the sea' is known today as San Francisco Bay. Impressed by its vastness, Crespi remarked, that 'it is a very large and fine harbour, such that not only all the navy of our Most Catholic Majesty but those of all Europe could take shelter in it'.[4] Realizing by now that they had gone well beyond – by about 150 km – the Monterey Bay they were seeking, a council was convened of the four officers (Portolá, Rivera, Fages, and Costansó) including the two chaplains (Gómez and Crespi). They decided 'that the expedition should return and that the port [of Monterey] should be sought for with greater care'.[5] On 11 November, Portolá and his men began retracing their steps southward; and a fortnight later they returned to the bay now suspected of being the real Monterey Bay.

Given their uncertain circumstances and lacking naval support, they decided that the expedition should return all the way south to San Diego. When they arrived there on 24 January 1770, they were greeted by a most distressing scene. The packet boat *San Carlos* remained at anchor exactly where she had been when Portolá set out to find Monterey Bay over six months earlier. She was essentially unmanned, serving only as Vila's headquarters and residence, guarded by a handful of sailors and soldiers. As for the other two vessels assigned to support the expedition, the *San Antonio* and *San José*, there was no sign of them at all. The failure of any supplies to reach the fledgling settlement at San Diego could be read in the faces of the colonists, many of whom were desperately malnourished. Some had died or were dying slowly, day by day, from the baneful effects of scurvy – or possibly some other more contagious affliction. The only encouraging sign was the progress Serra and his Franciscan missionaries, Fray Fernando Perrón and Fray Juan Vizcaíno, had achieved in the construction of 'some humble huts or sheds … one of them destined to serve as a chapel'.[6]

To compound the hardships the colonists endured, the natives at San Diego – the Spaniards called them Diegueño Indians – resisted giving up their native beliefs and customs. When Portolá's party returned from the north, not one Diegueño had been

---

[1] Smith and Teggert, eds, *Diary of Gaspar de Portolá*, p. 39. Bolton. *Outpost of Empire*, p. 22, places their vantage point on Montara Mountain. The site can probably be narrowed further to Montara Mountain's North Peak (el. 579 m). North Peak is located about 1.86 km due north of Half Moon Bay and 27 km due south of the Golden Gate entrance to San Francisco Bay.

[2] See Cabrera Bueno, *Navegación*, pp. 302–3, where these geographic features are cited as landmarks to assist Manila galleon masters on their voyages skirting along the North American coast to reach Acapulco.

[3] Smith and Teggert, eds, *Diary of Gaspar de Portolá*, p. 39.

[4] Bolton, ed., *Fray Juan Crespi*, p. 28.

[5] Smith and Teggert, eds, *Diary of Gaspar de Portolá*, p. 39.

[6] See Bolton, ed., *Historical Memoirs*, 2, pp. 267–8. The mission of San Diego de Alcalá was formally founded by Serra on 16 July 1769, the first of the Alta California missions.

converted to the Christian faith. In mid-August 1769, the natives, armed only with bows and arrows, launched an attack on the Spaniards. The uprising came uncomfortably close to succeeding, but it was eventually put down by the small garrison of leather-jacketed soldiers using firearms, steel-tipped lances, and whatever protection their leather jackets afforded them against stone-tipped arrows. After this incident, a stockade was put in place around the mission settlement to protect it from further attacks.[1]

Faced with such difficulties and growing doubt as to whether they would ever see a relief ship, the expedition's leaders fell to quarrelling about what ought to be done. Portolá convened a council at which it was decided that Rivera should lead a party back along the same overland route they had originally used to reach San Diego from the south. It was hoped, according to Portolá, that 'he [Rivera] might go to [Baja] California, and also bring back the herd of cattle which was intended for this mission'.[2] Portolá also seems to have set a deadline of 20 March 1770, at which time, if no relief ship had arrived, he would reluctantly order the abandonment of San Diego. Serra adamantly opposed this and enlisted the support of Vicente Vila, whose poor health had largely confined him to his cabin aboard the *San Carlos*. This potentially bitter dispute was cut short, however, with the sighting of the *San Antonio*'s sails on 19 March 1770 off-shore of San Diego, under Juan Pérez's command. For a second time, the expedition's success or failure hung in the balance, dependent upon an event conveniently saving it at the last moment.[3]

After provisions, other supplies, and reinforcements (including a new crew for the *San Carlos*) were put ashore at San Diego, Pérez and the *San Antonio* set sail on 16 April bound for Monterey Bay. In addition to her officers and crew, she had aboard Father Serra, Miguel Constansó, military engineer, and surgeon Pedro Prat.[4] On the following day, 17 April, Captain Portolá set out with a new overland party, including among its members Fray Juan Crespi and Lieutenant Pedro Fages, to go in search of Monterey Bay once more – this time with more realistic expectations of its virtues as a harbour. Unlike circumstances in 1769, when the ships of the expedition reached their initial objective (San Diego Bay) ahead of the overland parties, the situation in 1770 was exactly the reverse. Soon after departure, the *San Antonio* encountered winds so adverse to gaining latitude that she found herself driven as far south as the thirtieth parallel, at a considerable distance off the coast as well. But eventually winds developed

---

[1] See ibid., 2, pp. 269–71.

[2] Smith and Teggert, eds, *Diary of Gaspar de Portolá*, p. 51. Rivera set out for Vilicatá on 11 February 1770', escorted by twenty leather-jacket soldiers, two muleteers, and two of the California Indians, with ten horses, and eight mules to bring the provisions'. Rivera did in fact return to San Diego with provisions and a herd of 123 head of cattle soon after the arrival of Pérez and the *San Antonio*. See Bolton, ed., *Historical Memoirs*, 2, pp. 262–3.

[3] Portolá has been criticized for his premature willingness to admit failure, as opposed to Serra's determination that the expedition not abandon its hold on San Diego regardless of the circumstances. There is little question that Serra viewed the expedition with more zealousness than Portolá. But the latter had a practical reason to set a deadline beyond which the settlement's viability could not be sustained. It was simply Portolá's estimate that by mid-March provisions would be so depleted that the expedition would barely have sufficient food to sustain it on an enforced march back to Baja California. See Chapman, *Spanish California*, pp. 99–101.

[4] See Bolton, ed., *Historical Memoirs*, 2, pp. 278–9.

more favourable to sailing north, sending the *San Antonio* scudding up to the vicinity of Point Reyes (lat. 37°59.7′N at its Light). At that juncture, Pérez steered a course to the south-east following along the shore until, on 31 May, after forty-six days at sea, he located Monterey Bay and entered the harbour sheltered in the lee of Point Pinos. He was greeted by Portolá's overland party, which had reached the harbour eight days earlier on 24 May.[1]

With this accomplishment, the expedition had achieved its basic purpose to initiate the Spanish occupation and settlement of Alta California, anchored by presidios and missions at San Diego and Monterey. Serra and Crespi would formally found a mission at Monterey on 3 June 1770, called San Carlos Borromeo. The following year the mission was relocated about seven km south of Monterey Bay, on a small river called Carmel (or Carmelo) – considered by Serra a more favourable location.[2] Although the harbour at Monterey Bay continued to have its detractors, it was destined to become the focal point of Spain's colonial presence in Alta California – insofar as its civil or secular governance was concerned. The Mission San Carlos Borromeo, as relocated at Carmel, would perform a similar role in matters ecclesiastical. The overland route Portolá and his men had forged, stretching from San Diego north to Monterey and San Francisco, came to be known as El Camino Real (The Royal Road). It was along this route that Franciscan missions would begin to spring up. In addition to the missions at San Diego and Monterey, two new missions were founded in 1771: San Antonio de Padua (14 July) and San Gabriel Arcángel (8 September). The year 1772 saw another new mission founded called San Luis Obispo de Tólosa (1 September).

5. The Bucareli Expeditions

The time now seemed ripe to prepare for the expedition – as originally envisioned by José de Gálvez – to undertake maritime exploration of 'la costa septentrional de California' 'the northern coast of California'. In addition to geographic exploration, such an effort would aim to lay formal claim to the north-west coast up to the sixtieth parallel, learn something of its native inhabitants and natural resources, determine the extent of the Russian or other foreign presence in the north-eastern Pacific and whether it posed a threat to Spanish claims or interests.

Meanwhile, a new viceroy in Mexico City had taken office in September 1771, replacing the Marqués de Croix. He was Antonio María de Bucareli y Ursúa,[3] and he

---

[1] See ibid., 2, pp. 281–9.

[2] Serra seems to have favoured the Carmel site because, in Palóu's words, '[it] is surrounded by hills with good pasture for every kind of stock. It has an abundance of firewood, as well as timber for building'. Ibid., 2, p. 320. The nearby river, moreover, ensured a reliable supply of water for drinking and irrigation. It also may be that Serra sought to keep his Indian converts separated from what he considered the baneful influence of the presidio's soldiery. See Smith, *Architectural History*, p. 18.

[3] Viceroy Bucareli, a native of Seville, was related to noble families in Spain and Italy. On his paternal side he was descended from a distinguished Florentine family, boasting 'among its connections three popes, six cardinals, and other high offices; and on the maternal, the Ursuas were related to several ducal families … . He had distinguished himself in several campaigns in Italy and Spain, … he was called to be governor and captain-general of Cuba, where he again rendered valuable services to the crown, which were rewarded with a promotion to the viceroyalty of New Spain'. Chapman, *Spanish California*, p. 94.

was not as enthusiastic as Croix, Gálvez or Serra were about the California enterprise in general and maritime exploration of the north-west coast in particular. Resources immediately available at San Blas to undertake such sea ventures were slim. The packet boats *San Antonio, San Carlos* and *San José* were primarily supply ships not designed for lengthy voyages of discovery in the uncharted, bone-chilling waters of the north-west coast. Except for Pérez's superior seamanship in command of the *San Antonio*, the less-than-stellar performances of Vicente Vila and the other packet boats did not auger well for their continued use in exploration. Loss of the packet boat *San José* with her entire crew and cargo, not to mention the many officers and men who had perished from disease or deprivation in gaining the footholds at San Diego and Monterey, were legitimate reasons for the new viceroy's concern. He came on scene initially favouring efforts to find an overland route from Sonora through what is today the state of Arizona (then known to the Spaniards as Pimería Alta) to supply and reinforce the Alta California settlements primarily by land instead of by sea, while postponing any maritime exploration in the far north.[1]

Within two years, however, Bucareli's opinion would undergo a change regarding the urgency of maritime exploration on the north-west coast. Russophobic alarms were rife in Madrid at the highest levels of the government. Julian de Arreaga, minister of the Indies, had pointedly reminded Bucareli 'of the King's order to take appropriate measures to find out if these Russian explorations were going forward'.[2] Fray Junípero Serra, in his capacity as father president of the California missions, travelled from Monterey to Mexico City, arriving in March 1773. There he presented to Viceroy Bucareli a thirty-two-point petition airing grievances, placing requests, and seeking the viceroy's unequivocal support of the Alta California missions.[3] Although Serra made no explicit mention of plans for exploration of the north-west coast, his Franciscan missions would clearly be threatened if nothing was done to confront the Russians well to the north of California. Whatever the case, by the end of July 1773, Bucareli had initiated the process to assemble the resources necessary for an exploring voyage – including who should lead it.[4]

In the course of Serra's journey to Mexico City, he noted while passing through San Blas, early in 1773, that there were only three qualified masters on station there: Juan Pérez, José de Cañizares, and Miguel Pino.[5] Considering Pérez's meritorious service in the Portolá-Vila expedition as captain of the packet boat *San Antonio*, his service in three subsequent voyages to keep the missions supplied, and the fact that he outranked

---

[1] Finding an overland route linking New Spain (Mexico) with Alta California through Sonora and Pimería Alta (present-day state of Arizona), was of considerable interest to Bucareli. Such a route would diminish the need to sail across the Gulf of California. It would also reduce if not end the need to travel through the barren, inhospitable terrain of Baja California. Two expeditions led by Captain Juan Bautista de Anza, a veteran frontier army officer, with the assistance of Fray Francisco Garcés, a Franciscan missionary and explorer, succeeded in finding a way to reach Monterey and the other Alta California settlements overland through Sonora and Pimería Alta (1774–76). See Bolton, *Outpost of Empire*, pp. 27–130; 134–225; and Geiger, *Franciscan Missionaries*, pp. 92–7.

[2] Bucareli to Arriaga, 27 July 1773, *Publicaciones del AGN*, 30 (1936) 2, p. 221. Editor's translation of: ' … de orden del Rey tome las medidos que crea convenientes para descubrir si pasan adelante estas exploracieones de los rusos'.

[3] See Serra to Bucareli, 13 March 1773, in Bolton, ed., *Historical Memoirs*, 3, pp. 3–36

[4] See Bucareli to Arriaga, 27 July 1773, in *Publicaciones del AGN*, 30, 1936, 2, pp. 221–4.

[5] See Serra to Bucareli, in Bolton, ed. *Historical Memoirs*, 3, p. 4.

Cañizares and Pino, Pérez was the obvious choice to lead the exploring expedition. He had, in fact, taken the initiative to submit his request to lead it.[1] Cañizares, in Serra's opinion, was too youthful and inexperienced to entrust with sole command of a packet boat much less an exploring vessel.[2] As for Miguel Pino, he was seeking a leave of absence, according to Serra, 'to return to Cádiz, his native place, where he has his wife and mother, from whom he has been absent now for many years'.[3] An infusion of new, experienced naval personnel assigned to San Blas was essential if Spanish claims to, and settlements on, the North American west coast above Baja California were to have any prospects of being sustained.

Bucareli readily accepted Pérez's offer to lead the expedition, giving him authority to plan its details, select the ship to be used, and choose her crew members.[4] As for the ship's second in command, Bucareli had only a single choice to offer. He was Master Second Class Esteban José Martínez, a thirty-one-year-old native of the Andalusian city of Seville. Brought over from Veracruz, on Mexico's Atlantic coast, he was among the first new mariners sent to bolster the San Blas naval corps. A graduate of Seville's Colegio de San Telmo, a highly regarded navigational school, Martíinez was reputed to be well experienced in coastal navigation on the Atlantic and Pacific coastal waters of Iberian America.[5]

Recognizing that the packet boats were not well suited for exploration, construction of a new ship better designed for such duty was initiated in 1771 at the San Blas shipyard, under the urgings of José de Gálvez. He had expected, in his usual enthusiasm, the new vessel, a 225-ton frigate to be called the *Santiago* (alias *Nueva Galicia*), would be completed by year's end – an unrealized expectation. When Gálvez returned to Spain in 1772, he ceased to be directly involved in the affairs of the Department of San Blas. In his absence, progress on the new frigate began to lag, but Bucareli and Serra intervened to see that she was completed and launched as quickly as possible. Serra's interest in the *Santiago* had less to do with exploration than the ship's cargo capacity. Writing in his petition to the viceroy in March, 1773, Serra says:

> I consider it necessary to use all urgency in equipping the new frigate with all possible haste, for, in view of its great size (for I went inside of it and saw its spacious hold), a voyage by it, together with the two packets [*San Carlos* and *San Antonio*], can relieve the presidio and missions ... from suffering and hunger, and can make all the men happy and content.[6]

The *Santiago* slid down the ways at the San Blas shipyard late in December 1773, the largest ship ever to be constructed there. With little if any benefit of trial runs, she weighed anchor at midnight, 24–25 January 1774, on her maiden voyage bound for Alta California. Her officers and men numbered eighty-eight, including her captain, Juan Pérez. The purpose of the *Santiago*'s voyage was avowedly to deliver provisions, supplies and twenty-eight passengers – prospective colonists – to the presidio and

---

[1] See Bucareli to Arriaga, 27 July 1773, in *Publicaciones del AGN*, 30, 1936, 2, p. 223.
[2] Serra to Bucareli, 13 March 1773, in Bolton, ed. *Historical Memoirs*, 3, p. 4.
[3] Ibid.
[4] See Bucareli to Arriaga, 27 July 1773, in *Publicaciones del AGN*, 30, 1936, 2, p. 223.
[5] See Thurman, *San Blas*, p. 142; and Cook, *Flood Tide of Empire*, p. 55.
[6] Serra to Bucareli, 13 March 1773, in Bolton, ed. *Historical Memoirs*, 3, p. 5.

mission at Monterey.¹ But once Pérez had cleared Monterey, he had secret instructions to sail as far north as the sixtieth parallel, go ashore there and take possession; following which, he was to return southward, keeping close to shore, examining it in detail, learning what he could about its native population and natural resources, checking for Russian (or other foreign) establishments, and going ashore to take possession wherever possible – north of any foreign settlements, should they be encountered.² Pérez had been given an assignment that would prove exceedingly difficult to fulfill in all its many details.

The first instance of trouble came on 28 February, about a month into the voyage, when boisterous seas caused the *Santiago* to sustain some structural damage.³ Pérez decided to put in at San Diego to make repairs – a stop he was not authorized to make except in an emergency. Whether this situation qualified as an emergency is debatable. But the colonists at the San Diego settlement, who were down to half rations, welcomed the sight of the frigate's sails, which offered them hope of staving off famine.⁴ In any event, the *Santiago* was delayed twenty-five days before resuming her voyage northward. On 8 May she reached and cast anchor in the shelter of Monterey's Point Pinos. Another delay lasting twenty-six days ensued while offloading cargo for the Monterey settlement and re-stowing the provisions, supplies, and water casks sufficient to sustain the *Santiago*'s crew on her secret voyage of discovery. Pérez was well aware of the dire consequences of shipboard water shortages, and he clearly was determined to avoid an ordeal such as Vila and the crew of the *San Carlos* had suffered in 1769.

The voyage would first take the *Santiago* about 500 nautical miles westerly of Monterey before gaining winds favourable to sailing northward. From there, until 14 July at latitude 50°21′N, the frigate made good headway on a more or less northerly course. Some rain squalls and unnerving fog banks were encountered, though otherwise things seemed to have gone well. That day, however, Pérez called a meeting of his officers, expressing his concern over a dwindling water supply. He explained to them that he had

> determined to fall in with the coast, considering the short water supply we had, the uncertainty of a port in which to obtain it, and other reasons that are expressed in the instrument accompanying this [Pérez's] diary.⁵

On the morning of 15 July, the helmsman was ordered to steer a course north-east and within two days the coast hove into view. The *Santiago* was off the west coast of

---

¹ Documentation of the *Santiago*'s officers, crew and passengers, at the time of her departure from San Blas, January 1774, is in Mexico City's AGN, Ramo 'Division': Historia, 61: fojas [folios] 225–8. English translations of these documents may also be found in Beals, ed., *Juan Pérez*, pp. 151–7.

² The full text of Bucareli's Instructions, translated into English, is in Servin, ed., 'Instructions of Viceroy Bucareli', 3, pp. 237–48.

³ This damage, described by Kenyon in 'Naval Construction', p. 137, consisted of broken 'trestletrees on the fore and mainmasts'. Another source, Francisco Mourelle's narrative of Pérez's 1774 voyage, says 'the cross-beams of both masts were sprung'. Beals, ed., *Juan Pérez*, p. 104. Thurman, citing as his source 'Primera exploración de Juan Pérez 1774' (331), MS in MN, says: 'the vessel sprung one of its main body joints'. Thurman, *San Blas*, p. 128.

⁴ See ibid., p. 131.

⁵ Beals, ed., *Juan Pérez*, p. 73. The 'instrument' to which Pérez refers is no doubt a letter of 31 August 1774, he sent from Monterey to Viceroy Bucareli. See ibid. pp. 50–55.

present-day Graham and Langara islands, the former being the largest of the Queen Charlotte Islands and the latter the northernmost. Pérez searched for a sheltered harbour, only to come upon a huge opening in the coast – now called Dixon Entrance – through which powerful currents flow. These currents thwarted efforts to enter the opening to reach the leeward side of a point of land Pérez called 'Punta de Santa Margarita' (now Langara Point, the northernmost point of Langara Island). There would be two encounters with the native Haida people of the area, who came out in their impressively large and elaborate canoes to meet and barter with the Spaniards.[1]

In the end, having earlier abandoned his favourable northward track, Pérez now realized he had sacrificed all hope of reaching latitude 60°N, having to settle instead for a latitude estimated to be no higher than 54°40′N. He and his crew – not to mention their water supply – were in no condition to continue sailing north into the teeth of north-west winds, which prevail close to shore during the summertime; neither could they attempt to regain the longitude they had lost in seeking to replenish their water supply. And to make matters worse, once they reached the coast, the weather, currents and shoreline topography prevented them from going ashore for water or to take formal possession. Pérez summed up his unhappy situation in this way:

> Having reflected on the inconstancy and confusion of the weather, and also the uncertainty of finding a place farther northward where one could anchor and take on water, [I realized that] by cutting the daily [water] ration I would scarcely be able to have [enough] for returning. I [therefore] determined not to press on farther, and from this latitude follow the coast to Monterey.[2]

The expedition would now begin its voyage to return south. Thus far, Pérez and his men had little to show for their efforts – except for informative descriptions of the Haida Indians and their natural environment. Martínez, in his account, wrote that he

> noticed in their canoes some small plates of iron … . But what surprised me was to see among them [the Haidas] half of a bayonet and another [Indian] with a piece of a sword made into a knife.[3]

These objects gave rise to speculation by Martínez that they might have been relics of an earlier Russian presence, specifically the disappearance of a landing party of fifteen men Aleksei Chirikov sent ashore in July 1741 at Lisianski Strait, some 250 nautical miles north-west of Dixon Entrance.[4] Exactly what happed to Chirikov's men remains a mystery to this day. Martínez's report of metallic objects of probable European origin among the Haidas was the only evidence the expedition found to indicate a possible Russian connection anywhere along the coast they examined.

Returning south proved largely uneventful, except for an attempt to anchor, go ashore for water, and take formal possession in early August 1774. It was at a place Pérez called 'Surgidero de San Lorenzo' (St Lawrence's Roadstead), better known today as Nootka Sound on Vancouver Island.[5] For reasons not entirely clear, Pérez

---

[1] For Pérez's account of the events off the point he called Santa Margarita, including the contacts with the Haida, see ibid, pp. 73–81, and for Martínez's version, ibid. pp. 100–102.
[2] Ibid., p. 81.
[3] Ibid., p. 101.
[4] For details of this event, see Golder, *Bering's Voyages*, 2, pp. 295–7.
[5] See Beals, ed., *Juan Pérez*, pp. 88–9.

decided to look for a harbour in the vicinity of latitude 49°N. Not having the services of a smaller-sized escort vessel, he was forced to bring the *Santiago* close to shore to find an anchorage – an especially risky business in unknown and uncharted waters. On the evening of 7 August, he anchored the frigate temporarily on the exposed seaward side of Nootka Sound's Hesquiat Peninsula, planning to dispatch a launch in the morning to find a better place to anchor. Meanwhile, the *Santiago* received a number of native visitors in canoes, with whom the ship's company bartered. On the morning of 8 August, just as the launch was about to get underway, a wind came up from the west, threatening to dash the frigate and her launch on the peninsula's rocky and exposed shore. Pérez immediately ordered the anchor cable cut, and the *Santiago* narrowly managed to escape destruction with her launch still in tow. This event seems to have shaken Pérez's confidence; from then on he became increasingly wary, tending to remain a comfortable distance off shore. On 10 August, he sighted a snow-covered mountain, which he named 'Cerro de Santa Rosalía' (Peak of St Rosalie), probably what is known today as Mount Olympus, (el. 2424 m; lat. 47°50′N) in north-west Washington State. But he sailed past the entrance of the Strait of Juan de Fuca (lat. 48°23.5′N, at Cape Flattery Light) and later the mouth of the Columbia River (lat. 46°16.6′N, at Cape Disappointment Light) without notice or mention. He was usually too far off shore to identify such detailed shoreline features at or near sea level.[195]

Continuing south along the Oregon coast, Pérez made only one sighting of a coastal feature worthy of mention. He described it as 'a point that slopes steeply into the sea, … is massive and has a very prominent cliff that is between white and yellow in colour'.[196] Judging by the latitude he observed off shore in its vicinity, at noon on 14 August (44°35′N), and its general appearance, it must have been Yaquina Head (lat. 44°40.6′N at its Light), just north of Newport, Oregon. For some uncertain reason, Pérez failed to name this headland. Soon after its sighting, the skies clouded over and the coast became shrouded in fog, hindering observations of latitude as well as visual contact with the shoreline. It was not until the *Santiago* reached the vicinity of Cape Mendocino (lat. 40°26.4′N at its Light) on 21 August when breaks in the clouds and coastal fog began to allow Pérez to take hurried observations to determine his latitude and take bearings on coastal landmarks. Four days later they passed the Farallon Islands, off Drakes Bay, without any attempt to look for the seaward entrance to the bay discovered by the Portolá expedition in 1769 – present-day San Francisco Bay. By 27 August 1774, the *Santiago* returned to anchor once more in Monterey Bay. After rest and recuperation, the frigate and her crew set out on 9 October to return to San Blas, where they arrived without incident on 5 November 1774.

This first of the so-called Bucareli expeditions had mixed results. Most objectives spelled out in Viceroy Bucareli's detailed instructions were not achieved: Pérez failed to reach latitude 60°N; no landings were made anywhere on the coast; no possession-taking rituals were performed; and the sightings he recorded along the coast were rather meagre – with the notable exception of descriptions of the Haida and Nootkan natives. In fairness, however, Pérez had been sent on a task hugely difficult for an expedition consisting of a solitary frigate unassisted by a smaller-sized escort vessel.

---

[1] Ibid., pp. 90–91.
[2] Ibid., p. 93. The yellow and white colours Pérez reported are consistent with the similarly tinted sandstone cliffs at Yaquina Head.

The voyage might accurately and charitably be described as a reconnaissance of the north-west coast, which pioneered the way for others more daring, better equipped or with better luck. It is worth noting that only two members of the ship's crew perished in the course of the expedition.[197] By mid-August, scurvy had begun making its unwelcome appearance, seriously afflicting some fourteen to sixteen crew members.[198] But considering the size of the ship's company – eighty-eight officers and men upon leaving San Blas – less than a fifth were severely stricken with the symptoms of this dietary deficiency. Neither of the two deaths appear to have been scurvy related. The crewmen lived in crowded, unhealthy conditions, and according to Malcolm Kenyon, the overcrowding may have been intentional to the extent that it might guarantee enough replacements to match what was anticipated to be a high death rate.[199] Ironically, the low death toll that actually prevailed in the 1774 voyage only ensured overcrowding would persist throughout the voyage.

Upon returning to San Blas, on 3 November 1774, Pérez knew he had some explaining to do. Viceroy Bucareli would, no doubt, be anxious to learn why most of his instructions had not been achieved. He might also wonder why Pérez had made an unauthorized, twenty-five-day stopover at San Diego en route north. There would also be questions about the twenty-six days it took to offload provisions and supplies at Monterey. And then, there was his troubling failure en route south to stop and look for the seaward entrance to the expansive bay the Portolá expedition had discovered in 1769. Another question of cartographic import also arose: Why had Pérez failed to prepare a chart of the expedition's discoveries – however limited they may have been?

In a letter addressed to Viceroy Bucareli, dated 31 August 1774, written at and dispatched overland from Monterey, Pérez summarized what happened on the voyage.[200] He dwells on the various difficulties the expedition encountered, explaining how circumstances – the weather in particular – had conspired to thwart his earnest intentions to comply with the viceroy's instructions. As for the unauthorized layover at San Diego, Pérez could plausibly claim that it was responsive to an emergency. And the lengthy time it took to offload cargo at Monterey was attributed mostly to the crew's poor health and exhaustion. In the case of his failure en route south to seek out the passage known today as the Golden Gate, connecting San Francisco Bay with the Pacific Ocean, it was his crew's weakened condition and the swift onset of darkness at nightfall that dissuaded him from attempting to enter that now celebrated passage. But realizing that, as commander of the expedition, he could not avoid bearing responsibility for the expedition's obvious shortcomings, Pérez expressed his regret to the viceroy:

---

[1] On the eve of the *Santiago*'s departure from Monterey in June 1774, the ship's boatswain, Manuel López, fell ill and died. Whatever the cause of his death, it is unlikely to have been scurvy. He was buried in the chapel at the *presidio* at Monterey. See Cutter and Griffin, eds, *California Coast*, p. 141. The only death known to have occurred while the expedition was actually at sea was that of an apprentice seaman named Salvador Antonio, who died of uncertain causes off the Queen Charlotte Islands on 24 July 1774. He was buried at sea. See Beals, ed., *Juan Pérez*, p. 82.

[2] Ibid., p. 95.

[3] See Kenyon, 'Naval Construction', p. 136.

[4] See Beals, ed., *Juan Pérez*, pp. 50–55.

for not fulfilling Your Excellency's desire as I might have wished. But ... whether or not it is the will of God or that such success is reserved for someone else, the fact is that the way is opened and recorded for others who may be worthy of sailing it with better fortune.[1]

When the *Santiago* reached San Blas on 3 November 1774, Pérez dispatched a second letter to Viceroy Bucareli together with a copy of his Diario ('Journal') of the frigate's exploring voyage. In this letter, Pérez explains that it had

> not been possible to construct a map of the coast discovered because the rolling aboard ship rendered it impossible, as well as the inconveniences suffered in these quarters of which Your Excellency is not unaware.[2]

In an effort to correct this omission, Pérez explained that he was having a map drafted based on the expedition's observations, surveys and other records. If such a map was in fact completed, its existence and whereabouts long remained a mystery. But in 1988, just such a map, purporting to have been drafted by José de Cañizares, was unexpectedly discovered in the US National Archives, suggesting it was not just another unfulfilled promise.[3]

There were plenty reasons for the viceroy to be displeased with Pérez's performance in the expedition, had he been so inclined. Yet it appears that Bucareli was surprisingly sympathetic with Pérez's explanations and excuses, even to the extent of sometimes being in full agreement with them. In a letter dated 26 November 1774, to Julián de Arriaga, minister of the Indies, Bucareli not only had words of praise for Pérez's role in the voyage but recommended him for promotion from the rank of *alférez de fragata* 'junior ensign' to *teniente de fragata* 'junior lieutenant'.[4] Unfortunately for Pérez, he would not live to reap the benefits of such a promotion. But promotion or not, Pérez would continue to have an active, though somewhat diminished, role to play in the next naval operation to be mounted at San Blas. Bucareli seems to have realized early that a second and possibly additional voyages would be necessary to penetrate the secrets, geographic or otherwise, of the north-west coast. With this in mind, reinforcements of a few additional masters would not suffice; the need was plainly for a contingent of highly motivated, aggressive, commissioned officers of the line from Spain's Real Armada (Royal Navy).

And so, on 13 June 1774, when Pérez and the *Santiago* had barely been at sea a day since departing Monterey on their exploring voyage, a half-dozen Spanish naval

---

[1] Ibid., p. 55.

[2] Ibid., p. 57.

[3] The map must have been drawn by Cañizares sometime between Pérez's return to San Blas early in November 1774 and his departure from there with the second Bucareli voyage in mid-March 1775. It apparently was sent to Viceroy Bucareli's residence south-west of Mexico City at Chapultepec Castle, where it evidently was misplaced or ignored, only to fall into the hands of US Army troops in an assault on the Castle seventy-two years later during the US-Mexican War in September 1847. Later, the map and a number of other unrelated captured documents were transported to Washington, DC, where they have remained in the USNA's collection. For more about the provenance and authenticity of this map, see Beals, 'The Juan Pérez-Josef de Cañizares Map', pp. 46–55.

[4] In addition to the rank of *primer piloto* 'master, first class', Pérez held the temporary naval rank of *alférez graduado de fragata* 'brevet junior ensign'. As Pérez failed to survive the Hezeta-Bodega expedition of 1775, his promotion to *teniente de fragata* 'junior lieutenant', as recommended by Viceroy Bucareli, was never awarded.

officers were leaving the harbour at Cádiz in the south-west of Spain. All were graduates of the Spanish naval academy, the Escuela de Guardias Marinas.[1] Travelling in a group, they would make the Atlantic crossing to Veracruz on the Mexican east coast, after which they would travel inland to the former Aztec capital of Tenochtitlan, long since transformed into Mexico City, the capital of New Spain. Arriving in the city on 25 October 1774, they were the vanguard of a major reinforcement of naval personnel at San Blas. Three of the officers, Bruno de Hezeta y Dudagoitia, Miguel Manrique, and Fernándo de Quirós y Miranda, held the rank of *teniente de navío* ('senior lieutenant'); the other three, Juan Manuel de Ayala, Diego Choquet, and Juan Francisco de la Bodega y Quadra, held the rank of *teniente de fragata* ('junior lieutenant'). In addition to these officers of the line, the Naval Department of San Blas had assigned to it several masters, including Juan Pérez, José de Cañizares, Cristóbal Revilla, and Francisco Antonio Mourelle de la Rua. The last named was a native of Galicia in Spain's north-west corner, who was as highly motivated and aggressive as any line officer in any navy. It would be from these six line officers and the four San Blas masters that Bucareli would select the officers to lead his second exploring expedition to the north-west coast.[2]

The three senior lieutenants went directly to San Blas, where they were assigned their respective commands. As the most senior officer among them, Hezeta would go aboard the frigate *Santiago* as over-all commander of the expedition; Pérez would join him as first master, with Revilla as second master. Manrique was made captain of the packet boat *San Carlos*, with Cañizares as master. Quiros received command of the packet boat *San Antonio*, which would serve only as a supply ship and play no direct role in exploration. Meanwhile, the three junior lieutenants, Ayala, Choquet, and Bodega remained in the city of Tepic, about 50 km inland from San Blas, to await their orders. There seems to have been some last-minute uncertainty as to how many and which vessels would be employed. Despite the difficulties the first expedition had experienced without benefit of a smaller-sized escort, the need for such a vessel on the second voyage was not agreed upon without some dissent. It may have been because San Blas had no ship expressly intended for this purpose, which required that a suitable sized vessel would have to be found among the existing San Blas fleet to serve as an escort.

The ship finally selected to escort the flagship *Santiago* was a twin-masted, 59-ton *goleta* ('schooner'), officially named *Nuestra Señora de Guadalupe*, but informally called the *Sonora*. Her assignment as an escort on a lengthy, open-ocean voyage of discovery would be a challenge her designers or builders had not likely contemplated for such a small ship. Built in 1769 at the San Blas shipyard, she was originally intended for nothing more demanding than short coastal voyages or crossings between Sonora and the inner coast of Baja California, to deliver mail or transport personnel and small cargoes. She measured along her keel, about 10 m (34 ft); her deck measured roughly 12 m (40 ft); width abeam was about 3.5 m (12 ft); width of hold, 3.5 m (12 ft); depth of

---

[1] See Harbron, *Trafalgar and the Spanish Navy*, p. 105.
[2] See Bucareli to Arriaga, 26 November and 27 December1775 in *Publicaciones*, 30, 1936, 2, pp. 228–37; Beals, *For Honor and Country*, pp. 33–6. Mourelle was originally assigned to the *Santiago*, but he successfully requested a transfer to the *Sonora*. See Thurman, *San Blas*, p. 145, n. 12.

hold, 2 m (6.5 ft).¹ The planners of the expedition seem to have convinced themselves that this diminutive, cramped schooner had the capability of undertaking an extended voyage far into the North Pacific, It was not an opinion, however, shared by all of the mariners who would actually participate in the voyage. Nevertheless, as no better choice was available, the *Sonora* would have to do. After hurried careening and last-minute repairs Junior Lieutenant Juan de Ayala received word in Tepic that he would command the schooner with Francisco Antonio Mourelle as second officer.

When command of the *Sonora* fell to Ayala, Bodega was effectively passed over – potentially a serious blow to his prospects as a naval officer. Historian Warren Cook suspected that this decision stemmed from Bodega's Peruvian birth and was unrelated to considerations of his seniority or other qualifications.² Of the officers who were initially assigned ship-board commands – a key step for advancement of any naval officer – all were native to provinces in the Iberian homeland; they were in a word *peninsulares*. Regardless of his birthplace, Bodega was descended from an illustrious line of ancestors, who had distinguished themselves in service to the Spanish Crown.³ In any event, it was not in his character to shrink from what he conceived to be his destiny, and he was not about to remain ashore while his fellow officers were headed north on the high seas. He persuaded the expedition's commander, Bruno de Hezeta, to allow him to volunteer for service under Ayala, as the *Sonora*'s second officer, even though they both held the same rank. As Warren Cook has written 'it was truly a self-effacing gesture'.⁴ But it would also open the door on an unexpected opportunity for Bodega to prove his mettle.

The packet boat *San Carlos* was the third ship designated to play a role in Viceroy Bucareeli's second expedition. Sailing under command of Senior Lieutenant Miguel

---

¹ The linear dimensions of 18th-century Spanish ships were measured in *codos*, *pies*, and *pulgadas*, units of measurement long since gone out of use. The *codo* is more or less equivalent to the English cubit, which can vary between 18 and 22 English inches. The *pie* is 0.91 of an English foot; the *pulgada* is 0.91 of an English inch. An inventory of the *Sonora* conducted at San Blas in March 1777 gives her dimensions as follows: length stem to stern '*esclora a esclora*' 24 *codos*; keel length '*quilla limpido*' 20 *codos*; width abeam '*manga*' 6 *codos*, 15 *pulgadas*; floor timber '*plan*', 4 *codos*; depth of hold '*puntal*', 6 *codos*, 15 *pulgadas*. See Hernandez and Barrios, 'Inventario de la Goleta ... Sonora' (3 March 1777). Converting these figures into metric and English equivalents results in the following dimensions, first based on a 22-inch *codo*: length stem to stern, 14 m (44 ft.); keel length, 11 m (37 ft.); width abeam, 4 m (13 ft.); floor timber, 2 m (6 ft.); depth of hold, 4 m (13 ft.). Alternatively, based on an 18-inch *codo*, her dimensions would be: length stem to stern, 11 m (36 ft.); keel length, 9 m (30 ft.); width abeam, 3 m (10 ft.); floor timber, 2 m (6 ft.); and depth of hold, 3 m (10 ft.). These dimensions are all problematic to one extent or another. They all depend on what value the *codo* is assigned. The depth of hold appears to be too deep for a vessel of the *Sonora*'s kind. And Bodega himself alludes to the *Sonora*'s keel length as only 18 *pies* 'feet' long, when he must have intended to write *codos* 'cubits'. (See below, p. 86, n. 1). Whatever the confusion over the schooner's dimensions, there is no doubt that she was a very small, cramped ship for a crew of 14 seamen and two or three officers to spend months upon months together at sea.

² See Cook, *Flood Tide of Empire*, pp. 70–71, for discussion and speculation as to whether or not Bodega was passed over because of his Peruvian birth – a creole, as Spaniards born in Spain's American colonies were called, a label that could have some social disadvantage. Freeman Tovell (pers. comm. 29 January 2003), however, is inclined to believe that this and other similar decisions were based primarily on seniority.

³ For example, among Bodega's ancestral lineage, Alvaro de la Quadra seems to have been the Spanish ambassador in London when Mary Tudor (Mary I, Queen of England) was married to Prince Philip of Spain (later Philip II, king of Spain). See Menchaca, *De California a Alaska*, p. 7.

⁴ Cook, *Flood Tide of Empire*, p. 71.

Manrique, the *San Carlos* would deliver Alta California's annual provisions and supplies to the settlement at Monterey Bay, following which she was to continue northward along the coast to track down, once and for all, the elusive seaward entrance of Portolá's San Francisco Bay. And in the event the passage was successfully located, Manrique had instructions to enter and conduct a hydrographic survey of the bay. It had less potential for glory than the assignment to explore the north-west coast. But considering earlier reports of the bay's considerable magnitude, it promised to be an important, arduous task demanding the utmost in skill and patience.

By the evening of 16 March 1775, the expedition's three ships were ready to set sail from San Blas. Almost from the beginning the *Sonora*'s performance belied the optimistic information the expedition's officers had been given about her seaworthiness. Under full sail and even with the best wind and sea conditions she simply could not keep pace with the flagship *Santiago*. It would plague the expedition for many weeks to come. But on the third day out, 19 March, an incident took place that would briefly shift attention away from the expedition's hardware to its leadership. In the afternoon, the *San Carlos* suddenly and without apparent reason fired her two cannons, followed by a red flag run up the packet boat's main mast – standard distress signals. The distress, however, turned out to concern the captain himself, Miguel Manrique, who had become mentally deranged. Suffering from delusions that his fellow officers were trying to kill him, he armed himself with a half-dozen loaded pistols, and terrorized his crew, threatening to shoot at least one of them. Manrique was relieved of his command by Hezeta with the full concurrence of the other officers, including the surgeon and chaplains. The delusional Manrique was returned by launch to San Blas. This curious incident forced an immediate change in the expedition's command. The *Sonora*'s captain, Juan de Ayala, was ordered to replace Manrique as captain of the *San Carlos*. This vacancy would in turn be filled by Bodega, who while eager for advancement, took no pleasure in seeing a fellow officer stripped of command and sent ashore to an uncertain future. In any event, as of 20 March, Bodega assumed full command of the *Sonora*.[1]

After Ayala was transferred to the *San Carlos*, while he was inspecting firearms left in the captain's cabin, one of the loaded pistols accidentally discharged, wounding him in the right foot. Despite what must have been a painful injury, Ayala remained in command of the packet boat throughout her voyage to Monterey, her subsequent search for and discovery of the Golden Gate, the mapping of San Francisco Bay, and finally her return to San Blas. But his injury prevented him from personally conducting the hydrographic charting of the bay, a task that agreeably fell to *San Carlos*'s master, José de Cañazares.[2]

As for the flagship and her escort, their instructions remained essentially the same as those that had governed the Pérez expedition in 1774. Writing to Arriaga in November 1774, Bucareli explains that 'the instructions to this officer [Hezeta] will be

---

[1] For other accounts of this episode see below, p. 88; Beals, *For Honor and Country*, p. 52; Campa Cos, *A Journal of Explorations*, p. 16; and Wagner and Baker, eds, 'Fray Benito de la Sierra's Account', p. 209.

[2] See Galvin, ed., *First Spanish Entry*, an extract from the 'Journal of Frey Vicente Santa María', pp. 13–37; extract from the 'Journal of Juan Manuel de Ayala', pp. 79–87; and 'Report of José de Cañazares ... to Captain Ayala', pp. 95–8.

the same as those carried by D. Juan Pérez'.¹ There would, however, be at least two important differences. The *Santiago* and *Sonora* were expressly ordered not to stop at Monterey on the voyage north – with the usual exception in the event of an emergency. It would take a vigorous protest by Bodega to persuade Hezeta to reject Pérez's advice to do exactly that – stop at Monterey. Another change in Bucareli's instructions to the second expedition was to increase the latitude for which they were expected to strive, from 60°N to 65°N. Evidence of this change is found in a remark made by Bodega in his 1775 Journal acknowledging ' I am obligated to go up to 65 degrees'.²

As the voyage involved two ships instead of one, it was also necessary to make some pragmatic adjustments accordingly. What was to be done in the event the *Sonora*'s poor sailing qualities made her more hindrance than help? This problem, which arose early in the voyage, was temporarily solved by the *Santiago*'s taking the *Sonora* in tow, until a more permanent solution was found in Bodega's resourceful reconfiguring of the schooner's sails and rigging. Officers aboard the *Santiago* – notably Pérez and Revilla – had from the beginning of the voyage held serious misgivings about the *Sonora*'s abilities to navigate and survive in high northern latitudes. By mid-July, Revilla was convinced the schooner ought to be forthwith returned to San Blas. But his colleague, Pérez, had a change of mind, citing the schooner's improved performance under Bodega. Though somewhat grudgingly, Pérez agreed that the *Sonora* and her crew had earned the privilege to continue north with the expedition – a judgement supported by Hezeta and welcomed by Bodega and Mourelle.³

Another issue arose from the presence of two ships in the expedition, namely: What should be done in the event the ships became separated? This question surfaced shortly after the expedition's unfortunate encounter with natives on 14 July 1775 on the coast of present-day Washington State. Once the *Sonora*'s continued role in the expedition was decided, replacements for the crewmen killed and the boat lost on 14 July were transferred from the *Santiago*, together with additional weapons and ammunition.⁴ By the end of July rough seas were making it difficult to maintain visual contact between the ships. An alarming increase in scurvy among the *Santiago*'s crew led to pleas by Pérez, Revilla and the ship's surgeon, Juan Gonzales, to abandon voyaging farther north. Bodega and Mourelle were unsympathetic and increasingly disposed to part company with the flagship and strike out on their own. Both were convinced that Pérez and Revilla had given up reaching latitudes above the fiftieth parallel, much less latitude 65°N, and that they would soon prevail upon Hezeta to turn back far short of the expedition's goal – whether 60° or 65°N.

---

[1] Bucareli to Arriaga, 26 November 1774. in *Publicaciones*, 30, 1936, 2, p. 227. Editor's trans. of 'Las instrucciones que daré a este oficial [Hezeta] serán las mismas que llevó D. Juan Pérez'.

[2] See below, p. 108. Most if not all secondary accounts of the voyage, including Bancroft, *History* 1, p. 158; Chapman, *Spanish California*, p. 239, Cook, *Flood Tide of Empire*, p. 72, and Thurman, *San Blas*, p. 144, agree on the change from 60°N to 65°N. There is also a relevant remark by Bucareli in a letter to Arriaga dated 26 November 1775. In alluding to a proposal for a third expedition Bucareli writes: 'I propose going to spend winter at the Port of San Diego in order to sail next year [1776] to the latitude of 65 degrees'. Editor's translation of 'propuso ir a invernar al Puerto de San Diego para navegar el año próximo a la altura de 65 grados', in *Publicaciones*, 30, 1936, 2, p. 229.

[3] See Beals, ed., *For Honor*, p. 79; and below, pp. 108–9.

[4] See ibid.

And so, when the *Sonora* lost all contact with the flagship on 1 August 1775, her officers seemed to relish the opportunity to commence a solitary voyage taking them north into the waters of south-eastern Alaska to within two minutes of latitude of the fifty-eighth parallel. Details of this remarkable voyage – including when the *Sonora* was sailing in company with the *Santiago* – are described elsewhere in this work in Bodega's own account, so there is no need to repeat them here. Much as Bodega and Mourelle had expected, the flagship turned south after reaching a point off Vancouver Island at an estimated latitude of 49°40′N.[1] Although the *Santiago*'s officers – Pérez and Revilla in particular – were viewed by Bodega and Mourelle as lacking élan, the flagship's return south was far from devoid of accomplishments. Even with a crew ravaged by scurvy, Hezeta hugged the coastline as much as possible, which led among other things to the discovery on 17 August 1775 of the mouth of the Columbia River – called by the Spaniards Bahia de la Asunción 'Assumption Bay', or later Entrada de Hezeta, 'Hezeta's Entrance'.[2] With too few men able to work the ship properly, he was unable to enter the river to conduct a rudimentary survey, for which reason Spanish claims of prior discovery would later be clouded. The advantages the Spanish Crown might have garnered from this discovery were thus diluted, though not entirely lost.

On a more positive note, however, during the second expedition's northward progress, Hezeta and Bodega, both in company and apart, had made four landings and performed four possession-taking ceremonies from Trinidad Head in northern California to Sea Lion Cove, on Kruzof Island, in south-eastern Alaska. Ayala and Cañizares had found the seaward entrance to San Francisco Bay and charted its interior waters. On the *Sonora*'s return voyage to Monterey Bay, while diligently searching for the Golden Gate, Bodega found instead two previously uncharted inlets, now called Bodega and Tomales bays.[3] It all represented a substantial improvement over Pérez's efforts in 1774. The two voyages together added notably to the cartography of the North American coast stretching from Monterey Bay in California over 1,200 nautical miles (2,300 km) to Kruzof Island in south-eastern Alaska. In achieving this, they also dispelled, temporarily at least, some of Madrid's xenophobic concerns, having found practically no evidence of Russian or other foreign activity.

It is clear from Viceroy Bucareli's correspondence with Minister of the Indies Arriaga that the viceroy intended to launch a third successive expedition in 1776.[4] He was particularly interested in further investigation of the 'mouth or entrance seen by D. Bruno de Ezeta at 46 degrees'.[5] But these exploring expeditions had proven costly drains on the royal treasury, producing wealth mainly in the form of geographic knowledge with no immediate prospects of compensation in more tangible forms.[6] The existing San Blas fleet was also needed to keep supplies and provisions flowing to the Alta California settlements, which continued to show a distressing inability to

---

[1] See Beals, ed., *For Honor* pp. 82, 124. Hezeta's navigation tables indicate that on 11 August 1775 under 'Latitude by Dead Reckoning' the entry reads '49°40'; under 'Latitude by Observation', the entry is '49° [...]' indicating the minutes were missing or undecipherable.

[2] Ibid., pp. 86–9; see also Bucareli to Arriaga, 27 December 1775, in *Publicaciones*, 30, 1936, 2, p. 236.

[3] See below, p. 122, n. 5.

[4] See Bucareli to Arreaga, 26 November 1775, in *Publicaciones*, 30, 1930, 2, p. 229.

[5] Ibid., 2, p. 236

[6] The cost of Pérez's expedition in 1774 was 15, 455 *pesos*, 4 *reales*, 11 *granos*; and for the Hezeta-Bodega expedition in 1775, the cost was more than double at 36,740, *pesos*, 2 *granos*. The two expeditions taken together cost the royal treasury a total of 52,195 *pesos*, 5 *reales*, 1 *grano*. See Chapman, *Spanish California*, p. 242.

become self-sustaining. These circumstances worked to hamper and delay plans to return north quickly with a third exploring expedition. The pace of these preparations might have been speedier had Madrid or Viceroy Bucareli known in advance about the appearance in March 1778, off the present-day Oregon coast, of two British exploring vessels, *Resolution* and *Discovery*, under command of Captain James Cook. In a sense they were following where Sir Francis Drake had left off almost exactly 200 years earlier. Cook even described the shore as 'the long looked for Coast of new Albion',[1] using the name Drake had applied to the country he had visited in June 1579. Had Madrid or Bucareli possessed knowledge of when and where these two British vessels would reach the coast where the Spanish Crown had just sought through two costly maritime expeditions to affirm its sovereign claims, a Spanish attempt to intercept Cook's expedition or at least hinder its progress might have ensued.[2]

Lessons learned in the first two Bucareli expeditions would guide the planning for the third venture northward. Preparations for the voyage had begun as early as 1775 – but pursued somewhat leisurely. With a chronic shortage of ships at San Blas and especially those designed expressly for exploration, the keel of a new vessel was laid in April 1777, better designed and equipped to meet the exigencies of exploration in the North Pacific Ocean,[3] She would be a 189-ton, three-masted frigate, also sometimes referred to as a *corbeta* (corvette), similar in appearance to the *Santiago*, but slightly smaller.[4] Viceroy Bucareli would christen her officially the *Nuestra Señora del Rosario*, although informally she was called the *Princesa*. Her hull was launched on 15 December 1777 and by 20 April 1778 she was fully outfitted and ready for sea trials. According to Kenyon: 'The San Blas pilot [master] Joseph Camacho took her out on a seventeen-day shakedown cruise from 18 October to 3 November 1778, … Camacho reported that she handled well'.[5] The *Princesa* would be accompanied by an escort, but not a schooner such as the *Sonora*. The escort would instead be a ship nearly identical to the *Princesa*, with shallow-draught surveying to be conducted by means of launches carried aboard the *Princesa* and her yet-to-be designated escort. As there was no other ship available at San Blas fit for the task – or time and resources enough to build one – the escort vessel would have to be found elsewhere.

It was for this reason that Bodega departed San Blas on 28 December 1776, to travel overland to Acapulco, remaining there long enough to assist in the hydrographic mapping of the harbour. He then took passage on a merchant ship named *Fenis*, which took him to Paita, a port city about 900 km north-west of Lima. He covered the

---

[1] Beaglehole, ed., *Journals*, 3, p. 289.

[2] Through espionage, Madrid had learned of the departure of Captain Cook's third expedition from Plymouth, England, in July 1776, together with information that its mission probably involved exploration of the north-west coast of North America. Without knowing exactly when and where Cook's expedition would actually reach the north-west coast any Spanish countermeasures would have had very slim prospects of success. See Cook, *Flood Tide of Empire*, pp. 88–93, for discussion and speculation that the third Bucareli expedition may have been originally intended to intercept Cook.

[3] Information concerning this ship, *Nuestra Señora del Rosario* (alias *Princesa*), is based primarily on Kenyon, 'Naval Construction', pp. 148, 150–57, 153–9.

[4] The *Princesa*'s dimensions are given by Kenyon, 'Naval Constrruction', p. 153, as: deck length, 83.2 ft (25.4 m); keel length, 67.6 ft (20.6 m); width abeam, 22.8 ft (7.0 m). By way of comparison, the *Santiago*'s deck length was 84.8 ft (25.9 m); her keel length, 77.2 ft (23.5 m); and width abeam, 26.9 ft (8.2 m).

[5] Ibid., p. 158.

remaining distance by road to reach his Peruvian birthplace.[1] There he negotiated the purchase and refitting of a ship to serve as the *Princesa*'s escort. This ship, a 197-ton, three-masted frigate named *Nuestra Señora de los Remedios* 'Our Lady of Help' informally known as the *Favorita*, required a considerable overhaul to make her suitable for voyaging in high North Pacific latitudes. It took two and a half months, from 15 September to 30 November 1777, to complete the overhaul.[2] In addition to the ship, Bodega obtained a supply of spare anchors, cables and cannons for use by other ships at San Blas. Once the *Favorita* was ready with a crew assembled, Bodega sailed her north from Peru, arriving at San Blas in February 1778.[3] This addition to the San Blas fleet was similar enough to the *Princesa* to be her twin sister. In deck length, the *Favorita* was 25.8 m (84.5 ft), slightly longer than the *Princesa*'s 25.4 m (83.2 ft); in keel length the *Princesa*, at 20.6 m (67.6 ft), was a bit longer than the *Favorita*'s 20.1 m (65.8 ft). The widths abeam of both ships were the same, 7.0 m (22.8 ft).[4]

After many delays, the third and last of the Bucareli expeditions to the north-west coast finally set out from San Blas on 11 February 1779.[5] The *Princesa* was the designated flagship under command of Senior Lieutenant Ignacio de Arteaga, who was also in overall command of the expedition. He had arrived in New Spain in 1774, separately and sometime after the six naval officers – Hezeta, Manrique, Quirós, Ayala, Choquet, and Bodega – who had originally served in ship-board capacities. Arteaga, however, had been given a shore-bound post as commandant of the Naval Department of San Blas. He came to his new expeditionary assignment in 1779 lacking any experience exploring in the North Pacific. As for Bodega, his exploits in the 1775 voyage had earned him respect among his fellow officers and a valorous reputation among Crown officials reaching as high as King Charles III himself. On 16 March 1776 – the first anniversary of the second Bucareli expedition's departure from San Blas – Bodega was promoted to the rank of *teniente de navío* 'senior lieutenant'. He was also recommended for the title of Knight in the Order of Santiago. And in December,

---

[1] See Bodega, 'Mandos y Comisiones que seme han confiado', and Bernabeu, *Bodega y Quadra*, p. 33, for brief accounts of Bodega's travels to reach Lima.

[2] The *Favorita*, as purchased, lacked the shelter of a cabin for her officers; she had no separate quarterdeck, forecastle or poop deck, and probably was steered by tiller. The submerged part of her hull was badly damaged by marine worms, which required repair; her entire main deck needed re-planking, among other problems that had to be remedied if she was to accompany the *Princesa* north on a voyage of discovery. See Kenyon, 'Naval Construction', p. 140.

[3] See Bernabeu, *Bodega y Quadra*, pp. 112–13, for Bodega's account of his mission in Peru to secure the frigate *Favorita*, and sail her to San Blas.

[4] Dimensions of the *Favorita* are based on those given by Kenyon, 'Naval Construction', pp. 139–40.

[5] There are numerous first-hand accounts of the Arteaga-Bodega expedition of 1779. They include the journals of Arteaga, Bodega, Quiros, Camacho, Pantoya, Cañizares, Aguire, and Fray Riobó. Located in a variety of archives and libraries most have neither been published nor translated into English. See Cook, *Flood Tide of Empire*, p. 594, for a listing of these documents and their locations. The full Spanish text of Bodega's 1779 Journal was published in Bernabeu, *Bodega y Quadra*, pp. 111–58. The ABC (Victoria, BC) has an unpublished English translation of Bodega's account of the 1779 voyage. The text of an account by Fray Riobó. rendered into English by Thorton, is in 'An Account of the Voyage made by the Frigates "Princesa" and "Favorita" in the Year 1799 [*sic*, actually 1779] from San Blas to Northern Alaska', *CHR*, 4, 1918, 2, pp. 222–9. Useful secondary accounts are in: Thurman, *San Blas*, pp. 163–79; Cook, *Flood Tide of Empire*, pp. 93–100; and Wagner, *Cartography of the Northwest Coast*, 1, pp. 191–6.

1776, that honour was approved by royal order of King Charles III.[1] Whatever the reasons, Bodega's meritorious services were still insufficient to have assured him overall command of the expedition.

The two frigates, *Princesa* and *Favorita*, sailed on courses taking them much farther westward into the Pacific Ocean than either previous Spanish voyages in 1774 or 1775. Their officers were intent on gaining sufficient west longitude to find the most favourable winds to carry them north. This also minimized the temptation to stop at one of the Alta California settlements. Viceroy Bucareli's instructions to the expedition had increased the latitude they were instructed to reach from 65°N to 70°N.[2] They aimed for an initial landfall in the vicinity of latitude 58°N, placing themselves high enough, they thought, to reach this newest impossible task the viceroy had given them. It meant bypassing the entire coasts of present-day California, Oregon, Washington and British Columbia. The voyage was going well until 20 April when the ships became separated in a storm. On 1 May the *Favorita* came in sight of the symmetrical 975 m (3,200 ft) high volcanic cone of Mount Edgecumbe at the entrance to Sitka Sound (lat. 56°59.9′N at Cape Edgecumbe Light). Bodega would have immediately recognized this distinctive landmark from his previous encounter there in 1775.[3] From there they turned south-eastward to reach Bucareli Bay (lat. 55°13.5′N), also discovered by Bodega on his previous voyage. They were much relieved to be rejoined there by the *Princesa* after a fortnight of separation.

Bucareli Bay was their rendezvous point in the event of separation. Both frigates remained at that location for nearly two months, anchored most of the time in a sheltered cove they called Santa Cruz (Holy Cross). There a cross was erected, mass said, and a sermon delivered.[4] During the interval of their stay, they replenished their supplies of water and wood. Unexpectedly, much of the *Princesa*'s crew fell seriously ill with a mysterious malady, resulting in the deaths of several sailors.[5] Friendly relations with the indigenous Tlingit people prevailed at first, but the longer the Spaniards remained the more tensions heightened, which finally proved fatal to at least one and possibly more of the natives. Meanwhile, Francisco Mourelle, Bodega's stalwart and trusted master in the 1775 voyage, led a corps of three masters – José Camacho, Juan Pantoja y Arriaga and Juan Bautista de Aguirre – using the ships' launches, to chart the intricate tangle of channels and many small islands off the west side of Prince of Wales Island in the southern Alexander Archipelago.

---

[1] The military order of Santiago was an award and title for 'exceptional valour in a military assignment together with unimpeachable lineage of highest purity', according to Thurman, 'Juan Bodega y Quadra and the Spanish Retreat from Nootka', in *Reflections of Western Historians*, p. 51. The recommendation process was initiated by Viceroy Bucareli and the nomination seems to have proceeded through the usual channels, eventually reaching King Charles III himself. Apparently Bodega received the award as he later used the title in his official correspondence, indicating that the award had been officially authorized.

[2] See Bernabeu, *Bodega y Quadra*, p. 114; Thurman, *San Blas*, p. 172; and Wagner, *Cartography of the Northwest Coast*, 1, p. 192

[3] See below, pp. 109–10.

[4] The act of taking possession was apparently not performed, as that had already been accomplished in 1775. See Bernabeu, *Bodega y Quadra*, p. 125 and n. 50; see also Wagner, *Cartography of the Northwest Coast*, 1, p. 192.

[5] The nature of the illness that broke out aboard the *Princesa* is uncertain. A field hospital of sorts was established ashore, following which further deaths were limited to two. It could have been scurvy, but that term is not used by Bodega to describe the affliction whatever it was. See Bernabeu, *Bodega y Quadra*, p. 126.

On the first day of July, the expedition departed Bucareli Bay seeking winds favourable to continue their northward progress, hopeful of reaching the goal of latitude 65°N set for the 1775 expedition, or even the goal as now revised upward to 70°N. After a little more than a week's sailing, a very high, distinctive, snow-capped mountain hove into view roughly 120 nautical miles due north. It probably was 5,489 m (18,000 ft) Mount St Elias, and sighting it was an indication that their hopes of reaching latitude 65°N – much less 70°N – were unlikely to be realized. This mountain and other nearby high peaks are located where the west coast of North American begins turning onto an east–west axis, baring northward progress by sea. The summit of Mount St Elias is at latitude 60°15.5′N, and the shoreline due south of the mountain is at 59°45′N. These topographic barriers forced Arteaga and Bodega to veer westward, whereupon they encountered what was probably Cape Suckling (lat. 60°01′N). Soon after, they came upon Kayak Island and its southern extremity at Cape St Elias (lat. 59°48′N), the scene of Bering's first landfall in 1741, and which had not been visited since by Europeans until Cook's expedition in 1778, a year before Arteaga and Bodega arrived.

When the *Princesa* and *Favorita* doubled Cape St Elias, Arteaga and Bodega could see that the coast offered little promise of advancing much above latitude 60°N. Two days later, on 20 July, as they were becalmed close to shore, they encountered two natives each in separate 'canoes' – or more likely kayaks or *baidarkas* – who approached the frigates. The natives communicated by signs that a certain nearby inlet or passage would offer them a favourable anchorage. On this advice, Arteaga gave orders for both ships to enter and anchor in what is today called Port Etches, nestled on the west side of Hinchinbrook Island.[1] Bodega dispatched Cañizares and Pantoja in the *Favorita*'s launch to determine if the land off which they were anchored was an island or part of the mainland. By the next day they returned with the answer that it was indeed an island. They also reported the existence of a *gran ensenada* ('great bay') – probably present-day Prince William Sound – but they were doubtful if it offered any hope of finding a passage leading northward.[2] Meanwhile, with the *Princesa*, the ship's company was busy collecting wood and water, or planting a cross and performing the rituals required for taking formal possession of what they called Puerto de Santiago, present-day Port Etches. At latitude 60°17′N, it would prove to be the northernmost location at which Spanish explorers would perform such acts claiming sovereignty on the North American west coast.[3]

A meeting of the officers and masters was called by Arteaga aboard the flagship *Princesa* on 24 July to assess the expedition's circumstances. Scurvy was beginning to

---

[1] This anchorage was the same one in which *Resolution* and *Discovery* of Cook's third expedition had anchored briefly a year earlier in May 1778. The name Hinchinbrook was applied by Cook to the cape on the south-west point (lat. 60°14′N) of an island bearing the same name. However, John Gore, a lieutenant aboard *Resolution*, named the point Cape Hold with Hope, as it is shown on one of the maps drawn by Henry Roberts, master's mate on the *Resolution*. See Beaglehole, ed., *Journals*, 3, p. 343, n.1; see also, *Journals*, *Charts and Views*, map LI. Arteaga called the island Santa María Magnalena and the anchorage Puerto de Santiago, both of which names proved as ephemeral as Gore's name of Cape Hold with Hope. See also Bernabeu, *Bodega y Quadra*, pp. 147–8. The name Port Etches was applied to the anchorage several years later in 1787 by Nathaniel Portlock and George Dixon, in honour of the London firm for which they were sailing. See Beaglehole, ed., *Journals*, 3, p. 344, n. 2.

[2] See Barnabeu, *Bodega y Quadra*, p. 148.

[3] See Cook, *Flood Tide of Empire*, p. 97.

make its unwelcome appearance, and there seemed little hope of gaining latitude higher than 61°N based on Cañizares and Pantoja's reconnaissance of the expedition's surroundings at Puerto de Santiago (Port Etches). All finally agreed to press on westward, even if it meant losing some latitude, keeping the coast in sight, watching for a passage north, and retiring south only when the onset of autumn weather compelled it.[1] They now sailed along the seaward coast of Montague Island, continuing along the southern coast of the Kenai Peninsula. After being pummelled by heavy rain and high winds, the skies cleared, and the expedition came upon a group of islands off the Kenai Peninsula's southernmost point. There, on 1 August, they took refuge, anchoring on the south side of an island, which was one of three now collectively called the Chugach Islands. It was probably Elizabeth Island, western-most of the three.[2] But whichever island it was, the anchorage was an exposed one, and the *Favorita*'s launch was sent out to find a better location. They found such a place, which Bodega described as:

> [A] bay [*ensenada*] north-west of where we were anchored, where there is shelter and enough depth to anchor the two frigates, with more protection than we [now] have, with shelter from the winds of the first, second, and fourth quadrants, with depth of seventeen fathoms, an even and clear [bottom]; its shores surrounded by high hills populated with some small pines and an abundance of wild grasses.[3]

On 2 August Fernando Querós, second in command of the flagship *Princesa*, with Bodega, the chaplains, and a large contingent of both ships' companies – noticeably lacking the expedition's commander, Arteaga, who had fallen ill – went by launch to the newly found anchorage to take possession in the customary manner. They named it Ensenada de Nuestra Señora de la Regla (Bay of Our Lady of the Rule) and determined its latitude to be 59°08′N.[4] On the following day, the weather improved and with clearing skies they could see to the north-west a snow-covered mountain, which was

---

[1] See Barnabeu, *Bodega y Quadra*, p. 148.

[2] The three islands of this group are today named, east to west respectively, Chugach, Perl, and Elizabeth. Although any one of them could have been the one at which the expedition anchored, Elizabeth Island – the nearby location also of Cape Elizabeth – is considered to be the most probable location by Wagner, *Cartography of the Northwest Coast*, 2, p. 194.

[3] Barnabeu, *Bodega y Quadra*, pp. 151–2. Editor's translation.

[4] Cook, *Flood Tide of Empire*, p. 97, translates the bay's Spanish name Ensenada Nuestra Señora de la Regla as 'Our Lady of the Rule Bay'. Our Lady of the Rule was the patroness of several places in the Philippines and Cuba and was very popular in syncretistic folk religion (Santeria). It was originally a Spanish devotion, supposedly derived from the Rule of St Augustine. It is difficult to pinpoint the bay's location as the data concerning its whereabouts are contradictory. The latitude Bodega recorded for it, 59°08′N, falls directly on the south side of Elizabeth Island, where the two frigates cast anchor. But Ensenada de Regla cannot be located at the same latitude as the anchorage at Elizabeth Island, because Regla is described as being north-west an unspecified distance from where the frigates were anchored. Bodega's description of the bay they called Ensenada de Nuestra Señora de Regla suggests it was somewhere on the extreme south-west part of the Kenai Peninsula. There are two small inlets – one named Port Chatham and the other unnamed – on the peninsula across from Elizabeth Island (within 2 to 5 nautical miles; at lat. 59°12′N and 59°14′N, respectively) that are possible candidates for the location of Ensenada de Regla. Another inlet, now called English Bay, is farther away to the north-west (about 14 nautical miles; lat. 59°23′N at its mouth), but somewhat more commodious and better sheltered than the other two. Modern charts show the depth at its mouth to be 17 fathoms, consistent with Bodega's description. See Bernabeu, *Bodega y Quadra*, p. 152, and National Oceanic and Atmospheric Administration (NOAA) Chart 16013, *Alaska South Coast, Cape St. Elias to Shumagin Islands*.

particularly impressive by its height and having the appearance of a volcano. Any doubt of its volcanic nature was erased on the evening of 3 August, when the expedition witnessed the mountain erupt in a cloud of volcanic ash. It was one of two 3,000 m (10,000 ft) volcanoes, now called Iliamna and Redoubt, located on the Alaskan mainland on the west side of Cook Inlet. Whichever of these mountains it was, its Spanish discoverers called it 'Volcán de Miranda'.[1]

It appears that the two frigates were not moved to the better anchorage at Ensenada de Regla, remaining instead at Elizabeth Island. Once again the launches were sent forth to explore and chart numerous small islands, probably the Barren Islands and possibly as far away as outliers of Afognak Island. The Ensenada de Regla was mapped as well. But by 7 August, sickness aboard the flagship *Princesa* had reached serious proportions with the death of eight of her crewmen and many others – including the commander himself – too ill, mostly with scurvy, to function well if at all. Bodega and the *Favorita*'s crew seemed to have been more resistant to scurvy or whatever was afflicting their colleagues on the *Princesa*. Much as Bodega would have preferred to continue exploring farther westward for several more days, he realized the *Princesa*'s officers and men were hardly in condition to do so. He believed the expedition had sufficiently demonstrated that no north-west passage existed from Cape St Elias westward to the place they had dubbed Regla (Elizabeth Island and vicinity). Although they had not attained latitude 70°N or even 65°N, at Puerto de Santiago (Port Etches) they had finally reached the goal set in 1774 to go ashore and lay formal claim at latitude 60°N. As they had encountered neither Russian nor British interlopers – unaware that Cook's ships, *Resolution* and *Discovery*, had preceded them on basically the same track the previous year – they were led to conclude erroneously that the threat of foreign encroachment was slight or non-existent.

And so, on the evening of 7 August, as Bodega noted: 'We set sail from the Isla de Regla'.[2] They were south-east bound, headed non-stop for San Francisco, the nearest Spanish outpost where they could seek relief. In so doing, they bypassed the entire North American west coast from Alaska's Kenai Peninsula to Cape Mendocino on the northern California coast. It would bring to a close the Bucareli-planned efforts to explore, investigate, chart, and claim the north-west coast in the name of the Spanish Crown. By coincidence, news of Viceroy Bucareli's death on 9 April 1779 and Spain's declaration of war on Great Britain in June 1779 reached the settlement at San Francisco at about the same time the two frigates, *Princesa* and *Favorita*, arrived there

---

[1] This erupting volcano must have been either Iliamna or Redoubt. If the eruption was viewed from the mouth of English Bay, Iliamna would be more consistent than Redoubt with the bearing (NW7°N) and distance (13 leagues) that Bodega recorded. See Barnabeu, *Bodega y Quadra*, pp. 151–2. This evidence would seem to favour Iliamna as being the erupting volcano. But as it is not certain that the eruption was witnessed by Bodega from English Bay, it is difficult to be certain which of the volcanoes was the one actively erupting.

[2] Arteaga called the island where they anchored Isla de San Aniceto, while Bodega's name for it was Isla de Regla. Its present-day name, Elizabeth Island, is derived from a promontory on it, which was named Cape Elizabeth by Captain Cook on 20 May 1778, not realizing it was on an island. See Beaglehole, ed., *Journals*, 3, p. 356. It is close to where the *Princesa* and *Favorita* in all probability cast anchor on 1 August 1779; an ideal anchorage it is not, because of its open exposure in nearly every southerly direction. Although Bodega says they intended to remove the frigates to a more sheltered location, there is nothing to indicate it was actually done. In fact, when the expedition departed this area to return home, Bodega refers to the anchorage they had left as Isla de Regla. See Barnebeu, *Bodega y Quadra*, p. 155.

Figure 1: Oil portrait by Julio García Condoy of Juan Francisco de la Bodega y Quadra Mollinedo (1744–94). Reproduced by permission of the Museo Naval, Madrid.
Bodega is posed before the backdrop of Nootka Sound on the west coast of Vancouver Island. The scene depicts a twin-masted *goleta* (schooner) with a native canoe nearby. At the time of his death, Bodega held the rank of *capitán de navío* equivalent to a post captain in the eighteenth-century British Royal Navy.

in mid-September 1779.[1] These two events, so remote and seemingly unconnected, combined with the mistaken perception that neither Russian nor British designs on the north-west coast were of immediate concern, would merge to insure the suspension of further Spanish voyaging to that distant region for the next nine years.

6. Bodega y Quadra: *Marinero Limeño*

His full name was Juan Francisco de la Bodega y Quadra Mollinedo, and each surname was a reminder of his aristocratic ancestral heritage. All three names were squarely rooted in the Basque Country of northern Spain – particularly the village of Poveña in the province of Vizcaya. This village was the birthplace of Tomás de la Bodega y

---

[1] See Bolton, *Historical Memoirs*, 4, p. 186.

Quadra, destined to be Juan Francisco's father. Tomás and his brother, Juan, were the sons of Juan de la Bodega y Quadra and Agustina de los Llanas. The two boys descended from a long line of respectable landed noblemen. According to the late Antonio Menchaca, a collateral descendant of Juan Francisco, one of their ancestors, a certain Alvaro de la Quadra, served as the Spanish ambassador in London sometime during Philip II's reign as king of Spain (1556–98).[1]

How the two young brothers, Tomás and Juan, found themselves emigrants in a distant part of the Spanish overseas empire is a tale of some interest. The mother of the two boys, Agustina de los Llanas, died a premature death, following which their father, Juan, remarried. His second wife, Josefa de Llano, bore him four more children. But Josefa was apparently uncomfortable with the two stepsons she had acquired in her marriage to Juan, and it became necessary to find another place for them to reside. That place would prove to be with an uncle named José de la Quadra, in distant Lima, capital of the viceroyalty of Peru. The two young brothers would travel there, take up residence with their uncle, and grow to manhood. Tomás married Francisca de Mollinedo, a Peruvian creola, who was a member of Lima's colonial aristocracy and a descendant of the peninsular Quadras as well. Aided by such a well-connected partner in marriage, Tomás became a successful and wealthy man engaged in various enterprises, including trade, farming, and probably shipbuilding. It was into these circumstances that Juan Francisco was born on 22 May 1744, the second of seven children.[2]

Two years later, Lima suffered enormous damage from a powerful earthquake. It struck on 28 October 1746, with an estimated magnitude of 8.4 on the Richter scale. At least 5,000 people are believed to have lost their lives, many of whom were victims of a tsunami that swept the Peruvian coast.[3] The towers of Lima's main Cathedral were toppled, a statue of King Philip V collapsed, Pizarro's palace was destroyed, and the port city of Callao was nearly obliterated by the tsunami. Three large cargo-laden merchant ships anchored in the harbour were cast ashore and wrecked. The fabric of the city of Lima was badly disrupted, leading to a march on the Plaza de Armas by people rendered homeless in the earthquake. The disaster, however destructive and harrowing, spared Juan Francisco, who was then only in his infancy, and other members of his family seemed to have survived as well.

---

[1] This biographical sketch of Bodega y Quadra is based substantially on the following sources: (1) his service records in VICO, the Archivo-Museo Don Alvaro de Bazán, Viso del Marques, Ciudad Real; (2) Bernabeu, *Bodega y Quadra*, pp. 20–50; (3) Menchaca, *De California a Alaska*; and (4) Tovell, *Bodega y Quadra Returns to the Americas*; 'The Career of Bodega y Quadra', in *Spain and the North Pacific Coast*; and pers. comm. 29 January 2003 and 11 August 2003. The late Antonio Menchaca, retired Spanish naval officer and novelist, was a collateral descendant of Juan Francisco. Freeman M. Tovell, a member of the Royal Canadian Naval Voluntary Reserve in World War II, with 35 years in the Canadian Diplomatic Service, is a former Canadian ambassador to Peru and Bolivia (1962–5). He is also the author of an unpublished full-length biography of Bodega.

[2] Bodega's date of birth has been established conclusively through baptismal records located by Antonio Menchaca. These records, in the *Parroquia del Sagrario* 'Parish of the Sacrarium', Lima's main Cathedral, indicate that Juan Francisco, the twelve-days-old infant son of Tomás de la Bodega y Quadra and Francisca Mollinedo, was baptized on 3 June 1744. A handwritten document obtained by Menchaca testifies to this fact and is signed by parish priest Jeremias Revello Río, dated 3 October 1989. Subtracting the twelve days that elapsed between Juan Francisco's birth and baptism places his birthday on 22 May 1744.

[3] See Mallis ed., 'Earthquakes of Peru', pp. 8–11.

There is little to report about Juan Francisco's life as a child growing up in the aftermath of Lima's devastating earthquake. The Bodega family residence was in the heart of Lima in a large house next to the Plaza de Armas and behind the viceroy's residence.[1] It was a location likely to have received earthquake damage, but if so apparently nothing fatal to Bodega family members. His adult relatives and acquaintances were no doubt concerned for many weeks, possibly months or even years, with reconstruction of the two battered cities of Lima and Callao. Antonio Menchaca speculates that the youthful Juan Francisco 'would have grown up surrounded by generosity, good taste, and lordly pomp, as befitted the society of the time'.[2] In any event, the young Limeño began thinking about what direction his future might take. On 16 March 1761, at age sixteen, he entered the Real Colegio de San Martín, de la Universidad de San Marcos. It was an educational institution in Lima animated by Enlightenment Age ideals of learning, thinking and questioning. Although such things were not necessarily anathema to Juan Francisco, he decided to seek a career in the Spanish naval service. He and his brother, Manuel Antonio, travelled to Spain, where Juan Francisco entered the Academia de Guardias Marinas at Cádiz,[3] while his brother studied law at the Universidad de Alcalá de Henares, near Madrid. Juan Francisco entered the Academia in 1762 and graduated on 21 September 1765 as a *guardiamarina* 'midshipman'. He was eighteen years old and his service in the Armada Real Española (the Spanish royal navy) commenced at that time and continued throughout his life.

Midshipman Juan Francisco's first assignment was aboard a 74-gun *navío* 'warship' named *Terrible*, under the command of Post Captain Francisco Garganta, on a Mediterranean voyage. Next, our *marinero limeño* sailed on another 74-gun warship named *Princesa* – not to be confused with the frigate of the same name used in the third Bucareli expedition. The ship was commanded by another post captain named Francisco Espinola, on a voyage with ports of call including Naples, Italy, the Sicilian town of Palermo, and return to Cartagena on the south-east coast of Spain. On 21 April 1767, Juan Francisco went aboard a warship named *Garzota*, under Commander Francisco Saravia. During this Mediterranean voyage, on 12 October 1767, Juan Francisco was promoted to *alférez de fragata* 'junior ensign'. On 30 October, he was transferred to a light, shallow-draught type vessel called a *jabeque* (chebeck), named *Ibizenco*.[4] He disembarked the vessel on 1 December 1767, apparently at Cartagena, where he remained ashore until 15 October 1768. On that day, Juan Francisco embarked upon the most venturesome and longest voyage in his life at that time.

---

[1] See Tovell, 'The Career of Bodega y Quadra', p. 173.

[2] Menchaca, *De California a Alaska*, p. 10.

[3] This institution is also sometimes referred to as the Escuela de Guardias Marinas, or as by Menchaca, ibid., p. 12, the Real Compañia de Guardiasmarinas. In any event, it is today called the Academia de Guardias Marinas, the Spanish Naval Academy. See Blanca, *La Marina en Cádiz*, pp. 19–23, 121–30.

[4] The *jabeque* or 'chebeck', a vessel peculiar to the Mediterranean Sea, was propelled by both sails and oars. It was a highly manoeuvrable, three-masted, lateen-rigged, shallow-draught vessel armed with as many as 32 cannons. It is thought to have originated with the pirates of the Barbary States during the 17th century and been adopted by the Spanish Navy in the 18th century to combat Algerian pirates. See O'Scanlan, *Diccionario*, p. 324; Landström, *Sailing Ships*, pp. 158–9; and Harbron, *Trafalgar and the Spanish Navy*, p. 47.

He would sail aboard a 60-gun warship named *Septentrión*, which is thought to be the first such ship built at Cartagena, Spain (1751).[1] The ship, under command of Post Captain Antonio de Arce, sailed from Cartagena through the Strait of Gibralter, across the Atlantic Ocean to Buenos Aires, Argentina; then around Cape Horn to the Pacific Ocean and north to Concepción, Chile. The outbound voyage would continue northward bringing Juan Francisco to the city of his birth, Lima, Peru. He surely took the opportunity to visit his relatives and acquaintances there, as he had not seen them for about seven years. On the return voyage, the *Septentrión*, under Commander Manuel Bravo, reached Cartagena on 17 August 1772. Four months later, on 11 January 1773, Juan Francisco was promoted to *alférez de navío* 'senior ensign'.

A month after his promotion, while in Madrid, he wrote letters to Julián de Arriaga, *secretario de Marina e Indias*, and Andrés Reggio, *director general de la Armada*, requesting permission to travel to the port of Veracruz in the viceroyalty of New Spain, to resolve certain family business affairs.[2] His request was denied, although he was subsequently offered the possibility to be one of a group of six officers being assembled to go to San Blas by way of Veracruz to bolster the officer corps at San Blas. In this way Juan Francisco found himself on the list of possible candidates from which the officers would be selected. His initial request may not have been motivated entirely by patriotism; but when he learned he was among the chosen six he welcomed the assignment with enthusiasm. Departing Madrid, he arrived at Cádiz on 24 September 1773.

Juan Francisco's prospects now seemed ascendant, as he and his fellow officers prepared to cross the Atlantic Ocean. He was promoted to *teniente de fragata* 'junior lieutenant' late in April 1774, and by 13 June he and his five fellow officers departed the Cádiz harbour aboard an *urca* (stores ship) named *Santa Rita*, under command of Senior Lieutenant Juan Palacios, bound for New Spain. After touching at San Juan de Puerto Rico and Havana, they arrived at Veracruz on 26 August 1774. From there they travelled overland to Mexico City, where Viceroy Bucareli welcomed them on 7 September. There they received their instructions and forthwith departed the capital city by 13 December for the Naval Department of San Blas on the Pacific coast. It was dismaying to Juan Francisco that he and two of his fellow officers, Juan de Ayala and Diego Choquet, each holding the same rank of junior lieutenant had been neither assigned the command of a vessel nor even a role in the prospective expedition. They were ordered to remain at the inland town of Tepic, to await the organizing of another expedition. But suddenly everything changed when the expedition's planners in Mexico City finally decided – almost at the last minute – that the flagship *Santiago* needed an escort vessel with officers and a crew to sail her. That decision would prove a boon to Juan Francisco's naval career, although it would also expose him and those who would sail with him to enormous hardships fully described in his Journal of the schooner *Sonora*'s voyage from 16 March to 20 November 1775.

Juan Francisco never married, and it was likely that his lengthy absences at sea were most responsible for this. It is known that sometime during his stay at Tepic he met and became engaged to marry the daughter of a certain Lieutenant Colonel Miguel Marín del Valle, who commanded the militia garrisoned at San Blas. The prospective

---

[1] See ibid., p. 168.
[2] See Bodega to Arriaga and Reggio, 11 February 1773 (AGS, Valladolid, Marina, vol. 37). The full Spanish text of this letter is published in Bernabeu, *Bodega y Quadra*, p. 24.

bride, whose given name goes unmentioned, was a member of one of Tepic's principal families. On Juan Francisco's own admission, the difficulties of making arrangements at a distance and between voyages proved a serious obstacle to the marriage. Finally, his meagre financial resources led to wedding-day postponements which eventually became permanent.[1]

After returning to San Blas, Juan Francisco and his master aboard the *Sonora*, Francisco Antonio Mourelle, immersed themselves in developing a map of the North American west coast, bringing together what had been learned from the first two Bucareli expeditions.[2] Plans for a third expedition were also on the viceroy's mind, and Juan Francisco would be expected to assist in its planning and probably participate in it directly. His actions in the 1775 expedition had earned him high praise from the viceroy, and he was promoted to *teniente de navío* 'senior lieutenant' on 16 March 1776, exactly a year after that expedition had been launched. Viceroy Bucareli also recommended him for the title of Knight of the Order of Santiago.[3]

As part of the effort to prepare for a third expedition, Juan Francisco would have another opportunity to visit his birthplace of Lima, Peru, and once more meet his relatives and acquaintances in the Peruvian capital. On 28 December 1776, he departed San Blas travelling sometimes by land and other times over longer distances by sea, arriving at Lima on 5 May 1777. His purpose there, and more particularly in Lima's port city of Callao, was to negotiate the purchase and equipage of a Guayaquil-built frigate, generally known as the *Favorita* and later called the *Nuestra Señora de los Remedios* 'Our Lady of Remedies'. Juan Francisco completed his mission to secure the frigate, and by mid-December 1777 he sailed the newly acquired vessel north to San Blas, where he cast anchor in its harbour on 20 February 1778. During most of the following year, Juan Francisco would be at sea in command of the *Favorita*, on her Alaskan voyage of discovery in company with Ignacio Arteaga and the *Princesa*.[4]

On 4 February 1780, shortly after returning from the third Bucareli expedition, Juan Francisco was made commandant of the Department of San Blas by the new viceroy, Martín de Mayorga (Bucareli's replacement). This was soon followed by his promotion on 10 May to the rank of *capitán de fragata* 'commander'. Juan Francisco moved to strengthen San Blas' coastal defences by raising a battalion of provincial militia and training two companies of Indians. He further ordered construction of a fortress guarding the entrance to the harbour at San Blas. But his tenure as naval base commandant ended abruptly on 4 November 1780.[5] In February of 1781, Viceroy

---

[1] See Bodega to Antonio Valdés, 23 January 1787. In his service record, Bodega alludes to his abortive marriage proposal in this letter written at Cádiz, to minister of Marine, Antonio Valdés. The text of the letter has been translated into English and published in Tovell, *Bodega y Quadra Returns to the Americas*, pp. 9–11.

[2] See Wagner, *Cartography of the Northwest Coast*, 1, pp. 170–71; and 2, pp. 342–3, for the Bodega-Mourelle map of 1775.

[3] See above, p. 69, n. 1.

[4] See above, pp. 73–4, for more detailed discussion of Bodega's mission to Lima, the acquisition of the frigate *Favorita*, and the Arteaga-Bodega expedition to Alaska in 1779.

[5] Accusations of acts 'prejudicial to the royal treasury' made against Bodega by a certain Juan Sartorio, were forwarded to Martín de Mayorga, the new viceroy replacing the deceased Bucareli. Mayorga apparently did not give the accusations or the accuser much credence, and after an investigation all charges were dismissed. This incident, however, may have shortened Bodega's first tenure as commandant at San Blas. See Bernabeu, *Bodega y Quadra*, p. 40, and n. 51.

Mayorga, put Juan Francisco in command of the frigate *Santiago* – by now a venerable workhorse of the San Blas fleet. Her new mission would be to transport supplies of quicksilver from Peru for use in Mexican mines. So once again Juan Francisco would be returning to the sea and his birthplace as well. The *Santiago* departed San Blas on 5 June 1781 but did not reach Callao until 17 July 1782. This considerable delay was the result of stopovers along the way to conduct hydrographic mapping on the coast and to make repairs to the *Santiago* which necessitated a lengthy stay at Panama.

When the *Santiago* finally cast anchor at Callao that July day in 1782, up-country Peru was convulsed in a major rebellion of the indigenous Inca population. Initially, many Spaniards – *peninsulares* and creoles alike – were killed at the hands of rebels led by a descendent of Inca royalty, known by his Spanish name as José Gabriel Condorcanqui, but who had changed his surname to Túpac Amaru.[1] One of the early casualties in this anti-Spanish rebellion, was a *corregidor* 'magistrate' named Manuel de Bodega in the province of Paria.[2] Although this person bore the same name as Juan Francisco's brother, it is uncertain if he was a kinsman of any sort. Whatever the case, while Juan Francisco was in the Lima vicinity the Inca rebellion seems to have had little if any effect or hindrance on his activities. He departed Callao on 19 March 1783, returning aboard the *Santiago* with a cargo of war *matériel* and other tools or utensils to San Blas, arriving there fourteen weeks later on 20 June.

Considering that Spain's exploration on the north-west coast had ceased, there was now a surplus of naval officers in the Department of San Blas. Juan Francisco received orders to travel to Veracruz, from where he would embark on the warship *Santo Domingo* bound for Havana on 2 December 1783. Arriving there eighteen days later, he was ordered transferred to the warship *Dichoso*, a vessel designated for surplus officers, where he remained until 15 March 1784, when he was transferred to the 74-gun warship *San Cristóbal* (alias the *Bahama*). This ship was under the command of Post Captain Félix del Corral, with Juan Francisco as second in command. It was during this time, on 15 November 1784, that he was recommended for promotion to the rank of *capitán de navío* 'post captain'. The *San Cristóbal* set sail from Havana for Cádiz on 14 January1785, reaching her destination a month and a half later on 2 March.[3]

From the time Captain Bodega y Quadra returned to the Department of Cádiz until the resumption of Spanish exploration on the north-west coast in 1788, relatively little is known of his activities. He seems to have served as an adviser or instructor of midshipmen at the Academia de Guardias Marinas, where he had commenced his own naval training twenty-six years earlier. It also appears that he travelled to Galicia in the north-west corner of Spain to visit Francisco Antonio Mourelle, his master on the *Sonora*'s 1775 voyage. Juan Francisco may have also been seeking genealogical proof of his ancestry in Galicia and the Basque Country connection with his nomination as a Knight of the Order of Santiago.[4] Whatever the

---

[1] The events of this Peruvian rebellion are recounted in Fisher, *The Last Inca Revolt*.

[2] Ibid., pp. 72, 141, 152.

[3] Bodega appealed directly to King Charles III in a petition dated 18 October 1784 to be transferred to Spain, specifically the Department of Cádiz. The petition includes *Mandos y Comisiones que se me han confiado* 'Commands and Assignments entrusted to me', a handwritten summary of his activities during his naval service, a copy of which is preserved in VISO.

[4] See Bernabeu, *Bodega y Quadra*, pp. 42–3; and Thurman, 'Juan Bodega y Quadra and the Spanish Retreat from Nootka', pp. 52–3.

case, he was also giving serious thought as to what ought to be done if Spain were to renew and press on with its exploration and claims on the north-west coast. In a letter to Antonio Valdés y Bazán, ministro de Marina, written at Cádiz on 23 January 1787, Bodega outlined a proposal that would anticipate the Malaspina expedition in many respects, including: an emphasis on cartography and collecting scientific data; the use of a pair of ships specially designed and equipped for exploration; and a capability to navigate globally. In this same letter, there is a sense that Bodega is concerned that his naval career was all but over. He had been inactive for two years, and recognizing that he would not likely be given command of his proposed scientific expedition he would gratefully accept an opportunity to command once more the Department of San Blas.[1]

On 24 March 1789, Captain Bodega y Quadra was finally appointed to the post he had requested – a position he had previously held for nine months in 1780: Commandant of the Department of San Blas. It was an appointment made directly by the king himself, Charles III, and it was part of a major new effort to reassert Spanish claims of sovereignty on the north-west coast of America. New Spain would also be getting a new viceroy. He was Juan Vicente de Güemes Pacheco de Padilla, Conde de Revillagigedo, hand-picked for his support of Madrid's more assertive stance concerning the north-west coast of America. Viceroy-designate Revillagigedo, with his entourage, and commandant-designate Bodega y Quadra, with six junior officers, sailed aboard the 68-gun warship *San Ramón*, from Cádiz harbour bound for Veracruz, where they arrived on 26 May 1789. The six officers selected were: Jacinto Caamaño, Manuel Quimper, Salvador Fidalgo, Ramón Saavedra, Francisco de Eliza, and Salvador Menéndez Valdés. These were obscure names, but they would later find a place of honour in the maritime exploration of the north-west coast. Unlike the six junior officers who had made the same journey in 1774, this new group had the advantage of being led by Bodega y Quadra, who as a young lieutenant had himself participated in that earlier period of exploration, and who was well versed in the difficulties of San Blas and the hazards they faced on North Pacific waters.

When Conde de Revillagigedo and Bodega y Quadra arrived at Veracruz that May in 1789, events during the previous year on the Alaskan and north-west coasts had taken a decidedly worrisome turn. Spanish authorities had learned from the French explorer Jean-François de Galaup, Comte de La Pérouse, during his visits to Spanish settlements along the Pacific coast in 1786, of the substantial Russian establishments in Alaska. This had resulted in an expedition, mounted from San Blas in March 1788, to investigate such reports.[2] It was led by Esteban José Martínez, aboard the frigate *Princesa*, accompanied by Gonzalo López de Haro, second in command and captain of the packet boat *San Carlos*. The expedition had achieved mixed results. It made the first direct contact with Russian settlements in the Aleutian Islands, but because of a dispute between Martínez and López de Haro, the expedition failed to examine the coast stretching between Alaska and California for foreign activity. Martínez, however, had

---

[1] Bodega to Antonio Valdés, 23 January 1787, written at Cádiz. For an English translation see Tovell, *Bodega y Quadra Returns to the Americas*, pp. 9–11.

[2] Only one account by a participant in the Martínez-López de Haro 1788 expedition has been published in English translation. See McDowell, *José Narváez*, pp. 99–163. See also Cook, *Flood Tide of Empire*, p. 594, for a listing of the extant journals and related documents of this expedition.

been led to believe by Russian informants at Unalaska – Evstrat Delarov and Potap Zaikov – that a Russian effort to establish an outpost at Nootka Sound was impending. This information set off a second expedition, also under Martínez's command, to counter such a move by the Russians. Martínez, still sailing the *Princesa*, reached Nootka on 5 May 1789, with his escort, the packet boat *San Carlos*, under López de Haro, arriving twelve days later. They were surprised to discover that a place called Friendly Cove was sheltering British and American ships – and others flying Portuguese colours – but no sign of Russian ships or occupation. Martínez proceeded to take possession of the place, ordering the erection of some small structures, and fortifying a point overlooking the entrance to Friendly Cove. Martínez announced that the foreign vessels were trespassing on Spanish territory, and that they would have to vacate the anchorage. A confrontation ensued between Martínez and James Colnett, captain of the snow *Argonaut*, which would spiral into an international crisis, bringing Great Britain and Spain to the brink of war.[1]

It was into these unsettling circumstances that Conde de Revillagigedo and Captain Bodega y Quadra would now step to assume their respective new responsibilities in Mexico City and at San Blas. Revillagigedo was there to replace Manuel Antonio Flores, as viceroy of New Spain; Bodega would be filling a vacant post, in which Martínez had been serving informally as the de facto commandant of the Department of San Blas. The Colnett-Martínez confrontation at Nootka Sound was in fact occurring at roughly the same time the new Spanish leadership was going ashore at Veracruz and preparing to travel overland to their new posts. When Bodega reached San Blas, he learned that Martínez had arrested Colnett, placed his crews in irons, and seized two ships, the *Princess Royal* and the *Argonaut*. The ships, their officers and men, and cargos, were all taken captive and transported south for detainment at San Blas. The outgoing viceroy, Manuel Flores, had supported Martínez's actions, but his replacement, Conde de Revillagigedo, had some misgivings. Moreover, Martínez's treatment of the captives – who were mostly British – had been rather harsh and not much given to grant even the smallest favour. In contrast, the newly-appointed San Blas commandant, Bodega y Quadra, was singled out for praise in his conduct toward the prisoners. Colnett later remarked: 'To this officer [Captain Bodega] I am greatly indebted for his kind attention and obtaining permission for me to go to Mexico [City] to claim redress for our past treatment'.[2]

---

[1] Between 1785 and 1788 at least 16 British, 3 American, 3 Portuguese-flagged vessels, and an Austrian-flagged vessel scoured the Alaskan and north-west coasts for sea otter skins. See Cook, *Flood Tide of Empire*, p. 551. In the view of Madrid, they were all technically violating Spanish claims of sovereignty. As for the Portuguese- and Austrian-flagged ships, they were actually British, disguised as Portuguese or Austrian. And because they were not properly licensed by the South Sea Co. and the East India Co., they were in violation of British law as well. The confrontation between Martínez and Colnett, at Friendly Cove in Nootka Sound in May 1789, initiated what has come to be called the Nootka Sound Controversy, a hugely complex diplomatic episode too lengthy to be recounted in detail here. For what is still considered the most complete study of the controversy and its outcome, see Manning, 'The Nootka Sound Controversy', pp. 279–449. For more recent examinations, see Cook, *Flood Tide of Empire*, pp. 146–270; and Nokes, *Almost a Hero*, pp. 139–70.

[2] Quoted by Manning, 'The Nootka Sound Controversy', p. 353, and attributed to 'Colnett, Voyage, 96–102, note'. For other similarly favourable mention of Bodega's humane treatment of the captives detained at San Blas, see Howay, ed., *Journal of Captain James Colnett*, pp. 82, 89.

At first it seemed the issue could be resolved without further resort to force. It would require restraint on both sides – particularly from the two main antagonists, Colnett and Martínez. Efforts to gain the release of Colnett, his officers and men, the return of his ships, and indemnification for his losses, were progressing with good prospects. But hopes for a quick and quiet resolution were dashed when John Meares, one of the principal figures behind the enterprise to occupy Nootka Sound, travelled directly from Macao to London, where he placed the issue squarely before Parliament and the British public. It effectively took this small dispute on a remote shore to the highest levels of the British and Spanish governments, and further entangled the matter in national pride. Naval forces of both nations were mobilized and placed on alert in the event talks broke down without a diplomatic solution. While negotiations between London and Madrid slowly ground forward, Bodega remained patiently at San Blas waiting for whatever orders or directives diplomacy might generate. His wait would finally be ended on 28 October 1790, with the signing of the First Nootka Convention (also called *Tratado del Escorial* 'Treaty of the Escorial') at the Escorial palace, twenty-five kilometres north-west of Madrid.

Although an agreement had been reached in the Convention of 1790 between London and Madrid at the upper levels of their respective governments, many details and ambiguities would have to be addressed and ironed out on the site of the original dispute, namely: Friendly Cove in particular and Nootka Sound in general. Each side agreed to send a commissioner empowered to meet and resolve such issues on the very ground where the trouble had all begun. Not surprisingly, Viceroy Revillagigedo selected Bodega for this delicate task. No other officer in the Armada Real had a better understanding of the history of Spanish claims on the north-west coast of America, combined with diplomatic skills uncommon among naval officers. It would prove to be his third and final assignment on the north-west coast. In considerable contrast to Bodega's first arrival on that coast as a young lieutenant in command of the little schooner *Sonora*, he now travelled aboard a large flagship, the *Santa Gertrudis*, a frigate sent out from Spain. He was accompanied by six other ships of various kinds, which together formed what came to be called the *Expedición de Límites* (Expedition of Limits or Boundary Expedition).[1]

Bodega arrived at Nootka Sound on 29 April 1792, while his British counterpart, Captain George Vancouver, who was engaged in hydrographic mapping elsewhere, did not put in at Nootka until four months later at the end of August. During this time Bodega took the opportunity to question witnesses of the events leading up to the confrontation. They included Maquinna, the principal Nootkan leader; Francisco José Viana, Portuguese captain of the *Ifigenia Nubiana* (one of Meares's ships in 1789); and two American fur traders, Robert Gray and Joseph Ingraham. This may have caused Bodega to toughen his stance in the negotiations he would later have with Vancouver.[2] But it did nothing to prevent him from extending the most cordial of welcomes to his British guests. The officers were treated to lavish five-course meals,

---

[1] In addition to the frigate *Santa Gertrudis*, the ships included another large frigate from Spain, the *Concepción*; five San-Blas-built vessels, the frigates *Princesa* and *Aranzazú*, the brigantine *Activa*, and two schooners, the *Sútil* and *Mexicana*. They did not all sail in the same convoy at any given time, but they were a very substantial force at Bodega's disposal while he remained at Nootka Sound in negotiations to settle the Nootka Controversy.

[2] See Lamb, ed., *A Voyage of Discovery*, 2, pp. 664–5.

consisting of nearly every order of fresh food and drink imaginable.[1] Meetings and discussions were conducted in a friendly, relaxed atmosphere, even to the point of sharing cartographic information. But when the key issues they were there to settle came up for discussion they were unable to find a mutually acceptable compromise.

Bodega contended that whatever structures Meares may have erected at Friendly Cove they had vanished so that there was nothing tangible to return, except ownership of the raw land on which it had been located. As for claims to Nootka Sound and its vicinity, Bodega was only willing to turn over the site of Meares's improvements, together with structures and gardens Spanish personnel had built, occupied or cultivated on sites away from Meares's location. And finally, Bodega claimed he was unable to make or accept any proposal prejudicial to Spanish claims of sovereignty.[2] Vancouver, on the other hand, believed he was there to receive, on behalf of the British Crown, not only Meares's properties seized by Martínez at Friendly Cove, but Nooka Sound:

> *in toto*, and port Cox [an anchorage in Clayoquot Sound]. These I was still ready to receive, but could not entertain an idea of hoisting the British flag on the spot of land pointed out by Senr. Quadra, not extending more than a hundred yards in any direction.[3]

With this, negotiations had reached an impasse beyond which the parties were unable to proceed. They would now be required to seek new instructions from their respective governments, which in those pre-electronic times meant delays of many weeks or even months, and sometimes years. Despite the language barrier and their inability to reach an accord, Bodega and Vancouver came away from their negotiations with a deep and genuine respect for each other.[4] From the Spanish perspective, the Boundary Expedition's main diplomatic aim had been to fix a boundary or limits separating British and Spanish claims on the north-west coast. In this, it had proven unsuccessful and perhaps more importantly by admitting that there was some degree of validity to British claims, Spain's claims to sovereignty were effectively ended. For Bodega it was now time to return south, where he would report the outcome of his

---

[1] It has been suggested that Bodega's generosity may not have been purely a gesture of good will but part of his diplomatic strategy. As Vancouver's officers and men had been on reduced rations since missing their rendezvous with the storeship *Daedalus* in Hawaii, they might have been susceptible to such a gambit. See Fireman, 'The Seduction of George Vancouver', pp. 434–7. This view, however, is disputed by Freeman Tovell (pers. comm. 29 January 2003).

[2] Viceroy Revillagigedo and officials in Madrid had already decided that the Spanish occupation of Friendly Cove would have to be abandoned, but not necessarily all sovereign claims to it. Their plans called for establishing a boundary between British claims (north side) and Spanish (south side) at the Strait of Juan de Fuca and placing a settlement they called *Nuñez Gaona* on the south side entrance of the Strait, at present-day Neah Bay. It too would prove untenable and was abandoned before its construction could be completed. See Lamb, ed., *A Voyage of Discovery*, 4, pp. 1568–70.

[3] Ibid., 2, p. 676.

[4] Negotiations were conducted in part through an exchange of letters, which were translated by Thomas Dobson, a Spanish-speaking midshipman aboard Vancouver's storeship, *Daedalus*. Sometimes there were verbal exchanges in which Dobson served as an interpreter. Another bridge across the language barrier was provided by Lieutenant William Broughton, captain of the armed brig *Chatham*, Vancouver's escort, and José Mariano Moziño, a naturalist attached to Bodega's staff. Broughton and Moziño both understood French, which enabled them to help bridge the gap between those who spoke only English or only Spanish. See ibid., 2, p. 675; and Moziño, *Noticias de Nutka*, p. 69.

meeting with Vancouver and resume his responsibilities as commandant of the Naval Department of San Blas.

Meanwhile, a project was being planned at San Blas to colonize a bay just north of San Francisco.[1] It had been discovered in 1775 by Bodega and it bore his name. He was scheduled to direct the planning of this operation, but his failing health required that a younger officer take over. Bodega had earlier suffered from painful headaches at Nootka Sound.[2] In early spring 1793, he fell seriously ill, which induced him to write a letter dated 24 March to Viceroy Revillagigedo explaining:

> I have lived for many days with a pain that is impossible for me to endure. The physicians assure me it is curable if I take some time off from the affairs of the Department. I am deserving, Your Excellency, that you allow me to delegate [my duties] to a corresponding officer while I obtain some relief.[3]

Revillagigedo approved what amounted to a request for sick leave, but he required Bodega to remain close to his post because of strife in Spain. On 7 March 1793, revolutionary France had declared war on monarchist Spain, ironically resulting in an alliance of the latter with Great Britain – which a year earlier had been considered Spain's prime adversary on the north-west coast of America. Bodega's health, however, continued to worsen, and he was forced to turn over his command of the Department of San Blas to Manuel Quimper.

In the company of several attendants, Bodega travelled to Mexico City, pausing on the way at Guadalajara on 13 November to draw up his last will and testament. Farther along, in the city of Querétaro, they met Pedro Carvajal, a retired surgeon of the Spanish navy and Bodega's friend and physician, who attended him until his death. Upon reaching Mexico City, they took lodging in simple quarters, with room for two of his attendants, Juan Mesía, a native of Ferrol, in north-western Spain (Galicia) and José Antonio Navarrete, from Tepic, Mexico. Bodega was also attended by his personal chaplain, Padre Alejandro Jordán, and a woman named Manuela Atayde. At his death on 26 March 1794, he was two months short of his fiftieth birthday. He was buried three days later at the convent of San Fernando in Mexico City.[4]

When George Vancouver learned of Bodega's death he penned a touching tribute to his erstwhile diplomatic rival. It was at Friendly Cove in September 1794, when a change in the Spanish commandant brought word of Bodega's demise, to which Vancouver responded:

> The appointment of this gentleman [José Manuel Alava] as governor of Nootka had taken place in consequence of the death of our highly valuable and much esteemed friend Senr.

---

[1] The project to occupy and colonize Bodega Bay, led by Juan Bautista Matute, proved unsuccessful. According to Wagner, its abandonment had less to do with its difficulties 'as to the change in relations with Great Britain and a final amicable settlement of the disputes with her over the north-west coast. It was reserved for the Russians to take possession of the place [Bodega Bay] and become a thorn in the side of the Spaniards.' Wagner, *The Last Spanish Exploration*, p. 29.

[2] Tovell, pers. comm. 29 January 2003.

[3] Bernabeu, *Bodega y Quadra*, p. 49 (editor's trans.).

[4] This account of Bodega's last days relies primarily on Bernabeu, *Bodega y Quadra*, pp. 49–50. Others such as Thurman, 'Juan Bodega y Quadra and the Spanish Retreat from Nootka', p. 62, and Cook, *Flood Tide of Empire*, p. 415, add some details such as the possibility that Bodega may have suffered a seizure while in seclusion at Guadalajara.

Quadra, who in the month of March had died at San Blas [*sic* actually Mexico City], universally lamented. Having endeavoured, on a former occasion to point out the degree of admiration and respect with which the conduct of Senr. Quadra toward our little community had impressed us during his life, I cannot refrain, now that he is no more, from rendering that justice to his memory to which it is so amply intitled [*sic*], by stating, that the unexpected melancholy event of his decease operated on the minds of us all, in a way more easily to be imagined than described; and whilst it excited our most grateful acknowledgements, it produced the deepest regret for the loss of a character so amiable, and so truly ornamental to civil society.[1]

---

[1] Lamb, ed., *A Voyage of Discovery*, 4, pp. 1396–7.

THE JOURNAL

Year of 1775 Journal of the Exploration made on the northern coast of California undertaken by Juan Francisco de la Bodega y Quadra in the schooner *Sonora*

His Majesty [Charles III] having ordered the appointment of six officers from his naval corps to go on voyages of discovery to the northern coast of California; and to serve under the command of His Excellency, Knight Commander of Malta, Don Antonio María Bucareli, viceroy and captain-general of the kingdom of New Spain; the individuals selected [for the assignment] being Don Bruno de Hezeta, Don Miguel Manrique, Don Fernando Quirós, Don Juan de Ayala, Don Diego Choquet and myself; we value it highly for the details such an extended voyage offers us as a recommendation so singular that it fosters such a sense of superiority.

    We left Europe on the 13th of June, 1774, and we arrived at Mexico City on the 25th of October, of the same year, whereupon we presented ourselves to His Excellency, the Lord Viceroy, who wanted [accomplished] what the first [expedition under Pérez] went forth to do earlier in the year.[1] It was decided that Don Bruno de Hezeta, as the one with the most seniority, would embark in the frigate *Nueva Galicia* [alias *Santiago*] with the purpose of exploring up to sixty degrees or more, and to come [southward] along the coast, reconnoitering ports or inlets, recording their shorelines and taking possession of those lands. Don Miguel Manrique and Don Fernándo Quirós were named to command the packet boats *San Carlos* [alias *El Toison*] and *Príncipe* [alias *San Antonio*], with the aim of supplying provisions to the presidios at San Diego and Monterey, in support of their missions. This was to enable them to assist the frigate in case she put into port in distress. They also had orders to examine *Puerto de San Francisco*.[2] We left behind Don Ignacio Arteaga, who came later and was named commandant of the new Department of San Blas. [As for] Don Juan de Ayala, Don

---

[1] This expedition embarked from San Blas in January, 1774, under command of Juan Pérez. Its failure to reach the 60th parallel or go ashore to perform possession-taking formalities were in part responsible for the need to launch a second expedition in 1775. See Beals, *Juan Pérez*, pp. 33–41; and above, pp. 56–9.

[2] What is today called San Francisco Bay was first encountered by Spanish overland explorers in 1769, when the Portolá expedition came upon it by chance while trying to locate Monterey Bay. But its opening to the sea, the Golden Gate, had remained undiscovered from its seaward approach until 1775, when the packet boat *San Carlos*, under command of Juan de Ayala, sailing out of Monterey, made its discovery and entered to chart it. Both Hezeta and Bodega y Quadra were under orders to look for this elusive, frequently fog-bound passage as they returned from the north. Hezeta tried to find it but failed because of fog. Bodega y Quadra probably sighted it, although he was unable to gain entry. See Smith and Teggart, eds, *Diary of Gaspar de Portolá*, p. 39; Treutlein, *San Francisco Bay*, pp. 15–34; Beals, ed., *For Honor and Country*, pp. 93, 95; and above, p. 52.

Diego Choquet and me, we three were ordered to reside in Tepic (a place twenty leagues distant from San Blas) until another expedition [was readied].

On the 15th of December, we departed Mexico [City], and having arrived at Tepic, we learned that His Excellency had determined that the schooner [*Nuestra Señora de Guadalupe*, alias *Sonora*], which lay in port, be speedily careened and readied to make a voyage of discovery as the frigate's escort. She was to be placed under command of the senior officer, who was Don Juan de Ayala. With the aim of acquainting myself with Puerto de San Blas, its artillery and ships, I went to said Department. I realized the schooner's smallness and her incapacity for a voyage so lengthy and exposed might, in carrying only one officer, cause the expedition delay. I thus decided, attentive to the best interests of such an important mission, to request that I embark on the said schooner in the capacity of second [officer]. It was of no concern to me to go under the command of another [officer] of my same rank; or for me to agree to the immense voyages for which I am pleased to risk myself. Notwithstanding the discomforts and ever-present perils that are to be expected on a vessel eighteen feet [*codos* or cubits?] in keel length,[1] and in such extreme latitudes, where experience has shown me the rigors of the winds and swollen seas, which seldom are lacking, I saw fit to consider the assignment on which I have come exclusively as an honour, sacrificing [if necessary] my health and even my life for His Majesty. Sustained by my regard, desirous that my determination would be effective, I sent an official letter to the commander [Hezeta] (for there was no time to write to the Most Excellent Lord Viceroy), in which I requested that I might embark on the schooner. And attending to my reasons, it appearing that my determination had been successful, he responded that I was to remain aboard, and he rejoiced in my resolve, agreeing that it would be very advantageous to the service of His Majesty and agreeable to His Excellency.

The 16th of March, 1775, at ten o'clock in the evening, the schooner's list of all necessary supplies being now [complete], with provisions for a year, water for four months, the stern draft in six fathoms, and the bow in five and a half, the commander signalled for us to set sail, which we did with a wind out of the north (which is a land breeze). At the same hour, the packet boat *San Carlos* set sail under the command of Senior Lieutenant Miguel Manrique, who was to go to Monterey. We sailed the rest of the night under full sail striving to keep up with the frigate.

At sunrise on the 17th of March we found ourselves greatly separated, which led me to believe that some pattern of the currents might be the reason, because no diligence whatsoever had been spared maintaining ourselves in convoy [with the *Santiago*]; nor was it attributable to [the *Sonora*'s] poor speed, because the information we have says she is an exceedingly swift sailor. After sunrise it was entirely calm, and it remained so until three o'clock in the afternoon, when the north-west wind came up (which is the [sea] breeze). And with the same [wind] over the land until sunset, we cast anchor about one league south-east of Piedra Blanca (white rock),[2] and five [leagues] distant from San Blas.

---

[1] 'Dieciocho pies de quilla', 'eighteen feet of keel'. The *Sonora*'s actual keel length was somewhere between 30 ft, at a minimum, and 38 ft at a maximum, with the probability that it was closest to 34 ft. Bodega must have intended to write 'dieciocho codos de quilla' 'eighteen cubits of keel', which would have been closer to reality. See Landin, *Mourelle de la Rua*, p. 19, n. 14; and also above, p. 63, n. 1.

[2] Piedra Blanca refers to one of two reefs, today called Piedra Blanca del Mar 'white rock seaward'; and Piedra Blanca de Tierra, 'white rock landward', off the entrance to San Blas. See Thurman, *San Blas*, p. 22.

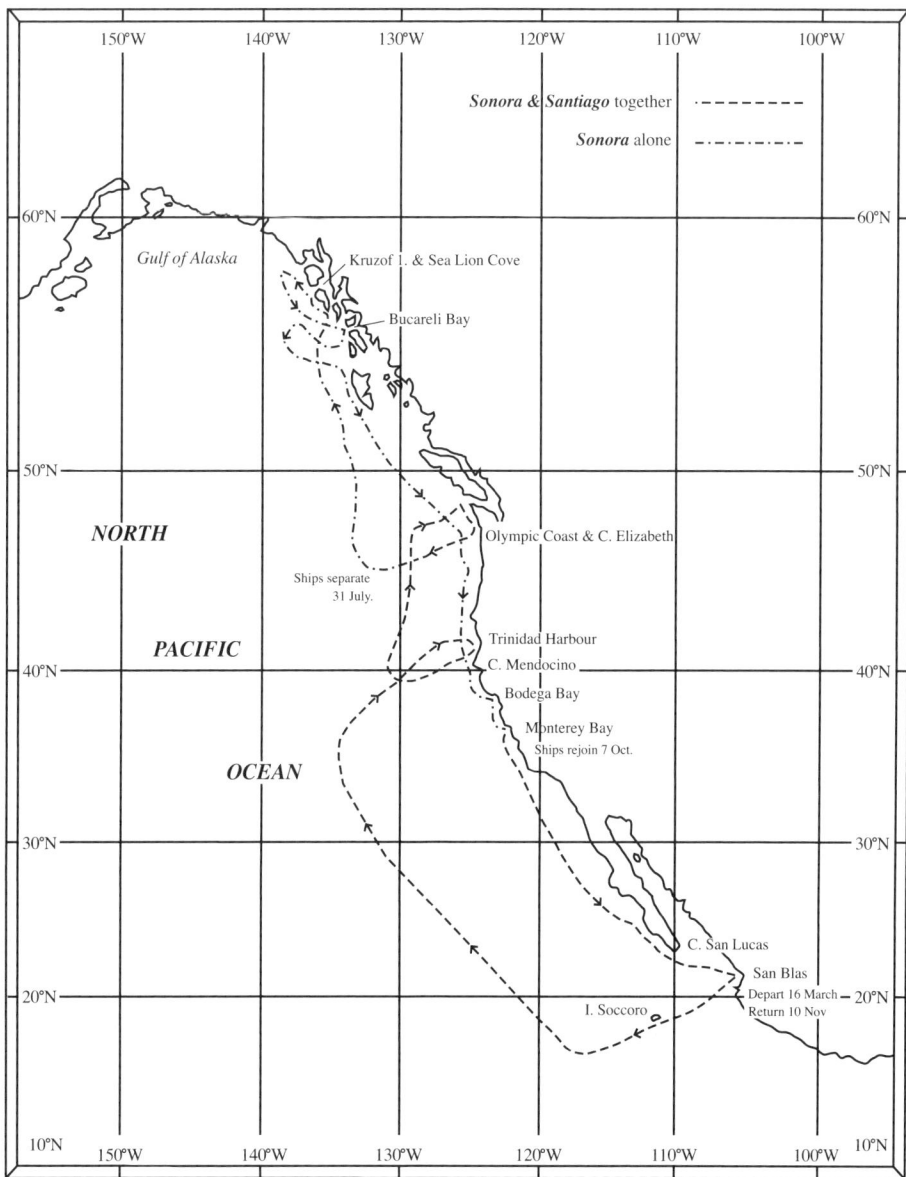

Map 1: Voyage of the *Sonora*, 1775: From San Blas, Mexico, to Sea Lion Cove, Alaska and return.

We have navigated in the same manner until the 19th [of March], sailing in order to get outside [of the Islas Tres Marías 'Islands of Three Marys'],[1] with the land breeze and with the landward [sea] breeze, until it went calm and we were obliged to anchor. On this day at three o'clock in the afternoon, the packet boat *San Carlos* signalled for help. And the commander [Hezeta] sent the launch to learn what had happened. At about nine o'clock at night the captain [of the *San Carlos*, Manrique] arrived at this schooner, who by his actions demonstrated that he was not in his right mind. After having quieted him down, insofar as his cavillings, they conducted him to the frigate, where neither bleeding nor other remedies that they applied met with improvement. On the contrary, he became more enraged and grew worse. Having been unable to get hold of himself, however much it was tried to quiet him, he endured the rest of the night crying a sea of tears, possessed by powerful manias and apprehensions.[2]

At dawn on the 20th [of March] the situation was the same. Seeing [Manrique's] inability to recover his normal senses and considering him unfit for command, the commander [Hezeta] convened a meeting in order to determine what ought to be done to resolve things. It was decided (although much to our sorrow) to send him ashore, for there was no other recourse; and to transfer command of the packet boat to the captain of this schooner, Don Juan Manuel de Ayala. Hence forth from this day, I was recognized as captain of the said schooner. I was handed a sealed envelope that I am required to open in case of separation, along with various instructions and signals.

The breeze freshened this day, and we weighed anchor. In a short while it was discovered that the frigate's fore-topmast was sprung, and that this damage had been noted after the previous voyage. But, finding nothing that might be able to substitute for it in its absence, because the one brought as a spare is extremely small and deteriorated, the commander determined to continue with it, repairing it and tying it together with gammoning, which was done in a day. Its sail was returned to [usefulness] navigating, and it will be able to carry a stiff sail provided precaution [is observed] to take in a reefing or always to lower it when the wind freshens.

Desirous to know how this schooner handled in a moderately fresh breeze, I ordered all sails set, and having hove the log line, it made only two knots headway. It was hard to believe that it made so little headway, when ashore it had been highly praised as a sailing ship. I judged that it was a matter concerning the rather mediocre foresail. I ordered a mainsail put in its place, which was rigged from stem to stern, but this endeavour gained little, for the increase in speed was about two or three yards [per hour]. I continued on, exercising redoubled diligence to increase speed. The major part of the entire cargo was brought astern; then returned to make it the same as the bow; it was stowed in unusual ways; the stays and rigging were loosened and hauled taut again; the masts were un-wedged. And finally, no task remained undone that practice had taught me; but there was no way I could succeed in surpassing three knots, it being impossible for me to stay with the frigate, even when she did not carry more than her main and foremast topsails. The commander [Hezeta], who saw that the

---

[1] The Islas Tres Marías includes four islands, three of which bear the names of biblical Marys, running NW–SE, between 60 and 80 naut. miles SW of San Blas. See Thurman, ibid., p. 45.

[2] The accounts of the *Santiago*'s two chaplains indicate that Manrique spent the night in the company of Fray Miguel de la Campa Cos, the only person in whom the deranged officer placed any trust. See Campa Cos, *Journal of Explorations*, p. 16; and Wagner, and Baker, eds, 'Fray Benito de la Sierra's Account', p. 209.

consequence of this situation was endlessly delaying the voyage, decided to extend a towline which was taken aboard [the schooner]. So much strength was demanded that it would be impossible for any stream cable to withstand [the strain]. I resolved to set two sails, which I ordered put in the tops. With this I have succeeded in maintaining myself, without the constant anxiety I was living with so disproportionately, or the risk it posed for me in countering my desire to continue without being a drag on the mission.

The 24th [of March] at twelve o'clock [noon] the southernmost of the Islas Marías [Isla Marías Cleofas] was surveyed at a distance of three leagues, which put the location of the ship in 111 degrees, 20 minutes, longitude west of Paris. Although by solar observation [her latitude] was 21 degrees 14 minutes, north, the French map[1] shows 21 degrees 40 minutes, with the result that said islands are situated on this map somewhat farther north of their true positions.

In this place we lost sight of the packet boat, and she began to follow a course south-west one-quarter west in order to get away from the coast. We are continuing to pursue our course routinely, except for having seen various birds. Some are black with a speckled white breast; the wings are large, the beak long, and the tail in the form of scissors. They are called frigate birds. Others are entirely white, with a tail seemingly of one feather only; they are called tropic birds. Others remain in the water like ducks, and they are called boobies.[2]

The experience has been that the winds slacken at night and freshen in the daytime, chiefly around the [lunar] conjunction, at which there have been several calms. The winds have run in the fourth [NW] quadrant from north-west up to north.

The 29th [of March] at sunset we saw an island called Socorro [help, succour].[3] The commander wanted to replenish the water supply, provided we were not squandering time away. Because of the small force of the lunar [tide], it was of no benefit to us in sailing, so it was decided to approach the island. We took a sighting of it, and it bore to the west a distance of nine or ten leagues, which we tried all night to close, luffing as much as possible so as not to fall off leeward or its parallel.

The 30th [of March] at sunrise its bearing from us was north-west one-quarter west to its midpoint, a distance of four leagues. Although we sailed the necessary distance, we were unable to put abreast of it, apparently because the currents carried us toward

---

[1] This French map was no doubt the work of Jacques Nicolas Bellin (1703–72), a hydrographer in the French Dépôt des cartes et plans de la Marine. Hezeta also had such a map aboard the *Santiago*, which he describes as being published in 1766, under the title *Carte reduite de l'Océan Septentrional*. Bodega apparently had two versions of this map aboard the *Sonora*, the editions of 1756 and 1766. See Bernabeu, *Bodega y Quadra*, p. 61, n. 6; Wagner, *Cartography of the Northwest Coast*, 1, p. 178; Beaglehole, ed., *Journals*, 1, n. 2; and Beals, *For Honor and Country*, pp. 57–9.

[2] The frigate birds Bodega y Quadra describes were most likely the Magnificent Frigatebird (*Fregata magnificens*). The tropical birds he mentions were the Red-billed Tropic bird (*Phaeton aethereus*). As for boobies, they were probably one of two species known to frequent the waters off Mexico's Pacific coast: the Brown Booby (*Sula leucogaster*) or the Blue-footed Booby (*Sula nebouxii*). See Udvardy, *Audubon Field Guide*, pp. 360–64.

[3] This island was apparently first encountered by Spanish explorer Hernando de Grijalva in 1533, who named it Santo Tomás. Today it bears the name Socorro ('help' or 'succour'). Considering its less than inviting bleakness, it is uncertain how it acquired such an unlikely name. See Mathes, *Vizcaíno and Spanish Expansion*, p. 4; and Wagner, *Spanish Voyages*, p. 6.

the south, as was manifest in the differences between [latitude] by observation and by dead reckoning. Despite the calm, we observed 11 minutes farther to the south than what we judged. Attesting to it was the survey made at twelve o'clock [noon] on the 31st [of March] which was north-west one-quarter north, a distance of three leagues to its midpoint.

The commander, seeing that the lack of wind was preventing us from approaching the island, sent to me his ship's boat equipped so that I might endeavour to tow myself close and inspect its vicinity. We set out, assisting the tow with the force of the schooner's oars, but throughout the day it could not be reached, and we succeeded in getting ourselves no closer than about one league.

At sunset the commander signalled for us to sail in company (so as to avoid any separation), and laying aside the superfluous oars, we thereupon put about. Although it was calm all night, at daybreak the island was about three leagues away.

The 1st of April, at about two o'clock [p.m.], we surveyed the island to the north-north-west a distance of three and a half leagues. And we observed the same latitude as on the previous day, from which it follows that all our effort for twenty-four hours was only to maintain ourselves against the impulse of the currents.

The 2nd [of April] at sunrise a slack wind began out of the north-west, and we returned to being towed. At eight-thirty o'clock [a.m.] we came about to a heading north one-quarter north-east, and we continued approaching the island, which at twelve o'clock [noon] bore to the north a distance of about four leagues, and at eight o'clock at night a distance of two [leagues]. It remained completely calm. We departed with contrary [winds] that came from one part or another of the island, sailing on a heading east-north-east and north-east one-quarter north, with the winds north and north-west.

The 3rd [of April] at eleven-thirty [a.m.] we came about on a heading west-south-west, [the island] being to the north-west five degrees west a distance of four leagues. We remained in its vicinity in the shelter of a large mountain, which is what kept us becalmed. With light winds that were running in different directions, we departed on a course west one-quarter south-west, with the aim of leaving it, because of its excessive harshness and no promise of finding the least refuge.

Description of this Island

Its shape is almost round, extending its lengthiest from north-east to south-east, and with little question it is five leagues between these two extremes. I consider it to be in 115 degrees 18 minutes longitude west of Paris, and in 18 degrees 50 minutes latitude [north] according to the observations and surveys while being delayed.

All around its circumference, particularly on the south-west point, some large, steep rocks stand out, which resembled trees at a distance. The island is capped by a prominent mountain, barren and craggy, splashed in parts by some white stains that look like houses. The sea dashes against it with great force, and it could not be examined closely as to whether some inlet was nearby; nor did we undertake to investigate it, seeing it was so rocky and sterile.

The name Socorro [help, succour], which is given this island, induces a favourable impression for those who sail to the presidios of San Diego and Monterey; but in all the information I have read of voyages, I have neither found by whom it was discovered, nor what they intended by attributing to it a name that has nothing in common with its appearance.

In [Venegas's] *History of California*, 1st volume, folio 152, there is mentioned a voyage that was made by Captain Hernando de Grijalva out of Tehuantepec in the year 1524 [*sic* actually 1533]. Having sailed 300 leagues, he encountered a deserted island, which he named for Santo Tomás (St Thomas), and the distance calculated from his departure leaves no doubt that this must be the same as that which is called Socorro. [Some] have sought to attribute its discovery to themselves, obscuring the truth and usurping this captain's claim to being its first discoverer.

The French chart also gives it the name of St Thomas, but greatly north of its true latitude, and greatly west of my longitude, which cannot be appreciably wrong.

The 4th [of April] at ten o'clock [a.m.] we lost sight of [the island], bearing to the north-east one-quarter east of us, a distance of seven or eight leagues. We endeavoured to haul the wind and return to our voyage, without having anything particular that happened.

The 6th [of April] the commander learned that [the *Santiago*'s] bowsprit was sprung, which he succeeded in having repaired with ropes and reinforcing sleeves. Thus, this damage, like that which was found with the fore-topmast following the previous voyage, has been noted.

After departing the island, I have noticed that the sky showed signs of clouding over; the horizons dark; the sun rarely allowing itself to be seen. The wind-driven spray, which has been strong since San Blas, has ceased completely. The beneficial winds are rather cold, and they continued with little force until the lunar opposition; [we are] experiencing a change in circumstances, a lessening of the currents that are always set in a southerly direction.

After the [lunar] opposition, the winds increased to the north-north-east, and they freshened powerfully, extending themselves. Occasionally, however, they were in the north-east, with their strength allowing me to try to get the schooner up to carrying a stiff sail and [testing] the spirit of my crew. I ordered sail to be crowded. So much fear was conjured up in seeing [the schooner] heeling over excessively that some [crewmen] feigned sickness as a reason to transfer to the frigate. But it was not surprising as we pressed ourselves to the extreme, for I recognized that the poor devils have ample reason to be fearful, as they see little refreshment, they endure continual drenching on the planks of the ship's deck, and the heavy seas that incessantly come in on the windward side offer no place to be dry. But for me it was necessary to carry a stiff sail, because otherwise the expedition would never be completed. I have tried to dissuade them [of their fear], but this strong effort created so much of an impression on them that they remain astonished when the frigate, with reefings taken in, is obliged to lower the main and foremast topsails, and they see that I maintain [the schooner] fully rigged.

Not only have these men been possessed with fear, but it has extended to the crew of the frigate, not one of whom having voluntarily offered to come as a replacement for any who (for lacking the care of a surgeon) transfer [to the *Santiago*] to recover from sickness. This has obliged the commander [Hezeta] to announce banishment [to the *Sonora*] for those who might commit some offence; and as they consider it the most rigorous punishment, it has successfully settled down the [*Santiago*'s] crew.

On one of those days in which the breeze was rather fresh, the master [Mourelle] and I, having gone below to rest from the constant alternation of the night [watches], leaving for the moment the boatswain in charge of the watch, a cabin boy advised me the wind had freshened, and that they had lowered the mainsail and taken in a reefing; and for the same reason, the frigate had done likewise. Wanting to rid them of their fear, I went out and ordered that the reefing be loosened and the sail hoisted. Showing annoyance, I told them that henceforth they were never again to lower an inch of sail without my permission; that it was a contemptible indignity for men to demonstrate such a cringing spirit; that how would I be able to sustain confidence or hope in them in higher latitudes where the seas and winds are certain to be of greater force and circumstances more unfavourable, if now with clear weather they were exhibiting such timidity. That upon reflection, I placed the same value on my life as on theirs; that we, the master and I, would exercise caution and anticipate contingent risks in advance; and that seeing the frigate shortening sail should not have frightened them, because [the *Sonora*] was not under the same compulsion that previously had necessitated [the *Santiago*] to lessen sail as much as possible. Finally, I reminded them of the glory that they would have should they endure from start to finish, and the expression of esteem that all of them would deserve; and that I was really a comrade in their hardships.

After this day, I recognized in them a new spirit and desire to please me. I have endeavoured to appear grateful, praising them for their thoughtful behaviour, and giving them some small presents as circumstances allowed. I have succeeded in making them contented, spirited and resolute participants in my own fate, even following me into whatever destiny may decree.

The 11th of May, the winds began to wheel around to the 2nd [SE] quadrant, but so changeable that later they rotated through the entire circumference of the horizon until returning finally to between east and south-east, with various squalls and clouds. What has been experienced is the sea currents have abandoned their southward direction and, if indeed there is some difference, it is customarily to the north; that the seas alternate with the north winds and those nearby, and they are quieted down by those of the 2nd [SE] quadrant.

On the 21st [of May] the commander determined to convene a meeting in which it was to be decided if we ought to continue our voyage, or put in at the presidio at Monterey. And finding myself sufficiently informed about the reasons that were being put forth, and it being impossible for me or my master to attend on account of the heavy seas and winds, I submitted my opinion by means of a safety line that delivered it in the ship's boat, to which I attested in these words.

Opinion

It seems to me that there is no excuse whatever for putting in at Monterey. The reasons obliging me to be of this opinion are solely a matter of dedication to His Majesty's service and the best means of accomplishing our assignment.

It cannot be denied that in the latitude where we find ourselves it will be less difficult to cast anchor in 43 degrees (the inlet discovered by Martín de Aguilar),[1] which

---

[1] The inlet (or river as some accounts call it) that Martín de Aguilar is believed to have found in latitude 43°N has not been conclusively identified. Davidson, *Pacific Coast Pilot of California, Oregon and Washington*, p. 385, suggested that it might have been the Smith River, lat. 41°56′N, or the Chetko, lat. 42°03′N. He

[favours] not going to Monterey. In accordance with this principle, there is no doubt we could have the advantages so well understood. In this place, according to [our] information, there is no lack of water. The few who are sick [will] improve merely by setting foot ashore. The damaged fore-topmast and bowsprit can hold out well with the lashings that have been applied to them, and if not, remedied as much as possible, for one cannot always sail with comfort. We will succeed in this manner to replenish the water supply, strengthen the sick, search for the passage they picture for us;[1] and leave with the first southeaster, running along the coast, examining the harbours and inlets, so that in case the wind veers to the north-west, which are the prevailing winds, to have shelter when being unable to endure them.

In opposition to putting in at Monterey, we have the season, which is greatly advanced. We are laying ourselves open to achieving nothing this year and being compelled to wait for the following one, resulting in a serious setback for the service and interests of His Majesty and failure to comply with the orders of the Most Excellent Lord Viceroy. And if the thought is to make a new mast and bowsprit, it is equivalent to deciding to winter in.

Considering that the sailing qualities of the schooner differ from what they say, having all of its futtock [curved] timbers split, completely useless and incapable of holding a nail, and that this damage, with strong and steady winds, could result in some bad consequences. In all, [however], it does not appear to me sufficient cause to delay the expedition. Command having been thrust upon me, I must perform with honour corresponding to one's birth.[2] I am mindful of the advantages that can accrue to one's commission, if I succeed as I hope to in convoying with the frigate to 61 degrees or more. The exposure to danger and the arduousness of the enterprise are no secrets to me, owing to [the schooner's] smallness and poor steering, her little ability to carry a stiff sail or sail swiftly; and the need for me to crowd sail beyond what normally would be required. All these reasons, together with one that will be added, ought to have been presented to His Excellency by those who planned [the expedition], perversely reporting [the schooner's] fitness, which is now an opinion subject to change. This distrust should not serve as an obstacle for those who are willing to sacrifice gladly. With me are my master [Mourelle] and the small crew I have, who are resigned to the risks without seeming to be reckless; work has not been neglected even though no amount of water has yet been found. Although her strength is slight and the sail excessive, this is remedied by the continual watch that the master and I keep so as to

---

ruled out the Rogue River, lat. 42°25′N, because of its dangerous offshore reefs. Henry Wagner favoured the Mad River or Humboldt Bay, both of which are in the vicinity of the 41st parallel, in large part because he believed that observations of latitude made in the 16th and 17th centuries were unreliable, being as much as two degrees too high. See Wagner, *Spanish Voyages*, pp. 255, 407, n. 186; 408, n. 205. The Coquille River, lat. 43°07′N, and Coos Bay, lat. 43°21′N, are much closer to the 43rd parallel, and although they discharge sizeable volumes of fresh water into the ocean, they seem to have been considered unlikely possibilities by both Davidson and Wagner.

[1] Probably referring to a feature on Bellin's map called 'Entrée decouverte par Martín de Aquilar en 1603'.

[2] Bodega y Quadra was born in Lima, Peru, which meant he bore the onus of being a creole, a name applied to Spaniards born in Spain's American colonies. The high standards of performance he set for himself may have been his way of overcoming whatever disadvantages stemmed from his creole birth. But he also had an aristocratic ancestral heritage, and the phrase 'honour corresponding to one's birth' could have been in reference to that background. See above, p. 74.

prevent accidents. And lastly, God smiles on deeds of greatness, and if my luck were so adverse that help was out of reach, it is a glory for posterity to die for the King, each and everyone at his station.

The commander, giving attention to these reasons, accepted my opinion, and we proceeded on the voyage with winds out of the north and north-west. They persisted strongly until the [lunar] conjunction, and with very choppy seas, which were diminishing the latitude and longitude in which we were finding ourselves.

At the conjunction that followed the 30th of May, slack winds began to appear from the north-west down to W. south-west.

On the 1st of June an apprentice seaman became drunk. And it appearing to me that this might result in some sort of harm to him, I transferred him to the frigate, but in six hours he died of an intestinal inflammation.[1]

From this day, a seaweed began to be seen of sufficient length that only the head, in the shape of an orange surrounded by a bunch of various leaves, is visible.[2] The tail is an elongated tube, and at the lower extreme it has roots which maintain its hold on the rocks in the vicinity of the coast. Torn loose from them, the currents and winds carry them out as far as 100 leagues.

On the 5th and 6th [of June], fish, ducks, wolves [seals or sea lions?] and a slender, greenish weed that they call *sacate*, were to be seen.[3] The sea exhibited a colouration such that with all these indications affirmed the coast to be a distance of thirty leagues, which on this latest day I considered to be Cape Blanco, and twenty-seven [leagues] from Cape Mendocino, according to the map of New France published by M. Bellin.

On this day, the sea swell having increased and [with] constant chafing, the towline parted; the same thing having happened on several occasions in the course of the voyage. Not only have I watched, obliged to endure this inconvenience, but also it has been a continual worry in order to avoid running afoul of the [other] ship, which notwithstanding the watchfulness of each other, we were unable to prevent ourselves in several instances from coming [close] enough to scrape, particularly in one case that we suffered some damage. But to me living exposed to these contingencies was unavoidable, for there was no delaying the expedition with the sluggishness of this schooner.

On the 7th [of June] at one o'clock in the afternoon we saw the entire coast directly at a long distance, bearing to the southern extreme south-east, and the northern extreme north-east. We set as much sail as possible with the intention of drawing ourselves near [the coast].

On the 8th [of June] we saw the land rather clearly at a distance, it having been calm all night long. The experience over twenty-four hours has been that there is a

---

[1] This unfortunate seaman may have suffered a ruptured appendix, which in those times meant certain death.

[2] This 'seaweed' is a bulbous-headed, whip-like alga, commonly known as 'bull kelp' (*Nereocystis luetkeana*). Both Hezeta and Pérez call it *porras*, which the latter says was its Chinese name. See Ricketts and Calvin, *Between Pacific Tides*, p. 415; Beals, *For Honor and Country*, p. 72; and *Juan Pérez*, p. 74.

[3] *Sacate* is a Spanish word meaning grass, herbs or hay. But the exact nature and scientific nomenclature of this 'greenish weed' is uncertain.

difference southward of 29 minutes compared with where we considered we had gone, as inferred from the direction of the [ship's] course [by dead reckoning].

On the same day the wind freshened, and the commander signalled for me to sail ahead of him, examining the coast closely, searching for some harbour where the ships' drinking water could be replenished and the men refreshed. I immediately carried out the order, sailing to the north-north-east, crowding sail so as to reach the vicinity of the land before nightfall. I succeeded at six o'clock in the evening, but I saw no harbour, only various small inlets offering little shelter and a rugged coastal shore. Heavily forested mountains were discovered, and they promised water in all their ravines. At the same hour, I began to determine the average bearings of the most prominent points. I passed from place to place, which were surveyed allowing for the schooner's movement, gaining knowledge of the coast's direction. It was found to be 18 degrees of the 4th [NW] quadrant, corrected, and 18 degrees of the 2nd [SE quadrant]; the said points forming a small inlet that receded about one league in the space of twenty [leagues], which was from one side to the other.

At eight o'clock at night, I found myself about two leagues from land; and from the frigate, which continued out beyond about three to four [leagues]. I came about, and with very little sail I passed the night with the purpose of not drawing away from the coast, in order to continue with the reconnaissance next day. But at dawn on the 9th [of June] the frigate signalled [us] to rejoin her. We returned to sailing in company at ten o'clock [a.m.], keeping a distance of one league from her bow, until finding other land, at which time I saw in the distance its beaches, rocks, inlets, points, rugged and forested ground, sounding constantly in thirty fathoms of water in black sand. I followed along the coast with the bow headed south seeking shelter and drawing myself near a large point that was giving indications of [shelter]. Notwithstanding the great number of rocks that were on the north side.[1] It extended seaward a distance of one mile, giving an impression of other submerged [rocks] that were capable of extending farther out. Nevertheless, I turned the bow, perceiving that to the south-east it had an inlet, and I signalled the frigate that she might come in to cast anchor; and once inside, sounding constantly so that it is done more securely. I was signalled to send out the launch, but the distance at which I found myself from the harbour, and the fresh wind, rendered this plan useless. So I resolved to enter with the greatest caution and vigilance possible. At this time we saw that two Indian canoes left from the north side, which overtook the frigate, she being with shortened sail. I continued sounding down to seven fathoms of water, at which place I anchored inside the harbour. The frigate made her entrance by [following] in our wake. Later, soundings were begun of much of the harbour's interior, and having found the depth ample for a moorage place a stone's throw from the beach, we warped ourselves in very tediously. The schooner came to a halt and was moored at four o'clock [p.m.], fully secure, as was also the frigate.

In the interval of these tasks, four canoes loaded with Indians came alongside, who with great docility mixed with the crew and in an affable manner began to exchange deer and otter hides for knives, beads and other such trifles. I gave them other trinkets, with which in this spirit they [the Indians] embarked and demonstrated infinite gratitude.

---

[1] The large point they had sighted was what is called today Trinidad Head, lat. 41°03.1′N, long. 124°09.0′W of Greenwich, at its Light.

We saw a small house or Indian settlement nearby the beach, but from this they sent out no canoes at all.

The 11th [of June] we had finished the tasks [of making] our entry, and it was decided to take possession of a high mount that forms the point's entrance. For this the men were divided into various detachments, which were posted at key places, allowing the rest to march in an orderly fashion without the danger of some sort of attack. The advance guards were placed so as to search out at a distance the pathways where the Indians travelled. A square body of troops was formed on the mount with the crews from the two ships, well armed. The cross was put in the centre, and the persons employed in the chapel service were drawn up nearby. The mass was celebrated with everybody in this formation. The sermon was preached by Father Fray Miguel de la Campa, Order of San Francisco, missionary from the Apostolic College of San Fernando, Mexico [City]. Possession was taken in the name of the King [Charles III], with all the details and formalities that were provided in the instructions given by the Most Excellent Lord, Frey Don Antonio María Bucareli. In the ceremony they fired various salutes, the clamour of which frightened the Indians. But as they learned that no harm came to them, they [returned] to visit us as the ceremony was concluding.

The 12th [of June] measures were taken to replenish water and wood, and I prepared to have the schooner heeled over in order to see if she could continue on successfully. With the same purpose, I undertook the cutting of two top masts, yards and masthead caps, with the intention of putting up main and foremast topsails.

In the days that we maintained ourselves in this harbour, more than 250 Indians have been seen, consisting of persons from various settlements. Although we all paid close attention to understand them, sympathetic to their rules and laws, it was impossible for us [to understand], despite their affable efforts to explain themselves. Thus I alone will be doing the communicating, and that by other signs they grasp.

Description[1]
They are exceedingly docile, of medium stature, very well made and good looking, without the homeliness of the other Indians of America. They live in square huts, well constructed of stout planks, which are made nearly level with the ground, with circular doors, through which they can only be entered one at a time. The floor is perfectly flat and very clean; in the center they have a square hole one yard deep for maintaining the fire, by which they heat the entire surroundings against the cold that they experience.[2]

When it is excessively cold, the men cover themselves with skins of deer, wolf, sea otter, et cetera, but they usually wear nothing and expose their flesh to the mercy of the elements. Some of them wear on their head a crown of fragrant grasses and feathers. They have long hair (except for the children), and it is made into a knot, although some of them wear it spread across the shoulder.

---

[1] Bodega's description of the native population at Trinidad Harbour, along with the accounts written by the other officers and chaplains of the expedition, are the earliest concerning the Yurok Indians by Europeans.

[2] For illustrations showing the Yurok houses at Trinidad Harbour, see Heizer and Mills, eds, *The Four Ages of Tsuri*, pp. 91–3.

They wear in their ears two perfect pins of bone; a leather strap serves as a waist band, and in the vicinity of the ankle a thread is tied quite tightly. They paint their face, back and chest, black or orange, and they wear on their arms tattoos forming different designs.

The women cover their head with a hat (which they call *coras*) woven of agave fibre and other grasses.[1] They have their hair parted in two braids in the Greek fashion; they ornament their ears the same as the men. They are distinguished from [the men] by three lines tattooed on their lower lip, one of which runs from the upper edge of the lip to the middle of the lower part of the chin, and the remaining two run from the corner of the mouth to the bottom of the chin, an imperfection that disfigures the features of the face. They wear around the neck various necklaces of small beads made of fruits, bone and shells. From the waist to the calf of the leg they wear a nicely worked netting the color of saffron. They wear the same skins as the men to cover their shoulders.

The daily routine consists of following the orders of a captain who is in charge of each settlement having been so designated by virtue of being the eldest. This is arranged by turns, so that some go out in search of the necessary sustenance for all, in which activity the women customarily collect firewood. He also saw to it that at dusk the beach was searched, and after this having been done diligently, we gather ourselves together, although all of them retire at sundown.

It appears that their domain extends only to a village and bordering woods and beaches, and that they go to war with other villages, because we understood they would have been delighted for our assistance in subjugating their enemies. Nevertheless, there are friends between various villages, because they have assembled in this small settlement some 300 Indians divided into various groups with their women and children.

The law that they apparently observe is none other than that of perfect atheists, because it was impossible to ascertain whatever idol they might have had, nor the slightest [indication of] days or hours set aside for sacrifices. Nevertheless, we were convinced that an Indian death made them weep copiously within the captain's house, where they refused admittance to some of our men. But having [later] gained entrance nothing particular was found.

The weapons they have for defence are arrows with stone points, knives of the same [material], and some of iron in the shape of machetes, with handles of wood. It is understood that they obtained these [iron] weapons from the north, and they wear them around the neck or tied to their hand.[2]

Of the things given them, what they desired the most were knives and hoops of iron, although they were delighted with beads, rings and earrings. They scorned [our] food

---

[1] An example of such headgear is illustrated in ibid, p. 15.

[2] The presence of these iron weapons among the Yuroks at Trinidad Harbour was a matter of considerable interest to their Spanish visitors. For one thing, such weapons conceivably were evidence that the Yuroks had been in contact with other Europeans, such as Russians whose supposed presence on the coast was a worrisome possibility for officials in Mexico City and Madrid. Bodega's account fails to speculate on the origins of the Yuroks' iron blades, but his commander, Bruno de Hezeta, became convinced that the Yuroks had obtained these metal weapons by trade with natives to the north, or as in one case, by hammering a blade from a spike that had washed ashore with other wreckage. See Beals, *For Honor and Country*, pp. 69, 141, n. 21.

and drink, but they politely took it, pretending that they were eating it, later throwing it away. They were much pleased with tobacco, and they have small plots of it.

By hunting they provide themselves with deer, bison [elk?], seals and otters. In the days we were present, there were neither indications of other [animals] nor were birds other than jays, ducks and sea gulls seen in this vicinity; all in scarce supply; although the impossibility of our penetrating inland hindered reporting on this and other points.

The fishery includes sardines, mackerel, and *morcillones*[?]. Not having seen other fish in this harbour, with which they might have been able to provide themselves, I believe it is this deficiency that controls the seal and otter population, or it may be the season that limits it.

We also have tried to get them to tell us if they have seen other ships, but these efforts were in vain because no one whatever was able to understand any sign. Nevertheless, I am persuaded that no one else has come upon this coast.

It has not been possible to find out if metallic ores or precious stones were present, but no indications of them were found, and all I know for sure is that if they exist nothing is known about them. Finally, as much as possible was done to try to investigate details that might be able to contribute to measures that the King might want to take, but it was given up, as nothing was said, nothing could be ascertained.

As much of the land as I have seen left me with no doubt that it is very fertile and capable of producing the same [crops] as grown in all of Europe. All of its ravines have exquisite water, and it is overgrown by wild grasses with a verdure and fragrance that make it agreeable to sight and smell, amidst which grow numberless roses of Castile exceedingly fragrant, wild marjoram greatly scented, lilies, plantain, camomile, celery, thistle, and an endless array of other common [plants] of the field; also strawberries, blackberries, sweet onions, and truffles are seen.

The mountains are covered with a species of pine, perfectly straight and very tall, among which some were judged to be sixty yards in length, and more than two yards in diameter at the base.[1] They are suited for masts, spars, decking, and beams, because they are in sufficient quantity for manufacture and very straight grained.

It has not been possible to see if there is wood well-suited for futtock [curved] timbers, knees, et cetera. Nevertheless, what is on the riverbanks I judge will not fail to be of this quality.

Description of the Harbour, the layout which was worked up geometrically by Don Bruno de Hezeta, Don Francisco Mourelle and me to represent its shape.
On the point that is westward, there is a mount of fifty *toesas* [in height].[2] Its northern side connects with the mainland, where there is another [mount] that reaches twenty

---

[1] The dimensions of the 'straight and very tall' trees Bodega saw at Trinidad Harbour convert to metres as follows: 50.3 m tall; and 1.68 m in circumference. If he was correct in believing the trees were pines, there are at least two species that more or less fit these dimensions. They include: Ponderosa Pine (*Pinus ponderosa*), 45.7 to 54.9 m tall; 0.9 to 1.5 m circumference; and Western White Pine (*Pinus monticola*), 30.5 to 53.3 m tall, 0.61 to 1.52 m in circumference. Another possibility is that these trees were Port Orford Cedars (*Chamaecyparis lawsonina*), which attain heights between 42.7 and 54.9 m; circumferences of 1.2 to 1.8 m. See Brockman, *Trees of North America*, pp. 22, 30, 56.

[2] Approximately 83.8 m (275 ft) although Trinidad Head is actually 116 m (380 ft) above sea level.

[*toesas*].¹ Remaining substantially sheltered from winds of the 3rd [SW] and 4th [NW] quadrants, and by the landward mountains from the 1st [NE quadrant] and part of the 2nd [SE quadrant], the coast continues, though distantly, to close the anchorage.

At the entrance there is a high island devoid of any vegetation, and along the shoreline there are many high rocks, which keep the sea between them and the land [calm] as a river.

In the inlet where anchor was cast, any ship can seek protection at the mount with minimal fear, as likewise on the sandy ground, with only the precaution of sheathing her cables.

We repeatedly observed the latitude of the harbour to the best of our ability, finding it to be 41 degrees, 7 minutes.² Its longitude according to my reckoning is 19 degrees, 7 minutes, west of San Blas.³

Tidal movements follow the same regularity that they have in Europe: in the course of twenty-four and four-fifths hours, two flood tides and two ebb tides. One low tide was observed to fall seven inches and the other five, which difference made us continue with repeated experiments. Not only did they confirm the first, but at the lunar conjunction, which was on the 13th [of June] the tides were observed to be greater than in Spain, on which day the first fell nine inches and the other six. It was also found on this day that high tide was at twelve o'clock [noon], which unfailingly is the same hour of the day of the new moon.

The waters present on the same beach, where they flow in various brooks, are crystal clear; firewood is also present and easily obtained.

South-east of the harbour a distance of three-quarters of a league, there is a river that is about ten yards wide and as much as two yards deep in the middle; the water is brackish. The master [Mourelle] of this schooner entered into the [river] for some distance. Rather large turtledoves were seen (for which reason it was given the name [Río] de las tórtolas (river of turtledoves).⁴ Various flowers and blackberry brambles [were also seen]).

Departure from Puerto de la Trinidad

On the 19th of June, having completed re-provisioning water and firewood, we set sail at eight o'clock in the morning, with a slack wind out of the north-west, the only one that had been experienced in the days that we were in the harbour. At twelve o'clock [noon], it went calm, and we cast anchor south of the island nearest to it, where a canoe of Indians was seen loaded with shellfish, which they offered to give me. Having [received] them, the favour was returned by some trinkets and a promise to return.

---

¹ Approximately 33.5 m (110 ft). The 'mount' is doubtless Little Head, which is actually 38.1 m (125 ft) above sea level, immediately north-east of Trinidad Head.

² The anchorage between Trinidad and Little heads, where it seems most likely the ships cast anchor, is actually at latitude 41°03′N.

³ The longitude of their anchorage, assuming its location was in the cove between Trinidad and Little heads, is actually 19°00′W of San Blas.

⁴ This river is today called Little River. The 'turtledoves', after which the Spaniards named it, were most likely mourning doves (*Zenaida macroura*), a species native throughout much of temperate North America. See Udvardy, *Audubon Society Field Guide*, p. 544.

They said goodbye, embracing everyone with a great demonstration of emotion, making signs that they would always weep upon seeing the holy cross.

The 20th [of June] in the afternoon, a north-west wind began to blow, whereupon we weighed anchor. We were sailing to the west-south-west and west, the wind persisting constantly from the north-west to north.

The 28th [of June] the lunar conjunction occurred, and the winds were variable in all directions, particularly in the west, with rain showers and fog that separated me continually from the frigate. It forced us to signal by muskets and swivel guns in order to mark the position in which we found each other to be.

This day the topmasts and yards were finished, whereupon without losing an instant, I ordered their sails hoisted and set, desirous of pressing on with the voyage so as to achieve success. For when the frigate was earlier unable to loosen its reefings, it was at once essential to sail with the four principal sails, the spanker and mizzen topsails, with mostly favourable winds and smooth seas, which in such cases requires the top gallant sails to be set. For me the day is one of great joy for having succeeded in making the voyage less unpleasant and putting an end to being towed. Nevertheless, the excessive sail caused me to proceed with the greatest caution. But attending to that, the weather neither permitted me to be taken in tow, nor was I able to sail without serious loss of time. It was essential for me to endure these inconveniences in order to prevent such an important assignment from suffering delay on my account.

The 2nd of July, the wind freshened in the west, whereupon we dismasted the fore-topsail mast, and we split the main topsail yard, the lack of which was promptly remedied with other poles that had been cut as a precaution for this .

The 9th [of July] at twelve o'clock [noon] I considered myself, according to the French chart, in the middle [of the] entrance discovered by Juan de Fuca,[1] a distance of seven leagues to the south-west of the other northward point. The same signs were seen of being near land as were noticed in the other landfall.

This day, the wind freshened and the sea changed so excessively that the water casks came unlashed and were floating on the deck, carrying away the chocks on which they rested, [along with] railings and various trifles of little value. Ultimately, it forced me to lay to, turned towards the west-south-west.

The 10th [of July] the sea and the wind having calmed down, we were navigating with the bow headed north one-quarter north-east, intent on inspecting the land. Although the signs were indicating it to be not very distant, at sunset it was not seen, and we continued sailing all night quite cautiously.

The 11th [of July] the sky dawned clear, the sea the colour of soundings, with much seaweed and a multitude of ducks. At midday, we discovered land, of which the commander was informed by signal.

This same day, at twelve o'clock [noon], I considered myself [having] entered into Fuca's Entrance forty leagues east of the most northward point, which difference I

---

[1] The entrance shown on Bellin's map of 1766 called 'Entrée descouverte par Jean de Fuca en 1592' is located roughly midway between the 40th and 50th parallels. An account of Juan de Fuca's supposed discovery of 'a broad Inlet of Sea, between 47° and 48° of Latitude', found in the only extant source concerning the voyage, places Fuca's Strait at a somewhat higher latitude than Bellin's map. For Michael Lok the Elder's account of Juan de Fuca's supposed exploits, see Purchas, *Hakluytus Postumus or Purchas His Pilgrims*, 14, pp. 415–21. Reprints of Lok's account are also in Greenhow, *Memoir, Historical and Political*, pp. 207–11, and *History of Oregon and California*, pp. 407–11; and Cook, *Flood Tide of Empire*, pp. 539–43.

attribute to the bad placement of the coast on the charts for lack of discoveries, because my reckoning from Trinidad cannot suffer such error in so short a number of days.[1]

In the afternoon, the wind and sea increased such that at three-thirty o'clock it was necessary to lay to. Finding the frigate rather far to the leeward, fearing an interminable separation, which would put me in much jeopardy by being cut off from all help in case of an emergency, I had to fall off leeward toward her at no little risk, as I saw the ship [*Santiago*] in rather serious danger. But having drawn near her, I determined to lay to, turned south-west, because the sea was extremely stifling.

The 12th [of July] at daybreak, [I was] a distance of five leagues from the frigate, and from the land a little less than four [leagues], the wind fair and the sea smooth. I tried to join [the frigate], which I successfully did at eleven o'clock in the morning, although I was drawing away from the coast. I turned toward the land, about a league ahead of the frigate, and at six o'clock in the afternoon, being about a league and a half [off shore], I came about turning south. I was signalled the same, which I carried out, having sounded in eighteen fathoms, black sandy bottom and small shells. Various points, islets and mountains were examined, and [I found] that the coast runs for the space of seven leagues through 20 degrees of the 2nd [SE] quadrant, with no inlet, only a flat island a half league in length, close to the coast.[2]

Wanting to inform the commander of what had been done, and that we will remain for the time it takes to tack, I signalled by hailing. But I do not believe it was clear to them, because neither so much as a single sail was lowered, nor was there a response to the signal. Although with a full press of sail, it has still been impossible for me to maintain proper distance, leaving me continually behind, for not being able to set the main topsail because on the 11th [of July] we had sprung the masthead cap. Thus, in such fashion at sundown [the frigate] was lost from view, and however many lanterns, rockets and cannon shots were fired during the night, no response was heard.

At three o'clock in the morning, doing my duty to continue on an outward bound track, it not only deprived us of approaching the land all day long, but it also promoted separation [from the frigate]. I tacked, turning towards land, [as I was] doing prior to the signal. At sunrise, the frigate was seen a very long way off headed towards land.

---

[1] Bodega's meaning here is obscure, but apparently he is saying that his dead reckoning navigation placed the *Sonora* in a location that the Bellin map shows as the Entrada de Fuca, some 40 leagues (120 naut. miles) eastward into the Entrance, where in fact no such strait exists. Bodega, unwilling to believe his longitude reckoning might be so much in error, concluded that the error must have been in the map's placement of the coastline. In fact, both were wrong, and his longitude reckoning quite possibly the more erroneous of the two. Lacking a chronometer, he had no reliable means to fix his longitude accurately.

[2] This 'isla rasa' 'flat island' is apparently today's Destruction Island, lat. 47°40.5′N; long. 124°29.1′W of Greenwich, at its Light. A belief has arisen that it was named 'Dolores' 'sorrows' by the Spaniards, because of the tragic event that would take half of the *Sonora*'s crew two days later. Bodega's account, however, mentions no name whatever for the island. The expedition's commander and captain of the *Santiago*, Bruno de Hezeta, states that later in mid-August 1775, he encountered an island that he named 'Dolores', situated at lat. 47°58′N very close to the latitude of present-day James Island, lat. 47°54.3′N at its Light. Hezeta's description of the island as 'being populated by a dense grove of pine trees', Beals, *For Honor*, p. 85, is consistent with the appearance of today's James Island, which is heavily forested. There is no mention in either of the chaplains' diaries of an island named Dolores. It is only in Mourelle's account (see Landin, *Mourelle de la Rua*, p. 160; or Mourelle, *Voyage of the* Sonora, p. 34), that this name is applied to the island to which Bodega alludes. Neither Hezeta's account nor Mourelle's offer an explicit reason for applying the name Dolores to either island.

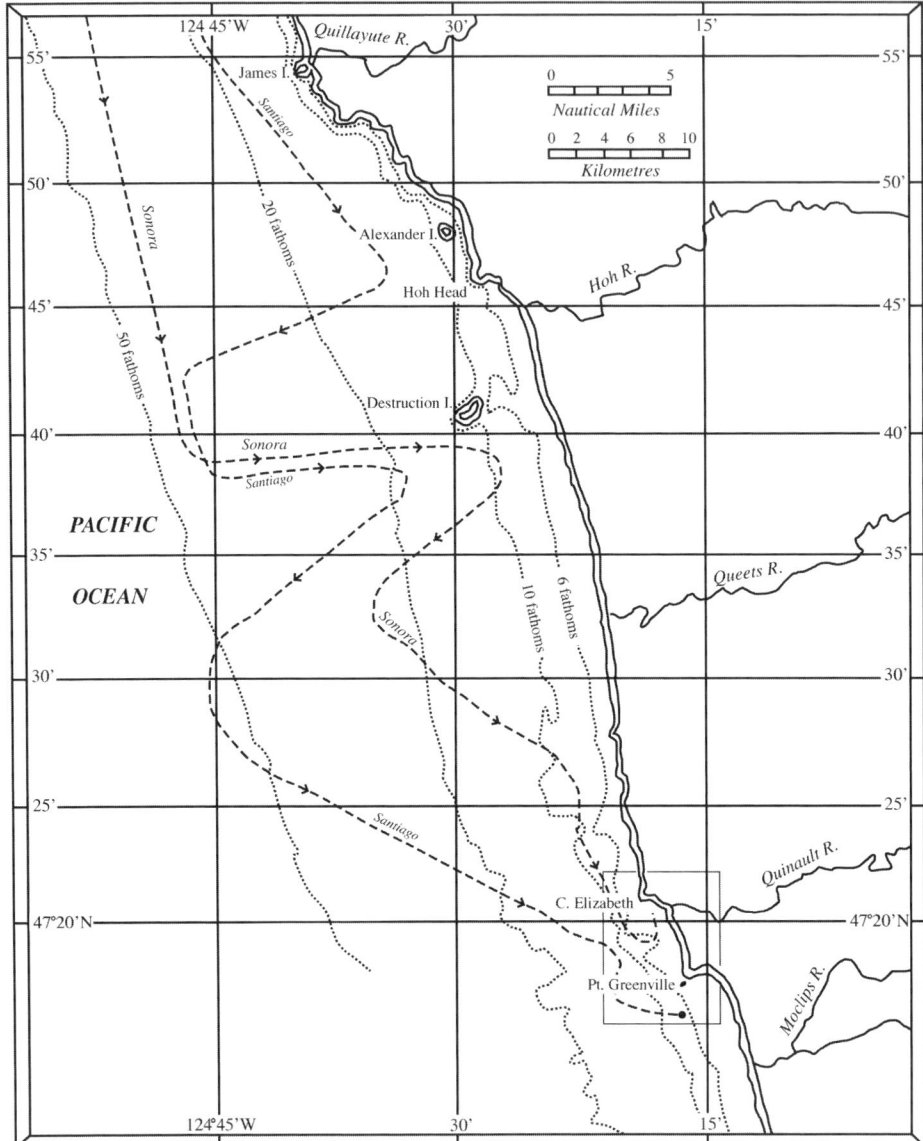

Map 2: Olympic Coast, Washington: Destruction Island to Cape Elizabeth.

Finding myself with very little wind, I sounded in thirty fathoms, waiting for [more wind]. It was observed that the currents went in the same direction as the coast, which was found by various surveys to be 19 degrees in the 2nd [SE] quadrant for a space of nine leagues, which began with the survey of the part farthest south, made on the 12th [of July], to the flat island, and by this, well known to continue in the 2nd [SE] quadrant.

At midday, the frigate was seen to the leeward rather distant, and that she came approaching land. In order for me to join [the frigate], I set sail, endeavouring to approach with the purpose of reconnoitring up to the place of rejoining her. Finding myself a distance of one mile from land, seeing clearly the rocks that break up the shore, and that there was neither inlet nor point that offered shelter, I fell off leeward to the south-east until six o'clock in the evening. Knowing that no harbour was to be had on the coast, and that to lose so much latitude was exceedingly painful, when so many risks had been taken in gaining it, I decided to anchor in the shelter of a point,[1] which appeared fit for the purpose of completing the re-supply of firewood and water; and after the matter of the Royal Service [possession-taking ceremony] is concluded, we ought to be able to pursue our voyage as soon as possible. At this time the frigate was found to be about a league [away]. I signalled that I was going to anchor, amidst soundings in as much as eight fathoms of water, finding myself in the part enclosed by this point. Suddenly, we were left in three fathoms, for which I immediately signalled the frigate of the shallowness, and although I tried turning to leave, it was impossible because of a large ridge of shallows that broke out, ranging all along the coast.[2] The same wind I used to anchor, being contrary to leaving through where we entered, and in order to report to the commander the situation in which I found myself – the impossibility at low tide of leaving the place where we would be able to anchor with safety that night, and to inform myself if there was a harbour sheltered by a point that was about a league leeward – I sent out the ship's boat with the master [Mourelle].

Already by this time, various canoes with rather tall, robust Indians had approached,[3] who, with offerings and flattery invited me to go to their dwellings. But despite their offerings and considering them even more docile than those at Trinidad, seeing that they had brought together nearly sixty Indians, and that I had no more than eight men, I ordered that they all be armed without [the natives] perceiving it. I then began to regale them with glass beads, pendants and handkerchiefs. In gratitude for these favours, they gave me some fish and promised to bring many more. They had no fear of approaching closely, and they commenced trading their pelts (with the sailors)

---

[1] This point, behind which Bodega sought temporary shelter, is almost certainly Cape Elizabeth, lat. 47°21′N, on the central Washington coast. Nevertheless there has been a protracted controversy about its identity. Robert Greenhow, translator-librarian to the US Department of State, wrote in his *History of Oregon and California*, p. 119, that this point 'is in the latitude of 47 degrees, 20 minutes, and on English maps is called *Grenville's Point*'. Greenhow's information about the latitude of Point Grenville (as it is named today) was faulty, however, as its latitude is actually 47°18.3′N at its Light. Over three decades later, George Davidson of the US Coast and Geodetic Survey, after visiting the locality, concluded that the *Sonora*'s anchorage was in the lee of Cape Elizabeth, and that it was the *Santiago* that had dropped anchor near Point Grenville. See Davidson, *Pacific Coast Pilot*, pp. 493–4. In 1933, Henry Wagner, a prominent figure in the historiography of Spanish voyaging off the California and north-west coasts, reopened the controversy by asserting that Bodega, as Greenhow had earlier thought, found shelter under Point Grenville, and that the *Santiago* had anchored some distance south in the open waters off the Moclips River. See Wagner, *Cartography of the Northwest Coast*, 1, p. 176. More recently, a re-examination of the evidence by Harry T. Majors in 1980 has convincingly revived the Davidson thesis. See Majors, 'The Hezeta and Bodega Voyage of 1775', pp. 212–21, 243–7, n. 18.

[2] This rocky shoal stretching southward from Cape Elizabeth is today called Sonora Reef. See Davidson, *Pacific Coast Pilot*, pp. 491–2.

[3] They were Quinaults, probably from their principal village near the mouth of a small river that bears an Anglicized form of their tribal name. According to ethnologist Ronald L. Olson, the name of the village was *kwi'nail*, from which the word Quinault is derived. See Olson, 'The Quinault Indians', 17.

for glass beads and knives. Finally, they took their leave with signs of their gratitude and friendship. I replied by presenting them some mirrors. They made such demonstrations of gratefulness that they tried to get us to go and sleep in their houses. They approached the cable, making signs that they would convey us yonder, but having responded to them by the same signs of affection, they went into the dusk, making a great clamour and demonstrating by actions and assistance that we were friends.

At nine o'clock at night, nine canoes were seen that came along side with a formidable outcry. Although I had no reason to doubt their trustworthiness, the inconvenient time of their arrival made me suspicious, and I quickly ordered my men well armed for any treachery. But it was quite the contrary to what I was thinking, because, obliged for the gifts that they had been made, without anything to return in interest, they came offering me immense numbers of all kinds of fish, such as: porgy, salmon, conger eel and sardines; whale meat and of other land animals; sweet onions and various containers of water. They drew near with this offering with much rejoicing. I showed them my pleasure, and that I greatly appreciated the attention shown. I went back to give them some handkerchiefs and glass beads, and to the women rings and earrings, with which they returned to their settlements. I remained amazed at the correctness of their conduct.

At twelve o'clock [mid] night, the ship's boat arrived aboard with the master, who informed me that the point to leeward had no shelter,[1] and that the frigate had anchored a league from my location. He also conveyed to me a verbal order, which the commander had issued, that I be informed that on the following day I am obliged to attend a meeting in which it would be decided if this schooner can continue the voyage to a higher latitude. [This was] because it was constantly feared we would not survive with the seas and winds that we had experienced on the 9th and 11th [of July]; and with this point decided, to set sail immediately.

This news was no surprise to me, because ever since the first days of departure from San Blas, they have done their best to dissuade me, eagerly urging my return to port, having me believe it was impossible for me to go up even to the latitude of Monterey. So I judge it of greater merit not to have made such dismal predictions, that the inconveniences and labours that I have come through I had expected on such a lengthy voyage.

At dawn on the 14th [of July] I prepared to set sail and anchor in the frigate's vicinity, but the sea was very low for this. The reefs preventing me from weighing anchor obliged me to wait for the flood tide. In this interval, the Indians returned and traded their pelts with the crew for old chests, showing the same friendship as the previous day and night. They placed themselves near the ship's boat, which was by the stern, signalling that they be given iron from the rudder. But having told them that it was impossible and given them some glass beads, they turned and withdrew, making known their affections in demonstrations of kindness.

---

[1] Referring to what is now called Point Grenville, lat. 47°18.3′N, long. 124°16.6′W of Greenwich, at its Light, a rugged promontory with several nearby rocky islets, one of which is in the form of a natural arch. The modern *US Coast Pilot* (10th edn, 1968, p. 223) confirms Bodega's opinion of its poor qualities as a place to anchor: 'An indifferent anchorage in northwesterly weather, ... the depths compel anchoring at such a distance from the beach that little shelter is afforded'.

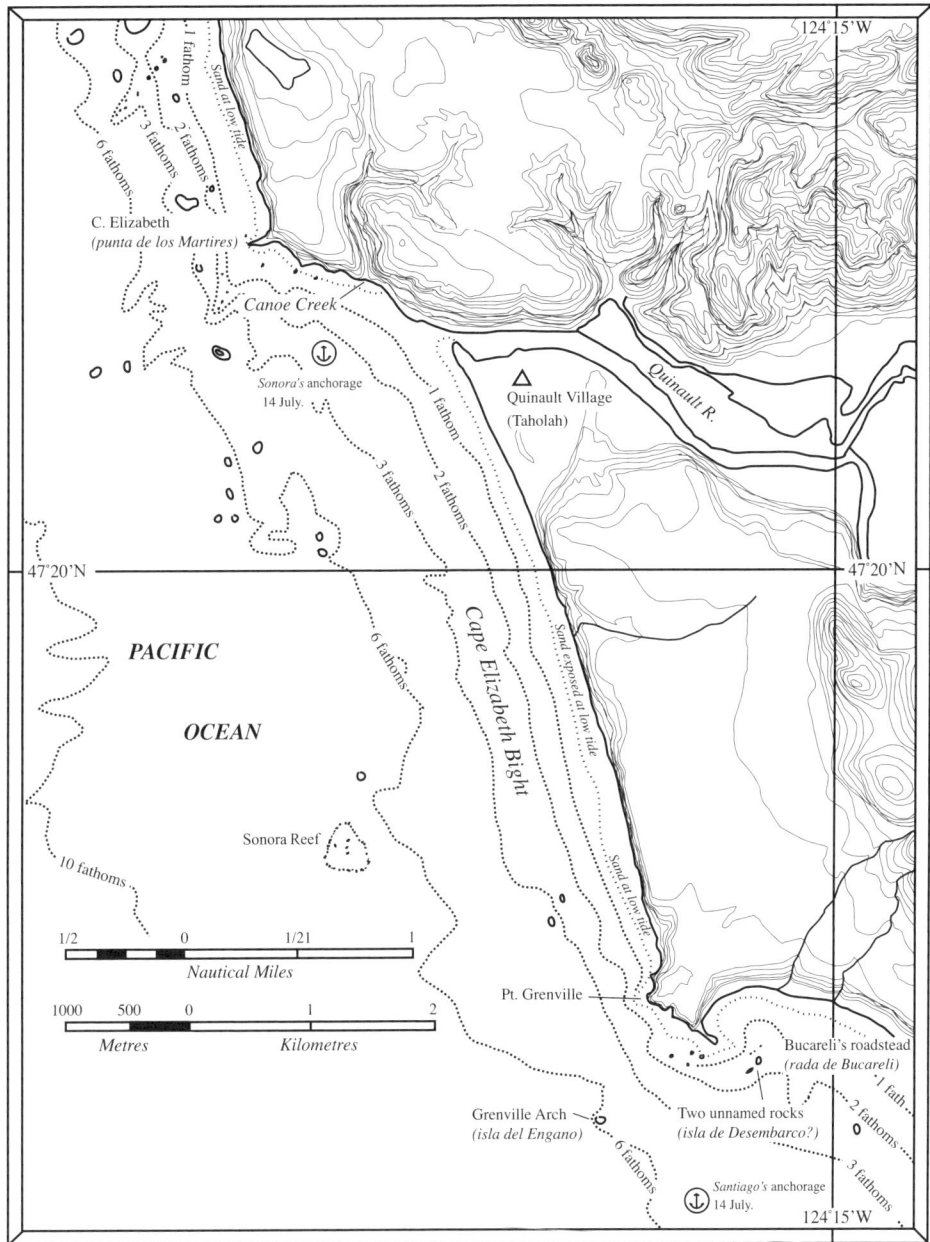

Map 3: Cape Elizabeth Bight: Cape Elizabeth to Point Grenville.

Anticipating that the commander was thinking to set sail that day in order to get out, and reflecting that, for a voyage so extended as we were expecting, a water shortage would have serious consequences after four months of continual hardships, and finding myself with something of a shortage after having it supplied in abundance from Puerto de la Trinidad in order to free my men of scurvy and other illnesses that are due to thirst, I decided to complete [the water supply], collect some firewood and cut a masthead cap in the time that the low tide made it impossible to leave. It was convenient for me to have on the same beach such an abundance of water and numberless [tree] trunks. For this purpose, I designated six men who were most sensible and capable for such instances, armed for all contingencies. Each of them was carrying his musket, a sabre and some [of them] two pistols. They were under command of the boatswain, Pedro Santana,[1] known for his worthy conduct and bravery among all of his class. Having given [the landing party] glass beads to offer as presents if any Indians got close, they departed with orders that, at the time of their landing, the ship's boat would be returned so as to come under [command of] the master and myself, to bring some water casks. Because of its small size, it did not permit us to have gone first, as was the case for a sailor who was supposed to have gone with them.

The sea being rather heavy, they were unable to reach the shore in perfect order. As soon as they began to alight, more than 300 Indians unexpectedly came out of the woods, attacking them in a defenseless moment. They were treacherously murdered, and finding myself without the means to help, I ordered the discharge of some muskets to see if it would frighten them. But as it could not do them harm, they paid little heed. Various signals that I made to the frigate requesting help were also fruitless, because the distance prevented them from being seen. In this way, finding myself isolated, without boat or men with which to go to their defence, I determined to weigh anchor, which was carried out at twelve o'clock [noon], the hour of high tide. [I was] moving myself toward the frigate in order that I be given the launch to take the most rigorous satisfaction; and to find out if any of the unfortunate ones had gotten free, because at the time of the attack two [sailors] were seen getting away, swimming to get aboard. But the coldness of the water prevented them from attaining their purpose, and it is not known if they turned back and died at the hands of the barbarians, or found refuge among the thickets of trees.

As soon as sail was set, nine canoes of Indians immediately came forth, each one of which had more than thirty in number. In a short time they assembled at a uniform distance, such that one of them with only nine of the most robust youths (and who on the previous day commanded each one of the rest) continued toward our side with all bows stretched taut and covered by some beautiful elk hides for defence,[2] trying at first to persuade us with the previous deceits. But though we were few and under sail amidst shoals, we were prepared in good order. Knowledge of those twisted intentions was very costly to me, so this time [I was] protecting myself from the assault they intended, urged on no doubt by [seeing] so few men who were left to us, an undeserved payment with which they returned the noble treatment that was given them.

They maintained themselves in this way for quite a while, and the other canoes were approaching the stern. I did not want to fire upon them until knowing what their

---

[1] In his 'Navegación ... 1775', Mourelle gives the boatswain's name as Pedro Santa Ana. See Appendix 1, p. 128, n. 1.

[2] Elk hides used as armour seems to have been a fairly widespread practice among north-west coast Indians. According to later fur traders such elk hides, which the natives called *clemons* (also spelled *clemans*), were exceedingly valuable in the purchase of sea otter pelts. See, for example, Howay, 'A Yankee Trader', pp. 83–94.

intentions were, in order to be sure of the first shots. [This was] because the only ones able to fire were the master, I and my servant, having one [sailor] at the helm, another sounding, and another at the masthead on the lookout for shoals, one making cartridges, a boy to fetch them, and five sick, with the seven unfortunate ones making up the remainder of the schooner's crew. In this condition, we set forth sailing with very little wind, observing that they lessened their paddling movement, maintaining the same distance, until they determined to board us at the bow, seeing that there were no men at it. But at the same instant, we opened fire with two swivel guns and three muskets, spraying them in such a manner that it did not permit them much space between each shot; and in which they, with supreme agility, gave a stroke of the paddle in order to flee. But their dexterity did not protect them, because the first shots downed the majority of them. Taking the precaution of covering themselves with hides, the remaining ones finally found refuge in their agility, carrying six less than the nine who came. I might have captured or killed those who remained, but I found myself without men and amidst shoals to which we were forced to give the most attention.

When the [natives] saw themselves out of range of musket fire, they were encircled by the remaining canoes, which in cowardly fashion decided not to approach. After considerable time in consultation, in which they will have reflected upon the difference between the first attack and the second, they withdrew to their settlements.

The commander, who having heard the last discharge, judged that there was need for the assistance of a stream anchor and cable, which aid was sent us with the launch; and which arrived, being already in the frigate's vicinity. After having anchored very close by, I went aboard with the master. I asked the commander if he would give me thirty men, with the aim of taking the most rigorous satisfaction for such a detestable abomination; and to assure myself of what became of those two unfortunate ones, whose misfortune would be [especially] lamentable if, getting safely away from the water, they remained in the hands of such vile oppressors. But having convened a meeting, the junior ensign and coastal pilot, Juan Pérez, and the second master, Cristóbal Revilla, were of the opposite opinion. Despite the reasons offered by my master and me, the commander resolved to leave without punishment for such effrontery.

This point decided, although not consistent with my opinion, the commander went on to discuss the seaworthiness of the schooner. Explaining that the seas and winds encountered have been exceedingly active, and that he believed on the 9th and 11th [of July] that they had considered that we might have perished. In order to avoid living with such sudden dread, it was decided that each should cast their vote as to whether it was appropriate to send [the schooner] back to port.

With attention to this statement and other reasons that he explained, it was the opinion of the second master of the frigate, Don Cristóbal Revilla, to return [the schooner] to San Blas. Next, Sr Don Juan Pérez gave his opinion, saying that although he could not deny the rough weather and the schooner's little strength, it appeared to him that whatever [happened] up to that latitude [47°23′N], she had come through happily. He had no doubt that she could attain other higher [latitudes].

In the opinion of my master and me, there was no doubt it could be done, when we were the ones responsible for the latitude having been gained from San Blas. Thus, it was agreed that we should continue [reinforced by] a boatswain's mate and five sailors, with whom I retired to my ship. And having been asked to send my opinion in writing the next day, I set forth [here] in the same form that I transmitted it.

Finding ourselves in latitude 37[47°?] degrees and longitude 27 degrees, 46 minutes west of San Blas, with attention to this schooner's slow speed, finding the futtock timbers completely useless as the carpenters affirmed, and finally, her limited seaworthiness, it was determined, Your Grace, to hold a meeting to consider turning back to the Department [of San Blas] or to put in at Monterey to examine the damage. There was no reason appearing to me sufficient to return or make a port stop involving the expedition's going ashore. I felt that neither pretence was agreeably consistent with His Majesty's service, the reasons for which I explained in writing. Don Francisco Mourelle, master of this schooner, being of the same opinion, it was decided, Your Grace, that we press on, and I have succeeded to attain latitude 49 degrees without having a reason that obliged me to seek a port. But with even more reason I am compelled to be of the same mind on this occasion, having experienced such strong seas and winds; and that I am certain [the schooner can] continue lying to without suffering the slightest damage, despite the fact that she heels over and ships water a lot. The first is impossible to prevent from happening on any small vessel. The second is not perceptible because of the weather board. Under the sail I have put aloft, her speed is not much different from the frigate's. And finally, I am obligated to go up to 65 degrees in the same manner as I have reached 49 [degrees]. Since the labours I have endured are, I expect, essential in all sea voyaging, especially for ships of discovery, they must be accepted with conformity and fortitude, because they are for the King's service and the nation's honour.

Desirous of accomplishing what corresponds to the charge that has been entrusted to me, and mindful that it will be more difficult for the frigate to reconnoitre the coast and its harbours without the schooner, it was the opinion in the meeting to press on. I am now putting it in writing, adding that it would have been an inappropriate rebuff to my spirit to abandon the mission without what is to me a grave reason impeding it.

Departure from the place called Rada de Bucareli
The 14th [of July] at 5:30 in the afternoon, we made sail from this roadstead, the latitude of which is according to my reckoning 47 degrees 24 minutes, and 21 degrees 19 minutes of longitude west of San Blas. Although the reckoning that it corresponds to according to the parallel on the French chart is the mouth of Fuca [Strait], situated between 47 and 48 degrees, it follows that if the aforesaid mouth exists, it can neither be in the latitude it is portrayed nor the longitude of its position.

With weak winds out of the north-west and north-north-west, we were separating ourselves from the coast, turning to a course west one-quarter south-west, without anything new in particular.

The 19th [of July] the commander sent over to me an official letter accompanied with a statement by junior ensign and coast pilot, Don Juan Pérez, and a certification by the surgeon, Don Juan Gonzales. Informed of its contents and considering that the statement and certification were saying that it was not possible for the frigate to pursue higher latitudes because of the advanced season, and likewise for the many sick men they had, I responded in the following terms.

Opinion

My Dear Sir: Informed of the official letter that Your Grace sends to me, and acquainting myself with the statement by junior ensign and coastal pilot, Don Juan Pérez, accompanied by the certification of the surgeon, Don Juan Gonzales, my feeling is that, despite its compelling, forceful reasons, we should continue on for sometime to see if in this interval we obtain more favourable winds, which will help us to avoid squandering away a voyage that has cost us so much discomfort. For although the evidence is obvious to me that the crew is small, the majority in weakened health and weary after four months of voyaging, to which they are unaccustomed, it appears agreeable to me to put ourselves at risk in whatever way, until the sick cannot summon any more strength. In such an event, we will be able to return, reconnoitring the coast, with the north-west winds that we have experienced prevailing in its vicinity.

With this opinion and that of my master, who agreed with me, despite being entirely contrary to all those on the frigate, the commander determined to continue, which we carried out without having experienced any winds other than north and north-west.

The 24th [of July] being in a calm, I requested of the commander his ship's boat, a cannon and a box of cartridges with shot, in order to have something for defence, which he immediately sent me.

The 30th [of July] at one thirty o'clock in the afternoon the wind freshened and the sea became very swollen. We laid to, turned to the west. At sunset the sea was increasing, and the wind was from the south-west. At nine o'clock at night, the frigate's lantern was barely visible, so that we repeated [firing] various rockets and a swivel gun in order to signal [our] position. All we saw in response (although obscurely) was a rocket very far to the leeward. In order to lessen the distance, I considered falling off to leeward, but the runaway winds and seas quickly obliged me to remain properly laid to.

The 31st [of July] at sunrise the horizons were cloudy, so that visibility was scarcely two leagues. The frigate was not to be seen on any side. I resolved to remain laid to, because the weather was going to lift, so that in clearing it would be possible that the frigate might have maintained the proper location and we might be seen. But this day and night passed with the same luck as before.

The 1st of August, it was overly dark, with more rain squalls, and the same sea and wind. The horizon was scarcely visible for half a league, but at midday it cleared, and the sea and wind were quieted. The frigate could not be seen. Under this assumption, I agreed with the master to follow a course west. Considering that the frigate was unable to continue to the east-north-east, because with the [magnetic] variation and the strong drift encountered, [she would be] keeping mostly to the east-south-east. When the southward currents produce a difference between dead reckoning and observed [positions] the course by which one would best succeed to reach land is 45 degrees; and that upon this reflection [the frigate] would navigate by the same return route as myself.

I remained on this tack until the 4th [of August] considering myself 170 leagues distant from the coast, after continuous rain squalls, fogs and winds variable between north-west and north-north-east, a little to the west-north-west. At eight o'clock in the evening, I tacked turning north.

Now that the winds commenced to be more favourable, and the frigate has not turned up, I decided to continue the [voyage for] discoveries as the instructions provide.[1] This, despite having foreseen that the season is advanced; that going up to higher latitudes in a ship so small could have an unhappy outcome, particularly if we were in need of medicines and a surgeon, and even of water, because after the last departure I got a measure of its great scarcity. To set an example, I was the first subjecting myself to the dismal water ration of a sailor; likewise for food. By having no other recourse, it was absolutely essential with me to press on [despite] all of these inconveniences, because it would be regrettable to run up new expenses to the royal treasury with successive expeditions, which by chance would suffer accidents of their own, and would never come to any good unless dominated by total resoluteness.

The 5th [of August] the winds began to blow from the south-west, and I put the bow to the north-west by the compass, with the aim of maintaining the proper distance from the coast so as not to expose myself to the north-west winds that run near it.

The 10th [of August] there was with the lunar opposition an eclipse, and a stiff, fresh south-west wind until the 14th [of August]. At one o'clock in the afternoon it veered to the east-south-east, exceedingly fresh, obliging me to lay to, turned to the east-north-east, taking in a reefing in the mainsail. But at six o'clock in the afternoon it was dry and [the wind] was observed returning from the north-north-west.

The 15th [of August] the sea was seen to be the colour of soundings, exceedingly vivid, immense trees, grasses, a great number of birds with red bills, feet and breasts,[2] and some whales, all indications of the land's proximity. But finding myself this day at about 56 degrees, 8 minutes [latitude], distant from the coast eastward 154 leagues and to the west sixty-nine [leagues] from an island that the [Bellin] map locates; the sailing master and I decided that we must be close to some unidentified island archipelago, which was carelessly not placed [on the map]. The wind came mostly in gusts, some unbearably strong. The fog was so thick that not even the ship's bow was visible. At six o'clock in the afternoon the wind changed to the west, strong. I immediately put the bow to the south, laying to until four o'clock in the morning of the 16th [of August] which grew calm and cleared up. [The wind] veered to the south-south-west at the same hour. I steered turning to the north one-quarter north-east.

At twelve o'clock noon, land was seen to the north-north-west a distance of six leagues. I steered toward it, and in a short time it was revealed from the north-west over to the north-east one-quarter east by the compass, forming some large inlets and some extensive mountains, one in particular that is also very elevated and perfectly shaped.[3] It is located on a prominent cape separated from the other mountain ranges; It [has] an abundance of snow forming some wide channels halfway down, which are beautiful and seen clearly at a distance of four leagues.

---

[1] These remarks suggest that the separation of the two ships was deliberate and agreed upon in advance. Bodega assumed the *Santiago*, with so many of her crew scurvy-ridden, would be unable to sail farther north. For this reason, in the interest of reaching the expedition's goal of 65°N latitude, Bodega seems to have purposely parted company with the flagship. Years later in 1792, Mourelle confirmed this and took credit for the idea. See Bernabeu, *Bodega y Quadra*, pp. 90–91, n. 23.

[2] This may be a reference to puffins, although the species most common on the north-west and Alaskan coasts, the tufted puffin (*Lunda cirrhata*) and the horned puffin (*Fratercula cornicula*), are not red-breasted. Another possibility is that Bodega was describing the red-breasted merganser (*Mergus serrator*), a duck that is a better match for his description than puffins. See Udvarty, *Audubon Society Field Guide*, pp. 390–91, 467.

[3] Unquestionably, this 'very elevated and perfectly shaped' mountain is today's Mount Edgecumbe. A volcanic cone rising some 975.6 m (3,200 ft) above sea level, it is a prominent landmark located near the south end of Kruzof Island, flanked on the east by Sitka Sound, and on the west by the Pacific Ocean.

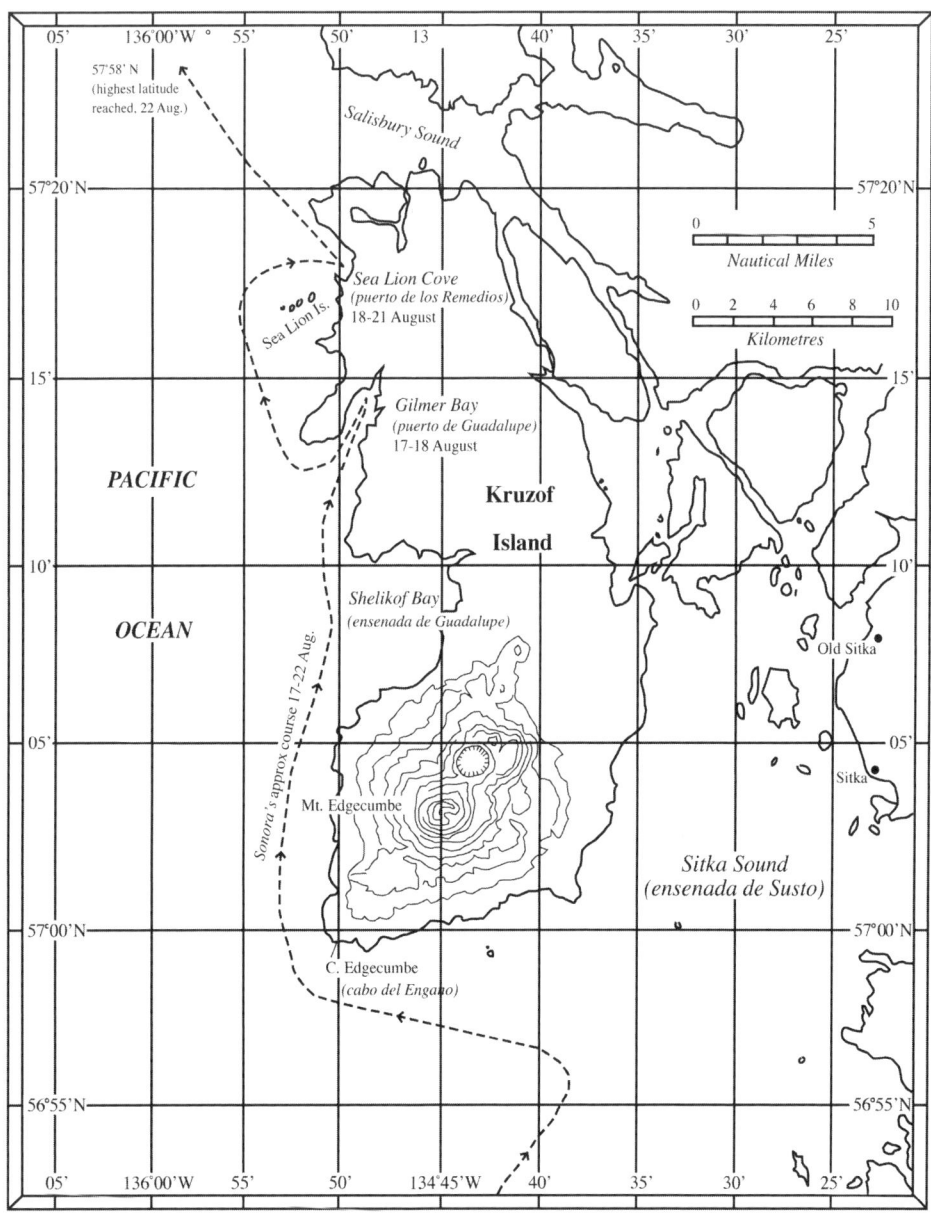

Map 4: Kruzof Island, Alaska: Mount Edgecumbe to Sea Lion Cove.

I named this mountain San Francisco,[1] and the cape Engaño (deception), which is located by two observations, made to the west from a distance of one mile, under latitude 57 degrees, 2 minutes. I consider it in longitude 34 degrees, 12 minutes west of San Blas.[2]

From this cape a survey was begun to demarcate the most prominent points to the east and west in order to perfect the location of the coast as it is shown in the illustration [a map of the combined discoveries of the first two Bucareli expeditions].

The 17th [of August] at sunrise the wind came on from the south, slack, and with it I was reconnoitring an *ensenada* [inlet or bay] the mouth of which is three leagues [wide] and extends inland the same [distance], which is formed by the cape to the west. In the point most northward I encountered a *puerto* [port or harbour] three-fourth of a league in length, and completely sheltered from winds, except from the south, which enters by its mouth. But however much we got near [the shore], no bottom was encountered less than fifty fathoms, nor a beach or level stretch discovered in which we might re-supply. Despite being night time, anchor was cast in fifty-six fathoms of water, over a mud of such character that it held the anchor securely.[3]

The 18th [of August] at sunrise I left this harbour, giving the bay and harbour the name Guadalupe. Being at the mouth, two canoes came out, and in each two Indian men and two Indian women were seen.[4] But they never wanted to come close and only urged for a short time that we go ashore.

They departed and I followed sailing to the north-north-east until nine o'clock in the morning, when I discovered a harbour.[5] Although not as large as the previous one, it had a beautiful beach, a river four or five yards wide and water eight fathoms deep. It was sheltered from the winds of the 1st [NE], 2nd [SE], and 4th [NW] quadrants; and protected from those of the 3rd [SW] quadrant by some islands that extended out as a group, one after another.[6] Anchor was cast in six fathoms of water, sandy bottom and

---

[1] Bodega's name for this mountain was written in an abbreviated form as Sn. Fran co, meaning San Francisco. See Bodega, 'Diario … 1775', p. 16(2), and Bernabeu, *Bodega y Quadra*, p. 92, n. 24 and p. 93. In Mourelle's account, however, the name is instead written 'Sn Jacinto'. See Mourelle, 'Navegación … 1775', p. 24(1), and Landin, *Mourelle de la Rua*, p. 200. Whatever reasons for this discrepancy, Mourelle's name for this mountain, San Jacinto, came to prevail on Spanish maps, but not on the mountain itself. Today it is officially named Mount Edgecumbe, a name it was given by James Cook in 1778 – although he spelled it Edgcombe. See Beaglehole, ed., *Journals of Captain Cook*, 3, p. 336; and *US Coast Pilot, Alaska*, p. 180.

[2] Bodega's Cabo Engaño 'Cape Deceit' is today called Cape Edgecumbe; its correct latitude is 56°59.9′N at its light, so that his observation was only two minutes too high. Its correct longitude west of San Blas is 30°35′, so that his dead reckoning overstated the cape's longitude by some 3°37′ – a discrepancy not particularly surprising as he had no reliable means to measure longitude.

[3] The *ensenada*, 'bay' or 'inlet', with a mouth three leagues (9 nautical miles) wide must refer to present-day Sitka Sound – called on Spanish maps 'ensenada de Susto' ('Bay of Fright'). A second bay or inlet was named by Bodega 'ensenada de Guadalupe', present-day Shelikov Bay. Next, he sighted, entered, anchored, and stayed overnight in a place he called 'puerto de Guadalupe', present-day Gilmer Bay. I am indebted to David Kroenke of Seattle, Wash., with first-hand sailing experience off Kruzov Island, for suggesting Gilmer Bay as the place the *Sonora* spent the night of 17–18 August.

[4] The Indians encountered at Gilmer Bay were most likely Tlingits.

[5] This harbour is Sea Lion Cove on the north-west shore of Kruzof Island. Bodega gives its latitude as 57°20′N, which is within two minutes of the actual latitude, 57°18′N, at the cove's centre.

[6] These circumstances can all be reconciled to Sea Lion Cove. The river is a small unnamed stream that flows into the cove on its south side, and the off-shore islands offering protection in the south-west quadrant are no doubt the Sea Lion Islands.

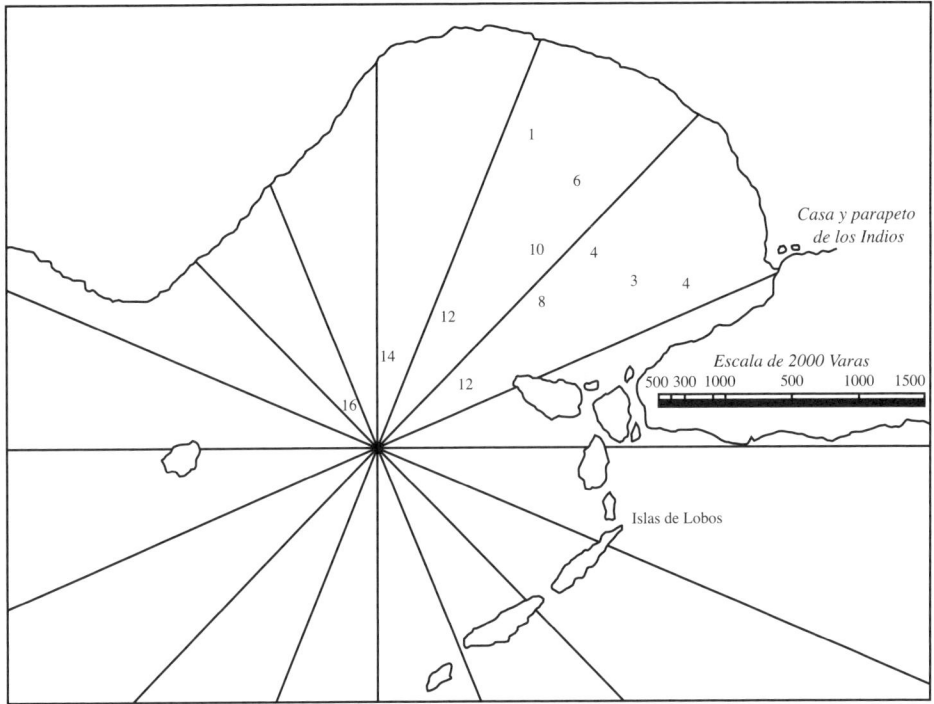

Map 5: Bodega's Map of Sea Lion Cove, Puerto de los Remedios.

seaweed. Distance to the land was a musket shot, and from the habitation of the Indians (which amounted only to a well constructed house and a parapet of poles for its defence) a little less than a quarter of a cannon shot.

I gave the name Remedios [remedies or help] to this harbour. It is situated in latitude 57 degrees, 20 minutes north, and in longitude west of San Blas, 34 degrees, 12 minutes.

On the same day I readied the crew, and at twelve o'clock noon I disembarked on land with fourteen men, well armed, with the purpose of taking possession, collecting water, firewood and a top mast, which was conspicuous in its complete absence. In the most defensible place in the harbour I made a stronghold with two swivel guns and some muskets. Leaving this place well fortified for the withdrawal, I went to take possession on a hill, where the cross was planted. Possession was taken of these lands, with all the requirements and formalities provided by the instructions.

This operation being concluded, I went with part of the men to reconnoitre the mouth of the river, in order to see where I would be able to obtain water most easily, without any of the Indians coming out from their parapet during that time.

Later when I had learned about the river, harbour and shore, I retired with my men on board, from where it was seen that the Indians pulled down the cross from the place in which I put it, and they fixed it to the front of their house in the best disposition one could want. They made various signals with their arms giving us to understand that they were keeping it there.

The 19th [of August] I disembarked with the same men on a point that projects farther out,[1] with the intention of gathering firewood and cutting the topmast. In this place, I made a stronghold as on the previous day, and I went to the river with six men so that water might be obtained. Being engaged in the work of filling casks, doubtless knowing that I was not going after them, about twenty unarmed Indians left their house and parapet, with only a kind of pennant [fashioned] from a white sheet hanging from a pole. Drawing near the [river] bank opposite from where I was, they spoke endlessly, but nothing I was able to understand. Later when they finished their babble, which lasted a great while, they remained in silence (apparently) waiting for a reply. I gave them to understand that we did not come to do them any harm; that we desired to be their friends; that they should have no fear; and that now we had come to obtain a small amount of water. One of them who appeared to be in charge ordered that they bring the water, and they brought to us a little from their house in a bucket or sort of reed vessel, which they handed over in the middle of the river. I treated them to glass beads and handkerchiefs, and they brought cured fish. All this being done by only one Indian and one of our men going into the middle of the river.

All the lengthy time that it took to fill the casks they were very cheerful, but when they saw that I was carrying them away, they insisted I should pay them for the water I had obtained, giving [me] to understand it was theirs. Not to offend them, I returned to present them various baubles that I still had, but without doubt they understood, because they were not content with anything. All of a sudden, running to their house, they came fully armed with some exceedingly large lances with stone points. They presented themselves, prepared to attack, making numerous skirmishes. I prepared myself and my men, and I gave orders not to fire until I told them to. Taking my gun and approaching the river bank, I gave [the Indians] to understand that if they took one more step I would have to fire upon them; and that therefore they should lay down [their] weapons; and that I would do the same if they wanted to be friends. As soon as I spoke to them, they retired to their house, and with my men, I finished collecting firewood, some water and the topmast.

In the river's vicinity there were fish in abundance and some were caught. It was found to be of the same taste as female haddock.

The Indians are neither as corpulent nor fair as I expected to encounter in that latitude. The pelts with which they cover themselves differ from those of Trinidad only in having a hood for more protection. They wear rings hanging from the cartilage of the nose, and some in the lower lip; the ears are covered with numberless small shells, very minute.

The temperature is exceedingly cold with continual rains and fogs, besides which the sun was not to be seen in the three days I was there. As a result, the men, with constant toil and little shelter, began to grow ill, causing me considerable agonizing over the lack of medicines or anyone to administer them.

Departure from Puerto de los Remedios
On the morning of the 21st [of August] I departed this harbour turning west-north-

---

[1] This point is an unnamed peninsula at the south end of Sea Lion Cove.

west, the wind south-east, with the purpose of ascertaining, to the extent of 1 or 2 degrees of voyaging, if I was on the coast situated to the east on the French map, or on the one to the west, from which I considered myself nearest.

The 22nd [of August] by the course taken and distance covered it was known that we were on the easterly coast, which was based on the observation made at twelve o'clock [noon] in 57 degrees, 58 minutes, north.[1]

The same day at two o'clock in the afternoon the wind came up from the north-west, fresh, with which in order to go up to a higher latitude it was necessary for me to gain longitude to the westward. Finding myself with the majority of the men sick, and the season such that it offered me no opportunity to search for favourable winds, I resolved to return on a course east-south-east, reconnoitering the coast at a distance of one mile. I contented myself with having gone up almost to 58 degrees, in a vessel from which one had no right to expect such success. For having been in situations countless times from which I might have retreated, I never failed in my primary intentions, even under pressure.

As for knowledge of the coast, I applied myself to do my best with orders to fix its location exactly, the results of which offer new objections to the map M. Bellin published in the year [17]66. Because his map locates the two tracks of the Russians, Bering and Chirikov, which they followed on similar voyages of discovery [17]47 [*sic*, actually 1741], it ought to have been able to locate their landfall. But from [the tracks] themselves, one may infer the error that [the coast] suffered in its placement, because if it were correct I would have had to encounter the westerly lands before those to the east.

It is possible to say that the currents could have carried me eastward, and the error could be in my reckoning and not in the map. But the chance of this is practically impossible, because it was necessary [for the currents] to stand with equal force toward the east, while I was being carried to the north. Yet it was never experienced, and to the contrary, the seas obliged me to trend westward, by coming continuously from the 2nd [SE] quadrant.

On the ships' tracks, M. Bellin supposes that their courses were executed toward the east-north-east and north-east, at the most. I intersected them, one at about 51 degrees, distance from land 157 leagues, and the other at 54 degrees, 15 minutes, distance 102 leagues, and I always kept to the north-north-west and north, at the least. It follows that, even conceding enough difference in my reckoning and that of my master, I would always be obliged to go up to the highest latitude in which that coast is shown on his map. But the truth is what presented itself across the bow, impeding me from making landfall at 62 degrees or more, just as I had believed.

On that map, he also supposes directions very different from what I surveyed. But I now consider that the map was constructed more upon his fantasy than upon truthful recollections; for whatever the way, it is the cause of my not having gone up to 65 [degrees], the purpose for which the voyage was undertaken.

I am putting the utmost care into the examination of the coast, hopeful of carrying out His Majesty's orders, and to locate the entrance that was supposedly discovered by

---

[1] This latitude (57°58′N) appears to be the maximum northward position reached by the *Sonora* in 1775.

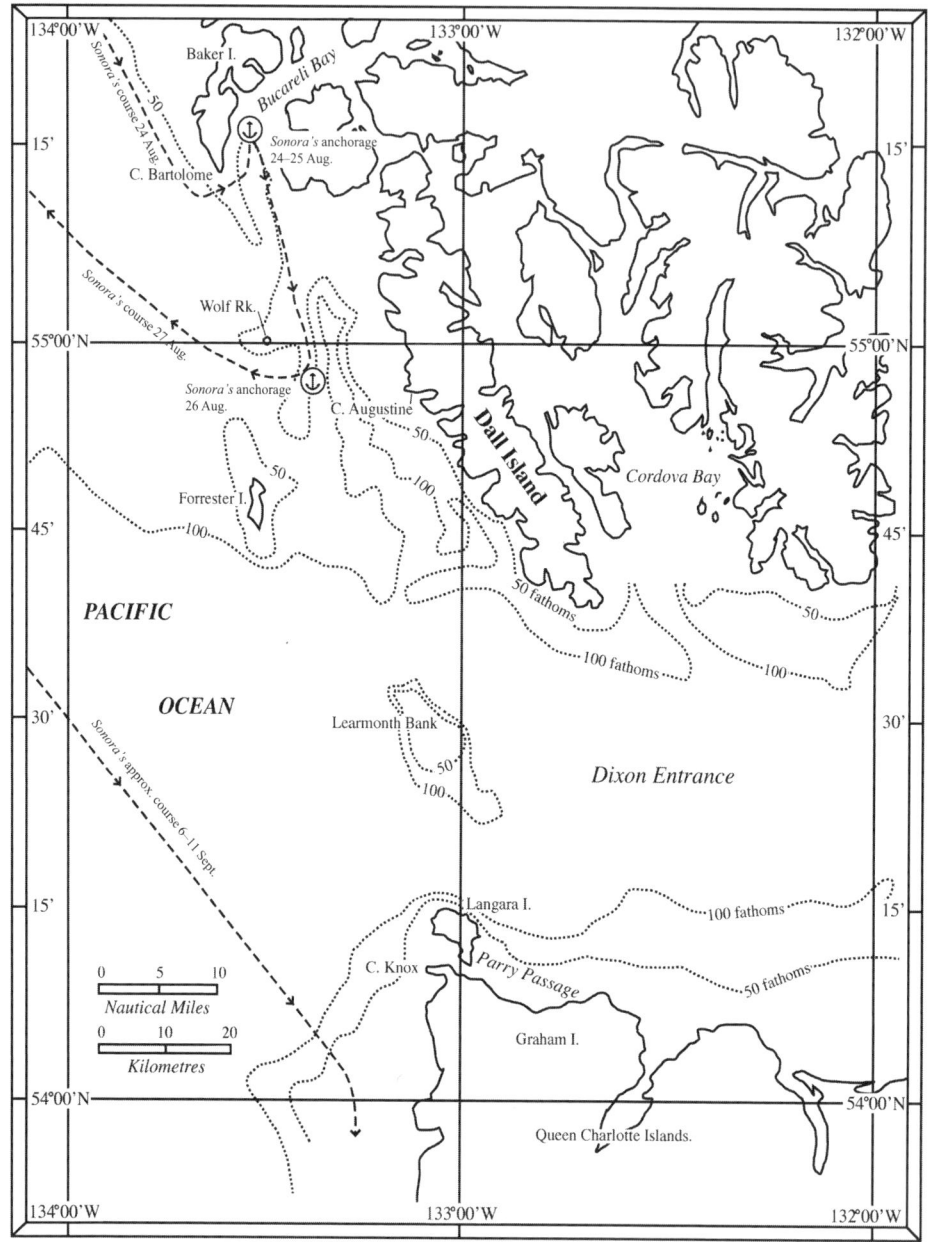

Map 6: Bucareli Bay and Dixon Entrance.

Admiral Fonte.¹ But I have not encountered that extensive archipelago of San Lázaro, into which he says he entered.²

With eagerness, I have stood down to the farthest ends of the many small inlets, doubling all the points present, and desisting from sailing at night so as not to miss the least measure of it. Still holding myself to the greatest accuracy, as such staunch determination requires, I have not the slightest doubt as to the trend of the coast and its location.

The 24th [of August] finding myself in the latitude of 55 degrees, 17 minutes, I doubled a cape and went into a large inlet, thus discovering an arm of the sea to the north, the terminus of which was not perceived. It was so sheltered from the winds that, entering with the sheets in hand, we remained completely becalmed. And with the oars and a tow by the ship's boat. I anchored within its mouth in twenty fathoms of water, mud bottom, distance from land two musket shots. Although I wanted to continue on to see its terminus or its direction, I was unable to accomplish this in the two days that I had, for there was neither any wind blowing nor a launch that could be embarked.

As soon as the ship was secured, finding myself indisposed, I gave orders to the master to go ashore with the preparations and precautions as before (although there was not a single Indian), in order to take possession of those lands. I gave [this inlet] the name Entrada de Bucareli [Bucareli's entrance],³ in honour of the Most Excellent Lord Viceroy. The [master] followed [my orders] carrying the cross in procession and with the prearranged formalities.

The latitude of this harbour and entrance, in the two days that the observation was repeated, was 55 degrees, 17 minutes, and I consider it west of San Blas, 32 degrees, 9 minutes, of longitude.⁴

This entrance is worthy of notice for the mildness of the climate, for the quietness of the sea, for the waters of rivulets and pools that were formed naturally, and the good

---

[1] Referring to a certain Admiral Bartolome de Fonte, whose historicity is doubtful, and whose claims to have made a voyage of discovery to the north-west coast in 1640 remain unauthenticated. An account of Fonte's supposed voyage first appeared in a London periodical called *Monthly Miscellany, or Memoirs of the Curious*, published by James Petiver in 1708. According to this account, Fonte sailed from Callao, the port town of Lima, Peru, in April 1640, with four vessels under orders of the viceroy of Peru to explore in the North Pacific and intercept certain vessels thought to be reaching the Pacific from Boston, Massachusetts, by mean of a north-west passage. It was claimed that Fonte sailed through a system of complex channels among a collection of islands, which he called the archipelago of St Lazarus. Then at the 53rd parallel he is said to have encountered a river that he named 'Río de los Reyes', 'River of the Kings'. Having sent his lieutenant Bernardo up the coast in one vessel, Fonte is supposed to have entered the river, ascending it far into the interior, where it is claimed he encountered a ship from Boston, Massachusetts. For more about Fonte's apocryphal voyage, see Greenhow, *History of Oregon and California*, pp. 82–4; Bancroft, *History*, 1, pp. 115–16; and Cook, *Flood Tide of Empire*, pp. 29–30.

[2] This archipelago of San Lazaro or St Lazarus, said to be located south of the 53rd parallel, might be seen as corresponding to the Alexander Archipelago of south-eastern Alaska, except that the latter group lies entirely north of the 54th parallel. Any apparent correspondence with geographic reality is far outweighed by many others that contradict it.

[3] What is today called Bucareli Bay is sheltered by and stretches north-east from Cape Bartolome, lat. 55°13.9′N; long. 133°36.8′W of Greenwich, at its light.

[4] Bucareli Bay extends from lat. 55°14′N up to 55°21′N Bodega's observation of 55°17′N is thus sufficiently accurate to be within that range. His estimate of its longitude west of San Blas, however, overstates its correct value of 28°20′W by 3°49′.

depth [of waters] and fish therein. Without doubt, I might have remained some days more had not the season been so advanced. But a brief description was drawn up, although hurriedly, as shown in illustration 8a [see Map 7].

The land is as fertile as anywhere on the coast. The Indians have not been seen, only some trails on which they customarily come to their lodges and catch fish.

The nights [in this place] are extremely clear and moderate, because of seven volcanoes of snow and fire, which with their steam illuminate and temper it.[1]

The 25th [of August] I replenished the water and firewood, which were running low, and I prepared to depart the following day with a mind to reconnoitre an island, quite large and fertile, which was to the south a distance of six leagues, and to which I gave the name San Carlos.[2]

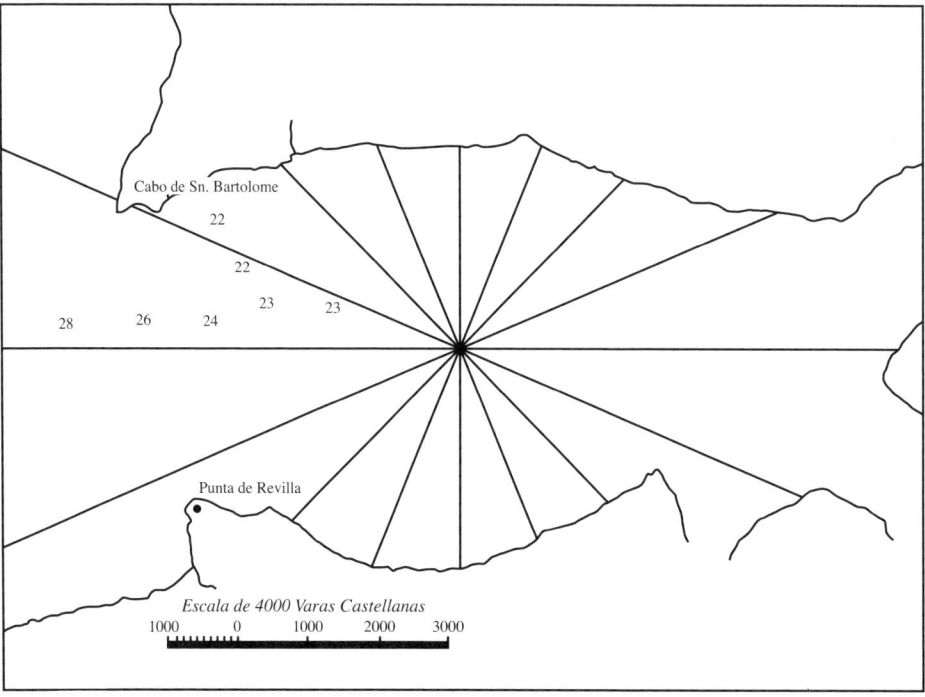

Map 7: Bodega's Map of Bucareli Bay, Entrada o Puerto de Bucareli.

---

[1] It is difficult to know what or where these seven volcanoes were. There are no volcanoes anywhere near the vicinity of Bucareli Bay. Mount Edgecumbe, a dormant volcano, is over 120 naut. miles to the northwest, and the main area of Alaskan volcanic activity is much farther away on the Kenai and Alaska peninsulas, and in the Aleutian Islands. It has been suggested by the Alaskan historian Dee Langenbaugh that a forest fire could have been mistaken for volcanic activity. See Olson, *Through Spanish Eyes*, p. 49, n. 25.

[2] This island must be present-day Forrester Island, lat. 54°48′N; long. 133°31′W of Greenwich, which lies due south of Bucareli Bay, a distance more like nine leagues, instead of six leagues as Bodega states. It is in all probability the same island Juan Pérez discovered in 1774 and named by him 'Isla de Santa Christina'. See Beals, *Juan Pérez*, pp. 80, 142, 144.

Departure from the Entrada and Puerto de Bucareli

The 26th [of August] at four o'clock in the morning I made sail with a slack north wind, and at twelve o'clock [noon] it stopped [and went] completely calm; being close to a small island nearly level with the sea, with some reefs to the east and west.[1] I anchored in twenty-two fathoms of water, distant about two leagues from the Isla de San Carlos.

From this location I saw a cape (which I named for San Agustín) to a distance of four to five leagues, but two [leagues] from it the coast receded to the east in such a manner that we made out the end of it.[2] Some currents experienced were so violent that they prevented the sounding lead from reaching the bottom. They have flood and ebb tides of the same duration, taking six and one-fifth hours to fall, and the same in rising. They move in an east–west direction, from which one might infer that that indentation is another entrance on the sea, where some copious river discharges, for the colour of its water is the same as at the mouths of rivers.

There is no doubt of the Cabo de San Agustín's location in 55 degrees of north latitude. Having heard that on the previous voyage made by Don Juan Pérez, he discovered in about the same parallel an entrance on the sea, I believe with good reason it may be the same.[3] Nevertheless, two sailors who went with [the Pérez expedition], and are present with me, do not recognize the land around the cape. They say, however, that they were southward where they saw the Indians.[4]

I was thwarted in the investigation of this place, with the conjunction of the moon, because the winds changed and held steady in the south-east.

The 27th [of August] I was coming to understand the winds would maintain themselves from the same direction during the flood tide at the least, and they were not only diametrically opposite to the ship's headway, but that they prevented the exploration of that mouth and its coast. Reflecting that I found myself in a latitude from which I would be able to go up to 60 degrees in a few days if the winds maintained themselves, when most of the men had regained strength, I resolved to make use of the time by going up to a higher latitude such as might be possible for me. It should be nothing difficult to do, assisted by the knowledge I have of the direction of the coast.

The same day in the afternoon, I set sail to the north-north-west, after having divided among the men four waistcoats, several yards of woollen cloth and small bales of goods that the commander had supplied me from the King's fourth to regale the Indians, and as well, some shirts, an overcoat of mine and the master's, because it was clear to me that the cold was discouraging and making them sick, as the poor devils had nothing with which to cover themselves.

---

[1] This small island is probably Wolf Rock, lat. 55°00′N; long. 133°26′W of Greenwich, although its distance from Forrester Island is closer to three leagues than Bodega's estimate of two.

[2] Probably present-day Cape Augustine on the west coast of Dall Island, lat. 54°57′N; long. 133°09′W of Greenwich.

[3] Pérez's discovery was almost certainly present-day Dixon Entrance, but this seems to refer to one of several indentations on the west coast of Dall Island. If Bodega meant two leagues south of his Cabo Agustín the most likely possibility is Gooseneck Harbour, lat. 54°53′N; long. 133°03′W of Greenwich. If he meant two leagues north, Sakie Bay, lat. 55°03′N; long. 133°1′W of Greenwich, would be the most likely candidate.

[4] The testimony of these two sailors seems to confirm that the small opening Bodega had in mind could not have been the entrance Pérez discovered in 1774, which is some 25 naut. miles wide and located between latitude 54°15′N and 54°40′N. See Beals, *Juan Pérez*, pp. 141–4.

The 28th [of August] the winds changed, forcing me to return toward land in 55 degrees, 50 minutes, where [the winds] resumed blowing from the south-east, and I continued sailing the same course.

On the 29th and 30th [of August] the winds veered to the south and south-west, coming in squalls with rough seas, which insensibly were pushing me toward the coast at 56 degrees, 30 minutes, which is most harsh and without bottom. With this difficulty, I was maintaining myself by tacking until the 1st of September, when God was willing to give us strong winds with which I separated myself from the coast, sailing to the south south-west.

On the aforesaid days of the 29th and 30th of August, seven men were discovered with scurvy, some in the mouth and others with aches that interfered with movement of the legs, as a result of which there remained only two men on each watch, it being essential that one was maintaining the helm.

Finding myself with men smitten by this evil contagion, without having anything with which to medicate them, and it is at the point of spreading to everyone for not having accommodations allowing for them to be separated, I knew it was impossible for me to exert any more effort to go up to a higher latitude. The return trip could even be called into doubt if the winds were to pick up, for not having men to work the ship. And so, I resolved to turn back, availing myself of the winds that were favourable to me, resigning myself to examining the coast as much as possible.

After the 1st day of September, the winds abated and were variable, until the 6th, when they blew steadily from the south-east with so much force that at midnight the mainsail with two reefings taken in was unable to endure it, and I lay to with the bow to the south-south-west. But the sea was increasing by the minute and the wind in such a manner that two o'clock in the morning I was unable to resist, despite making the utmost effort to remain lying to, on account of the short distance I was from the land. Being in these circumstances, a wave crashed over the entire vessel so powerfully it tore off rails, stanchions, clamps and boards of the roundhouse.[1] It carried away into the water everything topside. We believed ourselves about to founder at the wave's impact, because in the space of four minutes one saw nothing but sea foam throughout the ship. The cabin having been flooded with the rest of [the ship], and without knowing one from another until the scupper holes had drained off part of [the sea water] and some in the ship's hold through the hatchways. Moans and wails began to be heard, because the rails had dashed the boatswain against an anchor, leaving him motionless. Some of the sailors were also badly shaken up, although not dangerously so. But they were left useless for helping to work the ship, so that in such a critical hour I saw myself with only two sailors, the master, the watch, and my servant. Knowing that it would be impossible for us to remain without foundering, I gave the order to head the stern into the wind, continuing from a north-westerly direction, obliged to lose the little distance that I was from the coast; and with no other hope of deliverance except that the storm might abate before we met with the coast.

---

[1] This incident, which nearly put an end to the *Sonora* and the lives of her crew, occurred somewhere off the coast in the vicinity of Dixon Entrance. The term *toldilla* 'roundhouse' refers to the roof above the covered quarters at the ship's stern. It was essentially the *Sonora*'s *alcazar* (quarterdeck). See O'Scanlan, *Diccionario*, p. 525.

I carried out this procedure with only the flying jibsail. The bilge pumps needed to be checked, for in the course of the night they were of no use in pumping out the water, being already overflowing upon the stowage, because, despite having the hatchways and tarpaulins battened down, the water kept coming in. The crew was convinced that the ship was going to sink. No matter how much I encouraged the men, they believed she would founder before the weather let up. They were completely disheartened. Nevertheless, they worked with the greatest tenacity until the pumps were successfully repaired, and at eight o'clock in the morning of the next day there was not an inch of water.

We also had the misfortune to split a rudder gudgeon,[1] without having any spares. But it was remedied quickly with the utmost effort by making a spike, found by chance, into an eye.

The same day, the 7th [of September], the wind and sea were subsiding, and I returned to lying to from six o'clock in the evening until the morning of the 8th [of September]. When the north-west wind came on moderately, I continued to pursue my voyage with a mind to make landfall at about 54 degrees, relinquishing the reconnaissance that I was thinking about making in 55 [degrees], having remained with only two men for tending the helm, and the master and me, who watched over the working of the ship.

I continued sailing in this dejected state, and on the 11th [of September] the land was seen in 53 degrees, 54 minutes, at a distance of eight or nine leagues.[2] There was no committing myself concerning it, because the month is sufficiently blustery to make departure difficult, with the lack of men able to work the ship. I determined to run along [the shoreline] at a moderate distance, relinquishing by necessity inspection of its inlets, for which purpose it is necessary to travel in or through, as I have done when I found myself with opportunities for it. Nevertheless, I have endeavoured to examine [the coast] daily with the purpose of not leaving uninformed as to the direction it follows.

Being in latitude 49 degrees, I returned to bring myself within a mile's distance of the coast, proceeding as in the earlier reconnaissance of it.[3] Thus, by being satisfied with the favourable winds, as well as because the injured ones were in rather good spirits, they were all responding to working the ship, except for the boatswain and the eight men with scurvy, who I fear may die on the voyage.

In this fashion I sailed until the 21st [of September] when finding myself in 46 degrees, 20 minutes, slack winds came on from the south and south-east, which separated me from the coast.[4]

The 22nd [of September] the winds returned from the north-west. But in having been dealt the misfortune of the master and me overtaken at the same time by severe fevers, accompanied by the pains of scurvy, it was necessary to set a course directly for Monterey.

The few [healthy] sailors remaining and the sick ones, seeing that we were now ill, were so distressed that they imagined themselves lost. But having known its affliction,

---

[1] A rudder gudgeon (*hembra del timón*) is the part of a rudder's hinges in which the pintles or pins (*macho del timón*) are seated. See O'Scanlan, ibid, pp. 314, 347.

[2] Latitude 53°54′N places the *Sonora* off the coast of the Queen Charlotte Islands.

[3] Latitude 49°N places the *Sonora* off the coast of Vancouver Island between Nootka and Barkley sounds.

[4] Latitude 46°20′N places the *Sonora* off the Washington State coast, slightly north of the latitude of the Columbia River estuary. The winds that separated Bodega from the coastline thus deprived him of an opportunity to see the outlet on the ocean of North America's largest westward-flowing river.

we made every effort to present ourselves so as to conceal it, and in this purpose we had God's help.

The 24th [of September] finding myself somewhat relieved, land was sighted, which was found to be in 45 degrees, 27 minutes.[1] Thereupon I turned to engage in its reconnaissance with such promptness that I was hauling myself off scarcely a mile and a half from it, casting anchor at sunset in whatever place on the coast where we would not lose the slightest distance. On the lookout for the Río de Martín de Aguilar,[2] I did not encounter it between the aforesaid latitude [45°27′N] and 42 degrees, 50 minutes, where a cape that is shaped in the form of a table juts out into the sea.[3] Receding from it, the coast [extends] to the south-east, and there are various rocky isles to the south south-west.

According to available information, it is known that the discoverer of this river or entrance observed its mouth in 43 degrees; but from 44 [degrees] 50 [minutes] down to 42 [degrees] 50 [minutes], neither rock nor wood escaped my attention on the beach. It is true that the instruments with which observation [of latitude] used to be made in those times were not dependable, and they could have suffered some sizeable error, so I suppose it might be between 42 [degrees] and 42 [degrees] 50 min. For in the second latitude the southeasters and calms returned, and on the 26th [of September] they carried me on courses to the south. It being impossible for me to turn back to find it, nor yet being in 42 degrees, I came to make landfall on the 19th [of September, *sic* 29th?] at seven o'clock in the evening at about 40 degrees, 28 minutes, a distance of half a league.[4] But having discovered the landfall at Trinidad, from 42 degrees down to this place, it only left me in doubt (if it is certain that there is such a river) that it could be in those 50 minutes I was unable to examine.

From this last landfall, I continued with the same careful attention searching for Puerto de San Francisco, leaving no inlet unexamined. The 3rd [of October] finding myself in latitude 38 degrees, 18 minutes, I entered an inlet that is shown in Illustration 9 [see Map 8], in which I travelled until, in the south-east part, I discovered the mouth of an abundant river.[5] Entering it, I saw it was forming a large harbour, as represented in the same illustration, at the same time the tide was rising with extraordinary violence. Presently, I believed it might be the harbour that was being sought, so I went in and cast anchor facing Punta de Arena [Sand Point] sheltered from the coast by the Punta del Cordón [Cordon Point],[6] in seven fathoms of water. Expecting news of the people who had been charged with the responsibility of establishing themselves [at

---

[1] Latitude 45°27′N bisects Netarts Bay on the Oregon coast.

[2] This river, said to have been discovered during the expedition led by Sebastián Vizcaíno in 1602–3, took its name from Martín de Aquilar, captain of the frigate *Tres Reyes*.

[3] This cape is almost certainly present-day Cape Blanco, lat. 42°50.2′N; long. 124°33.8′W of Greenwich, at its Light, on the southern Oregon coast.

[4] The latitude Bodega gives for this landfall is within two minutes of Cape Mendocino, lat. 40°26.4′N; long. 124°24.4′W of Greenwich, at its Light, and it seems safe to conclude that this was the landmark he had sighted.

[5] The inlet is between Bodega Head, lat. 38°18.0′N, long. 123°03.2′W of Greenwich, at its Light, and the entrance to Tomales Bay, lat. 38°14′N; today it is called Bodega Bay. The 'abundant river' entering the bay from the south is Tomales Bay, a narrow body of water with an average width no more than a half mile, which stretches some 12 naut. miles to the south-east of Bodega Bay.

[6] The first point, which still bears the name Bodega gave it (though anglicized), Sand Point, is on the inland or east side of the entrance to Tomales Bay; the other point, which is today called Tomales Point, is on the west or oceanward side.

Puerto de San Francisco], but seeing no sign of them all day, I doubted that it was Puerto de [San] Francisco.

The Indians [Coast Miwoks] were there in great numbers. They were passing in tule [reed] canoes from one shore to the other, in order to come to a hill near where we were anchored.[1] After that, they joined into a large party and began crying out for nearly two hours without ceasing. At the end of this time, two [canoes] came alongside, and with the greatest generosity they presented plumage, necklaces of bone, a basket of white tallow with the taste of almond, and various baubles of this sort. I repaid their gifts with handkerchiefs, mirrors and glass beads, and they were very pleased.

The stature of these people is burly and robust, but the colour is swarthy. They are no different in any than other Indians that I have seen, except for the hairdo, which they wear in a manner that is the newest fashion worn by ladies in Spain.

The harbour is situated in latitude 38 degrees, 18 minutes, and in longitude west of San Blas, 32 degrees, 9 minutes. It is large, indeterminate towards the SE; and of such perfect shelter that all of it is an inner harbour. I did not inspect its bottom into the interior, because I wanted to put the sick ashore. It was agonizing to see them each in the final crisis without the least assistance. For this reason I endeavoured to put myself in readiness to sail, so that my departure might not be hindered. I came back to cast anchor near the point that I named Cordón.

The 4th [of October] at two o'clock in the morning the other tide began to rise, and at four o'clock in the morning the sea rushed through, entering against what was leaving so violently that some blows striking the ship covered us from stem to stern. One of them tore away the ship's boat, turning it into a thousand splinters. And at the same time one of the anchors was lost, it being impossible for me to return and stop to find it without the ship's boat.[2]

The mouth, not being sufficiently wide to handle that sea, makes the currents enter and leave in the aforesaid part with very great force, and the tides come rushing through the mouth whenever they are agitated by bad weather outside.

If I had been informed of this chance event (which does not always happen), I would have remained in the first location, or I would have gone farther out, in which locations I would have experienced nothing.

The bottom, in all that I have seen of it, is just as the map shows numbered in fathoms in the illustration [see Map 8].

Entrance into the aforesaid harbour is easy because the north-west winds are astern when entering. But departure, unless done with a southeaster, is forced to await low tide and setting sail with a tow, which is not so difficult because of the tidewater's force.

I observed that the tides on the day of [lunar] conjunction were high at twelve o'clock noon, and they kept regularly to six hours and twelve minutes for each ebb and flood tide.

The mountains are treeless, and it is the same in this vicinity. Only in the far interior does one see many trees, but the plains from the shores to the contiguous borderlands are verdant and pleasing, handsomely suitable for cultivation.

---

[1] This hill may be the same feature that is today called Sugarloaf Hill, el. 64.3m above mean sea level, which is separated from Sand Point by an expanse of sand dunes extending south-west from the hill.

[2] The *Sonora* had been struck by a tidal bore resulting from the narrowness of Tomales Bay. According to the *US Coast Pilot*, No. 7, p. 173, the current can reach six knots at the bay's entrance.

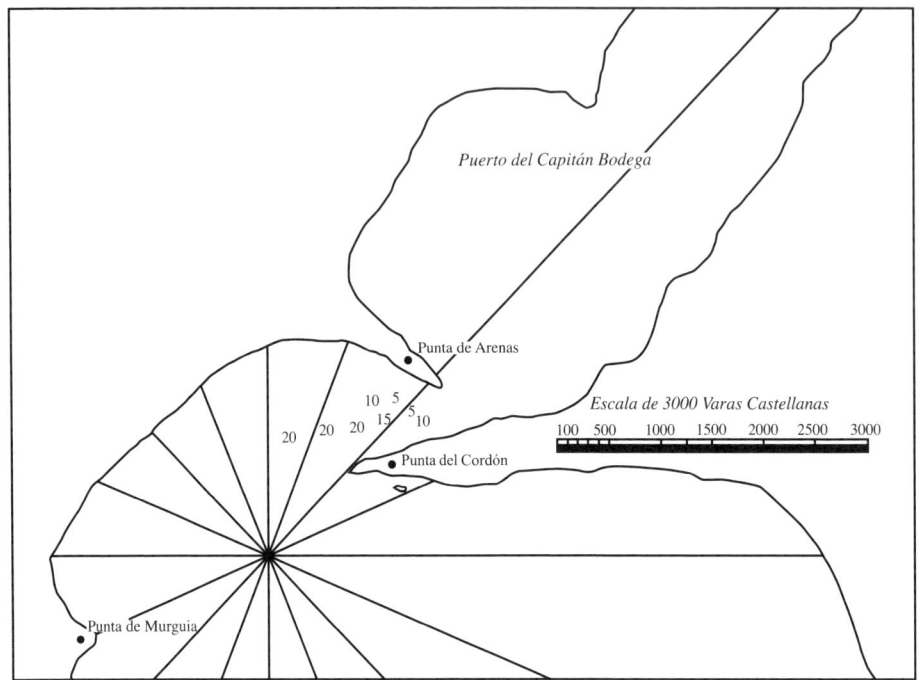

Map 8: Bodega's Map of Bodega and Tomales Bays, Puerto del Capitán Bodega.

At eight o'clock [a.m.] the dashing of the waves having ceased, two canoes came together alongside with another one present. They gave without seeking compensation, a liberality that I have not experienced among the others. I satisfied them with the best I could find, and they withdrew with signs of much pleasure.

Departure from this Harbour
This day, at nine o'clock [a.m.], I set sail and rounded Punta del Cordón. I steered to the south-south-west with a west wind in order to round a cape that was surveyed to the south a distance of five leagues.[1]

The 5th [of October] at dawn I was south-west of some small islands, which the charts place in the vicinity of the mouth of [Puerto de] San Francisco.[2] I knew of evidence that that was not it, and I put myself on a north-east course, entering into their midst, until being at the cape that I mentioned, from which the small islands are three leagues distant to the south south-west.

From this cape, I continued on a course east and east one-quarter north-east, sailing along the coast within a very short distance, and at six o'clock in the afternoon I found myself a distance of two-thirds of a league from a mouth I believed was, without the least

---

[1] This cape is no doubt present-day Point Reyes, lat. 37°59.7′N, long. 123°01.3′W of Greenwich, at its light, which shelters Drakes Bay.

[2] These are the Farallon Islands, located some 25 naut. miles west of the Golden Gate.

doubt, that of San Francisco.[1] But not having the ship's boat and being uncertain of the establishment [at San Francisco], I resolved to remain on course to Monterey, and I turned to a course south in order to round another cape that extends far out.[2]

At ten o'clock at night, it remained calm, and it continued so until the 6th [of October] at twelve o'clock noon, which with a slack wind, I put myself on a course south south-west, one league distant from the coast. But at eight o'clock at night, [the wind] veered to the north-west, fresh, with rain squalls and mists.

The 7th [of October] I considered myself on the parallel of Monterey, and I steered to the east-north-east so as to maintain it, although the mists did not permit me to see more than a quarter mile on the horizon.

At three o'clock in the afternoon, the coast was discovered to the south-east a distance of one mile, and proceeding on the same course, we coasted past a point that extended far out.[3] Finally, the frigate was sighted, together with the packet boat *San Carlos*, and having fired off some cannons, they came out with their launches to take us in tow as far as the anchoring ground. I cast anchor in three fathoms of water [over a] sandy bottom.

This harbour is situated in 36 degrees, 44 minutes, north, and I consider it west of San Blas, 17 degrees.[4]

The 8th [of October] I ordered that the sick be disembarked, those who were curable to the Presidio. The master and I went to live at a house that is near the anchoring ground, with the purpose of recovering our lost health with the land breezes, and to rest a few days from such an arduous voyage, mostly on a vessel of so little comfort it is a miracle that all of us have not arrived completely crippled. This, because we have come to be almost continuously drenched, as well as not having the least space in which to take two steps; obliged to be standing up, if they could endure it, or seated, whichever promised the greatest relief.

The 25th [of October] I found myself feeling better, and my men similarly so, because of the mildness of the climate, and likewise for the careful attention the reverend fathers of the Mission [San Carlos] have given us; and the Presideo commandant's great affability in making us generous offerings of vegetables, meat, milk, and as much as was possible. The commander [Hezeta] and I resolved to depart on the 1st day [of November].

Departure from Puerto de Monterey situated in North Latitude 36 degrees, 44 minutes, toward San Blas, under 21 degrees, 30 minutes [latitude]
The 1st day of November we got under sail with light winds out of the 1st [NE] and 2nd [SE] quadrants, and finding ourselves about two leagues outside of the points,[5]

---

[1] Bodega was probably correct in his supposition that this was indeed the entrance to San Francisco Bay, or as it is popularly known today, the Golden Gate.

[2] With no indication of this cape's latitude its identification is uncertain. It could be either modern Point Lobos, immediately south of the Golden Gate, or Point San Pedro, some 11 naut. miles farther south.

[3] Probably referring to the headland that runs from Point Cypress to Point Pinos on the south side of Monterey Bay.

[4] The correct latitude of Point Cypress is 36°35′N; Point Pinos Light is 36°38′N; and the harbour at Monterey is 36°36′N. The longitude of Monterey Bay is actually 16°45′W of San Blas.

[5] Referring either to points Pinos and Cypress, or possibly points Pinos and Año Nuevo, or even perhaps all three. In any event, the statement merely takes note that the ship had departed Monterey Bay.

they went calm so that we remained in view of the harbour until the 4th of November with winds out of the south and south-west.

The same day at twelve o'clock noon, the north-west wind came up, and we maintained courses south-south-east, east-south-east, and east one-quarter south-east, In accordance with what was needed to stay clear of the islands.[1] Landfall was made on the coast on the 14th [of November] at three o'clock in the morning, at about latitude 24 degrees, 15 minutes, from where I ran down the coast to Cape San Lucas.

I consider this cape under longitude 5 degrees west of San Blas,[2] and in north latitude 22 degrees, 49 minutes.

The 18th [of November] at six o'clock in the evening we saw the Islas Marias to the east one-quarter south-east. We anchored on the 20th [of November] in the Puerto de San Blas.

Despite the hardships I have suffered in the course of the voyage, I consider myself fortunate in having fulfilled, insofar as my part, the orders of the most excellent viceroy, Frey Don Antonio María Bucareli.

Juan Francisco de la Bodega y Quadra

---

[1] No doubt a reference to the Channel Islands off Santa Barbara. California.

[2] The longitude of Cape San Lucas, the southernmost point of Baja California, is actually 4°58′W of San Blas.

APPENDIX 1

Extract from Francisco Antonio Mourelle's 'Navegación … 1775' for 14 July

Figure 2: Anonymous oil portrait of Francisco Antonio Mourelle de la Rúa (1750–1820). Reproduced by permission of the Museo Naval, Madrid.
Mourelle was second in command under Bodega aboard the schooner *Sonora* on her 1775 voyage. He was among the first European mariners to reconnoitre the entire NW coast of America up to 62°N latitude. He is credited with discovery of the Vava'o Group of the Tonga Islands. Upon his death he held the rank of *jefe de escuadra* equivalent to the rank of rear admiral in the eighteenth-century British Royal Navy.

On the morning of the 14th, the sea was so low that the reefs were exposed along the coast, preventing us from making sail and obliging us to wait for high tide, which will occur at noon. In the interim, the Indians returned and offered our men various animal pelts in trade, making signs that they wanted pieces of iron, which they indicated by putting a hand on the rudder pintles. Our men managed to give them payment for their skins by pieces [of iron] from old chests. Finally, they went back, making the same signs that were made the previous day.

We, who had to sail out of there that day (according to orders the commander of the expedition had given me), were hoping, with the voyage being so long, that we would not be forced to leave with a deficient water supply. In order to prevent scurvy and other ailments, we had consumed it in abundance since [leaving] Puerto de la Trinidad. We determined to do all we could during the time the low tide prevented us from making our departure, including also the cutting of firewood and a mast. For this reason, six men were assembled, who were the most sensible, skillful and [well] prepared for such occasions. Each of them was given a musket, a pistol, some two [pistols], a sabre and a cartridge box. Boatswain Pedro Santa Ana, an individual who always manifested exemplary behaviour and valour among those of his rank, was in command.[1]

So that the aforesaid tasks could be accomplished, they took two axes and were ordered to direct the ship's boat so as to take the casks to a place they were to find that would be well suited to complete [the watering] in a brief time; and that when the ship's boat would return, we could go [ashore] with it.

They endeavoured to disembark near a river where there was somewhat more surf than elsewhere, which prevented them from making a completely trouble-free landing.[2] Following this, the Indians came out from those woods in excessive numbers, more than 300, attacking them unexpectedly in that defenseless instant. We think [the sailors] were done in, because from the instant that [the Indians] threw themselves upon them, we saw no more than one firearm discharged. Two of our sailors threw themselves into the sea in order to get away by swimming. The coldness of the water prevented them from reaching their goal, and we lost track of their whereabouts.

Lacking a boat, we were unable to render them assistance, so we tried firing the swivel guns and muskets. Inasmuch as these shots did not reach [the Indians], nor were they aware of the damage [firearms] could do had they been closer, they made neither the least movement nor desisted in their treacherous hostility, as it is necessary to infer. With no recourse remaining for us to render aid, we signaled the frigate for help, but the distance was such that they did not distinguish what we said. Finally, at 11:30 [in the morning], the Indians retired to their settlements, but we saw nothing of our men and little of the ship's boat.

---

[1] In Bodega y Quadra's account the boatswain's name is written Pedro Santana. See above, p. 106, n. 1.

[2] This river in all likelihood is the Quinault River, which empties into the Pacific Ocean in the bight formed by Cape Elizabeth. Neither Bodega's account, nor those of Hezeta and the two chaplains aboard the *Santiago*, explicitly say, as does Mourelle's, that this ill-starred landing took place near a river. The map prepared by the expedition of the vicinity in which this unhappy event occurred does, however, show a river or stream in more or less the same relationship as Cape Elizabeth and the Quinault River. See Beals, *For Honor*, pp. 75–8; Campa Cos, *Journal of Explorations*, pp. 41–5; Wagner, 'Sierra's Account', pp. 227–9; also pp. 104–7 above.

At 12:00 noon, the sea being at high tide, we weighed anchor. Accounting the few of us who remained, we set out to rejoin the frigate. For this, we put one man at the topmost head, another at the helm, and another sounding. Remaining for us to work the ship were a cabin boy and two seamen who, with four sick ones, completed the crew.

As soon as we were under sail, nine canoes of Indians came out. With increased numbers, they presently placed themselves a fixed distance from us, and while they maintained the same [distance], one of the canoes with only the nine captains who previously were in command [of the canoes individually], continued to within a very short distance of our side. Presenting themselves with their bows stretched taut [and] some handsome rawhides for defence, they at first [tried] persuading us with their previous deceit, making signs about the food they had given our men. But now we had weapons at the ready, and one of the sailors was hastily making cartridges. Nevertheless, we endeavoured to lure them by waving glass beads [and] handkerchiefs, but they were not enticed by attractive trinkets. They simply continued that means and knowing the few men we had, they made vehement demonstrations to attack us, taking their bows in hand. Although only three of us could handle muskets – the commander [Bodega y Quadra], his valet and me – we acquitted ourselves so well that in a brief time we killed 6 and shattered the canoe into splinters. Those who experienced the dire effects of our arms were stunned. When they were able to stop astern, giving us little time to fire on them, they went for the bow, going around the ship at very close distance, allowing an opportunity to do them harm. Although no one is agile enough to throw oneself down promptly at the noise of a trigger firing, there remains no doubt that if the side of their canoe had been more reinforced, not so much harm could have been inflicted on them. They also took the precaution of covering the dead with rawhides, and finally they withdrew beyond musket range, in which instance they went amidst the rest of the canoes, which in cowardly fashion had not come near to defend them.

There they had a brief consultation in which they must have reflected on the difference of the one attack from the other, and finally they retired to their settlement.

The commander [Hezeta], who heard the discharge [of our weapons], believed we needed a stream anchor and cable, with which foresight he sent us the launch with armed men so as to devastate that settlement and ascertain if some of our men, particularly those who threw themselves into the water, had been able to take refuge in the undergrowth of those woods.

A meeting was convened, and the commander [Hezeta] explained that [the expedition] found itself in a dangerous situation, that sea and winds could be threatening, that it was a long distance to the [native] settlement, and that also, the treachery being committed in plain view [argued against] a search of the landing site for any who might have been able to escape.

Don Cristóbal de Revilla and Don Juan Pérez were of the opinion that we should set sail, although my commander [Bodega y Quadra] and I openly urged that some revenge be taken for an act of this kind, especially when there had been bloodshed. [We argued further] that it was necessary to find out the manner in which those men had met their end; that it would be lamentable if those who had gotten away by swimming had saved themselves [only] to be left there with no further refuge from those barbarians; that it would be agreeable to His Majesty [Charles III] to make known to these people the superior power [of our arms], so that subsequently they would restrain themselves towards other [Europeans], who by similar chance might be in these circumstances; that [even] if the settlement was not nearby, we had sufficient time to wait for the following day to carry out the [punitive] operation; and that the winds have no force with the moon at conjunction, and that on that account a flood tide would be running. Nevertheless, the commander [Hezeta] went along with the opinions and votes of his officers [opposed to punitive action].

Now that this point had been decided, the commander took up the same matter he had warned me about yesterday concerning the schooner. After setting forth those same reasons he had communicated to me concerning her poor sailing qualities and the fear they [the *Santiago*'s officers] have for her, they offered their opinions. Of them, Cristóbal de Revilla was alone in feeling that she should go back; but those [opinions] of Don Juan Pérez, Don Juan Francisco de la Bodega y Quadra and mine agreed that she should proceed on. To this the commander [Hezeta] assented, and he asked me to put the opinions in writing to the schooner's captain, [including] mine, dated the 16th, [which] we delivered to him.

APPENDIX 2

The Schooner *Sonora*

Informally known as *Sonora*, but officially named *Nuestra Señora de Guadalupe*, this small, two-masted vessel was of a type called *goleta* in Spanish, the equivalent of 'schooner' in English. Built in 1769 at San Blas on the Mexican west coast, she was originally intended for mail delivery and transport of personnel and small cargoes in coastal waters and between Sonora and Baja California. In her first four years of service, the *Sonora* seems to have performed well in carrying out the tasks for which she had been designed. But in 1775 she was given an assignment that her designers had not contemplated, namely: open-ocean voyaging in North Pacific waters to accompany and assist a larger flagship on a voyage of discovery to what was then the remote and largely uncharted north-west coast of North America.

In March 1777, two years after the *Sonora*'s return from her voyage of discovery, an inventory of the ship was conducted, which included: her dimensions; her hull's seaworthiness; her masts, yards, rigging, sails, rudder, tiller, anchors, stores, water supply, ship's boat, and a great variety of equipage. It was in preparation for a voyage to transport two carriages from San Blas to Acapulco for use of the Inspector-General designate of the kingdom of Peru, Joseph Antonio de Areche. This Inventory is an important source of information concerning the *Sonora*'s dimensions and design.

The Inventory gives the ship's dimensions in *codos* (cubits, which can vary between 18 and 22 inches) and *pulgados* (0.91 inches) as follows: length stem to stern, 24 *codos*; keel length 20 *codos*; width abeam 6 *codos* 15 *pulgados*; floor timber, 4 *codos*; and depth of hold, 6 *codos*, 15 *pulgados*. For further discussion of the *Sonora*'s dimensions, see p. 63, n. 1. Her cargo capacity was fifty-nine 'Spanish tons',[1] based on the actual tonnage taken aboard at San Blas. As for her general design, the *Sonora* had a bowsprit 12 *codos* (about 6 m or 20 ft) long; two masts – a main-mast/top-main and a fore-mast/top-fore – each mast with a pronounced rake. She carried eight sails: main, main topsail, foresail, square sail, fore topsail, fore staysail, jibsail, and flying jibsail. She was steered by a tiller rather than a wheel. The officers' quarters consisted of two cramped cabins at her stern, the roof of which functioned as a quarterdeck. Two port openings were on the stern transom, but there is no mention of quarter galleries (windows) or any other port openings elsewhere. The 1777 Inventory says nothing about armaments, but it is clear from accounts of the 1775 voyage that the *Sonora* at that time was armed with several *pedreros* ('swivel guns'), muskets, pistols, ammunition, and sabres stowed in the

---

[1] The 'Spanish ton' or 'tonelada de arqueo' (ton of capacity) was based on 53.5 cubic feet per ton. The modern moorsom ton is based on 100 cubic feet per ton.

magazine – known in Spanish parlance as *Santa Barbara* – directly below the officer's quarters.

The 1777 Inventory carried out by Boatswain Santiago Hernando and Jossef Barrios del Castellon and the 1775 Journals of Bodega and Mourelle are the primary sources of information about the *Sonora*'s dimensions and design. If there were plans for her construction they apparently have not survived. Secondary sources include Malcolm Kenyon's 'Naval Construction and Repair at San Blas, Mexico, 1767–1797'; and two letters, dated 17 March 1990 and 28 September 1990, from Greg Foster to Herbert Beals, concerning efforts to build a life-sized cutaway replica of the *Sonora*'s interior for a museum exhibit. Timoteo O'Scanlan's *Diccionario marítimo español* has been helpful in understanding eighteenth-century Spanish nautical terminology.

BIBLIOGRAPHY

Manuscript Sources

Bodega y Quadra, Juan Francisco de la, 'Año de 1775, Diario de la exploración hecha de las Costas septentrionales de California hecha por D. Juan Francisco de la Bodega y Quadra en la Goleta la Sonora'. AGN, Historia, vol. 324, Mexico City, Mexico.

Bodega y Quadra, Juan Francisco de la, 'Navegación hecha por D. Juan Fran[ci]sco de la Bodega y Quadra Then[ien]te de Fragata de la Real Armada y Com[andan]te de la Goleta Sonora; a las Descumbrim[ien]tos de los Mares y Costas Septentrional de California'. MS 622, MN, Madrid,

[Bodega y Quadra, Juan Francisco de la], 'First Voyage to lat. 58° in a schooner of 18 cubits keel and 6 [a]beam, manned by a pilot, a boatswain, a mate, ten seaman, a cabin boy and a servant, 1775'. G. F. Barwick, trans. July 1912, ABC, Victoria, BC.

[Bodega y Quadra, Juan Francisco de la], 'Second Expedition up to 61° in the frigate "Nuestra Señora de los Remedios" [alias] "La Favorita" of 39 cubits keel and 13 [a]beam, stern draught 14 ft., and prow draught 13; year 1779'. G. F. Barwick, trans., July 1912, ABC, Victoria, BC.

Bodega y Quadra, Juan Francisco de la, 'Mandos y Comisiones que seme han confiados'. Havana, 8 de enero de 1784. VISO, Ciudad Real.

Bodega y Quadra, Juan Francisco de la, 'Relación de los merítos y servicios del Capitán de Navío Dn. Juan Francisco de la Bodega y Quadra'. VISO, Ciudad Real.

Hernándes, Santiago, and Jossef Barrios, 'Inventario de la Goleta de Su Magestad nombrada Nuestra Señora de Guadalupe, alias la Sonora, que sale despachada de esta Puerto para el de Acapulco conduciendo doz coches del usso del Señor Don Jossef Antonio de Areche Electo Visitador General del Reino del Perú, 3 March 1777. AGN, Provincias Internas 10, Mexico City.

Kenyon, Malcolm Hall, 'Naval Construction and Repair at San Blas, Mexico, 1767–1797', MA thesis, University of New Mexico, Albuquerque, 1965.

Mourelle de la Rua, Francisco Antonio, 'Navegación hecha por el Piloto segundo de la Armada Don Francisco Antonio Maurelle, en la goleta Sonora del mando del Teniente de Fragata Don Juan Francisco de la Bodega y Quadra, a las descumbrimientos de las Costas y Mares Septentrionales de Californias, que por orn. Del Exmo. Señor virrey Fr. Don Antonio María Bucareli y Ursúa, se hicieron en el año de 1775'. AGN, Historia, vol. 324, Mexico City.

Printed Sources

Aker, Raymond, and Edward Von der Porten, *Discovering Portus Albionis: Francis Drakes's California Harbor*, Palo Alto, Calif., 1979.

Baker, J. N. L., *A History of Geographical Discovery and Exploration*, New York, 1967.

Bancroft, Hubert Howe, *History of the Northwest Coast*, 2 vols, New York, n.d.

Bannon, John Francis, *The Spanish Borderlnds Frontier, 1513–1821*, Albuquerque, N. Mex., 1976.

Bawlf, Samuel, *Sir Francis Drake's Secret Voyage to the Northwest Coast of America, AD 1579*, Salt Spring Island, BC, 2001.

Bawlf, Samuel, *The Secret Voyage of Sir Francis Drake, 1577–1580*, New York, 2003.

Beaglehole, J. C., *The Exploration of the Pacific*, 3rd edn, Stanford, Calif., 1968.

Beaglehole, J. C. ed., *The Journals of Captain Cook on his Voyages of Discovery: The Voyage of the Resolution and Discovery, 1776–1780*, Hakluyt Society, extra ser. 36, Cambridge, 1967.

Beals, Herbert K., 'The Juan Pérez-Josef de Cañizares Map of the Northwest Coast', *Terrae Incognitae*, 27, 1995, pp. 46–56.

Beals, Herbert K., ed. and trans., *For Honor and Country: The Dairy of Bruno de Hezeta*, Portland, Ore.,1985.

Beals, Herbert K., ed. and trans., *Juan Pérez on the Northwest Coast: Six Documents of His Expedition of 1774*, Portland, Ore., 1989

Bernabeu Albert, Salvador, ed., *Juan Francisco de la Bodega y Quadra: El descubrimiento del fin del mundo (1775–1792)*, Madrid,1990.

Bishop, R. P., 'Drake's Course in the North Pacific'. *British Columbia Historical Quarterly*, 3:3, 1939, pp. 151–82.

Blanca Carlier, José María, *La Marina en Cádiz (Apuntes históricos)*, Cádiz,1987.

Bolton, Herbert E., ed. and trans., *Anza's California Expeditions*, 5 vols, Berkeley, Calif., 1930.

Bolton, Herbert E., ed. and trans., *Fray Juan Crespi: Missionary Explorer on the Pacific Coast, 1769–1774*, New York, 1971 (reprint of 1927 edition).

Bolton, Herbert E., ed. and trans., *Historical Memoirs of New California by Fray Francisco Palóu, O.F.M.*, 4 vols, Berkeley, Calif., 1926.

Bolton, Herbert E., ed. and trans., *Kino's Historical Memoir of Primería Alta*, Berkeley and Los Angeles, 1948.

Bolton, Herbert E., *Outpost of Empire, The Story of the Founding of San Francisco*, New York, 1931.

Bolton, Herbert E., *Rim of Christendom, A Biography of Eusebio Francisco Kino, Pacific Coast Pioneer*, New York, 1936.

Bolton, Herbert E., *Spanish Exploration in the Southwest*, New York, 1952.

Brockman, C. Frank, *Trees of North America*, New York, 1968.

Burrus, Ernest J. SJ, *Kino Reports to Headquarters, Correspondence of Eusebio F. Kino, SJ, from New Spain with Rome*, Rome, 1954.

Cabrera Bueno, José González, *Navegación especulativa, y práctica*, Manila, 1734.

Campa Cos, Fray Miguel de la, *A Journal of Explorations Northward along the coast from Monterey in the year 1775. Fray Miguel de la Campa Cos*, ed. and trans. John Galvin, San Francisco, 1964.

Chapman, Charles E., *The Founding of Spanish California, the Northwestward Expansion of New Spain, 1687–1783*, New York,1916.

Chapman, Charles E., 'Difficulties of Maintaining the Department of San Blas, 1775–1777', *Southwestern Historical Quarterly*, 19, 1916, pp. 261–70.

Chapman, Charles E., *A History of Spain*, New York, 1931.

Cook, Warren L., *Flood Tide of Empire: Spain and the Pacific Northwest, 1543–1819*, New Haven, Conn. and London, 1973.
Coxe, William, *Account of the Russian Discoveries Between Asia and America*, London, 1780 (Readex reprint, 1966).
Cummins, J. S., ed. and trans., *Sucesos de Las Islas Filipinas by Antonio de Morga*, Hakluyt Society, 2nd ser. 140, Cambridge, 1971.
Cutter, Donald C., and George Butler Griffin, eds and trans., *The California Coast: A Bilingual Edition of Documents from the Sutro Collection*, Norman, Okla., 1969.
Dahlgren, Eriik Wilhelm, *Were the Hawaiian Islands Visited by the Spaniards before their Discovery by Captain Cook in 1778?*, Stockholm, 1916 (New York, reprint, 1977).
Davidson, George, *Pacific Coast Pilot of California, Oregon and Washington*, 4th edn Washington, D.C., 1889.
Dmytryshyn, Basil, E. A. P. Crownhart-Vaughan, and Thomas Vaughan, eds and trans., *Russian Penetration of the North Pacific Ocean, 1700–1797*, Portland, Ore., 1988.
Fireman, Janet R., *The Spanish Royal Corps of Engineers in the Western Borderlands, Instrument of Bourbon Reform, 1764–1815*, Glendale, Calif., 1977.
Fireman, Janet R., 'The Seduction of George Vancouver: A Nootka Affair', *Pacific Historical Review*, 61, 3 August 1987, pp. 231–43.
Fisher, Lillian Estelle, *The Last Inca Revolt, 1780–1783*, Norman, Okla., 1966.
Fisher, Raymond H., *Bering's Voyages: Whither and Why*, Seattle and London, 1977.
Galvin, John, ed. and trans., *The First Spanish Entry into San Francisco Bay, 1775*, San Francisco, 1971.
Geiger, Maynard, OFM, *Franciscan Missionaries in Hispanic California, 1769–1848*, San Marino, Calif., 1969.
Gerhard, Peter, *Pirates on the West Coast of New Spain, 1575–1742*, Glendale, Calif., 1960.
Golder, Frank A., *Bering's Voyages, An Account of the Efforts of the Russians to Determine the Relation of Asia and America*, 2 vols, New York, 1968.
Golder, Frank A., *Russian Expansion on the Pacific, 1641–1850*. Gloucester, Mass., 1960.
Greenhow, Robert, *Memoir, Historical and Political, on the Northwest Coast of America, and the Adjacent Territories*. Washington, D.C., 1840.
Greenhow, Robert, *History of Oregon and California*, 4th edn, Boston, Mass., 1847.
Hakluyt, Richard, *The Principal Navigations, Voyages, Traffiques & Discoveries of the English Nation*, 8 vols, London and New York, 1910.
Hanna, Warren L., *Lost Harbor: The Controversy over Drake's California Anchorage*, Berkeley, Los Angeles and London, 1979.
Harbron, John H., *Trafalgar and the Spanish Navy*, Annapolis, Md., 1988.
Hayes, John D., 'The Manila Galleons', *US Naval Institute Proceedings*, Dec. 1934, pp. 1689–97.
Heizer, Robert F., *The Plank Canoe of the Santa Barbara Region, California*, Göteborg, Sweden, 1938.
Heizer, Robert F., and John E. Mills, eds, Donald C. Cutter, trans., *The Four Ages of Tsurai: A Documentary History of the Indian Village on Trinidad Bay*. Berkeley and Los Angeles, 1952.
Holmes, Kenneth L., 'Francis Drake's Course in the North Pacific, 1579', *Geographic Bulletin*, 17 (1979), pp. 5–41.

Howay, Fredrick W., ed., 'A Yankee Trader on the Northwest Coast, 1791–1795', *Washington Historical Quarterly*, 21, 1930, pp. 88–94.

Howay, Fredrick W., ed., *The Journal of Captain James Colnett aboard the* Argonaut *from April 20, 1789 to Nov. 3, 1791*, Toronto, 1940 (reprint, New York, 1968).

Humboldt, Baron Alexander von, *Political Essay on the Kingdom of New Spain*, 5 vols, John Black, trans., London, 1811 (reprint New York, 1966).

Inskeep, Edward L., 'San Blas, Nayarit: An Historical and Geographic Study', *Journal of the West*, 2, 1963, pp. 133–44.

Keen, Benjamin, ed. and trans., *The Life of the Adminral Christopher Columbus by his Son Ferdinand*, New Brunswick, NJ, 1959.

Kelsey, Harry, *Juan Rodríguez Cabrillo*, San Marino, Calif., 1986.

Kelsey, Harry, 'Finding the Way Home: Spanish Exploration of the Round-Trip Route across the Pacific Ocean', *Washington Historical Quarterly*, 17, 1986, pp. 145–64.

Kelsey, Harry, *Sir Francis Drake: The Queen's Pirate*, New Haven, Conn. and London, 2000.

Kushnarev, Evgenii G., *Bering's Search for the Strait: The First Kamchatka Expedition, 1725–1730*, ed. and trans. E. A. P. Crownheart-Vaughan, Portland, Ore., 1990.

Lamb, W. Kaye, ed., *George Vancouver, A Voyage of Discovery to the North Pacific Ocean and Round the World, 1791–1795*. 4 vols, Hakluyt Society, 2nd ser. 163–6, London, 1984.

Landin Carrasco, Amancio, *Mourelle de la Rua, Explorador del Pacifico*, Madrid, 1971.

Landström, Björn, *Sailing Ships*, New York, 1978.

McDowell, Jim, *José Narváez, The Forgotten Explorer, Including his Narrative of a Voyage on the Northwest Coast in 1788*, Spokane, Wash., 1998.

Majors, Harry T., 'The Hezeta and Bodega Voyage of 1775', *Northwest Discovery*, 1, 1980, pp. 208–52.

Makarova. Raisa V., *Russians on the Pacific*, ed. and trans. Richard A. Pierce and Alton S. Donnelly, Kingston, Ontario, 1975.

Mallis, Robert, ed., 'Earthquakes of Peru'. *Earthquake Information Bulletin*, US Dept. of the Interior Geological Survey, 7, March–April 1975, 2, pp. 8–11.

Manning, William Ray, 'The Nootka Sound Controversy', *Annual Report of the American Historical Association for the Year 1904*, Washington, D.C., 1905, pp. 279–449.

Masterson, James R, and Helen Brower, 'Bering's Successors, 1745–1780: Contributions of Peter Simon Pallas to the History of Russian Exploration toward Alaska', *Pacific Northwest Quarterly*, 38, 1, 1947, pp. 35–83; 38, 2, 1947, pp. 109–55.

Mathes, W. Michael, *Vizcaíno and Spanish Expansion in the Pacific Ocean, 1580–1630*, San Francisco, 1968.

Mathes, W. Michael, 'Apocryphal Tales of the Island of California and Strait of Anian', *California History Magazine*, 62, 1, 1983, pp. 52–59.

Menchaca, Antonio, *De California a Alaska, Vida y descubrimientos de D. Juan Francisco de la Bodega y Quadra*, Madrid, 1989.

Moziño, José Mariano, *Noticias de Nutka: An Account of Nootka Sound in 1792*, ed. and trans. Iris Higbie Wilson [Engstrand], Seattle and London, 1970.

Morison, Samuel Eliot, *Portuguese Voyages to America in the Fifteenth Century*, Cambridge, Mass., 1940.

Morison, Samuel Eliot, *Admiral of the Ocean Sea: A Life of Christopher Columbus*, Boston, Mass., 1942.

Morison, Samuel Eliot, *The European Discovery of America: The Southern Voyages, AD 1492–1616*, New York, 1974.

Mourelle de la Rua, Francisco Antonio, *Voyage of the* Sonora *in the Second Bucareli Expedition*, trans. Daines Barrington, ed. Thomas C. Russell, San Francisco, 1920.

Müller, Gerhard Friedrich, *Bering's Voyages: The Reports from Russia*, ed. and trans. Carol Urness, Fairbanks, Alaska, 1986.

Nokes, J. Richard, *Almost a Hero, The Voyages of John Mears, R.N., to China, Hawaii and the Northwest Coast*, Pullman, Wash. 1998.

Nowell, Charles E., ed. and trans., *Magellan's Voyage Around the World: Three Contemporary Accounts*, Evanston, Ill.,1962.

Nowell, Charles E., 'Arellano versus Urdaneta', *Pacific Historical Review*, 31, 1962, pp. 111–120.

Nuttal, Zelia, ed. and trans., *New Light on Drake: A Collection of Documents Relating to his Voyage of Circumnavigation, 1577–1580*, Hakluyt Society, 2nd ser. 34, London, 1914 (reprint Nendeln/ Liechenstein, 1967).

Olson, Ronald L., 'The Quinault Indians', *University of Washington Publications in Anthropology*, 6, 1936, pp. 1–190.

Olson, Wallace M., *Through Spanish Eyes, Spanish Voyages to Alaska, 1774–1792*, Auke Bay, Alaska, 2002.

O'Scanlan, Timoteo, ed., *Diccionario marítimo español*, Madrid, 1831 (Museo Naval facsimile, 1974).

Parr, Charles McKew, *So Noble a Captain: The Life and Times of Ferdinand Magellan*, New York, 1953.

Parry, J. H., *The Spanish Seaborne Empire*, New York, 1970.

Penzer, N. M., ed., *The World Encompassed and Analogous Contemporary Documents Concerning Sir Francis Drake's Circumnavigation of the World*, New York, 1969.

Publicaciones del Archivo General de la Nación, 30, 'La administración de D. Frey Antonio María de Bucareli y Ursúa, cuadragesimo sexto Virrey de Mexico', Tomo 2, Mexico City,1936.

Purchas, Samuel, *Hakluytus Postumus His Pilgrimes: Contayning a History of the World in Sea Voyages and Lande Travells by Englishmen and Others*, Glasgow, 1906 (AMS reprint, New York, 1965).

Purves, D. Laing, ed., 'A Voyage Round the World in the Years 1740–44 by George Anson', in *A Voyage Round the World by Sir Frances Drake and William Dampier*, London, 1880.

Ricketts, Edward F. and Jack Galvin, *Between Pacific Tides*, 4th edn, revised by Joel W. Hedgepeth, Stanford, Calif., 1969.

Rogers, Robert F., and Dirk Anthony Ballendorf, 'Magellan's Landfall in the Mariana Islands', *Journal of Pacific History*, 24, 1989, pp. 193–208.

Rose, Robert Selden, ed. and trans., *The Portolá Expedition of 1769–177: Diary of Vicente Vila*, Berkeley, Calif., 1911.

Schurz, William Lytle, *The Manila Galleon*, New York, 1939.

Servin, Manuel P., ed. and trans, 'The Instructions of Viceroy Bucareli to Ensign Juan Pérez', *California Historical Quarterly*, 40 Sept., 1961, 3, pp. 237–48.

Skelton, R. A., ed. and trans., *Magellan's Voyage: A Narrative Account of the First Circumnavigation by Antonio Pigafetta*, New Haven, Conn, and London, 1969.

Smith, Donald Eugene, and Frederick J. Teggart, eds and trans., *Diary of Gaspar de Portolá, during the California Expedition of 1769–1770*, Berkeley, Calif., 1909.

Smith, Donald Eugene, and Frederick J. Teggart, eds and trans., 'Diary of Gaspar de Portolá During the California Expedition of 1769–1770', *Publications of the Academy of Pacific Coast History*, 1, 1909, pp. 3–59.

Smith, Frances Rand, *The Architectural History of Mission San Carlos Borromeo, California*, Berkeley, Calif., 1921.

Thickens, Virginia E. and Margaret Mollins, eds and trans., 'Putting a Lid on California, An Unpublished Diary of the Portolá Expedition by José de Cañizares', *California Historical Quarterly*, 31, June 1952, 2, pp. 109–24; 31, Sept. 1952, 3, pp. 261–70; 31, Dec. 1952, 4, pp. 343–54.

Thorton, Walter, SJ, ed. and trans., 'An Account [by Fray Juan Riobó] of the Voyage made by the Frigates "Princesa" and "Favorita" in the Year 1799 [*sic*] from San Blas to Northern Alaska', *Catholic Historical Review*, 4 (1918), 2, pp. 222–9.

Thurman, Michael E., *The Naval Department of San Blas: New Spain's Bastion for Alta California and Nootka, 1767–1798*, Glendale, Calif., 1967.

Thurman, Michael E., 'Juan Bodega y Quadra and the Spanish Retreat from Nootka, 1790–1794', *Reflections of Western Historians*, ed. John Alexander Carroll, San Francisco, 1969, pp. 49–63.

Tovell, Freeman, M., *Bodega y Quadra Returns to the Americas*, Burnaby, BC, 1990.

Tovell, Freeman M., 'The Career of Bodega y Quadra: A Summation of the Spanish Contribution to the Heritage of the Northwest Coast', in Robin Inglis, ed., *Spain and the North Pacific Coast, Essays in Recognition of the Bicentennial of the Malaspina Expedition, 1791–1792*, Vancouver, BC, 1992.

Treutlein, Theodore E., *San Francisco Bay, Discovery and Colonization, 1769–1776*, San Francisco, Calif., 1968.

Udvardy, Miklos D. F., *The Audubon Society Field Guide to North American Birds*, New York, 1977.

United States Coast Pilot, No. 7, Pacific Coast: California, Oregon, Washington, and Hawaii, 10th edn, Washington, D.C., 1968.

United States Coast Pilot, No. 8, Pacific Coast Alaska, Dixon Entrance to Cape Spenser, 12th edn, Washington, D.C., 1969.

Venegas, Miguel, *A Natural and Civil History of California*, 2 vols, English version, trans. anonymous, London, 1759 (Readex reprint, 1966).

Wagner, Henry Raup, *Sir Francis Drake's Voyage Around the World: Its Aims and Achievements*, San Francisco, 1926.

Wagner, Henry Raup, *Spanish Voyages to the Northwest Coast in the Sixteenth Century*, San Francisco, 1929 (reprint Amsterdam, 1966).

Wagner, Henry Raup, ed., and A. J. Baker, trans., 'Fray Benito de la Sierra's Account of the Hezeta Expedition to the Northwest Coast in 1775', *California Historical Quarterly*, 9, 1930, pp. 201–42.

Wagner, Henry Raup, *The Last Spanish Exploration of The Northwest Coast and the Attempt to Colonize Bodega Bay*, San Francisco, 1931.

Wagner, Henry Raup, 'Urdaneta and the Return Route from the Philippine Islands', *Pacific Historical Review*, 13, 3, 1944, pp. 313–16.

Wagner, Henry Raup, *The Cartography of the Northwest Coast of America to the Year 1800*, 2 vols, Berkeley, Calif., 1937 (Mansfield Centre, Conn., reprint, 2 vols in one, n.d.).
Ward, Robert, 'Drake and the Oregon Coast', *Geographical Magazine*, July 1981, London.
Ward, Robert, 'Lost Harbour Found: The Truth About Drake and the Pacific', *Map Collector*, 45, Winter 1988, pp. 2–8.
Weber, David J., *The Spanish Frontier in North America*, New Haven, Conn. and London, 1992.

PART 2

THE JOURNAL OF HMS *BEAGLE* IN THE STRAIT OF MAGELLAN BY PRINGLE STOKES, COMMANDER RN 1827

Edited by
R. J. CAMPBELL

CONTENTS

List of Illustrations and Charts	144
Abbreviations	146
Note on the Text	146
Preface	147
INTRODUCTION	149
HYDROGRAPHIC HISTORY OF ESTRECHO DE MAGALLANES	155
THE EDITED TEXT OF PRINGLE STOKES' JOURNAL	179
APPENDIXES	
1. Geographical Position Book	242
2. Comparison of Geographical Positions	244
3. Canoes	245
4. Names in Estrecho de Magallanes given by Pedro Sarmiento de Gambóa still in use	247
Bibliography	249
Index	390

LIST OF ILLUSTRATIONS AND CHARTS

Figure 1: Grave of Commander Pringle Stokes. Photograph by Captain W. D. Barker. — 153

Figure 2: Olivier van Noort, chart, *Fretum Magallanicum*, 1602 (BL 455 b 10 Part 4). Reproduced by permission of the British Library. — 163

Figure 3: Joris van Speilbergen, chart, *Tijpus Freti Magellanici*, 1619 (BL 10026 df 17). Reproduced by permission of the British Library. — 164

Figure 4: John Narbrough, chart, *Streigthts of Magellan*, 1670 (published 1694) (BL Add. MS 5414 Art 29A). Reproduced by permission of the British Library. — 166

Figure 5: De Gennes, chart, *Detroit de Magellan*, 1698. Provided by the UK Hydrographic Office. — 167

Figure 6: François Frézier, chart, *Carte reduite de l'extremite de l'Amerique meridionnale dans la partie du sud*, 1717. Provided by the UK Hydrographic Office. — 168

Figure 7: John Hawkesworth, chart, *A Chart of the Straights of Magellan*, 1773 (BL 455 a 21). Reproduced by permission of the British Library. — 170

Figure 8: Louis de Bougainville, chart, *Carte du Detroit de Magellan*, 1772 (BL 4541). — 172

Figure 9: Antonio de Córdoba, chart, *Carta reducida del Estrecho de Magallanes*, 1788 (BL 302 K 15). Reproduced by permission of the British Library. — 173

Figure 10: Antonio de Córdoba, chart, *Carta reducida del Estrecho de Magallanes*, 1786 and 1789 (published 1793) (BL 302 K 15). Reproduced by permission of the British Library. — 174

Figure 11: Gerard Mercator, chart, *Exquisita & magno aliquot mensium periculo lustrata et iam retecta Freti Magellanici facies*, 1606 (BL Maps C 3 c 6). Reproduced by permission of the British Library. — 175

Figure 12: Chart 554, *The Strait of Magalhaens*, by Captain Phillip Parker King &c, 2 May 1832. Provided by the UK Hydrographic Office. — 178

Figure 13: Cape Froward bearing N25°W distance 2 Leagues with Cape Holland opening &c. Lieutenant R. H. Sholl. Provided by the UK Hydrographic Office. — 184

Figure 14: *Plan of Port Gallant*, HMS *Beagle*, 1827. Provided by the UK Hydrographic Office. — 188

Figure 15: *The Plan of Borja Bay*, HMS *Beagle*, March 1828 with *A View of the Eastern Entrance of Long Reach, in the Straits of Magellan.* Provided by the UK Hydrographic Office. — 192

Figure 16: El Morrion bearing S55°W and Cape Quod S75°W &c. Lieutenant R. H. Sholl. Provided by the UK Hydrographic Office. — 193

Figure 17: Cape Notch bearing WNW. Cape Upright bearing W by N &c. Lieutenant R. H. Sholl. Provided by the UK Hydrographic Office. — 197

Figure 18: Chart untitled [a Cove on the north shore 3 leagues to the westward of Cape Notch – Marian Cove] (from the Journal). Provided by the UK Hydrographic Office. — 201

Figure 19: *A Cove on the Northern Shore 3 Leagues to the Westward of Cape Notch*, HMS *Beagle*, 1827, [Marian Cove], with a *Plan of a Harbour on the So. Side East of Cape Valentine*, HMS *Beagle*, 1827. Provided by the UK Hydrographic Office. — 202

Figure 20: Chart of Tamar Harbour (from the Journal). Provided by the UK Hydrographic Office. — 204

Figure 21: *Plan of Tamar Isle and Harbour*, HMS *Beagle*, 1827. Provided by the UK Hydrographic Office. — 205

Figure 22: Indian of the Straits of Magellan with his sealing club. Provided by the UK Hydrographic Office. — 207

Figure 23: Fuegian woman and child with a native dwelling. Provided by the UK Hydrographic Office. — 208

Figure 24: Cape Pillar bearing SW^{b}S. Cape Cortado SE½S and Peak over Separation Harbour … S^{b}W. &c. Lieutenant R. H. Sholl. Provided by the UK Hydrographic Office. — 215

Figure 25: Evangelists bearing WNW &c. Provided by the UK Hydrographic Office. — 226

Figure 26: *Chart of the Western Part of the Strait of Magalhanes*, HMS *Beagle*, 1827. Provided by the UK Hydrographic Office. — 233

ABBREVIATIONS

BL   British Library, London
TNA   The National Archives (formerly Public Record Office), Kew
UKHO United Kingdom Hydrographic Office, Taunton, www.ukho.gov.uk

NOTE ON THE TEXT

Page references to the manuscript have been given within square brackets. The author's frequent use of dashes in the text has been standardized as follows:

Manuscript   *Published version*

. - [full point and dash]   . [full point]
; - [semi-colon and dash]   ; [semi-colon]
: - [colon and dash]   : [colon]
- [dash]   - [dash retained]

PREFACE

Note on Sources

HM Ships *Adventure* and *Beagle* were sent to survey the southern part of South America in 1826, under the command of Phillip Parker King, Commander[1] (who, between 1817 and 1822, had charted most of the north-west coast of Australia), with Commander Pringle Stokes in command of *Beagle*. On arrival they started in Estrecho de Magallanes (Strait of Magellan). In January 1827 Commander Stokes was directed to carry out a survey of the western part of the strait. The following journal was rendered to Commander King, by Commander Stokes on his return, who forwarded it to the Hydrographer, together with the charts made during the survey.

The journal is still in the possession of the United Kingdom Hydrographic Office. It is marked OD $\frac{SA2}{2}$ but is now bound in OD 18. It is 12.625 inches by 8 inches (321 mm × 204 mm), has a soft cover and contains 4 blank sides, 47 numbered text pages, 6 unnumbered pages giving advice on rounding Cape Horn (Cabo de Hornos), and meteorological tables with 3 final blank pages. Additionally there are 2 charts, 5 views and 2 pictures of Fuegians which have been inserted with blank versos; the views and pictures are full-size pages while the charts are smaller.

Names

Names are given using current conventions, i.e. those used by the country having sovereignty. Stokes used the English forms throughout his journal and these have been retained and where his remarks on a feature are referred to, the name he used is given with the national name in brackets, e.g. Cape Horn (Cabo de Hornos) or in a footnote. Both forms are indexed.

Acknowledgements

I am most grateful to the United Kingdom Hydrographic Office for permission to publish this journal and its accompanying charts, and to the British Library and The National Archives for their respective permissions to publish items from their collections. Crown copyright material is published by permission of the Controller of

---

[1] King was promoted Commander on 7 July 1821, and Captain on 25 January 1830; Stokes was promoted Commander on 26 December 1825. King was thus the senior officer of the expedition and a Captain when he rendered the results of the survey and published his account. In the 19th century it was customary to refer to the commanding officer of a naval vessel as Captain, regardless of his actual rank. Thus King frequently refers to Pringle Stokes as Captain Stokes.

Her Majesty's Stationery Office and the United Kingdom Hydrographic Office (www.ukho.gov.uk).

I am also grateful to the staffs of the above offices for their assistance in my research which has been invaluable, and to Lieutenant Commander Andrew David who first suggest the project and has provided considerable assistance and advice on the way; to Ms Jennifer Bryant and Ms Kathie Way of the Natural History Museum, London, and to Dr R. Prys-Jones of the Natural History Museum, Tring, who assisted with information on seaweeds, shells and birds respectively; to Captain W. D. Barker for the photograph of Commander Stokes's grave and permission to use it; to Mr T. N. Snow who translated a number of items from Spanish for me, and finally to Professor Michael Brennan and Professor Will Ryan who edited the work for the Hakluyt Society. I would add my apologies to anyone else I ought to have mentioned, but have not done so, and say that I am none the less most grateful for their help.

INTRODUCTION

In 1825 the Lords Commissioners of the Admiralty ordered that two ships be prepared for a survey of the south coasts of South America. These were the *Adventure* and *Beagle*. The senior officer, Commander Phillip Parker King, received the following orders:

> Whereas we think fit that an accurate Survey should be made of the Southern Coasts of the Peninsula of South America, from the southern entrance to the River Plata, round to Chiloé; and of Tierra del Fuego; and whereas we have been induced to repose confidence in you, from your conduct of the Surveys in New Holland; we have placed you in command of His Majesty's Surveying Vessel the Adventure; we have directed Captain Stokes, of His Majesty's Surveying Vessel the Beagle, to follow your orders.
>
> Both these vessels are provided with all the means which are necessary for the complete execution of the object above-mentioned, and for the health and comfort of their Ship's Companies. You are also furnished with all the information, we at present possess, of the ports which you are to survey; and nine Government Chronometers have been embarked in the Adventure, and three in the Beagle, for the better determination of the Longitudes.
>
> You are therefore hereby required and directed, as soon as both vessels shall be in all respects ready, to put to sea with them ....[1]

They sailed from Plymouth on 22 May 1826 and reached Rio de Janeiro on 10 August and, after a stay in Río de la Plata, sailed for the survey on 19 November the same year. King wrote:

> According to my instructions, the survey was to commence at Cape San Antonio, the southern limit of the entrance of the Plata; but for the following urgent reasons I decided to begin with the southern coasts of Patagonia, and Tierra del Fuego, including the Straits of Magalhaens.[2] In the first place, they presented a field of great interest and novelty; and secondly, the climate of the higher southern latitudes being so severe and tempestuous, it appeared important to encounter its rigours while the ships were in good condition – while the crews were healthy – and while the charms of a new and difficult enterprize had full force.[3]

They passed Cabo Vírgenes and having made contact with the local population proceeded to Puerto San Juan de la Posesión (Puerto del Hambre or Port Famine) where they anchored and 'Captain Stokes received orders to prepare the Beagle for examining the western part of the Strait'.[4]

---

[1] King, *Narrative*, I, pp. xv–xvi.
[2] [Note in original] 'Commonly called Magellan'.
[3] King, *Narrative*, I, p. 1.
[4] Ibid. p. 26.

The *Beagle* carried out necessary repairs, embarked wood and water and proceeded on her task on 15 January 1827. The journal published here was rendered to Commander King by Commander Stokes on his return.

It is difficult to assess the results of this survey. Captain Parry, Hydrographer,[1] wrote, in answer to a query about the work of the *Adventure* and *Beagle*, from Admiral Sir William Hope, one of the Council of His Royal Highness,[2] the Lord High Admiral: 'We have received a General Chart (in two sheets) of the Straits of Magellan, and Plans of Port Famine, Port Gallant and Tamar Harbour, the two latter accompanied an interesting Journal by Capt Stokes, together with a few views of lands and Natives.'[3] The plans of Port Gallant[4] and Tamar Isle and Harbour[5] and the Chart of the Western Part of the Strait of Magalhanes[6] are given at Figs 14, 21 and 26. A further plan of Borja Bay[7] (surveyed on the next expedition in 1828) is shown at Fig. 15 since it embodies a view directly based on Fig. 16 in Stokes's journal: two further plans of 'A Cove on the Northern Shore 3 leagues to the Westward of Cape Notch and Plan of a Harbour on the So. Side East of C. Valentine',[8] which were made during the 1827 survey are shown at Fig. 19.

The 'Chart of the Western Part of the Strait of Magalhanes' is marked 'Superseded by Capt King's (E 830a) last charts'.[9] Indeed Stokes himself produced a further chart of the same area the following year,[10] building on his chart of 1827, and both were used to produce the final version: Chart 554, *The Strait of Magalhaens* published by the Hydrographical Office of the Admiralty, 2 May 1832, which states that it was surveyed 'in His Majesty's Ships Adventure and Beagle by Captain Phillip Parker King, R.N., F.R.S., &c. 1826, 1830' (Fig. 12).

The geographical positions obtained were detailed in the Geographical Position Book[11] (Appendix 1) which has been marked in red to indicate that a shift in position was necessary due to a redetermination of the position of the Harbour of Mercy by back runs, since the time interval was shorter and there was less alteration in the rate of the chronometer. There is also a note at the end: 'Sir, The rate recalculated by Capt Fitzroy's latitude (52°44′57″)'. After the calculations the mean difference between Harbour of Mercy and Port Tamar is given. Captain King also rendered an Observation Book[12] at the end of the survey which gives details of chronometer errors, determinations of meridian distances between various places, determinations of latitude and longitude together with calculations etc. to indicate how they were obtained; relevant extracts are given together with modern accepted positions in

---

[1] Captain Sir William E. Parry, Kt., FRS, Royal Navy, Hydrographer 1823–29.
[2] William, Duke of Clarence (later King William IV) was Lord High Admiral from May 1827 to August 1828, advised by a Council of Admirals.
[3] UKHO Remark Book No. 1, November 1825 to 1832, p. 157.
[4] UKHO South America folio 2, G 232.
[5] UKHO South America folio 2, G 246.
[6] UKHO Shelf Fs E 497.
[7] UKHO South America folio 2, E 831a.
[8] UKHO South America folio 2, G 235.
[9] The UKHO Chart Catalogue describes this as 'Magellan Strait and part of Tierra del Fuego (no record but probably compilation by surveyors)'.
[10] UKHO Shelf Fs E 831.
[11] UKHO Geographical Position Book No. 80.
[12] UKHO Geographical Position Book No. 30.

Appendix 2, from these it can be seen that the mean difference in Latitude is 1′37″ and Longitude 3′19″ – accuracy of a very high standard indeed.

The Hydrographical Office published a set of sailing directions by Captain King in 1832. This draws on Commander Stokes's journal for a certain amount of its information (which is quoted in footnotes to the text of the journal). Stokes himself does not appear to have produced sailing directions to accompany his survey; no doubt this was planned for production when the survey was complete and never written.

The journal therefore covers a relatively modest part of a much larger survey. The work appears to have been well done and, with only minor positional adjustments, was embodied in the final product. Captain Beaufort, who had recently been appointed Hydrographer, in his report on 'Commander King's letter of 15 October, 1830', forwarding the results of the survey, said:

> The short sketch he has given (in enclosure No 1) of his transactions, and the general results of the voyage, appear to be more than borne out by the Charts, plans, views, and Chronometric Journals delivered to this office; – and there can be no hesitation in saying that when the survey is all put together, it will be acknowledged to be one of those which have eminently contributed to the credit of this country and to that of all the officers employed in it.[1]

The principal Admiralty Chart that resulted from this survey was numbered 554, titled *The Strait of Magalhaens commonly called Magellan*, published 2 May 1832 (Fig. 12). The last new edition of this chart was published in 1921 and its title states *The Main Strait from Cape Virgins to Cape Pillar from Surveys by; Captain Richard C. Mayne C.B., and the Officers of H.M.S. Nassau, 1867–8; Captain J. F. L. P. Maclear and the Officers of H.M.S. Alert 1880, Captain W. J. L. Wharton and the Officers of H.M.S. Sylvia, 1882–4. The remainder by Captain P. P. King and the Officers of H.M.S. Adventure and Commanders Pringle Stokes and Robert Fitz-Roy and the Officers of H.M.S. Beagle, 1826–30.* It was in use (with numerous corrections), with this title, until 1952, but it was not until 1964 that a new chart was published based on modern Argentinean and Chilean surveys.

A full account of the voyage is given in King and FitzRoy, *Narrative*, and an informal account of the first season in Estrecho de Magallanes in Macdouall, *Narrative*, from both of which quotations are given, so that no further details are necessary here.

Commander Pringle Stokes

Stokes was the son of Charles and Elizabeth Stokes. He was born 23 April 1793, and baptized at Chertsey on 2 May that year. He joined the Royal Navy 5 June 1805, as a Midshipman in HMS *Ariadne* (20), Captain Arthur Farquhar, serving in the North Sea; February, 1809, HMS *Orion* (74), Captain Sir Archibald Dickson, in the Baltic and Walcharen expedition; October 1809, HMS *Desiree* (36), Captain Arthur Farquhar, North Sea, during which period he passed for lieutenant at Sheerness, 5 August 1812; January 1813, HMS *Daedalus* (38), Captain Murray Maxwell, passage to India (the

---

[1] UKHO Minute Book No. 1, November 1825 to 1832, p. 316.

Daedalus was wrecked off Ceylon July 1813); July 1813, HMS *Minden* (74), Admiral Sir Samuel Hood, East Indies; he was promoted lieutenant and joined HMS *Leda* (36), Captain George Sayer, July 1815 remaining on the East Indies station; then as a supernumerary lieutenant in HMS *Alpheus* (36), Captain George Langford and HMS *Tyne* (26), Commander J. B. H. Curran, June 1816 to January 1817, for passage to Mauritius and thence to England. He was on half-pay until July 1821, when he joined HMS *Iphigenia* (42), Captain Sir Robert Mends, on the Africa Station, and was appointed to command HMS *Snapper* (9) on 13 March 1822, by Sir Robert Mends, and relieved by Lieutenant T. H. Rathery on 13 June the same year. *Snapper* was working off Sierra Leone, the Gambia and Senegal, while Stokes was in command.[1] He rejoined Sir Robert Mends, transferring with him to HMS *Owen Glendower* (42), 10 November 1822. He was promoted Commander in December 1823 when he again went on half pay until his appointment to HMS *Beagle*, (6), 7 September 1825.[2]

After the expedition which is the subject of the accompanying journal the ships returned to Rio de Janeiro returning to the Strait in January, 1828. Stokes was ordered to take the *Beagle* and survey the west coast of South America between the Strait and 47°S, with orders to 'return to Port Famine by the 24th July'.[3] The *Beagle* returned on 27 July, and King states 'but on passing our stern Lieutenant Skyring informed me that Captain Stokes was confined to his cabin by illness, and could not wait on me. I therefore went to the Beagle, and found Captain Stokes looking very ill, and in low spirits.' He appeared to improve to a state when, on 31 July, King was considering sending the *Beagle* off on another surveying expedition, 'the officers, however, knew more of the diseased state of his mind than I did; and it was owing to a hint given to me, that I desired Mr. Tarn to communicate with Mr Bynoe,[4] and report to me whether Captain Stokes's health was sufficiently restored to enable him to commence another cruize'.[5] It was while this report, which was very unfavourable, was being considered that the news was brought to King that 'Captain Stokes, in a momentary fit of despondency, had shot himself'.[6] He lingered in great pain and died on 12 August 1828, and was buried ashore. His grave is now marked by a simple white cross (maintained by the Chilean Navy) inscribed *In Memory of Commander Pringle Stokes H.M.S. Beagle Who died From the effects of the anxieties and hardships incurred while surveying the western shores of Tierra del Fuego.* † *12.8.1828* (Fig. 1). The original cross is in the Salesian Museum at Punta Arenas.

Lieutenant William G. Skyring

Skyring had served in HMS *Aid* (6), Captain W. H. Smyth, surveying in the Adriatic (1820), and with Commander Hewett, HMS *Protector*, on the east coast of England (1824) before his appointment to *Beagle*. He took command on Commander Stokes's

---

[1] Logs of HMS *Iphigenia*, TNA ADM 51/3457 and HMS *Snapper*, TNA ADM 54/3231.
[2] TNA ADM 107/44, f. 541a, ADM 9/14, item 5060, *Navy Lists*, Colledge, *Ships*.
[3] King, *Narrative*, I. p. 124.
[4] The surgeons of *Adventure* and *Beagle*. Bynoe had become surgeon of *Beagle* when Bowen was invalided home at the end of 1827
[5] King, *Narrative*, I. p. 152,
[6] Ibid. p. 153.

Figure 1: Grave of Commander Pringle Stokes. Photograph by Captain W. D. Barker.

death until relieved by Commander FitzRoy on the orders of the Commander in Chief, Rear Admiral Sir Robert Otway, KCB, in Rio de Janeiro, 13 November 1828, when he reverted to his former position and remained in *Beagle* until he was promoted commander, 25 February 1830. After a period on half-pay he took command of HMS *Ætna* (6), 20 September 1833, surveying off the west African coast where he was killed by the natives that same year.[1]

Lieutenant Robert H. Sholl

Sholl, the artist of Commander Stokes's journal, joined the *Adventure* as mate. When Lieutenant E. Hawes, of the *Beagle* was invalided, in September 1826, he was promoted in his place, and this appointment was confirmed by the Commander in Chief, Rear Admiral Sir George Eyre, KCB. Sholl died after an illness of ten days in January 1828 and was buried in Puerto San Julián.[2]

---

[1] Dawson, *Memoirs*, p. 107; King, *Narrative*, I, pp. 182 and 188; *Navy Lists*.
[2] King, *Narrative*, I, pp. xii–xiii and 121.

HMS *Beagle*

The *Beagle* was 235 tons, barque rigged – i.e. three masts, square rigged on fore and main, and fore-and-aft on mizzen – carrying six guns.

Ship's Company

Pringle Stokes	Commander and Surveyor
W. G. Skyring	Lieutenant and Assistant Surveyor
R. H. Sholl	Lieutenant
S. S. Flinn	Master
E. Bowen	Surgeon
J. Atrill	Purser
J. Kirke	Mate
B. Bynoe	Assistant Surgeon
J. L. Stokes	Midshipman
R. F. Lunie	Volunteer 1st Class
W. Jones	Volunteer 2nd Class
J. Macdouall	Clerk
Carpenter	

Sergeant and 9 marines and about 40 seamen and boys.[1]

---

[1] Ibid. pp. xii–xiii.

HYDROGRAPHIC HISTORY OF ESTRECHO DE MAGALLANES (STRAIT OF MAGELLAN)

> Apres allant et prenant la voye au cinquante et deuxiesme degre audit ciel antarticque le iour de la feste des onze mille vierges nous trouuasmes par miracle vng estroiet que nous appellasmes le Cap des onze mille vierges.
> (After going and taking the course to the fifty second degree of the said sky Antarctic, on the day of the festival of the Eleven Thousand Virgins,[1] we found by a miracle, a strait which we called the Cape of Eleven Thousand Virgins.)

Antonio Pigafetta of Vicenza, Knight of Rhodes, wrote the history of the voyage of Ferdinand Magellan and the discovery of the Strait which bears his name, from which the above is taken.[2] He sailed from San Lúcar on 20 September 1519, with five ships, only one of which, the *Victoria*, having circumnavigated the globe, returned, under the command of Juan Sebastian del Cano, arriving off San Lúcar on 6 September 1522.

> The land of this strait on the left hand side looked towards the Sirocco wind, which is the wind collateral to the Levant and South; we called the strait Pathagonico. In it we found at every half league a good port and place for anchoring, good waters, wood all of cedar, and fish like sardines, missiglioni, and a very sweet herb named appio (celery).[3]

There is little more description of the Strait and the chart shows a relatively straight east–west channel between two land masses, with a number of islands in the middle and at both ends. The account of a Portuguese companion of Odoardo Barbosa states: '… we found ourselves in a strait, to which we gave the name Strait of Victoria, because the ship *Victoria* was the first that had seen it: some call it the Strait of Magalhaens, because our captain was named Fernando de Magalhaens'.[4] Diogo Ribeiro[5] in his planispheres uses the name 'estrecho de fernam de magallanes' and, despite a

---

[1] 21 October, 1520. The feastday of Saint Ursula and The Eleven Thousand Virgins. See also p. 181, n. 4.

[2] There are other accounts extant, which, together with Pigafetta's account, are covered in Stanley, *First Voyage*. Pigafetta's text, in facsimile, from the manuscript in the Beinecke Rare Book and Manuscript Library of Yale University, together with a translation are given in Pigafetta, *Magellan's Voyage*.

[3] Stanley, *First Voyage*, p. 61.

[4] Ibid., p. 31.

[5] Diogo Ribeiro was a native of Portugal working for Charles V in Seville and involved in the preparation for Magellan's voyage. A letter dated 18 July 1519 from the Portuguese ambassador in Seville to King Manuel (quoted in Cortesão, *Portugaliae monumenta*, p. 87.) states 'he makes the compasses, quadrants and spheres, but he does not go with the expedition [Magellan's]'. He was appointed Cosmographer to the Casa de la Contratación 10 July, 1523 and died in Seville 16 August 1533.

subsequent attempt to change it,[1] the Strait has been known by Magellan's name ever since. The names given during this first passage that have remained in use are few – Cabo Vírgenes (east end), and Cabo Deseado (west end), while Tierra del Fuego is shown on Diogo Ribeiro's planispheres (as Tierra de los Fuegos) and apparently derives from an instance described in the letter of Maximillianus Transylvanus about the voyage – 'But one night a great number of fires were seen, mostly on their left hand, from which they guessed that they had been seen by the natives of the region.'[2] A mountain was named Campana de Roldan after the gunner, Roldan de Argote. This was identified by Sarmiento and shown on Admiralty charts,[3] adjacent to Bahía Campana (now Bahía Bell).

A second expedition was sent out by Spain under Comendador Garcia Jofre de Loaysa with Sebastian del Cano as his second in command and chief pilot, to relieve the small group of five Spaniards left by the first expedition to establish a factory at Tidore, in the Moluccas. They sailed on 24 July 1524, from La Coruña. The fleet was scattered after passing through Estrecho de Magallanes, both Loaysa and del Cano

---

The King ordered, by royal *cédula*, dated 22 March 1508 and again 24 July 1512 that a *padrón real* be kept up to date in the Casa de la Contratación. The information brought home by Sebastian del Cano in 1522, would have been included, probably by Diogo Ribeiro (Cortesão, *Portugaliae monumenta*, pp. 90–91).

There are five planispheres extant, all in the same hand, attributed by Cortesão to Diogo Ribeiro, two of which carry his name. These are all detailed and reproduced in Cortesão, *Portugaliae monumenta*, vol. I: 1525, pp. 95–8, Pl. 37, Archivio Marchesi Castiglioni, Mantua; 1527, pp. 99–101, Pl. 38, Thüringische Landesbibliothek, Weimar; 1529, pp. 101–3, Pl. 39, Biblioteca Vaticana, Rome; 1529, pp. 104–6, Pl. 40, Thüringische Landesbibliothek, Weimar; 1532 (western part only including America) pp. 107–9, Pl. 41, Herzog August Bibliothek, Wolfenbüttel. The names on these are all similar, but not identical.

The 1525 chart (Mantua) contains no legend along the top or bottom. The 1527 chart (Weimar) reads along the top and bottom: *CARTA VNIVERSAL EN QVE SE CONTIENE TODO LO QVE DEL MVNDO SE A DESCVB(IERTO)* [part torn away] *FASTA AORA HIZOLA VN COSMOGRAPHO DE SV MAGESTAD ANNO M.D.XX.VII. EN SEVILLA* (Universal chart in which is contained all that until now has been discovered in the world. A Cosmographer of His Majesty made it in the year 1527 in Seville.) The 1529 chart (Vatican) reads: *Carta Universal En que Se contiene todo lo que del mundo Se ha descubierto fasta agora. Hizola Diego Ribero Cosmographo de Su magestad: Año, de: 1529 e Sevilla:- La Qual Se deuide en dos partes conforme Ala capitulaçion que Hizieron los catholicos Reys de españa & el Rey don Juan de portugal En Tordesillas: Año de, 1494* (Universal chart in which is contained all that has been discovered in the world until now. Diogo Ribero Cosmographer of his Majesty made it, the year 1529 at Seville, which is divided into two parts according to the capitulation which took place between the Catholic Kings of Spain & King John of Portugal at Tordesillas in the year 1494). The 1529 Weimar chart has a very similar, but not identical legend. The 1532 chart (Wolfenbüttel) has no legend along the top or bottom, but adjacent to the southern tip of South America is written *TIERA DE PATAGONES./ Toda esta terra descubrio fernam de magallaes el año/ de: 1520: Donde allo estrecho por donde paso a ma/luco: toda esta tierra es muy esteril & de negum pro/ Recho los Indios son hombres de grandes cüerpos/ casy gigantes* (All this land was discovered by Ferdinand Magellan in the year 1520 where he found the Strait through which he passed to the Moluccas. All this land is very bare and of no profit at all: the Indians are men of big bodies, almost giants.).

[1] See below pp. 158–9.

[2] Stanley, *First Voyage*, p. 196.

[3] Lat. 53°58'S, long. 71°45'W. Height 2780 feet. King, *Narrative*, I, p. 27n, believed the name was originally given to the mountain which he had called Mount Sarmiento (lat. 54°27'S, long. 70°50'W). It is shown on Diogo Ribeiro's planisphere opposite the southernmost point of South America (Cabo Froward), on the south shore. King showed it as Roldan's Bell on his chart and it appeared on Admiralty charts until 1964. It is now represented on Admiralty Chart 1281 by a spot height of 2,772 feet. King calculated the height of Monte Sarmiento to be 6,800 feet (now given as 2,234 metres). It is permanently covered in snow and frequently enveloped in cloud.

died before their ship reached the Moluccas. They suffered various losses fighting the Portuguese, who were not prepared to surrender their monopoly of the spice trade. In 1529, when the news reached them that Charles V had sold his rights in the Moluccas to the Portuguese, the remnant of the crews surrendered and were returned home. The principal achievement of this expedition, with regard to Estrecho de Magallanes, was a comprehensive set of sailing directions written by the pilot Martin de Uriarte.[1]

In 1539 the Bishop of Plasencia, brother in law of the Viceroy of Peru, Don Antonio de Mendoza, equipped an expedition of three small vessels, under Don Alonso de Camargo, to open up communication between Spain and the Chilean ports. They entered the Strait, but Camargo's ship was lost at the entrance to the first narrows. He transferred to another vessel and managed to reach Callao. No record of his transit of the Strait appears to have survived. The third vessel returned to Spain having been separated from her consort and spent the winter in the south of Tierra del Fuego, possibly in Canal Beagle (Beagle Channel), apparently passing through Estrecho de Le Maire (Strait of Le Maire). A fragment of a journal survives.[2]

In 1554 Don Garcia de Hurtado became governor of Chile. In 1557 he despatched Juan de Ladrilleros with two ships to examine the southern coast of Chile and Estrecho de Magallanes – one returned to Valdivia with few of her crew remaining while Ladrilleros pressed on. He reached the Strait and wintered there but when he returned to Chile he had only two men left.[3]

The next significant voyage was that of Francis Drake who sailed round the world (1577–80), passing through Estrecho de Magallanes. On 30 January 1578, Drake captured a Portuguese vessel off the SW coast of São Tiago bound for Brazil. Her crew, except the Master, Nuño da Silva, who proved to be an experienced pilot for the coast of Brazil, were set ashore. Da Silva was retained and not set ashore until April 1579, at Guatulco. His expertise appears to have been a major factor in Drake's successful passage through the Strait.[4] The sailing directions available to Drake[5] give little more information than the courses and distances between 'Cape de las Virgenes', 'Cape del Estrecho' (the southernmost part of the Strait in the vicinity of Cabo Froward) and the 'mouth of the Strait' (Cabo Pilar).[6] No new charts or directions appear to have survived from this expedition.

The result of Drake's depredations in the Pacific was the mounting of a force of two vessels under Pedro Sarmiento de Gambóa, which was sent in pursuit, but failed to find him. It appears to have been generally believed that Drake would winter on the coast and then try to return through Estrecho de Magallanes. The Viceroy of New Spain, Don Martin Enriquez, wrote to King Philip II, 26 April 1579:

> it will be necessary for Your Majesty to give orders to have that passage safeguarded. I am told that, in this Strait, there is a place where a fortification could be built without great

---

[1] Translated and published in Markham, *Early Spanish Voyages*, pp. 90–101.
[2] Ibid. pp. 157–68.
[3] Burney, *Chronological History*, I, pp. 247–9.
[4] Nuttall, *Drake*, p. 114. Report by Captain Juan Solano Lieutenant Governor of the Province of Costa Rica, to the Licentiate Valverde, Presidente of the Audiencia of Guatemala, dated 29 March 1579.
[5] BL Harley MS 167, ff. 39–72. Given in Taylor, *Troublesome Voyage*, pp. 298–306.
[6] Taylor, 'The Dawn of Navigation', pp. 284–5.

difficulty. This was told me by an Augustine friar who had much intercourse with Urdaneta, a friar of his order, who was a mariner and passed through that Strait.[1]

The Viceroy of Peru, Don Francisco de Toledo, then sent Sarmiento to intercept Drake in Estrecho de Magallanes and to examine it and search for a place to fortify it.[2] He set out on 11 October 1579, in two ships with Juan de Villalobos as his second in command. They examined the islands and channels north of the entrance to the Strait in three boat journeys after which Villalobos became separated in a storm and returned while Sarmiento continued through the Strait and went on to Spain, where he arrived on 19 August 1580. There the king agreed to an expedition to fortify the Strait and appointed Don Diego Flores de Valdes in command: Sarmiento was to be Governor General of the forts and settlements in the Strait. The fleet finally departed Cádiz on 9 December 1581. After great difficulties, Valdes returned to Spain from Rio de Janeiro taking most of the stores and food with him but leaving Diego de la Ribera and Sarmiento with five vessels. They assembled what stores they could locally, to replace those taken back to Spain, and sailed for the Strait on 2 December 1583. They entered it on 1 February 1584, landed and established Ciudad del Nombre de Jesus on the north shore fourteen leagues from the first narrows. At this stage Diego de la Ribera departed with three ships (one, the *Trinidad*, had already been wrecked) leaving only one, the *Maria*. Sarmiento sent her into the Strait to the point of Santa Ana and took a party round by land and there founded Ciudad del Rey Don Felipe. Sarmiento then embarked in the *Maria* to return to Ciudad del Nombre de Jesus but on arrival a terrific storm drove him out to sea so that it was with great difficulty that he finally got back to Rio de Janeiro. It was by now July 1584. There followed a number of abortive attempts to relieve his men, loss of stores, mutiny and shipwreck. Finally the Spanish South American authorities refused to provide further assistance without direct orders from the King, and Sarmiento decided to returned to Spain. He left Bahia on 22 June 1586 and on 11 August was captured by Sir Richard Grenville and taken to Plymouth. He was captive in England until September when he was taken to Queen Elizabeth who provided him with a passport to return to Spain and 1,000 escudos. He departed on 30 October 1586, and was in Paris in November; thence to Bordeaux, where he was captured by the French and held to ransom. He finally returned to Spain where he submitted his report to the King dated 15 December 1589.

In his report he states that his charts and papers were thrown overboard before his capture by the English and that his remaining papers were seized when he was taken by the French, however the journal of his first passage through the Strait survives[3] and contains a detailed description of the passage. His charts have not been located if they are still extant. He collected native names where he could and recorded them and gave his own names to other features, about forty of which are in use today.[4] He altered the

---

[1] Nuttall, *Drake*, p. 222.

[2] Markham, *Narratives of the Voyages*, p. xxiv, states that Sarmiento's journals are in the Royal Library in Madrid, but that no trace could be found of his charts. The journal, published in 1768, in Madrid, contains a map dated 1769. See also note below.

[3] Published as *Viage al Estrecho de Magallanes Por el Capitan Pedro Sarmiento de Gambóa en los años de 1579 y 1580 y Noticia de la Expedicion Que despues hizo para poblarle*, Madrid, 1768; translated and edited, Markham, *Narratives of the Voyages*.

[4] See Appendix 4.

name of the Strait to Estrecho de la Madre de Dios,[1] which has not been retained and Cabo Vírgenes to Cabo Virgen María.[2] This latter (as Cape Virgin Mary) was used by most English mariners for many years until Magellan's original name was restored in the early nineteenth century.

The next voyage was that of Thomas Cavendish[3] (Candish), who sailed from Plymouth in the *Desire*, 120 tunnes, *Content*, 60 tunnes and the *Hugh Gallant*, 40 tunnes, with 123 men, on 21 July 1586. They reached Estrecho de Magallanes on 6 January, 1587 where they found the last survivors of Sarmiento's settlers. Cavendish went ashore in a boat and picked up Tomé Hernandez saying he would return and pick up the remaining seventeen Spaniards[4] (Hakluyt has twenty-three[5]), but when the wind turned fair he continued on his way and left them. The ships stopped at Ciudad del Rey Don Felipe, which was deserted and recovered the Spanish ordnance. He renamed it Port Famine. The ships anchored at night and proceeded by day, sounding with a boat ahead.[6] They named Cape Froward and a number of other places of which it would appear only Port Gallant (after the *Hugh Gallant*) has been retained.[7] They entered the Pacific on 24 February and, having pillaged and burnt their way along the coast, captured, among other vessels, a great ship of the king which was said to contain 122,000 pezos of gold together with 'silks, satins, damasks, with much and divers other marchandize'.[8] Only one ship returned to Plymouth by way of the Cape of Good Hope, on 9 September 1588.

Hakluyt prints a set of sailing directions written by M. Thomas Fuller, the master of the *Desire*, from this voyage. They include a section on Estrecho de Magallanes giving the latitudes of Cape Joy (Cabo Vírgenes), Port Famine (Puerto San Juan de la Posesión or Puerto del Hambre), Cape Froward (Cabo Froward) and Cape Desire (Cabo Deseado), together with notes on the courses and distances between the salient points.[9]

---

[1] Sarmiento de Gambóa, *Viage al Estrecho de Magallanes*, pp. 239–40.

[2] Ibid, p. 274.

[3] Candish is the form of the name used by Hakluyt who published an account of the voyage. Hakluyt, *Principal Navigations*, III, pp. 803–39. He himself seems to have used Caundyshe, Caundysh and Caundysshe. Quinn, *Last Voyage*, pp. 3 & 5.

[4] Markham, *Narratives of the Voyages*, pp. 364–6. Hernandez remained on board until the fleet arrived at Quintero, 30 March, 1587, when a foraging party was intercepted by some Spaniards with whom he acted as interpreter. He was then sent a second time to communicate with the Spaniards and escaped. He subsequently lived in Peru and gave evidence before the Viceroy of Peru, Don Francisco de Borja, Prince of Esquilache, on 21 March, 1620.

[5] Hakluyt, *Principal Navigations*, III, p. 806.

[6] Markham, *Narratives of the Voyages*, p. 367. Statement by Hernandez.

[7] Hakluyt, *Principal Navigations*, III, pp. 806–7, states they named Muscle Cove, probably Caleta Agua Dulce; Elizabeth Bay now Bahía Isabel; the Channel of Saint Jerome is also mentioned.

[8] Ibid. p. 837.

[9] Ibid. pp. 825–9.

	[Latitude]	[Modern Latitude]
Cape Joy	52°40'S	[52°20'S]
Port Famine	53°50'S	[53°38'S]
Cape Froward	54°15'S	[53°54'S]
Cape Desire	53°10'S	[52°44'S Cabo Pilar]
		[52°45'S Cabo Deseado]

Item, Comming from the Northwards, you shall see before you come to Cape Joy, a very long beach, about the length of 8 leagues, being 5 leagues short of the cape unto the Northwards. Also unto the Southwards of the cape, you shall see another beach, about a league long, adioyning hard under the cape; about which beach

There are also extant two manuscript charts, circa 1588, with insets of the Strait.[1]

The *Delight* of Bristol (Captain Andrew Merick) sailed for the Strait on 5 August 1589, and having been separated from her consorts, entered Estrecho de Magallanes about 1 January. They reached Puerto San Juan de la Posesión (Puerto del Hambre or Port Famine) where they picked up a survivor of Sarmiento's people who had been living alone there, and progressed ten leagues westward of Cabo Froward before being forced to return by the weather on 14 February. They got back to Monville de Hage, eight miles west of Cherbourg, in Normandy with only six of their company left alive.[2]

Thomas Cavendish attempted to repeat his successful circumnavigation leaving Plymouth on 26 August 1591, with three tall ships, the *Galleon Leicester*, *Desire* and *Roebuck* together with two barks. Having entered Estrecho de Magallanes on 8 April 1592, they reached a small bay on the south shore four leagues westward of Cabo Froward on 21 April and remained there until 15 May when, being unable to get westward and reluctant to winter in the Strait, they returned and were separated on the way towards Brazil. Cavendish decided to return home and died on the way.[3] The *Desire* (Captain John Davis), having wintered in Port Desire and expecting Cavendish would return, sailed for the Strait on 7 August 1592. He was driven by a storm off shore and on 14 August discovered the Falkland Islands.[4] He returned to the Strait on 18 August and by 25 August had managed to get to a harbour within fourteen leagues of the Pacific where he waited two weeks for Cavendish and then decided to enter the Pacific

---

is the entrance of the Streights of Magellan, the which Straights are in breadth six leagues over, from the cape unto the South shore, lying South and by East.

Item, From Cape Joy, being the entrance of the streight of Magellan, unto the first narrow passage of the said streight; the course is West and by North, and East and by South, and are distant 18 leagues; the land being in breadth from the one side to the other one league.

Item, From the first narrow unto the second narrow passage, the course is West & by South, and East and by North; and the distance is 12 leagues: and in breadth the one side is from the other about two leagues over.

Item, From the second narrow unto the islands that be called Elizabeth, Bartholemew, and Penguin islands, the distance is 5 leagues, and the course is Southwest and Northeast: the islands being distant a league and halfe the one from the other. [Names given by Drake.]

Item, From the sayd islands unto Port Famine is 16 leagues: the course is Southsouthwest, and Northnortheast. Moreover, from Port Famine unto Cape Froward, the course is South and by West, and North and by East: and they are distant 8 leagues asunder.

Item, From Cape Froward unto S. Jeromes river, is 16 leagues: the course is Northwest and Southeast. Also from S. Jeromes river unto the uttermost land on the South side, the which is called Cabo Deseado, the course is Northwest & somewhat to the Northward, and are distant 30 leagues. So the whole length of the streight of Magellan is 105 leagues.

[1] One in the Biblioteca Nazionale Centrale, Florence (Portolano 30), and the other in the Algemeen Rijksarchief, The Hague (Leupe Inventory 733). These charts are described and reproduced in Quinn, *Last Voyage*, pp. 150–59. The names on the latter appear to be in Thomas Cavendish's own hand. The two charts are not quite the same scale, but are about 1:1,000,000, the insets are about 1:260,000 (Florence) and 1:300,000 (The Hague). The latitudes are in broad agreement with Thomas Fuller's sailing directions. Professor Quinn remarks that both are based on lost Spanish originals, and probably date from 1588 or shortly thereafter.

[2] Hakluyt, *Principal Navigations*, III, pp. 839–42.

[3] His account of the voyage, in his own hand, written at sea before he died, is published in facsimile with transcript in Quinn, *Last Voyage*. It was also published by Purchas (*Hakluytus Posthumus*, XVI, pp. 1190–1201, and reprinted, ibid., 1906, XVI, pp. 146–77), who states 'Some passionate speeches of Master Candish [Cavendish] against some private persons not employed in this action, I have suppressed ...' (p. 148).

[4] Known over the years as Davis's Southern Land, Hawkins's Maiden Land, The Sebaldines, The Malouines and Isles Nouvelles.

Ocean and make for Isle of Santa Maria.[1] Having sailed on 13 September they thrice reached the open sea but were forced back by gales, losing their accompanying pinnace. On 10 October a providential break in the clouds enabled them to take an altitude of the sun and determine the course back into the Strait. They proceeded under bare poles with a following gale so that in six hours they were twenty-five leagues within the Strait.

> The storme growing outragious, ... we were constrained ... to guide the ship in the hell-darke night, when we could not see any shore, ... But our captain, as wee first passed through the Streights drew such an exquisite plat of the same, as I am assured it cannot in any sort be bettered: which plat hee and the Master so often perused, and so carefully regarded, as that in memorie they had every turning and creeke, and in the deepe darke night without any doubting they conveyed the ship through that crooked channel.[2]

Unfortunately it appears this chart has not survived. The vessel finally reached Bear Haven on the west coast of Ireland on 11 June 1593, with only 16 men left alive of whom five were able to work the ship in some sort.

In 1593 Sir Richard Hawkins set out on a voyage for Japan, the Philippines and Moluccas and China by way of Estrecho de Magallanes. He sailed from Plymouth on 13 June 1593, in the *Daintie*, with a pinnace which deserted him before he entered the Strait on 10 February. They had with them some of the men who had accompanied Thomas Cavendish on his second voyage who were able to assist with details of the harbours used on that occasion. They came within sight of the western entrance but were then blown back and ended up four leagues west of Cabo Froward which they named English Bay.[3] In due course a change in the weather allowed them to pass out of the Strait and by 15 April they were off Valdivia. Subsequently Hawkins attacked Spanish shipping but a Spanish fleet, under Don Beltran de Castro, caught up with him in June, off Atacame,[4] and on the 22nd he was forced to surrender. His journal[5] contains brief remarks about the Strait, but no sailing directions as such, nor do charts appear to have survived.

Thereafter the passage of the Strait was undertaken by a number of Dutch expeditions.[6] In 1598 Jacob Mahu sailed with five ships of Rotterdam for the East Indies through Estrecho de Magallanes. Part of the plan was to attack the Spanish

---

[1] Isla Santa Maria in Golfo Arauco, lat. 36°59′S, long. 73°32′W.

[2] Hakluyt, *Principal Navigations*, III, p. 849. 'The last voyage of the worshipfull M. Tomas Candish Esquire ... Written by M. John Jane, a man of good observation, imployed in the same, and many other voyages.' pp. 842–52.

[3] Possibly Bahía Wood, lat. 53°49′S, long. 71°37′W, into which Río San José flows. Hawkins mentions a goodly river flowing into the bay.

[4] Lat. 00°53′N, long. 79°51′W.

[5] Printed in Purchas, *Hakluytus Posthumus*, IV, pp. 1367–1415, and reprinted ibid., 1906, XVII, pp. 57–199, and in Williamson, *Observations*.

[6] The Low Countries, a confederation of counties and principalities, combined under the rule of the Burgundian dukes in the second half of the fourteenth century, were inherited by the Spanish crown when Philip the Handsome died in 1506. As a result of the repressive regime of the Duke of Alva, the Spanish Governor, and the execution of the Counts of Egmont and Hoorne, they revolted against Spanish rule in 1568 initiating the Eighty Years War. The southern portion was brought back under Spanish domination but the United Provinces of the Free Netherlands continued the war and built up an army and navy which formed one of the most effective fighting forces in Europe. Following on from the combination of the Spanish and Portuguese crowns in 1580, war was also waged against Portugal. There was a truce, 1609–21, and the war with Spain was concluded by the Treaty of Münster (1648). This did not include Portugal and, despite the

possessions on the west coast of South America in order to obtain booty and supplies before crossing the Pacific. Mahu died on the passage across the Atlantic and the command was assumed by Simon de Cordes in accordance with the orders given by their employers. On 6 April 1599, the fleet entered the Strait and anchored in a bay on the north side which was named Great Bay, then Green Bay and subsequently Bay de Cordes. This bay was probably Caleta Gallant,[1] where they subsequently wintered. They sailed on 23 August having buried over 120 men who had died due to the hardship suffered, and entered the Pacific on 2 September – now six vessels, having set up a shallop during the winter. They were almost immediately scattered by a storm. One ship reached Japan with twenty-four men on board (seventeen sick) who were succoured but their ship was not allowed to proceed;[2] one reached the Moluccas after a little successful operation off Peru, and was taken by the Portuguese; one was taken by the Spanish at Valparaíso: and one, under Sebald de Weert, returned through the Strait to Holland having made contact with the next expedition under Olivier van Noort, in Bay de Cordes. They arrived home with only thirty-six men left, on 13 July 1600. There is no record of what happened to the other two vessels.[3]

Olivier van Noort left Holland in 1598 with two ships, the *Mauritius* and *Hendrick Fredrick*, two yachts and 248 men. He finally sailed on 13 September. One of the yachts was condemned and burned on 21 June 1599, and they passed Cabo Vírgenes on 22 November. Having rounded Cabo Froward, they named and anchored in Bay d'Olivier, on the east side of Cabo Holland. They moved to the vicinity of Cabo Gallant and made contact with Sebald de Weert's ship, which, however, was unable to keep up with them when they moved on to Mauritius Bay (which they also named). The next stopping place was named Guesen (Beggarly) Bay and it was here that they court-martialled the Vice Admiral and, having found him guilty of having a 'tendency to excite mutinies', put him ashore and left him. On 29 February 1600, van Noort with two ships and a yacht entered the Pacific. The two ships separated in a fog on 12 March. Van Noorte cruised with the yacht off the west coast of America taking a number of Spanish vessels and then crossed the Pacific to the Spice Islands where he engaged in further conflict and lost the yacht. He finally returned to Rotterdam, 26 August 1601, the only one of the nine ships which set out in 1598 (apart from Sebald de Weert's ship) to do so.[4] As James Burney wrote:

---

separation from the Spanish crown in 1640, the war continued. The Dutch captured the Portuguese settlements on the Malabar Coast in 1663 but a settlement was not finally reached until six years later.

The Netherlands determined to break the Portuguese domination of the spice trade, and in 1594, '95 and '96 attempts were made to find a route to the Spice Islands north-east round Europe and Asia. In addition in 1595 a fleet, organized by The Far Lands Company, sailed round the Cape of Good Hope; it was supported by patents from the Stadtholder, Prince Maurits, however the profits from the voyage barely covered the cost. Another sailed through Estrecho de Magallanes in 1598. Other fleets followed, and by 1601 eight companies had despatched a total of 65 ships in 15 different fleets. In order to maintain prices and bring some order to the situation, the *Verenigde Oostindische Compagnie* (VOC), known in English as the United Dutch East India Company, was formed in 1602, with a monopoly of Dutch trade east of the Cape of Good Hope and through Estrecho de Magallanes. They also had authorization to wage war and make treaties with Asian rulers, build fortifications and trading posts, enlist soldiers and appoint administrators.

[1] Burney, *Chronological History*, II, p. 189.
[2] Some were allowed to build a ship and depart but W. Adams, the pilot, was retained and lived the remainder of his life in Japan. Ibid., p. 197.
[3] Ibid., pp. 186–204.
[4] Ibid., pp. 205–34.

Figure 2: Olivier van Noort, chart, *Fretum Magallanicum*, 1602 (BL 455 b 10 Part 4). Reproduced by permission of the British Library.

> The voyage of Olivier Van Noort contributed little to Geography; and, impartially considered, neither this nor the Voyage of the Five Ships of Rotterdam which preceded it, can give an advantageous opinion of the maritime knowledge and management of the Hollanders at that time. Both the expeditions are full of interesting events, but that of Olivier Van Noort is stained with many instances of shocking barbarity.[1]

Van Noort published an account of this voyage in Amsterdam in 1602[2] from which Fig. 2 is taken.

Joris van Speilbergen[3] sailed from Texel on 8 August 1614 with four ships, *Groote Sonne*, *Groote Mane*, *Æolus* and *Morgenster* and two yachts, *Jager* and *Meeuwe*, for the East Indies via Estrecho de Magallanes. The fleet arrived off the Strait on 7 March 1615, but was unable to enter it until 28 March due to adverse winds. The ships were separated in a storm, but reassembled in Bay de Cordes on 16 April, which they left a week later and sailed out of the Strait on 6 May. In general they tried to sail by day, with the boats going ahead to seek for anchorages, and anchor overnight, although, with a following wind, they were prepared to sail by night as well. A reasonably

---

[1] Ibid. p. 234.

[2] Noort, *Description*.

[3] Speilbergen's journal of this voyage was first published at Leyden, 1619, and is translated in de Villiers, *East and West Indian Mirror*. Speilbergen had been in the East Indies in 1601. He had sailed round the Cape of Good Hope, prior to the formation of the VOC, which he heard about while at Acheen, in January 1603, and where he sold one of his ships to the Company before returning home. He fought under Admiral Jacob van Heemskercke against the Spanish in the Bay of Gibraltar, in 1607, when he reported 12 Spanish ships and 14 smaller vessels were destroyed. de Villiers, *East and West Indian Mirror*, pp. xxxiv–lv.

Figure 3: Joris van Speilbergen, chart, *Tijpus Freti Magellanici*, 1619 (BL 10026 df 17). Reproduced by permission of the British Library.

accurate chart of the Strait was produced on this voyage and published with Speilbergen's journal (Fig. 3). They proceeded up the South American coast, fought a number of actions against the Spanish, crossed the Pacific and refitted in Jacatra (renamed Batavia, 4 March 1621) and with two richly laden vessels returned to Zeeland on 1 July 1617.

The next significant voyage (although it did not use Estrecho de Magallanes) was that of Schouten and Le Maire, 1615–17, through Estrecho de Le Maire and round the southern extremity of the land.[1] The account of this voyage was first published in Amsterdam in 1618 (and in England the following year)[2] and when the news reached Spain it was determined to send an expedition to ascertain the truth of the account and to survey the area. Bartolomè Garcia de Nodal was ordered to undertake this task and chose his brother Gonzalo de Nodal as his second in command. They sailed from Lisbon in *Nuestra Señora de Atocha* and *Nuestra Señora del Buen Soceso* on 27 September 1618. The ships arrived in Rio de Janeiro on 15 November, refitted, and sailed again on 1 December. They went south, anchoring for a day off Cabo Vírgenes on 17 January 1619, passing through Estrecho de Le Maire on 28 January, and proceeded round Cabo de Hornos (which was fixed in Latitude 56° less a sixth, and named Cabo de San Ildefonso), 5 February. Islas Diego Ramírez were discovered and named after their cosmographer, and the ships arrived off the western entrance to Estrecho de Magallanes on 25 February. They anchored two leagues from Cabo Deseado (probably the point known as Cabo Pilar today) in what would appear to have been Puerto Misericordia to rest and take observations for position, sailing again two days later. They anchored 'in a bay which they named San Joseph, being in the middle of the strait. On Friday the 1st March, at dawn, we entered a cove off this bay, which is dead water.'[3] This sounds like Caleta Gallant (lat. 53°41'S, long. 72°00'W), which is in the

---

[1] They arrived at Jacatra on 31 October 1616, where Jan Pietersz Coen, the VOC Governor General, confiscated their ship and sent the people home with Speilbergen's fleet, for infringing the VOC monopoly – refusing to believe that they had not passed through Estrecho de Magallanes. Jacob le Maire died on the way, but after two years' litigation the VOC was forced to return the confiscated vessel and its cargo to the owners and to pay all costs and interest since the illegal seizure.

[2] Schouten, *Relation*.

[3] Markham, *Early Spanish Voyages*, pp. 252–3.

right area and is the only anchorage with an enclosed bay off it. They observed the sun with an astrolabe; 47°, declination 6°52′ giving a latitude of 53°20′.[1] They beached the caravels, greased their hulls, wooded and watered, and sailed at dawn on 5 March anchoring in Bahía San Nicolás[2] (36 miles east of Caleta Gallant) the same evening. They passed Cabo Vírgenes again on 13 March and, calling at Pernambuco, returned home, arriving at San Lúcar de Barrameda on 9 July 1619, having completed their task. A chart and sailing directions were produced as a result of this voyage.[3]

In 1669 King Charles II sent Captain John Narbrough (Narborough)[4] in the *Sweepstakes*, 300 tons with a crew of eighty men and boys, and the *Bachelour*, pink, 70 tons with nineteen men and a boy, to the South Sea. The object of this expedition appears to have been to gather geographical information on the coasts of South America south of Río de la Plata, in Estrecho de Magallanes, and up the west coast, and to investigate the prospects of trade. They sailed on 26 September 1669, and Port San Julian 'which lies in about 49d. 20m. South Latitude' was nominated as the rendezvous in case the ships should be separated.[5] Approaching the coast of South America, towards the end of February, they lost sight of the *Bachelour*, and although they wintered in Port San Julian she did not rejoin.[6]

The *Sweepstakes* entered the Strait round Cabo Vírgenes[7] on 22 October. A number of names were given to the various features and a large chart produced[8] (see Fig. 4). They left the Strait on 19 November 1670[9] and sailed up the coast of South America arriving off Valdivia on 15 December, where communication was opened with the

---

[1] Ibid. p. 253. Presumably the angle measured was the zenith distance at noon, the declination at this date would have been south which would give a latitude of 53°52′S. There would therefore appear to be an error.

[2] This bay is described in considerable detail and there can be very little doubt that it was the bay that now bears the same name.

[3] Markham, *Early Spanish Voyages*, pp. 169–270. The chart is reproduced facing p. 188.

[4] This voyage was published in London, 1694 as the first part of *An Account of several late Voyages & Discoveries to the South and North towards The Streights of Magellan, the South Seas, the vast Tracts of Land beyond Hollandia Nova &c. also towards Nova Zembla, Greenland or Spitsberg, Groynland or Engronland, &c.* by Sir John Narborough, Captain Jasmen Tasman, Captain John Wood, and Frederick Marten of Hamburgh. It was also published as 'Captain Wood's voyage thro' the Streights of Magellan, &c, part III' (pp. 56–100) in *A Collection of Original Voyages*, edited by William Hacke, London, 1699.

Narbrough went on to have a very distinguished career in the Royal Navy. He served as Flag Captain to Rear Admiral Thomas Butler, Earl of Ossory, in the *Saint Michael*, in the English fleet under Prince Rupert at the battle of Texel, 1673. On Lord Ossory's retirement he was promoted Rear Admiral of the Red and Knighted at Whitehall, 13 September that same year. He went on to be Commander in Chief Mediterranean and a Commissioner of the Navy, dying in 1688. Dyer, *Life*.

[5] Narborough, *Several Late Voyages*, p. 12.

[6] In fact she had returned to England. Burney, *Chronological History*, III, p. 237. An unpublished journal of this voyage, under Captain Humphrey Fleming, signed by William Chambers, the mate (sold at Christies, 25 September, 2002), indicates that the pink made every effort to keep the rendezvous at Port San Julian but was unable to do so and subsequently returned home.

[7] Referred to by Narbrough as Cape Virgin Mary: Narborough, *Several Late Voyages*, p. 59.

[8] Narborough, *Several Late Voyages*, p. 71: 'but for a more exact Situation of the several Promontories, Bays, Ports, Rivulets, Soundings &c. I refer the Reader to the large Draught of the *Magellan Streights*, drawn by my own Hand on the place'. The original of this chart is now BL MS 5414 Art 29A.

There are also sailing directions under the title *A description of the Streights of Magellan Anno 1669 Capt John Wood taken in His Majesties Ship Sweepstakes Capt John Narbrough then Commander*, in Admiralty Library, Taunton, MS 4, and BL Sloane MSS 46A and 46B. Text is virtually identical in all three volumes.

[9] Hacke, *Collection of Original Voyages*, p. 92.

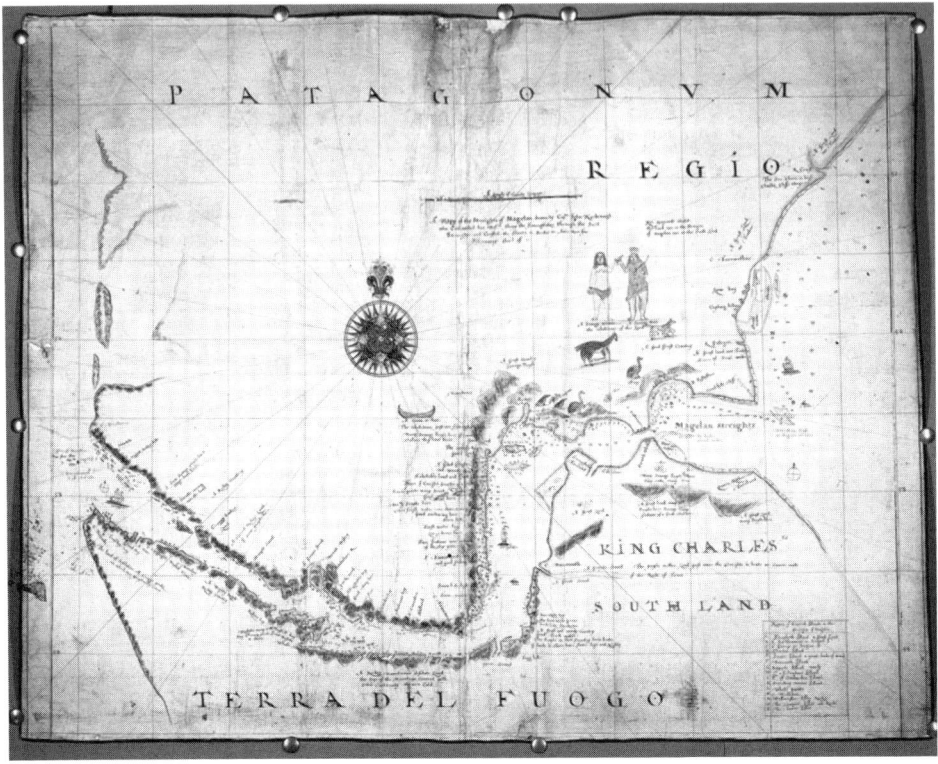

Figure 4: John Narbrough, chart, *Streigthts of Magellan*, 1670 (published 1694) (BL Add. MS 5414 Art 29A). Reproduced by permission of the British Library.

Spanish garrison. Lieutenant Armiger and John Fortescue with two seamen were detained (along with Don Carlos who had landed secretly on 15 December but did not contact the Spaniards until the ship had sailed). Narbrough departed on 21 December (leaving his men behind) returning to enter Estrecho de Magallanes on 6 January 1671. They stopped and refitted at Puerto San Juan de la Posesión (Port Famine) and departed from the Strait on 14 February and arrived 'upon the Coast [of England] about the middle of *June*, when we understood the *Spanish* Ambassador at Court had resented our Voyage into the South-Seas, but without any notice taken of it'.[1]

The French, with the approval of Louis XIV, sent out a privateering fleet in 1695 under Monsieur de Gennes, which arrived off Cabo Vírgenes 11 February 1696 and entered the Strait. They passed Cabo Froward but were blown back again and anchored in Bahía San Nicolás, where they named Río de Gennes which flows into the back of the bay. After a spell in Puerto San Juan de la Posesión, they reached Caleta Gallant on 20 March. However, they had continuous contrary winds, and at a council decided that, if the wind was not favourable in two days, they would return to La Isla Grande to take in provisions and seek their fortune elsewhere. After two days the winds

---

[1] Ibid. p. 100. See also p. 186, n. 2.

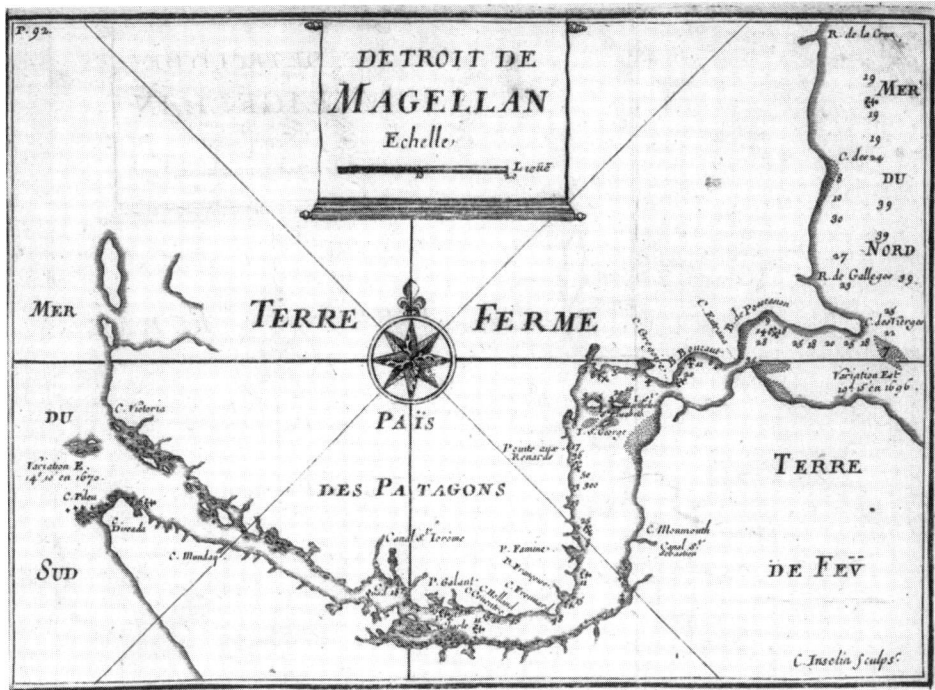

Figure 5: De Gennes, chart, *Detroit de Magellan*, 1698. Provided by the UK Hydrographic Office.

were still contrary so they prepared to return, which, after various failed attempts and difficulties, they did, finally returning to France on 21 April 1697.[1] This voyage was published by François Froger, in 1698, with a chart of Estrecho de Magallanes[2] (Fig. 5).

In 1698 a company was formed in France with the aim of settling colonies in those parts of South America not already occupied by Europeans. The *Phelippeaux* and *Maurepas* together with a frigate and bark sailed on 17 December 1698, under the command of a French naval captain, M. Gouin de Beauchesne. They reached Estrecho de Magallanes on 24 June 1699, by which time the two smaller vessels had parted company and returned to France. The winter was remarkably mild but the wind remained obstinately in the west and it was not until 21 January 1700, after 'nearly seven months spent in a most fatiguing and harassing navigation in the *Strait*',[3] that they entered the Pacific Ocean. In the Strait they gave names to a number of features, none of which have been retained, but otherwise do not appear to have added much to

---

[1] Froger, *Relation of a Voyage*, pp. 70–87

[2] Froger, *Relation d'un voyage*, contained plans of Port Famine, Port Galant (*sic*) and Baye Francoise et Embouchere de la Riv. De Gennes (as well as plans and views of other places visited). The London edition contained only the chart of the Straight (*sic*) of Magellan and The French Bay with the mouth of the River Gennes (as well as a much reduced number of the other plans).

[3] Burney, *Chronological History*, IV, p. 380.

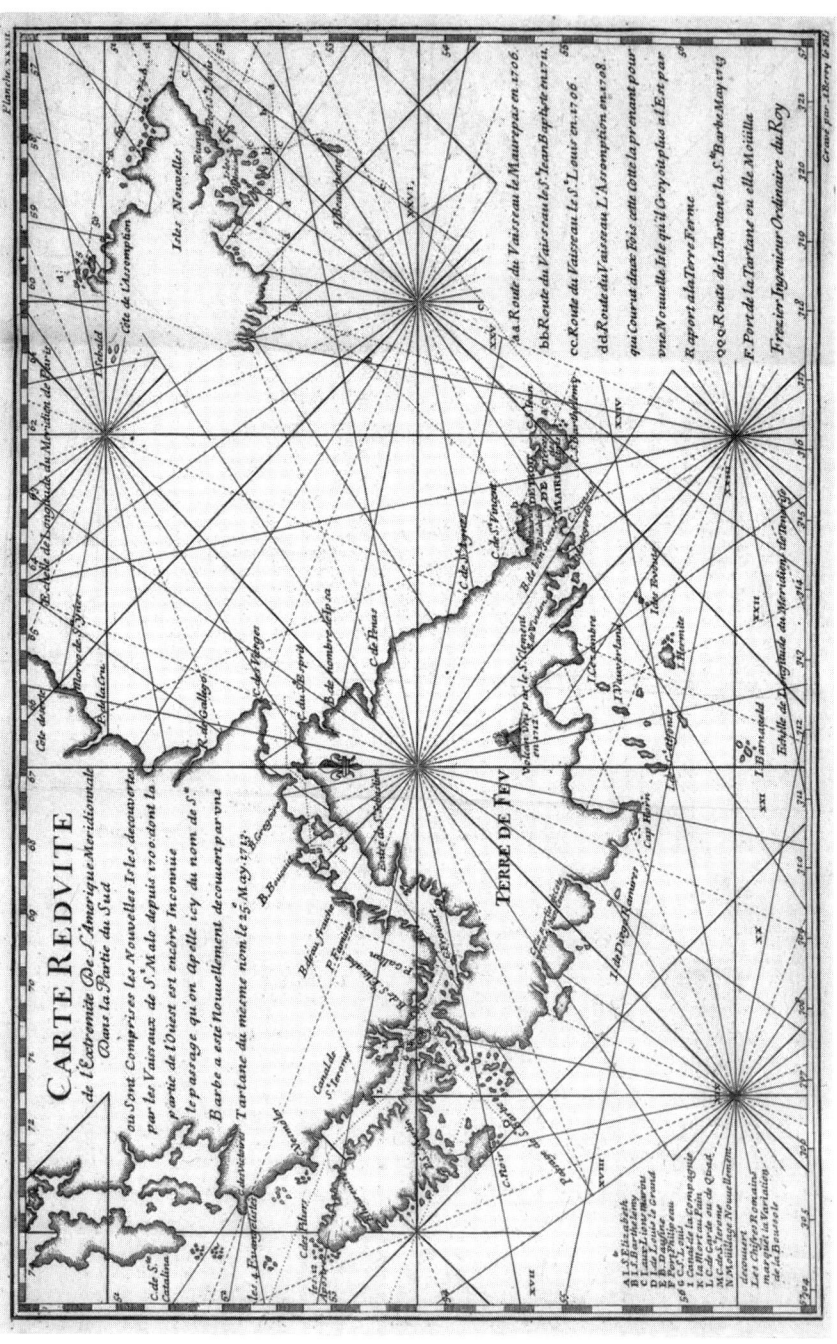

Figure 6: François Frézier, chart, *Carte reduite de l'extremite de l'Amerique meridionnale dans la partie du sud*, 1717. Provided by the UK Hydrographic Office.

the hydrographic knowledge of the area. They managed a little trade on the coast of South America. Returning home round Cabo de Hornos they discovered Beauchêne Island (south of the Falkland Islands) on 19 January 1701, and reached France on 6 August the same year.[1] A large number of French vessels continued travelling to the South Sea, and in 1712 M. Frézier, an officer in the King's service, travelled to Chile, passing round Cabo de Hornos both going out and returning, and writing up his travels.[2] In this work he includes an account of the tartane *S. Barbara*'s discovery of the Canal Bárbara.[3] Frézier also produced a chart of Estrecho de Magallanes on which this canal is shown as Passage du S. Barbara (Fig. 6).

The *Wager*,[4] under Captain D. Cheap, after rounding Cabo de Hornos in 1741, was wrecked on an island off the south-west coast of South America, and a party, having constructed a vessel, sailed through Estrecho de Magallanes from west to east. An account of this voyage was published by John Bulkeley and John Cummins, the gunner and carpenter respectively.[5] They suffered great hardship and loss of life but virtually no hydrographic information is given and it is only included here since it is referred to by Commander Stokes in his journal (see p. 195, n. 1).

In 1745 Richard Walter published *A Voyage round the World in the years MDCCXL., I, II, III, IV. by George Anson, Esq, &c.* In this he strongly advocated continued British interest in the Pacific and the establishment of bases at Pepys's Island[6] or the Falkland Islands.[7] When Anson was appointed to the Admiralty plans were put in hand to implement this policy and an expedition was planned in 1749. This came to the ears of the Spanish Ambassador at the Court of St James, who protested, and since at that time endeavours were being made to establish good relations with Spain, the plan was dropped. However the situation changed and in 1764 the *Dolphin* (24) and *Tamar* (16) were fitted out in great secrecy for a voyage under the command of the Honourable John Byron.

Commodore Byron, who had been in the *Wager* as a midshipman and returned to England with Captain Cheap, commanded the *Dolphin* on her voyage round the world, 1764–6.[8] They searched for Pepys's Island, and concluded it did not exist;[9] visited the

---

[1] This voyage is described in Burney, *Chronological History*, IV, pp. 375–85, where references are given to accounts in Woodes Rogers, *A Cruising Voyage round the World*, London, 1718, p. 117 et seq., and *Navigation aux terres australes*, and *Noticia de las expediciones al Magalhanes*.

[2] Frézier, *Voyage to the South Sea*.

[3] North end lat. 53°48′S, long. 72°08′W. Frézier, *Voyage to the South Sea*, pp. 285–7. M. Marcanil, in the tartane (normally a small Mediterranean single-masted vessel with a lateen sail) missed his way in the Strait sailing SW and SW by S from Elizabeth Bay and found himself in the Pacific Ocean on 15 May, 1713.

[4] A supply ship of the squadron under the command of Commodore Anson in his voyage round the world, 1740–44.

[5] Bulkeley and Cummins, *Voyage to the South Seas*, 1743; 2nd edn., 1757.

[6] 'Discovered by Cowley between the Latitude of 47°, and the Latitude 48° South, about Eighty Leagues from the Continent of South America'; Byron's instructions quoted in Gallagher, *Byron's Journal*, p. 4. See also ibid, pp. 166–7, for an extract from Captain Cowley's voyage round the world.

[7] Walter, *Anson's Voyage*, Chapter IX.

[8] Byron's journal of this voyage is given in Gallagher, *Byron's Journal*. The *Tamar* was commanded by Captain Mouat until he transferred to take command of the *Dolphin* on 30 April, 1765 (on Byron becoming Commodore), when the first Lieutenant of the *Dolphin*, Lieutenant Cumming, took over the command of the *Tamar*. At the same time Lieutenant Carteret transferred to the *Dolphin*.

[9] Gallagher, *Byron's Journal*, p. 42.

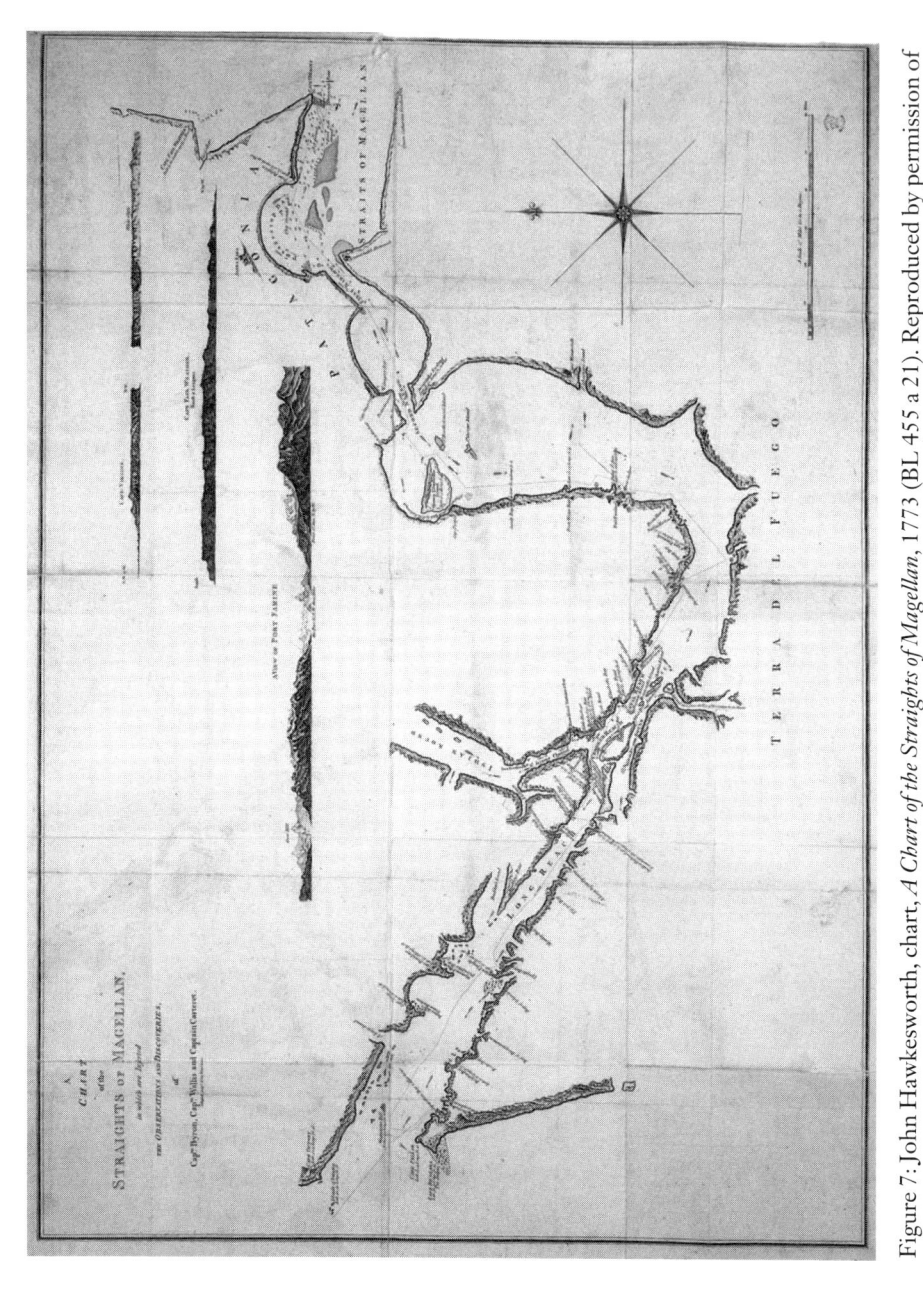

Figure 7: John Hawkesworth, chart, *A Chart of the Straights of Magellan*, 1773 (BL 455 a 21). Reproduced by permission of the British Library.

Falkland Islands which they claimed and named Port Egmont[1] before passing through Estrecho de Magallanes and crossing the Pacific.

This voyage was followed by that of Captain Wallis in the *Dolphin*, 1766–8, and Captain Carteret in the *Swallow* (14), 1766–9.[2] It was originally intended that these vessels should go round the world together but they became separated after passing through Estrecho de Magallanes.

As a result of these three voyages, charts of Estrecho de Magallanes and a number of other anchorages were published by Hawkesworth in the first volume of *An account of the Voyages undertaken by order of His Present Majesty for making Discoveries in the Southern Hemisphere, &c.* 1773[3] (Fig. 7). This also includes directions for the Strait and a number of anchorages by Captain Wallis who would appear to have been the first navigator to use lunar distances for longitude in that area[4] (Captain Carteret did not use lunar distances).[5]

Louis-Antoine de Bougainville also sailed round the world in 1766–9,[6] passing through Estrecho de Magallanes and producing a chart together with a number of plans. He had established a colony in the Falkland Islands in 1764, which he visited again the following year, and while in Estrecho de Magallanes to obtain wood sighted Byron's ships on passage to the Pacific. Due to complaints from the Spanish, Louis XV agreed to turn the colony over to Spain and Bougainville went to Madrid to arrange the terms. His first duty on his voyage round the world was to turn the colony at Port Saint Louis[7] over to the Spanish authorities, taking off any of the settlers who did not wish to

---

[1] Ibid. pp. 60 and 156. The Earl of Egmont was First Lord of the Admiralty.

[2] Carrington, *Discovery of Tahiti*, and Wallis, *Carteret's Voyage*.

[3] *A Chart of the Straits of Magellan in which are inserted the Observations and discoveries of Cap^n. Byron, Cap^n. Wallis and Captain Carteret*. Plans of: – 'Port Famine', 'Woods Bay', 'Port Gallant and Fortescue Bay', 'Cordes Bay and Harbour', 'St David's Cove', 'Island Bay', 'Swallow Harbour', 'Puzling Bay', 'Cape Upright Bay', 'Dolphin Bay', 'Cape Providence with the Bay and Anchoring Place to the N.N.E. of it', 'Cape Tamar', 'Elizabeth Bay', 'A Bay under the Islands Opposite York Head', 'St David's Bay', 'From York Bay to Three island Bay and Harbour'. There are several original surveys from these voyages in the United Kingdom Hydrographic Office, the British Library and the Dixson Library, Sydney, details of which are given in Gallagher, *Byron's Journal*, Carrington, *Discovery of Tahiti*, and Wallis, *Carteret's Voyage*.

[4] In Hawkesworth, *Account of the Voyages*, I, Directions are given at pp. 410–18. Two positions are given where longitude was obtained by observation, pp. 381 and 384, together with the position of Port Famine p. 412. In his 'Table of the Latitudes and Longitudes west of London etc.', p. 520, he records two values for Longitude, the first 'supposed' and the second 'observed by Dr Maskelyne's method' (Nevil Maskelyne, *The Mariner's Guide containing Complete and Easy Instructions for the Discovery of the Longitude at Sea and Land &c.*, London, 1763). The latter were observed during December 1766, and include: Cape Virgin Mary, lat. 52°24′S, long. 69°06′W (Cabo Vírgenes, lat. 52°20′S, long. 68°21′W); Point Possession, lat. 52°30′S, long. 69°50′W (Cabo Posesión, lat. 52°18′S, long. 68°58′W); Point Porpass, lat. 53°08′S, long. 71°30′W (Cabo Porpesse, lat. 52°56′S, long. 70°47′W) and Port Famine, lat.53°43′S, long. 71°32′W (Puerto San Juan de la Posesión, lat. 53°38′S, long. 70°56′W). There are no longitudes given for positions west of Port Famine, except Cape Pillar (April, 1767) 'Supposed Longitude 76°06′W' (Cabo Pilar, long. 74°40′W).

Wallis noted in his journal (now in the Alexander Turnbull Library) that the accuracy of his longitudes was due to his purser, John Harrison, a mathematician 'thro' whose means we took the Longitude by taking the Distance of the Sun from the Moon and Working it according to Dr. Masculines Method which we did not understand': Beaglehole, *Journals of Captain James Cook*, p. 119n1.

[5] Wallis, *Carteret's Voyage*, I, p. 123n4.

[6] Bougainville, *Voyage autour*; Bougainville, *Voyage*; Taillemite, *Bougainville*; Dunmore, *Pacific Journal*.

[7] Now Port Louis in Berkeley Sound at the eastern end of East Falkland Island.

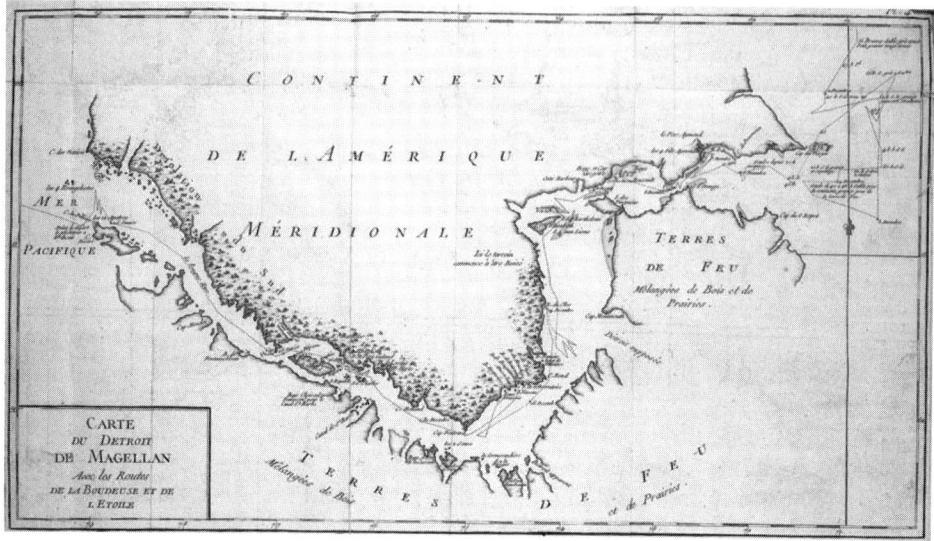

Figure 8: Louis de Bougainville, chart, *Carte du Detroit de Magellan*, 1772 (BL 454 a 1). Reproduced by permission of the British Library.

remain under Spanish rule. Thence, after visiting Rio de Janeiro and Montevideo, he entered Estrecho de Magallanes in early December 1767 and passed Cabo Pilar on 26 January 1768. During this passage his astronomer, M. Véron, together with the other officers, took observations for latitude in the normal way, and also made use of lunar distances to determine the longitude.[1] Bougainville's voyage contained a chart of the Strait (Fig. 8), with plans of the area round Caleta Beaubasin, opposite Cabo Froward, and that round Bahía Bougainville.

In 1785–6 an expedition in the fragata de guerra *Santa María de la Cabeza* (36) was sent out from Spain under the command of Capitan de Navío Don Antonio de Córdoba to survey the Strait. Although a number of the other expeditions mentioned above had carried out surveys, this was the first expedition since the Nodal brothers in 1618 that had had this specific aim. Córdoba made a second voyage in 1788–9, in the paquebotes[2] *Santa Casilda* (16) and *Santa Eulalia* (16) to complete his work by

---

[1] Having sailed before the first edition of *The Nautical Almanac* and *Tables Requisite* were published by Maskelyne for 1767, he used the method of Abbé Nicolas-Louis de Lacaille.

The longitude of Cabo Vírgenes was given as 71°25′20″ west of Paris and the eastern entrance of the second gut (Cabo Gregorio) 73°34′30″ (Taillemite, *Bougainville*, p. 292), which were the only longitudes based on lunar distances in the Strait, due to the bad weather. Paris (P) is 2°20′09″ east of Greenwich (G).

	Bougainville			Modern	
	Lat.	Long. (P)	Long. (G)	Lat.	Long.
Cabo Vírgenes	52°21′	71°25′20″	69°05′11″	52°20′S	68°21′W
Cabo Gregorio	52°40′	73°34′30″	71°14′21″	52°39′S	70°12′W

[2] Snows. A snow was described as a two-masted vessel, the masts of which were similar to the fore and main masts of a ship, but close abaft the main mast a trysail mast was mounted in a block on the deck with its head secured to the after side of the main-top. The sail, a loose-footed gaff trysail, was hooped to this mast and extended to the after part of the vessel: Falconer, *Dictionary*, and Burney, *Dictionary*.

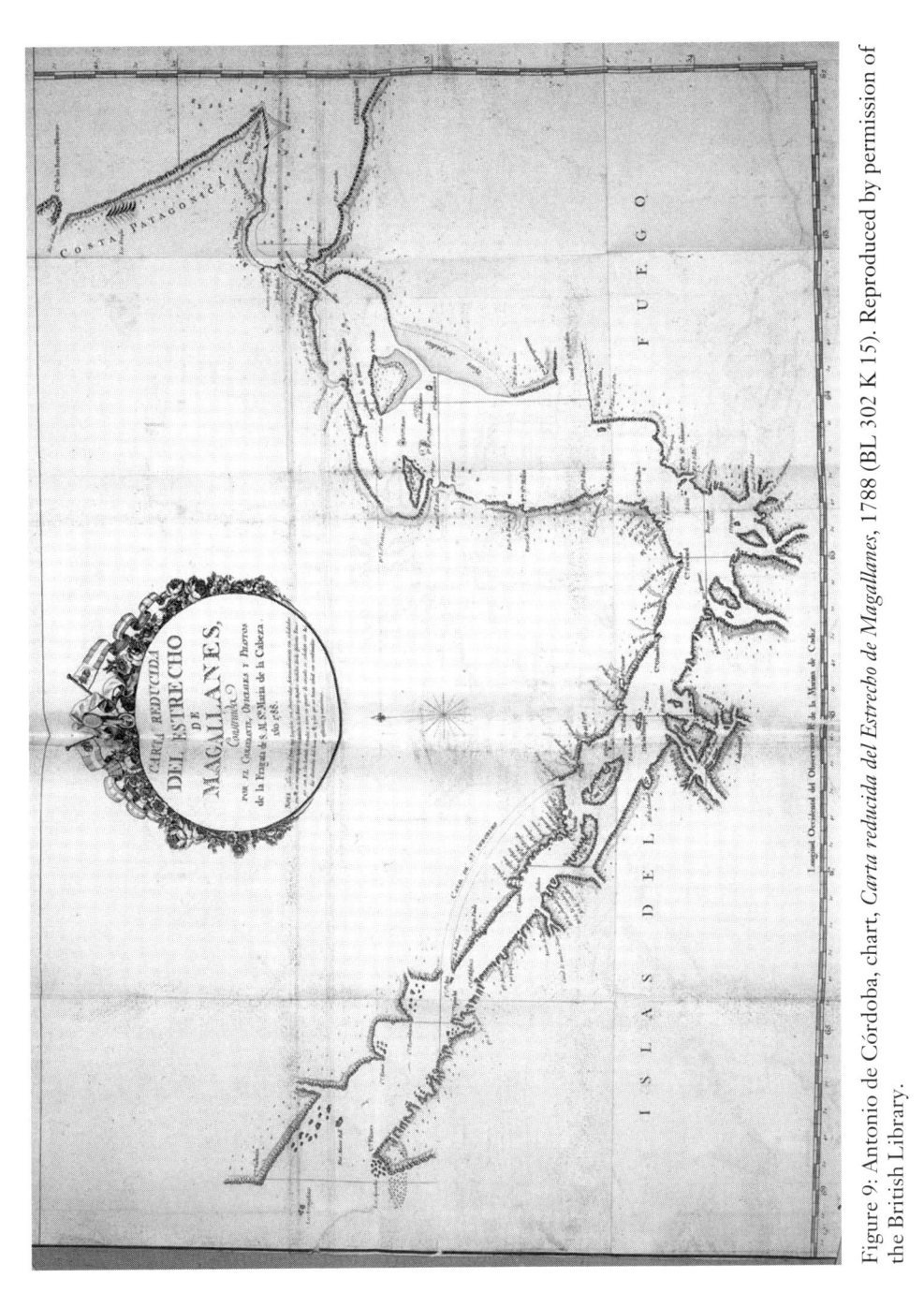

Figure 9: Antonio de Córdoba, chart, *Carta reducida del Estrecho de Magallanes*, 1788 (BL 302 K 15). Reproduced by permission of the British Library.

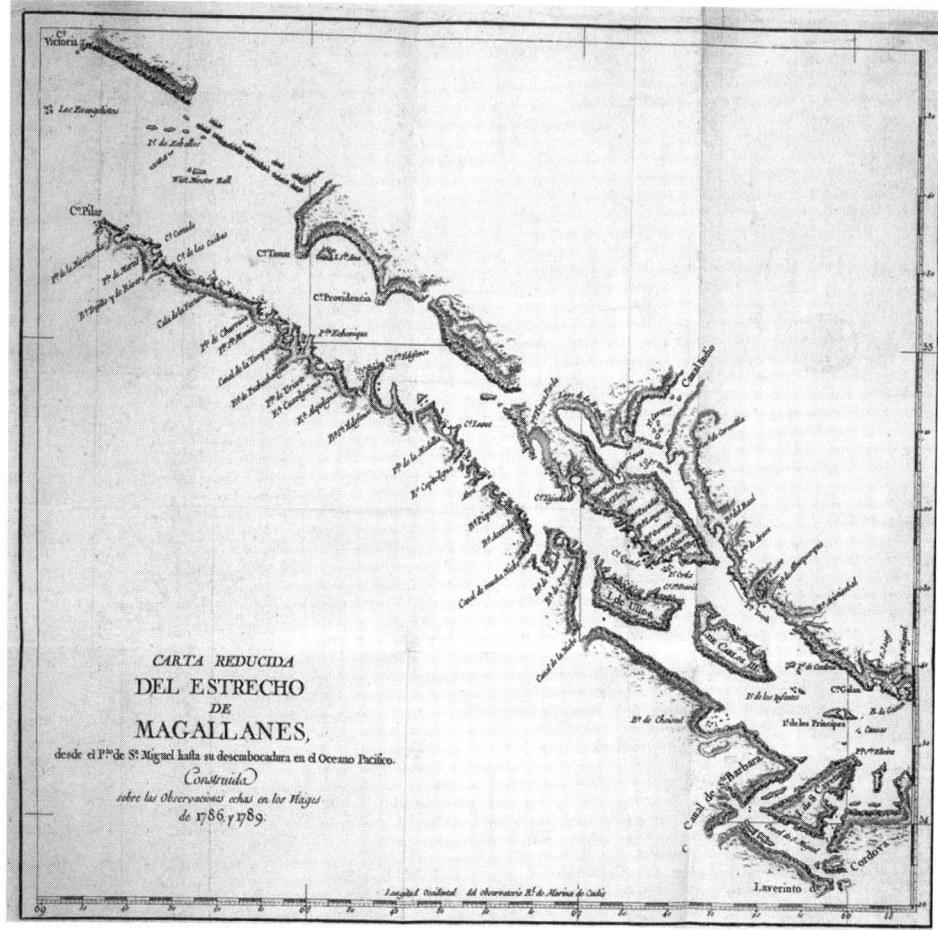

Figure 10: Antonio de Córdoba, chart, *Carta reducida del Estrecho de Magallanes*, 1786 and 1789 (published 1793) (BL 302 K 15). Reproduced by permission of the British Library.

examining the western part of the Strait and fixing Cabo Pilar and Cabo Victoria.[1]

This survey was the first to make use of timepieces,[2] which were combined with lunar distances to provide the control. There were also two *tenientes* (lieutenants) on board, on each voyage, who had been trained in hydrographic work under Don Vicente Tofiño.[3] They produced a chart of the Strait (Fig. 9), and another on a larger

---

[1] *Relacion* and *Apéndice a la Relacion; Voyage of Discovery*.

[2] Cordoba carried three timepieces, Berthoud Nos. 15 and 16 from the Observatory at Cádiz and Arnold No. 71, a pocket chronometer, which belonged to Teniente de fragata Belmonte, on his first voyage ([Vargas y Ponce], *Relacion*, p. 4), but apparently only Arnold No. 71 on the second navio, *Apéndice a la Relacion*, p. 2.). Arnold No. 71 was subsequently carried by Malaspina on his expedition of 1789–94.

[3] Tenientes de fragata Dionisio Alcalá Galiano and Alexandro Belmonte in *Santa María de la Cabeza*, *Relacion*, p. 2), and Teniente de navío Cosme de Churruca in *Santa Casilda* and Teniente de fragata Ciriaco Cevallos in *Santa Eulalia*, *Apéndice a la Relacion*, pp. 2–4).

scale of the Strait westward of Caleta Gallant (Fig. 10), together with a number of plans[1] which were all published in the anonymous account of the voyage attributed to José Vargas y Ponce, and based on their journals.[2]

Among the earliest charts of the Strait to be published is one by Mercator, in 1606,[3] which names Capo Forward (*sic*) and shows a few banks and soundings (Fig. 11). It appears to owe a certain amount to Thomas Cavendish's chart (see p. 160, n. 1) but is not entirely based on it.[4] The chart also names Cordes Bay, C. Mauritio and Capo

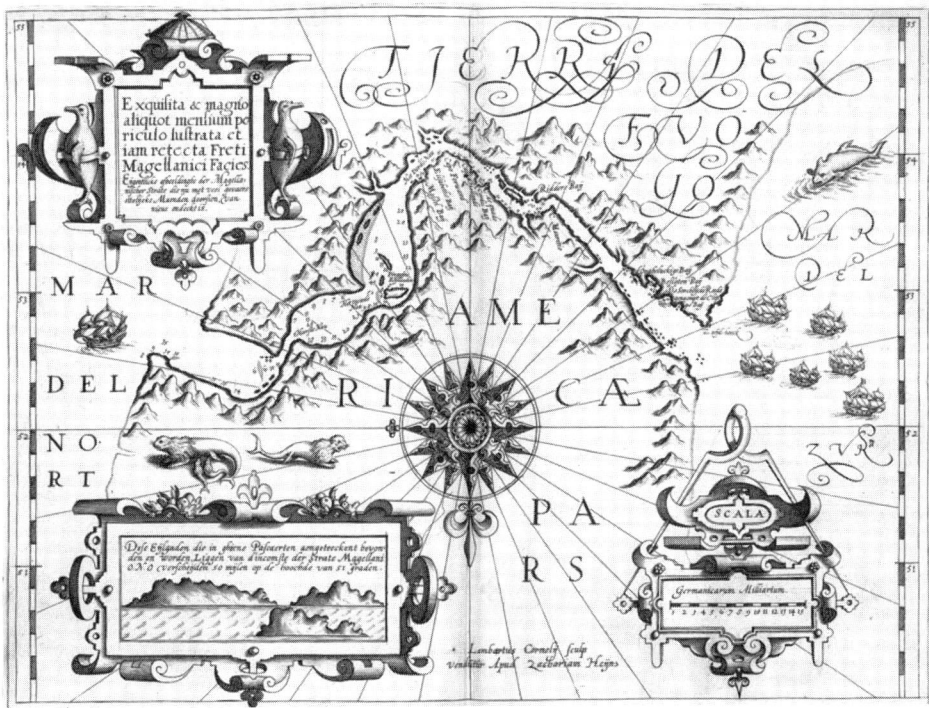

Figure 11: Gerard Mercator, chart, *Exquisita & magno aliquot mensium periculo lustrata et iam retecta Freti Magellanici facies*, 1606 (BL Maps C 3 c 6). Reproduced by permission of the British Library.

[1] Where they are different, modern names are given in brackets. Bahía de San Nicolas; Tres Bahías (directly east of San Nicolas including Bahía Bougainville); Bahía Valcarel (Bahía del Aquila); Puerto de la Hambre (Puerto San Juan de la Posesión); Puerto de San Miguel y Bahía de Gaston (Bahía Cordes); Puerto de San Antonio y Bahía Valdés; Laberinto de Cordova (name not currently in use) y Canal de Santa Barbara; Puerto Galan (Puerto Gallant) y Bahía de Fuerte Escudo (Bahía Fortescue); Bahía del Swallow ò Baronesa; Rada de Vacaro (Rada York); Bahía de Aristizabal (Bahía Isabel); Bahía de Solano (Bahía Woods).

[2] *Relacion* and *Apéndice a la Relacion*.

[3] Published in Mercator, *Atlas* (between ff. 353 and 354). The atlas also contains a map of America Meridionalis (between ff. 351–2) which marks 'P Famin', named by Cavendish. This atlas was translated into English by W. S. Generosus and published in London in 1635 and 1637 (Mercator, *Historia Mundi*, pp. 925–30). The text has additions and the chart is reduced with soundings and sand banks omitted.

[4] One of Cavendish's charts appears to have been in the Hague since it left England, while that now in Florence has a Dutch inscription on the back – *Brasilsche Caarte*. It is suggested in Quinn, *Last Voyage*, pp. 150–59, that both charts reached the Netherlands together, possibly with a collection of documents sent by Richard Hakluyt in 1594–5, or with pilots who had served with Cavendish and sailed with Mahu or van Noort.

de Holland, indicating that information from the voyages of Seebald de Weert and Van Noort (in the *Mauritius*) has also been used.[1]

James Burney reviewed the charts of the Strait in 1813,[2] and wrote:

> The Chart made by J. Cornelitz May who sailed with Admiral Spilbergen [sic], is to be reckoned the first published of the *Strait* which merits being called a Plan. The next improvement of consequence in the geography of the *Strait* is the Chart of Captain Narbrough. The published Charts of the *Strait* which are entitled to notice appeared in the order following:
>
> 1. The Chart of John Cornelitz May, shipmaster to Admiral Spilbergen [sic]; published 1619 [Fig. 3].
> 2. By Captain John Narbrough. Not published till 1694 [Fig. 4].
> 3. By Monsr Froger; Published 1698 [Fig. 5].
> 4. By M. Frezier. 1716 [Fig. 6].
> 5. A Chart of the *Strait*, published with M. de Bougainville's Voyage round the World. 1771 [Fig. 8].
> 6. A Chart, comprehending the observations and discoveries made in the *Strait* by Captains Byron, Wallis, and Carteret; engraved by Mr. J. Russel under the direction of our great navigator Captain James Cook; 1773 [Fig. 7].
> 7. A Chart constructed from a survey taken in the Spanish frigate Sta Maria de la Cabesa, in 1785–6; published in Madrid 1788. [Fig. 9. The chart from the second voyage is at Fig. 10.]

Admiral Burney goes on to examine the merits of the various charts and their deficiencies. He concludes 'The Spanish Chart of the *Strait* is to be esteemed the most accurate in the position of the principal Capes and Points of land … It is to the credit of the earlier Charts, that they have generally a very close agreement with the Spanish Chart.'[3]

The latitudes of the entrances to Estrecho de Magallanes and the various features in it had been known with a reasonable degree of accuracy from its discovery, but until the advent of lunar distances and timepieces, their longitudes had remained a matter of conjecture based on various navigators' reckonings (see comparative table at * on p. 177). Although Córdoba had carried out the best surveys practicable in the time available to him, he had very few soundings on his charts and most of the coastline was sketched in by eye between the headlands he had fixed. The Commissioners for executing the Office of the Lord High Admiral of the United Kingdom of Great Britain and Ireland etc. therefore instructed Captain Phillip P. King Esq, Commander of HMSV *Adventure*, with Captain Stokes of HMSV *Beagle* under his command to

---

[1] The text accompanying it mentions only the voyages of 'Franciscus Dracus' (Francis Drake) and 'Thomas Candischius' (Thomas Cavendish) in addition to that of Magellan himself. The English edition states that 'In the first entrance into the Magellan Sea, there is a new Castle which *Philip* the second King of *Spaine* commanded to be built there for the defence of these Straites in the yeere 1582'. Mercator, *Historia Mundi*, p. 930.

[2] Burney, *Chronological History*, III, pp. 376–82.

[3] Ibid, p. 378.

proceed to the entrance of the River Plata, to ascertain the longitudes of the Cape Santa Maria and Monte Video: you are then to proceed to survey the Coasts, Islands, and Straits; from Cape St Antonio, at the south side of the River Plata, to Chiloé; on the west coast of America; in such manner and order, as the state of the season, the information you may have received, or other circumstances, may induce you to adopt [dated 16 May 1826].[1]

The results of this survey were published as Chart 554 in 1832 (see Fig. 12, p. 178).

[1] King, *Narrative*, I, pp. xv–xix.

* Table of latitudes and longitudes

Name	Cabo Vírgenes		Cabo Pilar	
	Latitude	Longitude	Latitude	Longitude
Narbrough*	52°26′S	65°42′ W of L	53°05′S	72°49′ W of L
		70°54′W of G		78°01′ W of G
Bougainville**	52°21′ S	71°26′ W of P	52°50′ S	78°56′ W of P
		69°05′ W of G		76°36′ W of G
Córdoba***	52°20′ S	62°10′ W of C	52°45′ S	68°46′ W of C
		68°22′ W of G		74°58′ W of G
Modern	52°20′ S	68°21′ W of G	52°44′ S	74°40′ W of G

L=Lizard 5°12′W; P=Paris 2°20′E; C=Cadiz 6°12′W; G=Greenwich.
Values have been rounded off to the nearest minute.
*Narbrough, *Several Late Voyages*, pp. 59 and 78. Note Narbrough's chart shows longitudes East of Lizard.
**Geographical positions taken from Bougainville's chart. See also p. 172, n. 1 above.
***Geographical position of Cabo Vírgenes from *Relacion*, pp. 79–82, Cabo Pilar from *Apéndice a la Relacion*, p. 106.

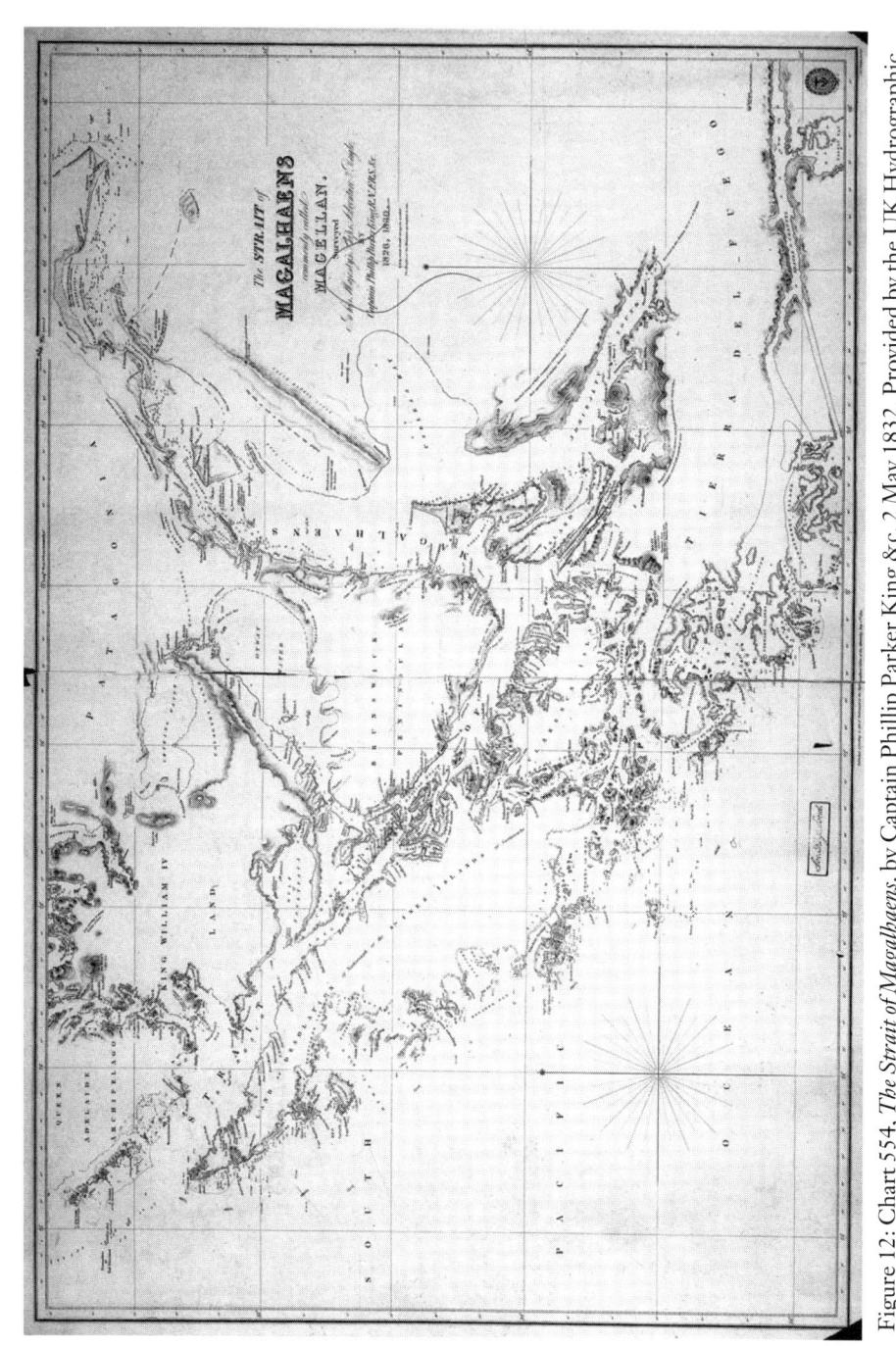

Figure 12: Chart 554, *The Strait of Magalhaens*, by Captain Phillip Parker King &c, 2 May 1832. Provided by the UK Hydrographic Office.

THE JOURNAL
of HMS *Beagle* in the Strait of Magellan
by Pringle Stokes, Commander RN 1827

Received
June 8th. 1827
[signed] Phillip P. King
Comm^d.

[p. 1]

The Journal of H.M.S.
Beagle
in the Straits of Magellan
between January 15th. & March 10th. 1827.
by Pringle Stokes, Commander R.N.
addressed to
Phillip Parker King Esqre. F.R.S. &c.
Commander of H.M.S. Adventure &
Senior Officer of an Expedition for the Survey of
The Southern Extremity of America &c.

With a Chart of the Western Division of the Straits, Plans of anchorages; & Drawings of the Coasts, & of the Natives —

The rates of my chronometers having been ascertained, the Beagle's victualling completed to three months, and water and fuel taken on board to the utmost capacity of her stowage, I sailed from Port Famine[1] on the morning of Jany. the 15th. for the purpose of carrying into execution your orders "to explore the Straits of Magellan westward of Cape Froward,[2] & particularly to fix the position of Cape Pillar,[3] the rock called Westminster Hall,[4] and the Isles of Direction":[5] your orders likewise restricted

---

[1] Now known as Puerto San Juan de la Posesión or Puerto del Hambre, lat. 53°38′S, long. 70°56′W.

   In 1581 Philip II of Spain sent an expedition to fortify Estrecho de Magallanes, as a result of which Don Pedro Sarmiento de Gambóa, established Ciudad del Rey Felipe at Punta Santa Ana. Thomas Cavendish (Candish), in 1587, anchored there and found the place deserted, most of the inhabitants having died of starvation. He named it Port Famine. Hakluyt, *Principal Navigations*, III, p. 806. See pp. 158–9.

[2] Cabo Froward, lat. 53°54′S, long. 71°18′W (Fig. 13).

[3] Cabo Pilar, lat. 52°44′S, long. 74°40′W (Fig. 24). The name was apparently given by Narbrough, although his account does not say so. It appears on his chart (Fig. 4). See also Markham, *Early Spanish Voyages*, p. 100n2 and p. 101n2.

[4] Isla Westminster Hall, lat. 52°37′S, long. 74°22′W. Name given by Narbrough, Narborough, *Several Late Voyages*, p. 77.

[5] So called by Narbrough, Narborough, *Several Late Voyages*, pp. 79 and 114; now known as Islotes Evangelistas, lat. 52°24′S, long. 75°06′W. See also p. 225, n. 2 below (Fig. 25).

my absence to "about the 21st. of February," and enjoined me to render to you on my return "a detailed account of my proceedings": It is in obedience to that injunction I lay before you the following Journal, which is, I am aware, little more [than] a mere log book account of hard beating to windward, and must owe whatever of interest it may be found to have entirely to its being the narrative of a successful attempt, under every adverse circumstances of wind and weather, to pass through these famous Straits, which from their having foiled the efforts of many – been cleared by few of any Nation without much of difficulty, danger, and disaster – and pronounced by some of our older Voyagers too dangerous for navigation – have acquired, even among Naval Men of the present day, a rather formidable name.

January the 15th

During the forenoon the wind was light & variable, with intervals [p. 2] of calm, it freshened up in the afternoon, and blew from the SW, a moderate breeze and fine weather. At 2 PM we doubled a Cape at the north shore to which various names have been given; our English Voyagers called it Cape Shut up, the French Cap Redond, and the Spanish Cabo San Isidor.[1] This is a headland of which the extreme southern point makes like a round island, within which is a high round topt hill wooded from its base to its summit, uniting itself with a much higher mountain in-land. The north shore of this Reach of the Straits is agreeable enough in the prospect: the mountains are green high up their sides; in the vallies and lower lands are woods of tall & well spread trees, and copses of lively coloured jungle; & the coast line presents several excellently sheltered coves and bays. In one of these, called by the Spaniards Bahia de San Nicolas,[2] we anchored for the night (at 8 PM) in 12 fathoms over a bottom of sand & pebbles: the centre of the small island in the middle of the bay (which forms a capital leading mark for it) bearing SW½W (magnetic) distant three cable-lengths. In coming in here sail should be kept on the ship to the last that she may be enabled to shoot into a good berth; otherwise as she soons gets becalmed by the high land on the western side of the bay, the stream of the river which discharges itself at the bottom of it will quickly carry her over to the eastern side where the water is too deep for good anchorage. The best berth is with the centre of the small island (already referred to) bearing about NE, distant a quarter of a mile, in 10 fathoms, a sandy bottom.[3]

I manned my boat and employed the remaining daylight in examining the place in which, as mentioned in Hawkesworth edition of Byron's voyage,[4] the celebrated Bouganville had, in the year 1765, hauled his Vessel for the purpose of procuring

---

[1] Now known as Cabo San Isidro, lat. 53°47′S, long. 70°58′W. It is shown as C. S. Isidro on the chart of The Strait of Magalhaens, 2 May, 1832 (Fig. 12) and on the chart in King, *Narrative*. This name was originally given to the point by Sarmiento.

[2] Bahía San Nicolás, lat. 53°51′S, long. 71°06′ W. The name was given by Bartolomè and Gonzalo Nodal in 1618; Sarmiento had called it Bahia de Santa Brigida y Santa Aqueda; de Gennes called it French Bay (King, *Sailing Directions*, p. 50).

[3] King, *Sailing Directions*, p. 51: 'Captain Stokes recommends in his journal, in coming in, to keep sail upon the ship, in order to shoot into a good berth, on account of the high land of Nodales Peak becalming the sails; and, to avoid the drift of the stream of the river setting the ship over to the eastern side of the bay. I do not, however, think that the stream of the river can affect a ship in any position between the islet and the peak. In taking up an anchorage, much care is necessary to avoid touching the bank. Less than 10 fathoms is not safe, but in that depth the security is perfect, and the berth very easy to leave.'

[4] Hawkesworth, *An Account of the Voyages*, I, pp. 62–3.

wood for the use of the French Settlement which he had then just formed on the Falkland Islands.[1] After seeing the abundant supplies of that article which Fresh Water Bay[2] and Port Famine afford, I had shared in the surprize which Byron expresses that any one should have come so far up the Straits to get it;[3] but on examining the spot I found that a happier selection could not have been made. It is a little Cove, just round the eastern point of the Bay of S^t. Nicolas, about a hundred yards [p. 3] wide and three times as long: here, moored to the shore, a ship may lie in 8 fathoms, perfectly sheltered from every wind, the water as smooth as in a wet dock; shapely trees of all dimensions growing within a few yards of the shore, and the wood when felled may be hoisted on board at once from the beach by tackles from the yard-arms: here, too, with a very little trouble, a supply of good fresh water may be got from the many little streams that will be seen making their way through the underwood of the flowering shrubs that skirt the margin of the beach. As we pulled up this sequestered Nook the unusual sounds of our oars and voices put to the wing multitudes of birds, and the surface of its waters was broken from time to time by shoals of fine fish. Some very eatable geese were shot, but we arrived too late to admit of our hauling the seine; my boat's crew, however, contrived to half fill the boat with the excellent muscles & limpits that are found here in great plenty – We could discover no traces of inhabitants.

Should any Voyager enter these Straits from the <u>eastward</u>, for a purpose similar to that which brought Bouganville hither, I would recommend to him this little Cove as better suited to it than any thing he will find between Cape Virgin Mary[4] and Cape Froward: I shall add that he may without hazard place confidence in the descriptions

---

[1] Bahía Bougainville (also known as Puerto Jack), lat. 53°50′S, long. 71°04′W.

Bougainville, *Voyage*, p. 158: 'I had there, without difficulty, in 1765, taken a loading of wood for the Malouines, and the crew of the ship had given it my name.'

Gallagher, *Byron's Journal*, pp. 67–8: 'Tuesday Feby 19th ... The French ship steering after us. I imagine she is either from the Malivin Islands (as the French are pleased to call them) to get Wood, or else upon a Survey of these Straits, I should rather think the latter, for if the first had been his intention it certainly was not necessary for him to come up as high as they are now to procure it. ...

Tuesday Febry 26th ... At 3 passed by the French Ship in a little Cove about 2 Leag^s to the S^{o}ward of Point Shutup, she had hauld her Stern close into the Woods, & we could see large Piles of Wood which they had cut laying on each side of her, so that I think there can be no doubt but she is here on purpose to get Wood for their New Settlement.'

[2] Bahía Agua Fresca, lat. 53°23′S, long. 70°58′W.

[3] Hawkesworth, *Account of the Voyages*, I, p. 63.

[4] This is Cabo Vírgenes, lat. 52°20′S, long. 68°21′W. It was originally named by Magellan for the eleven thousand virgins who, according to medieval legends, accompanied Saint Ursula from Britain to Rome, and were martyred with her on her return journey, at Cologne, on whose feast day, 21 October, he discovered the cape, in 1520. Sarmiento called it Cabo de la Virgen Maria and it is shown with that name on the chart (dated 1769) in Sarmiento, *Viage al Estrecho de Magallanes*. Sir Francis Drake, *World Encompassed*, p. 34, states 'August 20. we fel with the cape; neere which lies the entrance into the straight, called by the Spaniards *Capo virgin Maria*.' It is not named in Hakluyt's account of Drake's voyage, Hakluyt, *Principal Navigations*, III, pp. 730–42, or in the account of the voyage by Nuño da Silva, ibid., pp. 742–8; John Winter's account refers to it as Cape Victorie, ibid., p. 752. In the account of the voyage by Thomas Cavendish, notes by Thomas Fuller, Master of the *Desire*, it is referred to as Cape Joy, ibid., pp. 825, and 828–9. It is called Cape Virgin Mary by Narbrough and shown with that name on his chart (Fig. 4), and appears to have been known by this name to most English mariners. It is shown on the chart of 'The Strait of Magalhaens', published in 1832 (Fig. 12), and that published with the account of the voyages of the *Adventure* and *Beagle*, (King, *Narrative*) as C. Virgins.

and plans of these parts given in the Spanish Account of Don Antonio de Cordoba's Voyage to the Straits in 1786 in the Frigate Santa Maria de la Cabeza.[1]

January 16th

We weighed and made sail at daylight; but being becalmed by the high land had to tow the ship for a couple of hours ere we could clear the bay. During the fore part of the day the wind was westerly, sometimes light and sometimes squally; we made the best of our way to windward: In the course of the afternoon it fell calm, and the current was setting us rapidly to the eastward, when about 8 PM, a light breeze sprung up which enabled us to close the northern shore, and with the assistance of the boats ahead towing, we reached a little bay about a league and a half to the eastward of Cape Froward; in which, at 9.30 PM, we anchored in 11 fathoms, sandy bottom; about two cable lengths off a sandy beach.[2]

17th.

Our little bay had screened us so effectually from the wind that though when we weighed at 5 in the morning, we had the breeze so light as scarcely enabled us with all sail set to clear its entrance, no [p. 4] sooner were we outsides than we were obliged to treble reef the topsails. Wind at WSW, dead foul down the Reach. We continued to beat to windward under a heavy press of sail, our object being to double Cape Froward, and secure if possible an anchorage ere night fall under Cape Holland[3] six leagues further to the westward. At starting we extended our boards right across the Straits to within a third of a mile of each shore; making however, little progress we tried whether by confining our tacks to either coast we might discover a tide by which we might profit: for that purpose I began with the north side, for though we were there more exposed to the violent squalls that came down the vallies, I sought to avoid the indraught of the various channels that intersect the opposite coast: having made several boards without any perceptible advantage, I tried the south shore, and with such success that I was induced to keep it aboard during the remainder of the day's beat.

And here let me insert a remark – the result of my subsequent experience – that in consequence of the westerly winds which blow throughout the <u>Western</u> Division of the Straits of Magellan with the unvarying constancy – as regards their <u>direction</u>, not their <u>force</u> – of a tradewind, a perpetual current setting to the eastward, commonly at the rate of a mile and three quarters an hour, will be found in mid-channel: the <u>tides</u> exert scarcely any influence beyond a mile and a half from each shore; these tides, too, appear to set up one side of the Straits & down the other; the <u>weather</u> tide is in general sufficiently indicated by the ripple it causes.[4]

Heavy squalls off Cape Froward, obliging us repeatedly to cap the topsails[5] and clew

---

[1] *Relacion*.

[2] This may have been Bahía Rosa, lat. 53°53′S, long. 71°12′W.

[3] Cabo Holland, lat. 53°47′S, long. 71°41′W.

[4] The current is east going in the Strait at a rate of ½ to 1 knot, but it is much influenced by the prevailing W and NW winds. The tidal flood stream enters the Strait from both ends meeting in the vicinity of Cabo Crosstide (lat. 53°33′S, long. 72°26′W); the ebb flows outwards. *South America Pilot*, II, p. 223.

[5] The practice on the approach of a squall was to get the upper sails down which involved hauling the weather brace to bring the yard clear of the lee rigging, letting go the halliards and hauling the clewlines, buntlines and reef-tackles to get the yard down onto the cap, the clewlines then had to be hauled taut and

up the courses:¹ by day, however, their approach is announced in abundant time for the necessary precautions by their curling up and covering with foam the surface of the water and driving the spray before them like a cloud of smoke.

At 1 PM, we doubled Cape Froward. This Cape; called by the Spaniards El Morro de Santa Aguada² – the southernmost point of all America – is a bold promontory of dark-coloured slaty rock, its outer face nearly perpendicular, & within a [p. 5] quarter of [a] mile of it there is no bottom with 80 fathoms of line, it makes (both coming from the eastward & the westward) as a high round topt hill, connecting itself – by a ridge of land which has an uneven outline, convexes downwards, & runs by compass N by E – with a range of lofty craggy snow-topt mountains inland. The annexed drawing by Lieut^t. Sholl of the Beagle is a correct representation of this remarkable headland.³

By a little after nine in the evening we anchored under Cape Holland in 10½ fathoms, clay bottom; the low point, which forms the eastern extremity of the Cape, bearing E by N (true) distant ¾ of a mile.

Bougainville in his "Voyage autour du Monde" in the year 1767, observes that "Cape Froward has always been a point very much dreaded by Navigators."⁴ To double it & gain the anchorage under Cape Holland certainly cost the Beagle as tough a sixteen hour's beat as I have witnessed of a long time: for the number of tacks we had to make – no less than thirty-four – and the squally state of the weather which kept us constantly on the alert making or shortening sail, scarcely allowed the crew to have the ropes out of their hands the whole time; throughout the day, too, we were never once favoured with a slant;⁵ as soon as we had doubled Cape Froward & got into the next reach, we had the wind right down it: But what there is to inspire a Navigator with dread, I can't make out; for the coast on both sides is perfectly clear & a ship may work right across from shore to shore without hazard: There are squalls, now & then, it is true: but they give, as I have already noticed, such abundant warning of their approach, that mere common vigilance will render them harmless.

---

secured to prevent the yard mastheading itself again: J. Harland, *Seamanship in the Age of Sail*, London, 1984, pp. 221–2. It has not been possible to find the expression 'to cap the top-sails' in the standard nautical dictionaries of the period, but it would appear to be a reference to bringing the yards down onto the caps (the block of wood at the top of the lower mast through which the upper mast was hoisted), as described above.

¹ Truss the clews (the lower corners of the sail) up to the yards by the clew lines: Burney, *Dictionary*, pp. 89–90. The courses are the sails set on the lower yards.

² King, *Sailing Directions*, p. 52: 'The hill that rises immediately above the Cape, was called by Sarmiento, the Morro of Santa Agueda*.

* Sarmiento [de Gambóa, *Viage al Estrecho de Magallanes*], p. 218.'

³ Fig. 13.

⁴ Bougainville, *Voyage*, pp. 167–8: 'I here add a particular chart I have made of this interesting part of the coast of Terra del Fuego. Till now, no anchoring place was known on it, and ships were careful to avoid it. The discovery of the three ports which I have just described on it [port of *Beaubassin* (Caleta Beaubassin), bay *de la Cormorandiere* and bay and port of the *Cascade* (Caleta Cascada)], will facilitate the navigation of this part of the straits of Magalhaens. Cape Forward [*sic*] has always been a point very much dreaded by navigators. It happens but too frequently, that a contrary and boisterous wind prevents the doubling of it, and has obliged many to put back to Bay Famine. Now, even the prevailing winds may be turned to account, by keeping the shore of Terra del Fuego on board, and putting into one of the above-mentioned anchoring places, which can be done almost at any time, by plying in a channel where there is never a high sea for ships. From thence all the boards are advantageous, and if one takes care to make the best of the tides, which here begin to have more effect again, it will no longer be difficult to get to Port Galant.'

⁵ A wind of which advantage may be taken: Smyth, *Sailor's Word-Book*, p. 631.

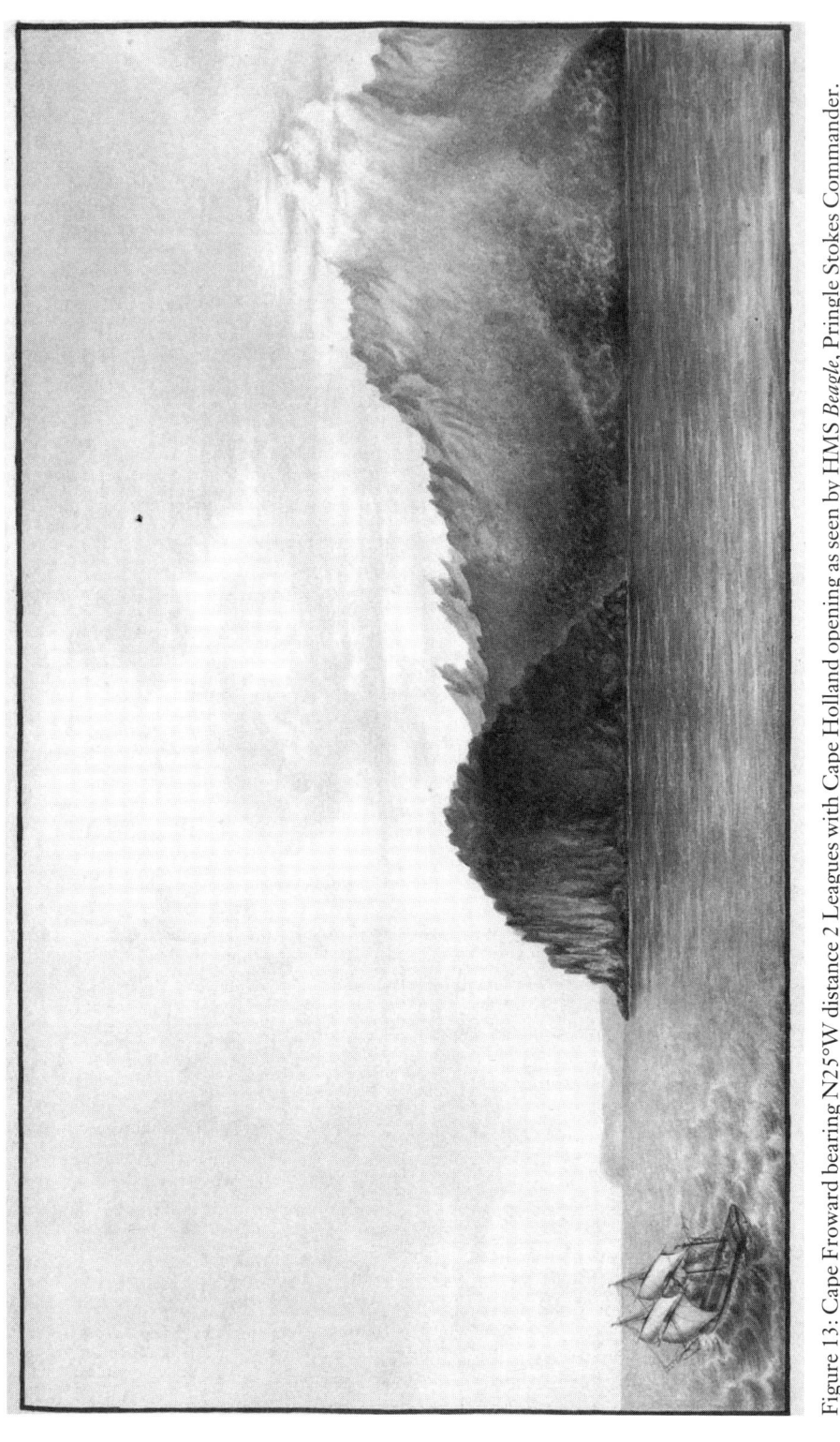

Figure 13: Cape Froward bearing N25°W distance 2 Leagues with Cape Holland opening as seen by HMS *Beagle*, Pringle Stokes Commander. Drawn by Lieutenant R. H. Sholl. Provided by the UK Hydrographic Office.

January the 18th.

Early in the morning I went on shore (with the necessary instruments) accompanied by the Assistant Surveyor Lieut[t] Skyring, for the purpose of determining the Latitude & Longitude of Cape Holland, and making a plan of the anchorage and adjacent coast. The weather throughout the day was showery & unsettled, the wind steady at WSW; strong and squally.

In the intervals of our surveying occupations, I tried to make some progress in-land, but a thick jungle composed [p. 6] of the stunted growth of the birch, the Winter's bark tree, & <u>arbutus & berberis</u> bushes, effectually prevented my advancing in any direction beyond the third of a mile from the beach.

The arbutus and berberis were plentiful here and good; though not as yet quite ripe: they are an excellent, & very pleasant, anti-scorbutic. The Winter's bark tree grows in this place to no considerable size; but its bark[1] had all the pungency of the hottest pepper, and was much used and relished by my people, as a capital spice. Wild celery abounds along the skirts of the jungle, and we found it very good, both as a pot herb & a sallad. Brant-geese and ducks of a tolerable flavour, may be shot: Mullet and Smelt, of the finest & largest kinds, are to be seen here in large shoals; and the seine may be hauled – without hazard of injuring it, along the sandy beaches on either side of the river to the eastw[d]. of the Cape. Fuel may be cut with ease, and in abundance; & good fresh water procured, at halftide, from the river: The landing I should deem to be very rarely either dangerous or difficult. With the exception of one forlorn hut, which, from the appearance of the muscle & limpit shells which strewed its area, seemed to have been long deserted; we saw not a trace of Inhabitants.

There is good anchorage along the coast for nearly four miles to the eastw[d]. of Cape Holland at the distance of ½ a mile from the shore, for the most part in 12 fathoms, with a capital holding ground of tough clay; we found it so tenacious, that on this, as on many similar occasions, Captain Charles Phillip's patent capstan[2] (with which the

---

[1] [Note in original] 'The "Winteranus Cortex magellanicus" of pharmacopology'.
 The bark of this tree was first brought to England by Captain John Winter of the *Elizabeth* (who sailed with Drake in 1577 for the South Seas and having been separated from him in a storm after passing through Estrecho de Magallanes returned to England) in June 1579. It was called, after him, Cortex Winteranus, by Carolus Clusius (Charles de l'Ecluse, 1526–1609), a Flemish botanist who visited England on a number of occasions and with the help of Sir Francis Drake obtained a number of plants from the New World: Woodville, *Medical Botany*, IV, p. 648, Blunt and Raphael, *Illustrated Herbal*, p. 156–7. It is described by Sir Richard Hawkins: 'This Tree carrieth his fruit in clusters like a Hawthorne, but that it is greene, each berry of the bignesse of a Peppercorne, and every of them contayning within foure or five granes, twice as bigge as a Musterd-seed, which broken, are white within, as good as Pepper, and bite much like it, but hotter. The barke of this Tree, hath the savour of all kinde of Spices together, most comfortable to the stomack, and held to bee better then any Spice whatsoer. And for that a learned Countriman of ours Doctor Turner, hath written of it, by the name Winters Barke, what I have said may suffice. The leafe of this Tree is a whitish greene, and is not unlike to the Aspen leafe.' Purchas, *Hakluytus Posthumus*, XVII, p. 125. The reference may be to Dr Peter Turner (c.1542–1614) who was a friend of Revd Thomas Penny (1530–89) an entomologist who corresponded with Clusius: Desmond, *Dictionary*, pp. 488, 623.

[2] Captain Charles Phillip was granted Letters Patent for his improved capstan, 20 September 1819 (TNA C 66/4214, item 11) under the terms of which, provided he submitted a specification within 6 months, he and his heirs had sole rights for fourteen years. He submitted the specification, 18 March, 1820 (TNA C54/9908, item 3, containing very detailed drawings), which states: 'My invention of an improved capstan consists in an arrangement of wheels, cogs and other machinery as hereinafter described by which an increased velocity of motion is given to the whelps [the projecting parts that rise out of the barrel of the capstan to increase its diameter and

Beagle was fitted) proved to us of special advantage. There is a Harbour, immediately under Cape Holland, called by the Spaniards, Puerto Salano,[1] but, as will be seen from our plan, it is altogether too confined to be eligible for any but very small vessels. On the whole, I take the best berth in the neighbourhood of Cape Holland to be that in which the Beagle was anchored, the bearings &c for which are in the preceding page. [p. 7]

January the 19th.

But no one who desires a spacious secure port will be under Cape Holland, if he can reach the Harbour of Cape Gallant;[2] for which place at seven in the morning, we weighed and made sail. The wind westerly, dead foul down the reach, fresh breezes & squally, the rain incessant, and the weather so thick that scarcely at any one time could we see both sides of the Straits.

In standing in to the north shore when about a league to the westward of Cape Holland, we struck soundings several times in 20 fathoms on a bottom of coarse sand, the nearest shore distant two thirds of a mile: this might be of use to a Vessel working to the westward as an anchorage in light winds to prevent her from being swept away by the tide or current; but it affords no shelter from the prevailing breezes. Between Cape Holland & Cape Gallant[3] there is likewise anchorage in what is called, by the English Cordes' Bay, by the Spaniards Bahia de San Miguel;[4] and within it is a very fine Harbour; but the entrance of the bay is so beset with ledges of sunken rocks, that if the weather be bad & a ship cannot reach Cape Gallant Bay, and does not wish to remain under sail during the night, (a measure which I deem scarcely advisable in any part of the Straits under any circumstances of weather) her best plan will be to gain with all expedition the anchorage under Cape Holland.

By 7 PM we had fetched into Cape Gallant Bay, in which there is very good anchoring ground; but the weather being bad, with every appearance of becoming worse, I worked at once into the Harbour, & having warped the ship up so as to be completely land-locked, moored in 4 f^{ms}, over a bottom of oaze, ½ a cable each way and an open hawse to the southward.

---

allow a greater portion of the cable to be in contact with it] and other parts of the capstan connected with them at the expense of a certain degree of power or an increased power by a diminution of velocity.'

[1] This would appear to be Bahía Wood (Woods), lat. 53°49′S, long. 71°37′W, shown on the chart as Woods Bay. Named by Narbrough for his mate on November 4. 1670. Narborough, *Several Late Voyages*, p. 71. and Hacke, *Collection of Original Voyages*, Captain Wood's Voyage, p. 89.

[2] This is now known as Bahía Fortescue, lat. 53°42′S, long. 72°00′W, through which Caleta Gallant, lat. 53°41′S, long. 72°00′W, is entered. Hacke, *Collection of Original Voyages* (Captain Wood's Voyage), p. 89, states 'This Bay is to the Eastward of Cape *Gallant*, to which we gave the name *Fortescue's* Bay, and within which is a fair sandy cove for small ships called by our Captain Port *Gallant*.' (This was Narbrough's voyage on which John Wood served as mate. Narbrough refers to Fostcues Bay, Narborough, *Several Late Voyages*, p. 71, in this vicinity, which may be a misprint.) John Fortescue, after whom the bay would appear to have been named, and Lieutenant Thomas Armiger were with the party detained by the Spanish at Valdivia and left there by Captain Narbrough. Narborough, *Several Late Voyages*, pp. 101–12, See also Bradley, 'Narborough's Don Carlos', in *Mariner's Mirror* for more details of their subsequent detention. Burney, *Chronological History*, III, p. 355, suggests that Port Gallant received its name during Cavendish's first voyage after a small ship in his squadron, the *Hugh Gallant*. See also p. 159.

[3] Cabo Gallant, lat. 53°42′S, long. 72°02′W.

[4] Puerto San Miguel, lat. 53°41′S, long. 71°53′W, is approached today through Bahía Cordes, lat. 53°43′S, long. 71°45′W. Bahía Cordes was named after Simon de Cordes, see p. 162.

All hands having been exposed during the whole of the day to an incessant drenching rain, I thought it proper, as soon as the ship was secured to splice the main brace by causing to be served out to every man an extra gill of spirits: adopting the precaution to direct that before it be issued, every one should present himself in dry clothes before the Officer of his Division. [p. 8]

January the 20th.
Contrary to our expectations, on the preceding evening, the day proved fine and clear; and Lieut^t Skyring & I availed ourselves of it, to fix the position of Cape Gallant and survey the Harbour, Bay & adjacent coast.

The Latitude I determined by circum-meridional altitudes of the Sun's upper & lower limbs,[1] observed alternately in a roofed quicksilver horizon, with an excellent Sextant (by Jones[2]) graduated to 10 seconds; The difference of Longitude was deduced from the mean result given by three admirable Chronometers:[3] and the Angles & Azimuths were measured by a very capital Theodolite (by Jones) of 6 inches diameter, graduated to 30 seconds.

I have been thus minute in mentioning our instruments &c, in order that, when I report hereafter having fixed the position of any place astronomically, it may be understood (without further explanation) that the instruments & methods mentioned above, were always employed.

... 21st.
The succeeding day, as well as the 20th. was clear and fine, and consequently favourable to our surveying operations: so we were enabled, in this instance, to get through our work satisfactorily & pleasantly. The plan of the Harbour, Bay &c. which we formed, and which accompanies this Journal,[4] will convey at a glance, much better than the most lengthened description, all the necessary information respecting the capacity of the Harbour, Bay &c. the depth of water and the shelter [p. 9] it affords from the prevailing westerly breezes.[5]

---

[1] A circum-meridian observation consisted of a series of timed altitudes of an heavenly body taken while the body crossed the observer's meridian. From these it was possible to obtain a latitude which was more accurate than that calculated from the meridian altitude alone.

[2] Probably Thomas Jones (1775–1852), who described himself as 'Astronomical, Mathematical, Optical and Philosophical Instrument Maker of 62 Charing Cross.': Taylor, *Mathematical Practitioners*, pp. 342–3. There is correspondence with this firm in the UK Hydrographic Office, from about this date, although nothing about the instruments for this expedition appears to have survived.

[3] [Note in original]. 'Note ... The Chronometers were N°. 228 & N°. 254 by Messrs. Parkinson & Frodsham, – (these two time pieces had under gone a most satisfactory trial in two arctic Voyages with Captain Parry); – our third Chronometer was N°. 134 by James McCabe, of London: – the abundant opportunity afforded (during a lapse of more than twelve months, and through very great and unusual changes of temperature, that they had been subject to while in my charge) of becoming thoroughly acquainted with their going – warrants me in saying that rarely have been constructed, even by British Artists, three such admirable pieces of mechanism.'

[4] Fig. 14.

[5] King, *Sailing Directions*, pp. 56–7: 'FORTESCUE BAY is the first best anchorage to the westward of St. Nicholas Bay. – It is spacious, well sheltered, easy of access, and of moderate depth. The best berth is to the south-east of the small islet, outside of Wigwam Point, in 7 or 8 fathoms. Having the entrance of PORT GALLANT open, small vessels may sail into the port, but the channel is rather narrow. The banks on the

Figure 14: *Plan of Port Gallant*, HMS *Beagle*, 1827. Provided by the UK Hydrographic Office.

In the Spanish books and charts the Harbour of Cape Gallant, is called, El Puerto de San Josef, and the Bay, Bahia de Forte Escudo.[1]

A ship working through the Straits, & merely wanting an anchorage for the night when hereabouts, will of course anchor in the Bay of Cape Gallant: a reference to the accompanying plan, and to the bearings given in the Margin[2] will enable her to select the best berth in the bay.

If it be required to get a supply of water & fuel, or to repair any damage in spars, hull &c. the Harbour Cape Gallant will be found safe and commodious.

The entrance of the Harbour is so confined, & the squalls that come off the circumjacent high land, often, so various and violent; that a square rigged vessel will rarely succeed in working into it: If a ship have not a leading wind – her shortest plan will be to stand in as far as she can fetch, into the entrance; and then down anchor, run out warps, and warp her up to the berth that may suit best: It is a capital harbour for any Vessel drawing less than <u>three fathoms & three quarters</u>: she may lie in about that depth (indeed, as the plan will immediately shew, a squadron may lie in that depth) perfectly sheltered from sea; land locked – a <u>soft</u> and <u>tenacious</u> holding ground. Excellent fresh water may be procured, at $1/3^d$ tide, (from the river marked in the plan) at the distance of a ship-length: Fuel abundant, and quickly procurable: and the Birch trees will afford spars of <u>sufficient size</u>, and <u>suitable straightness</u>, to replace a Frigates's Jib-Boom or Top Sail Yard. On both sides of the river (already referr'd to) the sandy beach affords places where the seine may be hauled with success, and with little hazard of tearing. <u>One boat</u> in <u>one hour</u> procured as many fish (mullets & smelts) as served us all on board (60 persons) more than two days.

The vegetable productions here are the same [p. 10] as I have mentioned having found under Cape Holland.

We found not a trace of a quadruped of any kind. The Birds were exactly such as were found at Port Famine, with the exception that no Parrots were seen here,[3] and that one of a species of the feathered tribe which a naturalist would hardly expect to meet in these latitudes, namely a <u>humming bird</u>,[4] was shot on the beach of the Harbour

---

western side, off Wigwam Point, are distinguished by kelp. When within, the shelter is perfect; but Fortescue Bay is quite sufficiently sheltered, and much more convenient to leave.'

[1] *Relacion* has a plan of 'Puerto Galan ò de S^n. Josef y Bahía de Fuerte Escudo'. These names are also used elsewhere in *Relacion* and *Apéndice a la Relacion* and on Córdoba's charts.

[2] [Note in original]
The outer point of Cape Gallant WbS½S.– (true)
The eastern extreme of the Bay E¼S.– (true)
The white patch on Charles's Island SW½S.– (true)
13 fathoms, oazey bottom.

[3] The only parrot commonly found in Estrecho de Magallanes is the Austral Parakeet, *Enicognathus ferrugineus*, specimens of which from King's voyage are held in the Natural History Museum's bird collections at Tring. King, *Narrative*, I, pp. 38 and 534: 'In the former the "parroquet" is said to have been sighted in Port Famine, and in both instances it is referred to as *Psittacus smaragdinus*, a synonym for *Enicognathus ferrugineus*.'

[4] *Relacion*, p. 318: 'Los Indios regalaron al Comandante D. Antonio de Córdoba un paxarito mosca ó calibre muerto y desecado con quasi todas sus plumas, enteramente semejante á los que se crian en los paises mas cálidos' – translated at *Voyage of Discovery*, p. 81, 'The Indians presented to the commander, Don A. Cordova, a small humming-bird, dead and dried up, but with almost all its feathers, in every respect similar to those found in the hottest climates.'

of Cape Gallant by one of my young midshipmen.[1] It has been extremely well preserved, and I send it to you as an interesting specimen to be added to the admirable Museum of the Expedition which you have formed on board the Adventure.

Just before we left Cape Gallant some papers were brought to me, which had been found by some of my Officers on the top of one of the lofty mountains that surround the Harbour; fragments of the glass bottle in which they had been originally deposited, were lying on the spot in which they were found – it had most likely been burst by the frost:[2] The papers had suffered somewhat from the weather; but they were legible. They proved to be two latin inscriptions, one by Bougainville in 1767 & the other by Cordoba[3] in 1789, stating the objects of their respective voyages & the names of the principal officers of the ships. I forward then to you herewith.

My observations place the Easternmost point of Cape Gallant in Latitude 53°.42′.11″ South; and in Longitude 1°.05′.05″ West of Cape Virgin Mary, the Northeasternmost headland of the Straits.[4]

The Variation of the compass is 23°.35′ Easterly.

The only hummingbird species commonly found in this area is the Green-backed Firecrown called by Vigors *Mellisuga kingii*, from a specimen sent to him by King, and now known as *Sephanoides sephaniodes*. King, *Narrative*, I, p. 535, records having seen it in a snow storm in May: 'It seemed regardless of the cold, and so long as the *fuchsia* and *veronica* were in blossom, so long did this hitherto supposed to be delicate little bird, remain to cull their sweets, or rather to prey upon the insects which buried themselves in the flowers; for, innocent as it seems, it is insectivorous.' This hummingbird feeds predominantly on nectar but will also take small insects.

[1] There was only one midshipman listed on board, J. L. Stokes, who was to remain with the ship until he succeeded to the command in March 1841 during the survey of the Australian coast where he remained until 1843. It could be that Commander Stokes was including the mate, J. Kirke, also a midshipman, and the two volunteers, R. F. Lunie and W. Jones, in this category, in which case it might have been any one of them.

[2] Macdouall records that he read a note in Wallis's account that the Master of the *Swallow* [George Robertson] had left a bottle containing a shilling and a note under a pyramid which he had constructed on one of the higher mountains [Hawkesworth, *Account of the Voyages*, I, p. 388] and that Captain Stokes had ordered a search to be made for it. The first attempt was unsuccessful, but on the second, a day later, the surgeon, Dr E. Bowen and clerk, John Macdouall, located a four-foot pyramid which proved to contain messages left by Bougainville and Córdoba: Macdouall, *Narrative*, pp. 90–6.

[3] [Note in original] 'Cordoba commanded two Expeditions, at different periods, for the survey of the Straits in neither of which did he ever reach to the westw^d. of Cape Gallant.'

This is not entirely true. Córdoba's ships, the *Santa Maria de la Cabeza* on the first voyage (1786), and *Santa Casilda* and *Santa Eulalia* on the second (1789), anchored in Caleta Gallant, and did not get further westward, but he sent surveying parties away on both occasions which did. In 1789 he sent two boats with a chronometer, Arnold No. 71, which examined the western part of the Strait and fixed Cabo Pilar and Cabo Victoria. See also pp. 172–4.

The finding is recorded at *Relacion*, p. 48 – translated at *Voyage of Discovery*, p. 48, 'In one of our excursions to the mountains which surround Port Galan, we found a bottle sealed up, containing a long Latin inscription, and placed there by M. de Bougainville, as he went this way, on his voyage round the world in 1768. In imitation of his example, we left another of the same kind; and gave to the mountain on which these monuments were left, the name of Cerro de la Cruz.' In February 1789, the papers were replaced and a new cross erected on the hill: *Apéndice a la Relacion*, p. 86.

King, *Narrative*, I, pp. 70–71, details the inscriptions and states that the originals were placed in the British Museum and that copies were made on vellum and replaced on the same spot.

[4] This difference of longitude is in error. The current positions give a difference of 3°41′. The figure quoted is the difference from Port Famine as given in the Geographical Position Book, Appendix 1.

January the 22nd.

At 4 in the morning we weighed[1] and made sail, with a light breeze from the ESE which freshened up in the course of the forenoon. We pursued our way to the westward with studding sails set low & aloft, until Noon when the ship lay to for an hour, [p. 11] while I landed (with the requisite instruments) upon York Point,[2] accompanied by the Assistant Surveyor, and determined its Latitude,[3] and took various angles with the theodolite for the Survey; that done, we returned on board, and the Ship continued her course with the same favouring breeze & lovely weather.

About this part, a heavy tide ripple will be observed, caused by the meeting of the waters of Jerom's Channel[4] with those of the Straits.

We were soon abreast of Jerom's Channel: Between it and Cape Gallant, the northern shore of the Straits exhibits for the most part a very agreeable landscape; in the distance, are lofty peaks and craggy mountainous ridges covered with snow, in the near are seen – oftentimes in extremely pleasing combinations – mountains hills & vallies with green sloping sides, well-grassed plains, woods & copses, water falls, rivers, and little streams, along the coast-line are several sandy coves and bays affording good anchorages; this part, too, appears to be a favourite haunt of the various water-foul of the Straits. But it is only to the northern shore this description applies; for the opposite coasts, as well as the islands that lie between, are rocky and in general devoid of verdure. Beyond Jerom's Channel to the westward, the land on both sides of the Straits wears throughout the remainder of their extent that aspect of rugged barrenness which led Sir John Narborough (in 1669) to call that part, with great propriety, "South Desolation, it being as he says, so desolate land to behold."[5]

Jerom's Channel opens most when bearing [ ] and appears to be [ ][6] miles at its entrance. I had fain explored it but that the fine fair wind we now enjoyed afforded me the prospect of soon accomplishing the points of more importance at the Straits western mouth to which my attention had been specially enjoined. This channel might, by possibility, cause a mistake (indeed it has already in more than one instance, particularly in that of Bougainville)[7] in as much as the northern shore of this Reach of

---

[1] 'Captain Stokes, in his haste to get through the straits, had forgotten to take on board several of the boarding-pikes which had been arranged along-shore, and to which were affixed silk handkerchiefs, for the purpose of survey. They were consequently left there, though he intended to remove them on his return to the harbour.': Macdouall, *Narrative*, pp. 96–7.

[2] Punta York, lat. 55°35'S, long. 72°17'W, at the eastern end of Rada York, named by Narbrough: Narborough, *Several Late Voyages*, p. 75.

[3] [Note in original] 'The Latitude of York Point is 53°.32'.35" South'.

[4] Canal Jerónimo, lat. 53°33'S, long. 72°23'W.

[5] Narborough, *Several Late Voyages*, p. 78: 'From *Cape Quad* to *Cape Desseada*, it is about twenty eight Leagues; and the Streight lies near North-west, and by West from *Cape Quad* into the South-Sea, and near in one Reach, which I call'd *Long-Reach:* and some of my Company call'd it *Long-Lane*. This part may properly be call'd the *Streights*; for it is high land all the way on both Shores, and barren Rocks, with Snow on them; and indeed from *Cape Quad* into the South-Sea, I call'd this Land *South-Desolation*, it being so desolate land to behold.'

[6] Spaces left blank in original. Canal Jerónimo opens most bearing 325° and is one mile wide at its entrance.

[7] Bougainville, *Voyage*, pp. 189–90: 'The entrance of this channel seems to be half a league broad, and in the bottom of it, the lands are seen closing in to the northward. When you are opposite the river du Massacre, or Bachelor, you can only see this false strait, and it is very easy to take it for the true one, which happened even to us, because the coast then runs W. by S. and W.S.W. till Cape *Quade*, which stretching very far, seems to close in with the westerly point of the isle of Louis le Grand, and leave no outlet.'

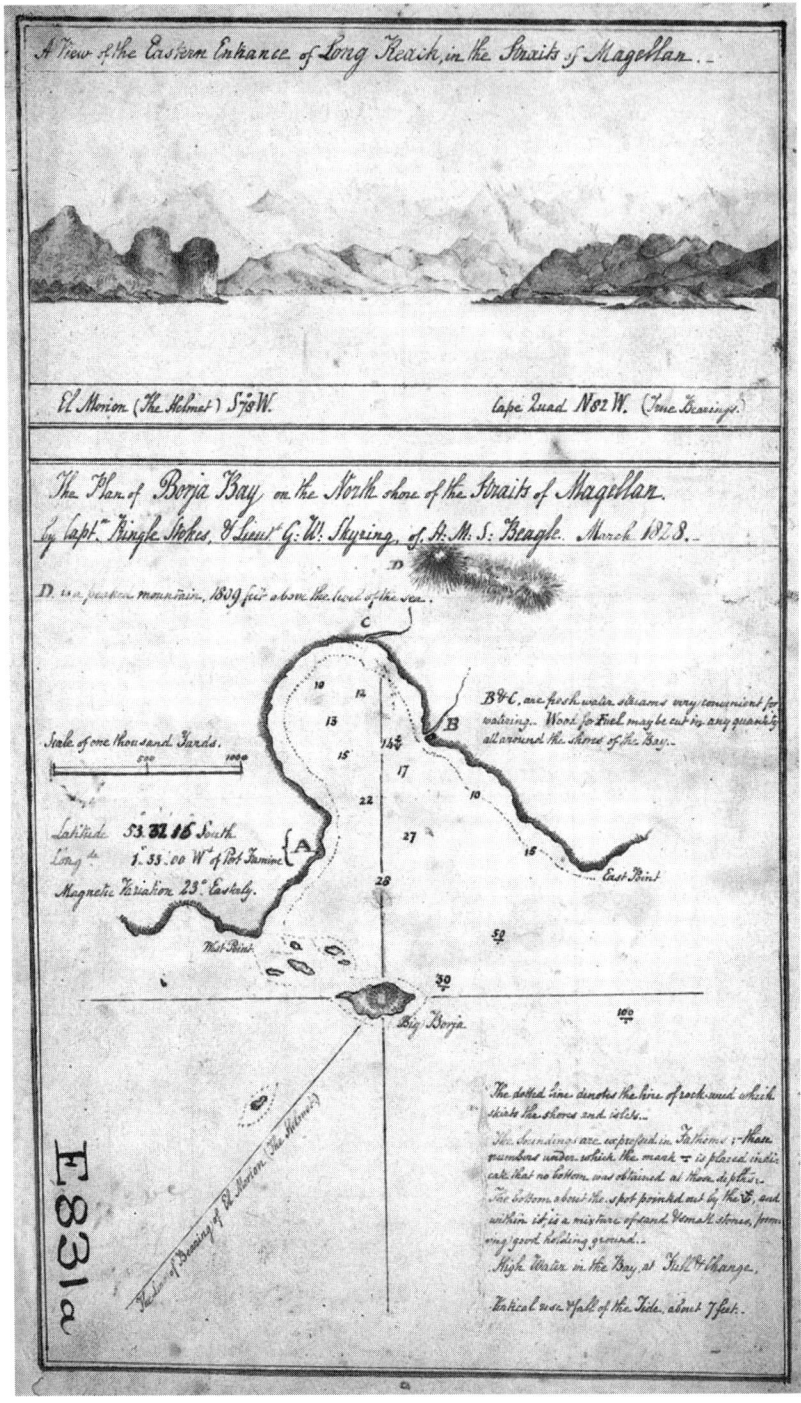

Figure 15: *The Plan of Borja Bay*, HMS *Beagle*, March 1828 with *A View of the Eastern Entrance of Long Reach, in the Straits of Magellan*. Provided by the UK Hydrographic Office.

Figure 16: El Morrion bearing S55°W and Cape Quod S75°W as seen by HMS *Beagle*, Pringle Stokes Commander. Drawn by Lieutenant R. H. Sholl. Provided by the UK Hydrographic Office.

which Cape Quad[1] is the northernmost[2] head-land, projects itself in such a manner on the southern shore as apparently to shut all further passage in [p. 12] the direction of the true channel of the Straits: but let the chart be consulted, and no one will ever deviate into Jerom's Channel. Sir John Narborough's remark is quite to the purpose: 'At Cape Quad the lands shut one with another as if there were no further passage, but as you make nearer to it you will see the opening more and more as the Straits round there more to the northward again.'[3]

At 3.30 PM, I landed at the base of a remarkable head land on the south side, from which Cape Quad bore N17°W & Cape Notch N47.50W; and got altitudes of the sun for settling its diff^{ce}. of Longitude by chronometer with the design of fixing its position astronomically as it forms a capital point in the triangulation of the two adjoining reaches. To this head-land, the Spaniards – from some fancied resemblance to that piece of armour – have given the name of "El Morrion" (The Helmet).[4] It is a lofty granite rock of which the outer face is perpendicular & bare & of greyish colour (this light coloured part is distinguishable several miles both to the east & to the west, forming an excellent leading-mark) on its summit grow dark coloured moss and shrubs, and at its base, very close to it, lie some low rocks. In this we are to see, I suppose, a Gigantic Casque & its dark plumes:[5] should Horace Walpole's romance happen to be suggested to a man's fancy by the Spanish appellation of the Helmet, he will hardly fail to see in Cape Quad (particularly when looking from the westward) the Castle of Otranto.[6] Of Cape Quad Sir John Narborough's description is very good, so I shall use his words: 'Quad is on the north shore and it is a steep up Cape of a rocky greyish face of a good height; before one comes at it, it makes like a great building of a Castle: it points off from the other mountains so much into the Channel of the Straits that it makes shutting in against the South Land & maketh an elbow in the Straits.'[7]

The ship passed Cape Quad. It was a lovely summer's afternoon – clear and warm, the breeze as favourable as we could wish. On several points of the southern shore were seen the fires kindled by the Natives – their customary mode, when [p. 13] a ship is descried, of inviting communication: we were, however, too anxious to make the most of the present prosperous breeze to admit of availing ourselves of their friendly invitation. We made out with ease Cape Notch,[8] Cape Buckly,[9] Cape Upright[10] and the head-lands of these rocky and desolate shores.

---

[1] Cabo Quod, lat. 53°32′S, long. 72°33′W (Fig. 16), projects well into the Strait from the northern shore.

[2] 'Southernmost?' inserted in the journal over northernmost, in a different hand.

[3] Narborough, *Several Late Voyages*, pp. 75–6.

[4] Cerro El Morrión, lat. 53°34′S, long. 72°32′W (Fig. 16).

[5] King, *Sailing Directions*, p. 65: 'Captain Stokes describes the MORRION, or ST. DAVID'S HEAD, to be a lofty granite rock, of which the outer face is perpendicular and bare, and of a light grey colour, distinguishable from a considerable distance both from the east and the north-west, forming an excellent leading mark to assure the navigator of his position.'

[6] *The Castle of Otranto, a story translated by William Marshall, Gent, from the original Italian of Onuphrio Muralto &c.* (pseudonym Horace Walpole, Earl of Oxford), London, 1765.

[7] Narborough, *Several Late Voyages*, p. 76.

[8] Cabo Notch, lat. 53°26′S, long. 72°48′W (Fig. 17).

King, *Sailing Directions*, p. 70: 'Captain Stokes remarks, that, the mountains of this part (Cape Notch) spire up into peaks of great height, connected by singularly sharp saw-like ridges, as bare of vegetation as if

Cape Quad, Cape Notch, Cape Upright &c these are homely names; but to us they were, nevertheless, highly interesting objects, associated as they are with circumstances of difficulty, danger & disaster, in the engaging narratives of Byron, Buckly,[1] Wallis, & Carteret.[2]

Here I tried the current in mid channel, and found it setting to the eastward, at the rate of nearly two miles an hour. No bottom, close to the shore, with 100 fathoms of line.

Towards nightfall the breeze became so faint as scarcely enabled us to stem the current: so I sought an anchorage, and at 9 PM anchored in 17 fathoms, oazey bottom, in Swallow Harbour[3] (so called by Carteret in 1767, after the name of his ship). There is a very fair plan of this harbour given in the Admiralty copy of Wallis & Carteret's Voyage:[4] one rock, indeed, in the centre of it is not correctly laid down, but danger can hardly arise from the inaccuracy, inasmuch as the rock is a[s] good as '<u>buoyed</u>' by the large patch of sea-weed that grows upon it in two fathoms water.

---

they had been rendered so by the hand of art. About their bases there are generally some green patches of jungle, but upon the whole nothing can be more sterile and repulsive than this portion of the strait.'

[9] Bulkeley's Island and Channel are shown on Commander Stokes's chart (Fig. 26), and can be identified as those known today as Islas Santa Ana, lat. 53°08'S, long. 73°20'W (of which Isla Pillolco is the western-most) and Golfo Xaultegua leading eastwards from the islands. The islands lie in the entrance to the gulf off the northern shore about midway between Cabo Notch and Cabo Upright. Isla Santa Ana with Punta de Bukley, now Punta Havannah (the south-west entrance point of the gulf), are shown on Córdoba's chart, 1788 (Fig. 9). Buckley's Sound (Golfo Xaultegua) and Buckley's Point are shown in the same place on the chart in Hawkesworth, *Account of the Voyages*, I (Fig. 7).

[10] Cabo Upright, lat. 53°05'S, long. 73°36'W.

---

[1] John Bulkeley was the Gunner of HMS *Wager*, one of the vessels which sailed with Anson on his voyage round the world in September 1740. The *Wager* was wrecked on Isla Wager (47°45'S, 75°00'W), 14 May 1741. Discipline was not maintained ashore and the party split. The ship's longboat was converted into a schooner and launched, 12 October, and, with the cutter and barge in company and a total of 81 men, sailed for Rio de Janeiro, with Lieutenant R. Beans, the Gunner, Mr Bulkeley, and the carpenter, Mr Cummins, on board. The barge returned to rejoin Captain Cheap who eventually managed to surrender to the Spanish, at Chiloé, with three others (one of whom was Byron) – all that survived of his party. They returned to England via Valparaíso. Meanwhile the schooner, named *Speedwell*, lost the cutter (6 November) and entered Estrecho de Magallanes, 10 November. After considerable hardship and loss of life they reached Rio de Janeiro, 28 January 1741/2, with only 30 men left, and eventually returned to England. At the subsequent court martial all were honourably acquitted of the loss of the *Wager*, except Lieutenant Beans who was acquitted, but with a reprimand from the court. The account of the *Speedwell*'s voyage was written by Bulkeley and Cummins, *Voyage to the South Seas*, 1743; 2nd edn, 1757.

From Bulkeley's account, the party in the *Speedwell* do not appear to have carried out any surveys or named any features in Estrecho de Magallanes. Byron refers to 'some remarkable hummocks on the north [in the vicinity of Point Possession], which Bulkeley, from their appearance, has called the Asses Ears': Hawkesworth, *Account of the Voyages*, I, p. 33. Bulkeley and Cummins, *Voyage to the South Seas*, 1743; 2nd edn., 1757, p. 148, only says: 'in Land lie two Peaks, exactly like Asses Ears'.

[2] The narratives of Byron, Wallis and Carteret are given in Hawkesworth, *Account of the Voyages*; Byron, *Narrative*, gives his account of the loss of the *Wager*.

[3] Bahía Swallow, lat. 53°30'S, long. 72°45'W, situated between Isla Carteret and the coast SE.

[4] Plan given in Hawkesworth, *Account of the Voyages*, I, 'To front Commodore Byron's Voyage'.

And here let me remark, that in the Straits of Magellan, <u>fixed</u> sea-weed is a never failing indication of foul ground.¹ Where patches of it occur will be found, sometimes, indeed, 10 fathoms, sometimes less than two, but invariably rocky bottom and consequently bad anchorage. Spots in which they are seen should be approached cautiously – the sounding lead in hand – until a more exact acquaintance warrants a bolder approach. This sea-weed is of two kinds: one, having <u>notched</u> edges is found growing from the depths above mentioned, the other,² having <u>smooth</u> edges, grows only on rocks that are awash at some time of tide.³ In the navigation of the <u>Western</u> divi-[p. 14]sion of the Straits (that is to say, from Cape Froward to the Isles of Direction) not only may the presence of fixed seaweed be deemed an excellent warning, but the absence of it an infallible assurance of freedom from danger. I have

---

¹ This indication of danger had been recognized from the early stages of navigation in the Strait. It is given in a 'Ruttier from the sayd river of *Plate* to the Streight of *Magelane*' printed in Hakluyt, *Principal Navigations*, III, pp. 724–6: 'And if you see beds of weeds, take heed of them and keepe off from them' (p. 726). The directions first appear in Hakluyt, *Principall Navigations Voiages*, II, pp. 806–8, which omits the remark about weed. It contains remarks on the existence and positions of 'a Towne called Iesus' and 'a towne called the Towne of King Philip', the two settlements established by Sarmiento in the Strait in 1584, which suggest that the rutter may be based on information obtained from Sarmiento after his capture by Sir Richard Grenville, in 1586. The 1600 version has corrected these remarks to indicate where the towns stood. It would appear therefore that the correction is either from Thomas Cavendish's expedition, 1586–8 (printed in Hakluyt, *Principall Navigations Voiages*, II, pp. 808–13, and Hakluyt, *Principal Navigations*, III, pp. 803–39), Captain Merick, 1589–90 (ibid., pp. 839–42), or Captain John Davis, 1591–3 (Thomas Cavendish's second expedition), ibid., pp. 842–52. Hawkins who sailed in 1593, having been captured by the Spanish did not return in time. The date of this caution is therefore between 1589 and 1600, and would appear to be from before 1593. Sarmiento has a similar caution in his published work 'whenever beds of sea-weed are seen they should be avoided. Some may have six, others ten fathoms, others much less under them.': Markham, *Narratives of the Voyages*, p. 104.

² [Note in original] 'a much larger kind'.

³ The seaweed with notched edges is probably the Giant Kelp, *Macrocystis pyrifera*, that with smooth edges a species of *Durvillea*, or possibly *Lessonia*. Hooker, *Botany of the Antarctic Voyage*, I, pp. 454–66, describes these seaweeds and says of 'D'Urvillæa' 'this, the *Lessonia* and *Macrocystis* are the three most remarkable *Algæ* of the Antarctic regions; the present exceeding any sea-weed, except *Lessonia* and the *Ecklonia buccinalis* of the Cape of Good Hope, in bulk; while the *Macrocystis*, to which we shall afterwards allude, is the longest vegetable product known' (ibid., p. 455). He also says that all three are abundant in Tierra del Fuego and the Falkland Islands, that *Durvillea* grows from the half-tide mark down and is invariably accompanied by the *Macrocystis pyrifera*: that *Lessonia* grow far beyond low-water mark, are 'arborescent, dichotomously branched trees' with trunks 5 to 10 feet long and as thick as the human thigh, that they grow alone or in miniature forest and are totally submerged during high-water or even half-tide but at low-water the branches project above the surface: the *Macrocystis* is notched and grows in a mean depth of six fathoms or more and that, between the Crozet Islands, he estimated the length of the largest specimen to be of the order of 700 feet (a figure which is still accepted today).

King, *Narrative*, I. p. 13, describes the weed thus: 'It is rooted upon rocks or stones at the bottom of the sea, and rises to the surface, even from great depths. We have found it firmly fixed to the ground more than twenty fathoms under water, yet trailing along the surface for forty or fifty feet. When firmly routed it shows the set of the tide or current. It has also the advantage of indicating rocky ground: for where there are rocks under water, their situation is, as it were, buoyed by masses of sea-weed (usually called by seamen kelp) on the surface of the sea, of larger extent than that of the danger below. In many instances perhaps it causes unnecessary alarm, since it often grows in deep water; but it should not be entered without its vicinity having been sounded, especially if seen in masses, with the extremities of the stems trailing along the surface. If there is no tide, or if the wind and tide are the same way, the plant lies smoothly upon the water, but if the wind be against the tide, the leaves curl up and are visible at a distance, giving a rough rippling appearance to the surface of the water.'

Figure 17: Cape Notch bearing WNW (*above*); Cape Upright bearing W by N (*below*), as seen by HMS *Beagle*, Pringle Stokes Commander. Drawn by Lieutenant R. H. Sholl. Provided by the UK Hydrographic Office.

restricted this remark to the Western division, for in the Eastern, the dangers are sand banks on which sea-weed does not grow. Floating sea-weed, detached & carried about by the action of the tides and currents, will occasionally be met with in various sized patches in different parts of the Straits at all depths of water. I recalled hearing before I left England on this expedition (and from Naval Men, too) that the navigation of the Straits of Magellan was rendered almost impracticable by reason of immense beds of sea-weed: how such a notion originated, I can't think; nothing, however, can be more erroneous. A ship's progress is as little impeded by seaweed in the Straits of Magellan as in the Straits of Dover. That which skirts the rocky coasts, or grows on rocky ledges, certainly presents an impediment to boats: but, as regards a ship, not only is she not impeded by it, but it affords her a capital advantage by supplying an ever faithful warning of danger.

Swallow Harbour is well sheltered from wind & sea, and can contain a large squadron of any draught of water; it is easy of access; the holding ground good, and it affords in abundance good fresh water and fuel. For the more minute hydrographical information, the plan and sailing directions, which will [be] given hereafter with the charts of the expedition, must be consulted.[1]

January the 23rd

At daylight a very light air from the eastward; we weighed and made all sail, and, with the assistance of all our boats ahead towing, soon cleared the harbour, and got into the fairway of the Straits.

At 8 AM, I landed at the base of one of the most remarkable head-lands in the whole Straits, called Cape Notch,[2] and got altitudes of the sun for settling its difference of Longitude chronometrically. It is a promontory of grey coloured granite, of middling height,[3] having a deep cleft in its summit [p. 15] this cleft has suggested to the Voyagers of different nations – English, French & Spanish, who have seen the Cape, the equivalent names of Cape Notch, Cap Fendu, and Cabo Tajado.

---

[1] King, *Sailing Directions*, pp. 67–8: 'SWALLOW HARBOUR is one mile and a quarter to the westward of Snowy Sound. It is a better anchorage for ships than any in the neighbourhood. The plan of it is a sufficient guide, the dangers being well buoyed and pointed out by kelp. It was first used by Captain Carteret in the Swallow; and Cordova gives a short description of it.

The anchorage is under the east side of the island which separates the harbour from Cordova's Condesa Bay, which forms it west side. Wallis describes the harbour to be "sheltered from all winds, and excellent in every respect. There are two channels into it, which are both narrow, but not dangerous, as the rocks are easily discovered by the weeds that grow upon them."*: Cordova's account of it runs thus: "To the westward of Snowy Sound are two bays, formed in a bight by an island. The eastern, Swallow Harbour, has in its mouth three islands and a rock; besides being strewed with kelp, which serves to point out the dangers in entering. Within, it is very well sheltered from all winds. The depth is from 40 to 16 fathoms, stones, and in some parts ooze. This bay is to the south of Cape Notch; and to recognize it, there is a cascade falling down the centre of a mountain at the bottom of the port, to the westward of which are two higher mountains; the summit of the eastern being peaked, and the western one rounded."**

*Hawkesworth [*Account of the Voyages*], I, p. 401 [Wallis gives a longer description at pp. 415–61].
**Relacion, p. 146.

[2] Cabo Notch, see p. 194, n. 8 above. King, *Sailing Directions*, p. 70: 'CAPE NOTCH is a projecting point of grey coloured rock, about 650 feet high, having a deep cleft in its summit. It is a conspicuous headland and cannot be mistaken.'

[3] [Note in original] '650 feet'.

The ship had scarcely passed Cape Notch, when the favouring easterly breeze died away, and it was succeeded by one directly adverse – light at first, but soon strong and squally. We commenced beating up towards our destination the western mouth of the Straits. By a little after sunset, we had fetched a deep Sound in the north shore three leagues to the westward of Cape Notch. I sent a boat to sound it, following her in the ship: but it fell dark without our having struck anchorage soundings. I now stood in towards a part of the eastern side of the Sound, where the sloping form of the high land seemed to promise a shalowing sea-board. We came suddenly in to 5 fathoms, with a sandy bottom; in which, having sounded in the boat the space of a cable-length and a half around without finding a less depth, I anchored the ship for the night, struck topgallant masts, pointed the yards to the wind, and veered away 80 fathoms of chain cable.

January the 24th

The day broke with fresh breezes from the WNW, thick weather, rain and hard squalls: under such circumstances nothing being practicable in the way of surveying, I thought it better to employ the day in beating to the westward – We found that we had taken up a berth much nearer to the rocky shore astern of us than I should have chosen had I anchored with the advantage of day-light: we were less than three cable lengths off it, the wind right on shore. However, the stream anchor was soon carried out, and the ship warped sufficiently off shore to enable us to make sail without risk;[1] & by 6.30 AM had commenced a series of tacks, which we continued for more than twelve hours without intermission, and with very little profit – for it was two in the afternoon ere we had realised a single league to windward.

In the afternoon we had the wind, sometimes light, and sometimes in hard squalls; but always right in our teeth. We observed that whenever the wind slackened, the current swept us back, & [p. 16] we quickly lost our westerly earnings. Desirous of getting an anchorage I sent a boat to sound a likely looking bight on the south shore; she brought back the information that it afforded no anchoring ground. I went in her and tried the current and found it setting E by N, at the rate of a mile and two thirds an hour.

In stretching over close to the north shore – trying for an anchorage, a little bay or cove was descried having at the bottom of it a light coloured beach apparently of sand: into this we stood, with a boat ahead sounding, and [in] it at 9 PM we anchored in 21 fathoms, sandy bottom. This little bay is altogether unnoticed in any preceding chart or book.

All the labour of this day's hard beating had not furthered us a league towards our destination.

---

[1] A vessel at anchor, unless there is a strong tide, will ride with its head to wind. In order to get under way the ship is hove up to the anchor which is then weighed, the sails are backed and a sternboard made until the ship pays off from the wind sufficiently to allow the sails to fill and the vessel to move forward out of her berth. In this case the stream anchor was laid out to windward to allow the ship to be hove sufficiently far off shore to make it safe to carry out this manoeuvre.

January the 25th.

Squally, thick weather with incessant rain. I sent the Master in a boat to learn the state of the weather outside & to try the current; his report on his return on board was that it blew a perfect hurricane from the westward, with weather so thick that nothing could be discerned at the distance of half a mile, & that the current was setting exactly as found on the preceding afternoon. I veered away the cable I had shortened in ready for a start had the report been in any way favourable, and determined to lie fast here during the day.

However, that this forced detention might not be altogether unprofitable, Lieut^t. Skyring and I, employed ourselves in forming a plan of this little anchorage; but to connect it with the adjacent coast must be left as work for weather very different from what we had while forming it. I have annexed the plan, from which will at once be learnt its capacity, depth of water, and the shelter it affords from the prevailing westerly breezes:[1] so I shall only add that in going in the western side should [be] kept aboard, that the holding ground is good, and that a supply of good fresh water & fuel may be got here. We found no traces of Inhabitants.

Throughout this day the Barometer rose gradually; but without the least amendment of the weather: at 8 PM [p. 17] it was at 29.96 Inches, which was higher than it had risen since the ship had been in the Straits; at the same time the weather was as bad, and looked as unpromising, as any we had yet experienced.

January the 26th.

We weighed early in the morning & recommenced beating up towards the mouth of the Straits; wind WNW, squally rainy and thick. We could see the land of either coast only by snatches through the mist; from what I did see of it, commend me I say to old Sir John Narborough for a description: I have already quoted him with applause; but in his description of this part he unites the excellencies of his customary correctness with extreme brevity, so I shall avail myself of his very expressive language: 'A cursed rocky land this.'[2]

---

[1] See Figs 18 and 19. This plan subsequently appeared on Chart 557, *Ports in the Strait of Magalhaens by Commander Pringle Stokes H.M.S. Beagle 1828*, dated April 10th 1835, under the name of Marian Cove. It is now known as Caleta Marión, lat. 53°18'S, long. 73°04'W. The cove is actually 13 miles from Cabo Notch or rather over 4 leagues.

King, *Sailing Directions*, p. 72: 'MARIAN'S COVE, one mile and a half to the west of Playa Parda, is a convenient anchorage; at the entrance it is about one third of a mile wide, and more than half a mile deep; a plan was made of it, which will be a sufficient guide. Captain Stokes, observes, that it affords shelter from the prevailing winds; the anchorage is 22 fathoms, good holding ground; but less water may be obtained, if required, there being 8 fathoms within sixty yards of the beach, at the bottom of the bay. In entering, the west side should be kept aboard.'

[2] This remark does not appear in the published accounts of Sir John Narbrough's voyage. His chart, however, has the following legend on the north shore west of Cape Quad 'A Cursed Rocky land I saw people here' which is no doubt the origin of the quotation. On the south shore between Cape Diseada and the longitude of Cape Froward are the following remarks 'The land of desolation all Craggy on which is perpetuall Snow' and 'A Rocky Mountainous desolate land the tops of the Mountains covered with Snow continually, the Aire Cold'.

On his return through the Straits, Narbrough says: 'The north side of the Streights from *Cape Victory* all along to the Eastward to *Cape Froward*, is all a ragged, rocky, mountainous, desolate Country; many high rocky Islands, and small Rocks, and sucking Rocks lie on the North-side of the Streights ...', Narborough, *Several Late Voyages*, p. 118.

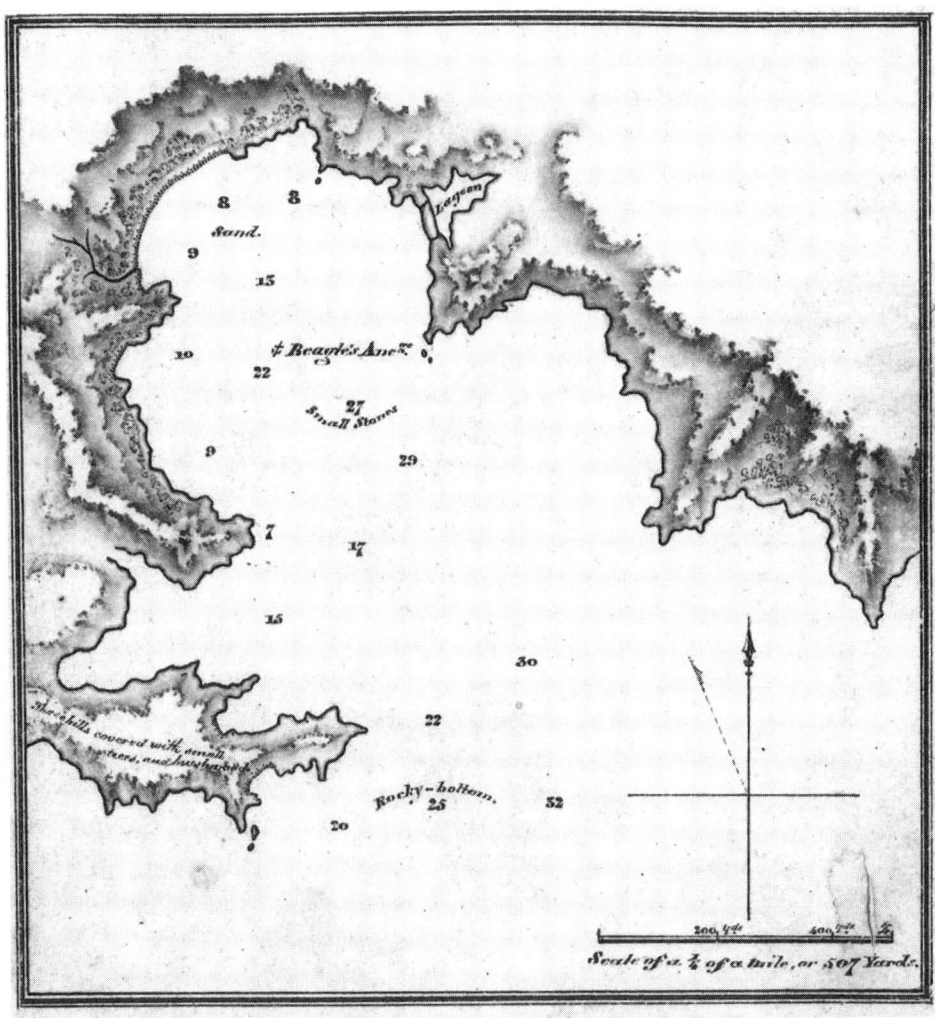

Figure 18: Chart untitled [a Cove on the north shore 3 leagues to the westward of Cape Notch – Marian Cove] (from the Journal). Provided by the UK Hydrographic Office.

By nine in the morning we had worked up to within a league of Cape Upright,[1] and just as it fell dark stood close in under the lee of a large island to the eastward of the Cape: here I tried for soundings with the hope of finding anchorage, but getting no bottom with 60 fathoms of line, stood off again and continued working to windward during the night. Wind WNW, squally & rainy. The coast on both sides is bold, during the night (which was a dark one) our boards were directed by view

---

[1] [Note in original] 'Cape Upright is on the Coast del Fuego, the southern side of the Straits'. Cabo Upright, lat. 53°05′S, long. 73°36′W.

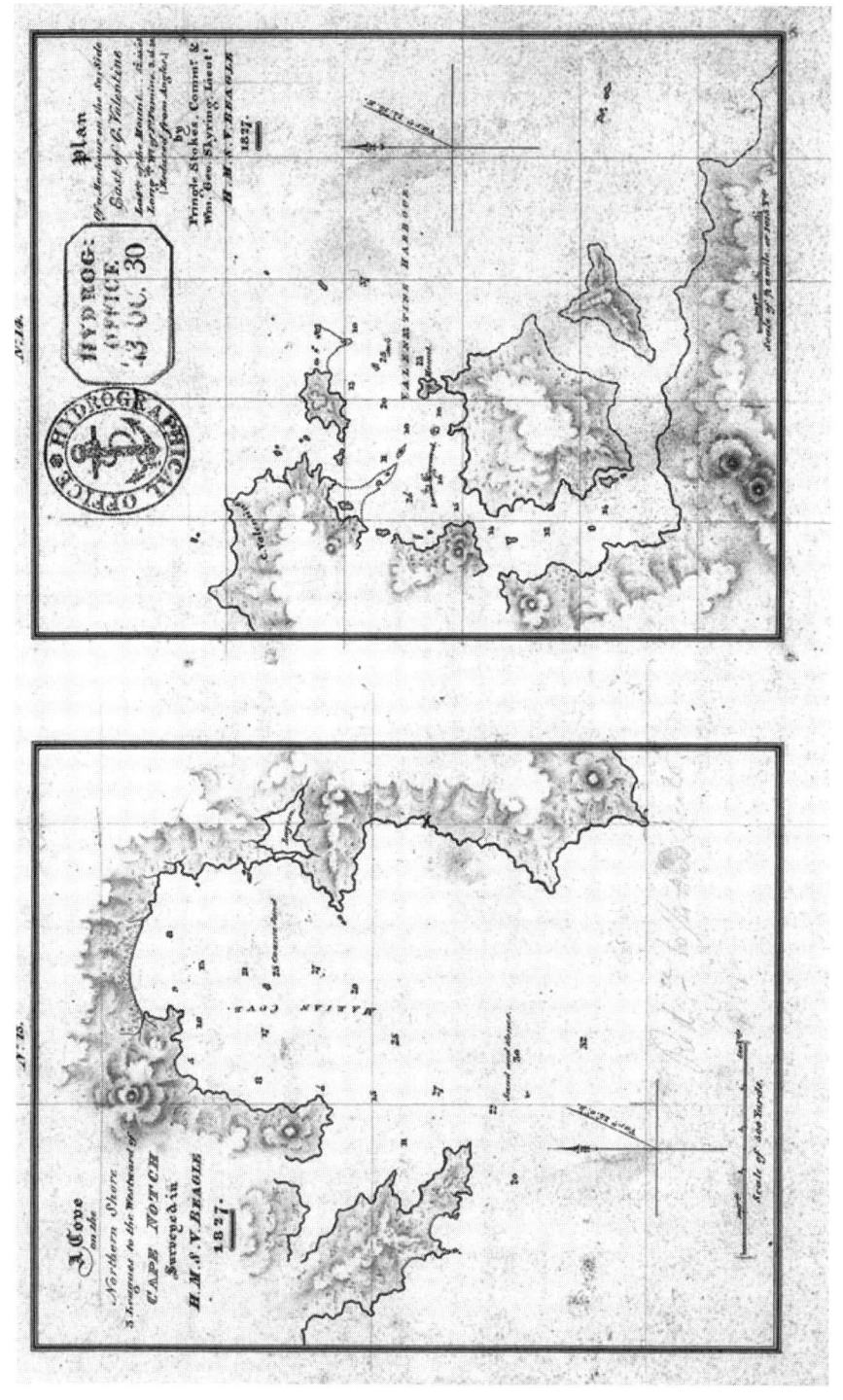

Figure 19: *A Cove on the Northern Shore 3 Leagues to the Westward of Cape Notch*, HMS *Beagle*, 1827 [Marian Cove], with a *Plan of a Harbour on the So. Side East of Cape Valentine*, HMS *Beagle*, 1827. Provided by the UK Hydrographic Office.

of Cape Upright when on one shore, and Cape Providence[1] on the other: we commonly tacked off either shore when at the distance of a mile or two thirds of a mile.

... 27th.

About 8 AM we doubled Cape Providence: this is a bold head-land (on the north side of the Straits) of which the summits are two arched masses of rock of considerable height; the adjacent is rugged, rocky, &, for the most part, barren. Cape Providence bears from Cape Upright, due North distant [   ].[2]

Of Cape Upright the drawing annexed to page 14th. is a correct representation;[3] neither to any one coming from the eastward or the westward, does it present the perpendicular appearance that might be expected from the name which has been given to it by some English Voyager – old Narborough, I believe.[4] It seems to have an upright structure, only in one point of view, namely, when right abreast of it, & pretty close. [p. 18] Hitherto the various Reaches (or bendings) of the Straits, had secured us from being troubled with any considerable sea: but the reach we had now entered is open to the whole fetch of the South Sea; hence, the prevailing westerly winds create such a confused, hollow, mountainous sea as I rarely witnessed in any part of the world in which I have served.

By about half past two in the afternoon, we had fetched close in under the lee of Cape Tamar.[5] I sent the Master in a boat to examine a bay which lies immediately under the east side of the Cape, and I directed him if he found a practicable passage, to lead in the boat, and I would follow with the Ship: he performed his duty on this, as on every occasion, like an admirable seaman; and about half past three in the afternoon, we anchored in the Bay in a very good berth in 18 fathoms, a sandy bottom.

---

[1] Cabo Providencia, lat. 53°00′S, long. 73°35′W. King, *Sailing Directions*, p. 75: 'Captain Stokes, upon this occasion, writes: "We continued beating to windward, the wind squally and weather rainy. The coast on both sides is bold. Our boards were directed during the night, which was very dark, by the sight of Cape Upright when on one shore, and of Cape Providence when on the other. We commonly tacked at the distance of a mile from either shore."'

[2] Distance left blank – it is 5 miles. King, *Sailing Directions*, p. 77: 'CAPE UPRIGHT bears due south five miles from Cape Providence. It has a rocky islet a quarter of a mile off its east extremity, surrounded by kelp, which also extends for some distance from the cape towards the islet, at the end of which there are 7 fathoms.

CAPE PROVIDENCE is a rugged rocky mountain, higher than the adjacent coast; it is deeply cleft at the top, and, when bearing about north, the western portion of its summit appears arched, the eastern lower and peaked. When the cape bears E. by S. (*mag.*) distant about one league and a half, a little round rocky islet will be seen open of it, about a quarter of a point of the compass more southerly. Stokes' MSS.'

[3] Fig. 17.

[4] Cape Upright is named on Narbrough's chart. He does not say that he named it, however, he does say that he named Cape Monday (Cabo Monday, lat. 53°11′S long. 73°24′W) and adds 'Here at *Cape Monday*, the Streight grows broader and broader to the Westward, but keeps all one Course, North-west and by West to *Cape Upright*; which is a steep upright Cliff on the South-side, and is distant from *Cape Monday* four leagues.': Narborough, *Several Late Voyages*, p. 77.

[5] Península Tamar, lat. 52°55′S, long. 73°46′W. The cape is mentioned by Byron but not by name and he does not state that he named it, Gallagher, *Byron's Journal*, p. 77. His ships were the *Dolphin* and *Tamar*, and it presumably was named after the latter. C. Tamar, Tamar Bay and Tamar Harbour are shown on Edward Leigh's chart of the Strait of Magellan, 1766, Dixson Library, Carteret MSS, illustrated in Wallis, *Carteret's Voyage*, I, facing p. 122.

During the thirty four hours that we had been under sail, since leaving our last anchorage, we had made upwards of fifty tacks, and carried through all a heavy press of sail; yet all this laborious beating had furthered us only [   ]¹ towards our destination, the western mouth of the Straits.

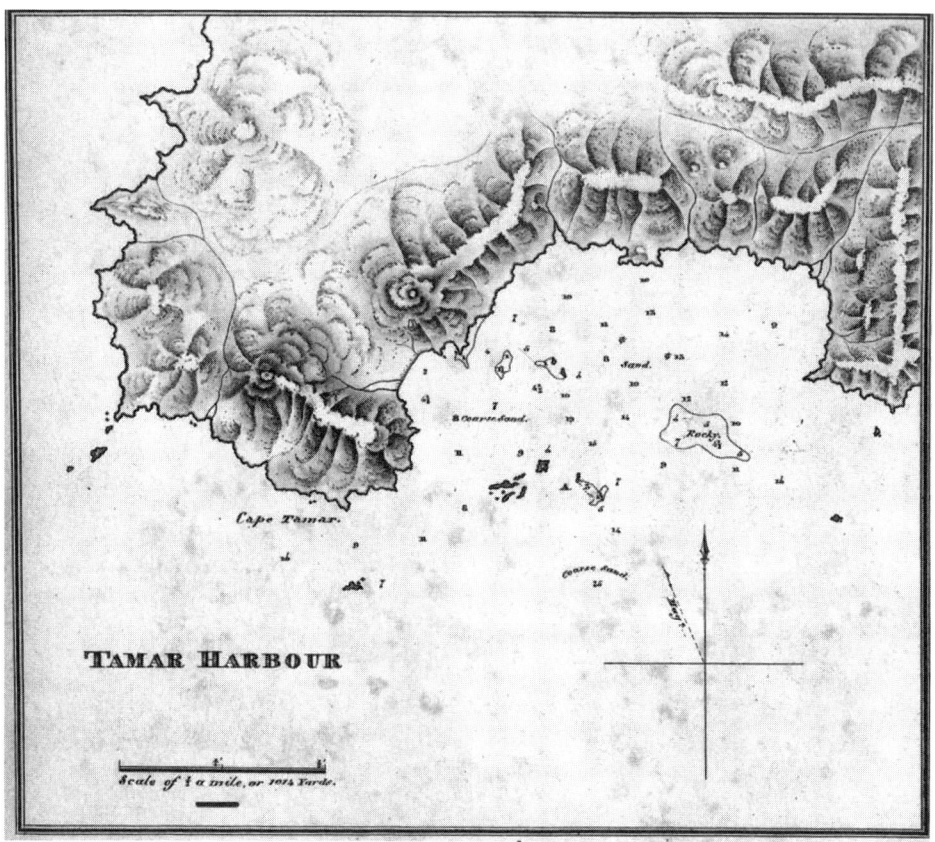

Figure 20: Chart of Tamar Harbour (from the Journal). Provided by the UK Hydrographic Office.

January the 28th. 29th. & 30th.

In Cape Tamar Bay² we remained during the next three days. This interval the Assistant Surveyor and I, employed – whenever the state of the weather permitted – in taking observations for the determination of the latitude and longitude of the Cape, and in making plans of the bay and adjacent coast.³ Parties of my people were allowed to go on shore occasionally for recreation, and my Officers often formed shooting parties; a jungle of stunted and weather-worn growth, confined, however, their

---

¹ Space left blank in original. It is about 34 miles from Caleta Marión to Cabo Tamar.
² Puerto Tamar, lat. 52°56′S, long. 73°46′W.
³ See Figs 20 and 21.

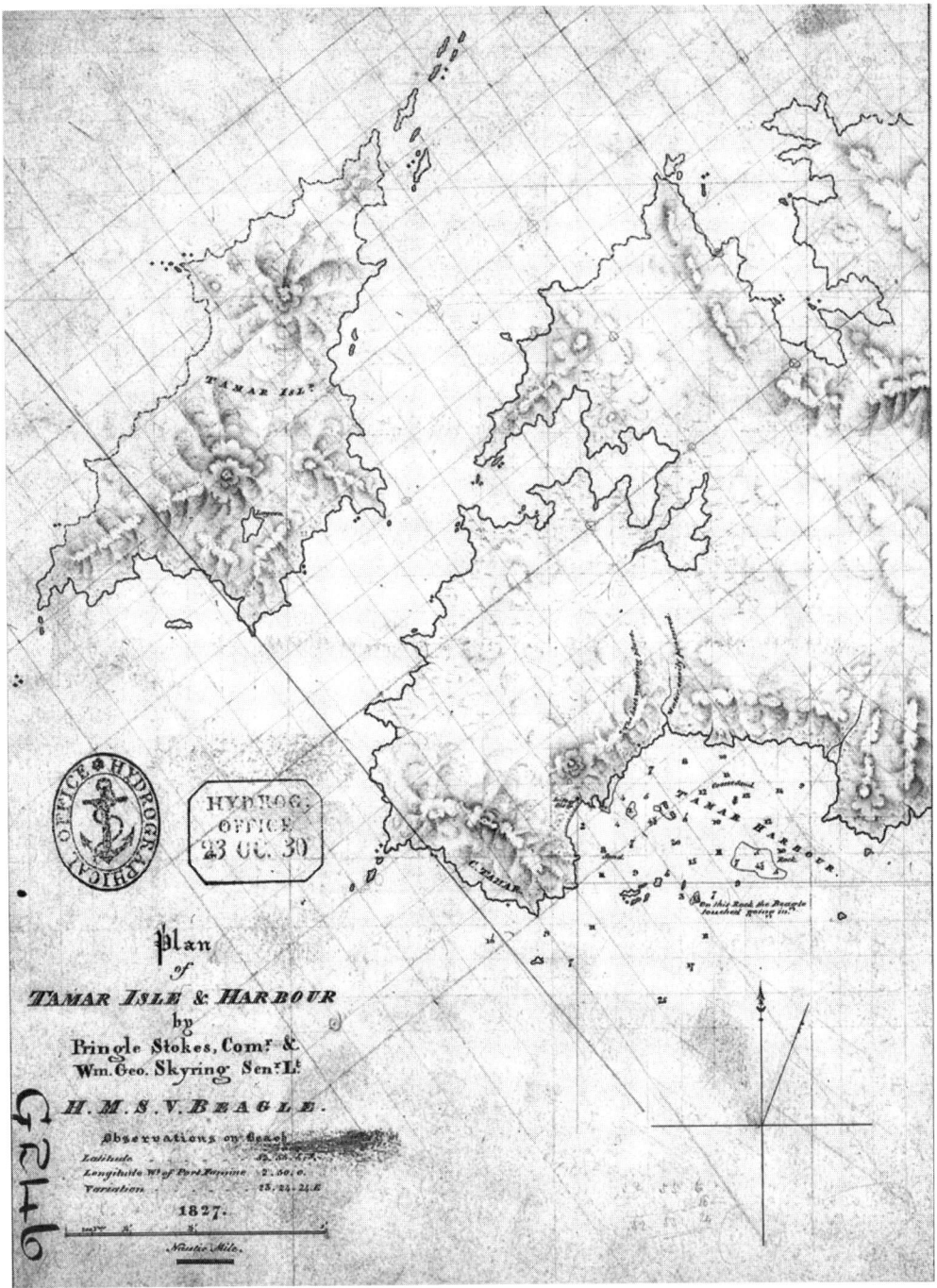

Figure 21: *Plan of Tamar Isle and Harbour*, HMS *Beagle*, 1827. Provided by the UK Hydrographic Office.

excursions to the margin of the shore, with a rare exception or two of success in reaching with great difficulty a mountain top. the prospect certainly did not reward the labour of the ascent – a more extensive view of rugged barrenness was all that was obtained by the merely curious; to us[1] the ascent was useful, by enabling us to take from [p. 19] a commanding point of view a correcter and more extensive series of angles.

The anchorage under Cape Tamar[2] will be found exceedingly useful to a Vessel working through the Straits. Although there are anchorages on the opposite southern shore, yet the prevailing <u>thick weather</u> would, <u>very often</u>, render it, to a Stranger to the Straits, let him have all the advantages that accurate charts & sailing directions can supply – exceedingly difficult to find for his ship a berth for the night. The <u>southern</u> coast is not indented by such deep and broad bays as the <u>northern</u>; hence the head-lands, islands &c. of the former are less distinguishable by their relative position; and their apparent form varies so much with every change of the point of view, that, in most instances, it is impossible for anyone, to give such a characteristic description of them, as would enable a stranger to recognize a particular point among the many that resemble it, and thereby secure the anchorage that may be in its neighbourhood. But the position of Cape Tamar, and consequently that of the anchorage which lies immediately under the east side of the Cape – is sufficiently indicated by its being the <u>westernmost</u> head-land of the deep bay of which Cape Providence is the <u>easternmost</u>. Cape Tamar bears [ ] of Cape Providence distant [ ][3] the intermediate land of the bay is much lower than either of the Capes, & is altogether a mass of rugged granite, with scarcely soil enough to give growth to a stunted and weather-worn jungle. A deserted dwelling or two (that bore the appearance of having been abandoned by its tenants), were the only indications that the human foot had ever trod these dreary shores. To the westward of Cape Tamar is an island much higher than the Cape, having at its summit a very deep and remarkable cleft; which, together with the trend of the channel that separates it from the land on which the Cape is situated, will enable a stranger to make it out.

A careful inspection of the annexed plan of this Bay (or Harbour) of Cape Tamar will suffice to enable a ship to place herself in the best birth: Fuel is to be cut here; & from the streams that trickle through the underwood along the sandy beach[4] good fresh water may be procured.[5] [p. 20]

---

[1] [Note in original] 'The Assistant Surveyor and I.'

[2] [Note in original] 'My observations place the Easternmost Extremity of Cape Tamar, in Latitude 52°55'6" South, & Longitude [ ] West of Cape Virgin Mary.'

Commander Stokes has left the longitude blank, however, Cape Virgin Mary (Cabo Vírgenes) is in long. 68°21'W and Cape Tamar (Cabo Tamar) in lat. 52°56'S, long. 73°46'W, and so it is 5°25' west of it.

[3] Spaces left blank in original. Cabo Tamar bears 301° from Cabo Providencia, 8¼ miles.

[4] [Note in original] 'On the eastern shores of the Bay.'

[5] King, *Sailing Directions*, pp. 77–8: 'On the east side of the promontory of Cape Tamar, is the useful and excellent anchorage of TAMAR HARBOUR. It is scarcely two miles wide, and rather more than half a mile deep. Its entrance is not exactly free from danger, but, with attention to the following directions, none need be apprehended. There is a sunken rock between a group of rocky islets, one-third over on the western side, and a patch of kelp, one-third towards the eastern side of the bay. With a westerly wind it would be advisable to give the outer rock a berth of two cables' length to avoid this danger, on which there are only 9 feet of water, and upon which the Beagle struck.

Figure 22: Indian of the Straits of Magellan with his sealing club. Provided by the UK Hydrographic Office.

January the 31st.

The hands were turned up at daylight 'up anchor', but the heavy squalls that came off the high land of the harbour rendered it too hazardous to attempt to start it[1] until about 8 AM, when availing ourselves of a lull, we weighed and made sail & recommenced beating up to the mouth of the Straits. Wind WNW dead foul, thick weather, rain, and hard squalls, a cross turbulent sea.

In the afternoon the squalls became more frequent and more violent, obliging us repeatedly to cap the topsails and clew up the courses.

"An excellent leading-mark for this shoal, is a whitened portion of bare rock, looking like a tombstone, about one-third of the way up the green side of the mountainous land that forms the coast of the bay. This stone bears N.76°W. (by compass) from the rocks to be rounded on entering the anchorage."*

The least water found among the kelp on the east side of the channel was 4½ fathoms, and near and within the edge towards the rocky islets there are 7 fathoms; so that with the lead in hand, and a look out for kelp, which should not unnecessarily be entered, there is no real danger to be apprehended. The Beagle anchored at about one-third of a mile from the back of the bay. The plan will shew what is further necessary to be known of the anchorage.* Stokes' MSS.'

[1] To move it, i.e. the anchor.

Figure 23: Fuegian woman and child with a native dwelling. Provided by the UK Hydrographic Office.

This squall gives, by daylight, abundant warning; it is preceded, commonly for half an hour by dark dense clouds which are seen forming on the weather horizon and gradually expanding upwards until their upper line attains the altitude of about forty degrees, then comes very heavy rain (& oft-times hail) and immediately follows the squall in all its fury. It lasts in general fifteen or twenty minutes.

In working to windward we frequently extended our boards to within a third of a mile of the south shore – not without some risk considering the state of the weather – with the hope of making out some anchorage: but the coast was hung with so thick and heavy a mist that not a single point mentioned by preceding Voyagers, could we recognize.

About seven in the evening we were assailed by a squall which burst upon the ship with a fury far surpassing all that had preceded it; had not sail been shortened in time, not a stick would have been left standing, or she must have capsized: as it was, the squall hove her so much on her broadside that the boat which was hoisted up at the starboard quarter was washed away and totally lost.[1]

I now stood over to the north coast to ascertain whether there was any anchorage under the lee of a Cape[2] bearing from Cape Tamar N 15° West[3] dis.^t 11 miles. On closing it, however, the weather was so thick at times that scarcely two ship-[p. 21] lengths could be discerned around: this circumstance, together with the time of the evening and the squally state of the weather, certainly did not tend to form a conjuncture particularly propitious to exploring unknown bays: and to think of passing such a night as was in prospect under sail if a harbour could be reached, had been silly in the extreme: so I was obliged to cast to the winds all the hard gained earnings of this day's beat and run for the anchorage whence I had started in the morning.

It was nearly dark ere we reached it; & in going in – desirous to keep well up to windward in order to fetch into the best anchorage in the bay – I gave the outer islot (marked A in the annexed plan [Fig. 20]) rather too close a birth, and the ship struck, in consequence, on a short rocky ledge that runs off the eastern end of it; fortunately she did not hang a moment – it was only touch & go – and at 9.30 PM she was anchored in 15 fathoms, a sandy bottom.[4] [p. 22]

A Boat excursion in search of Harbours

To lose, by running back, all the fruits of a laborious day's beat, was of course abundantly annoying; yet – considering the prevalent wind and weather – we were liable to a recurrence of it upon every fresh attempt to get to the westward, unless we could acquire such a knowledge of the coast as might enable us to stand boldly in and secure an anchorage – let the circumstances of wind and weather be what they list: So I manned a ten oared boat with a picked crew and victualled her for a week; and leaving the Beagle with two anchors down in Tamar Bay in charge of the Senior

---

[1] Macdouall, *Narrative*, pp. 99–100: 'The evening began to close in; still we dashed on, and still were driven back; the sea broke over the vessel many times, and she laboured very much, and oftentimes was buried in the deep trough of the sea. At length we were struck on the larboard side by a heavy sea, which threw us nearly on our beam-ends, the hammock-nettings on the larboard side, towards the forecastle, being completely under water, and the men who were there immersed above their knees: it broke the stanchions and carried away the first gig. The "idlers," who were battened down (and more than ten men were then on the sick-list), seemed to be quite unconcerned at the din and noise overboard – "nothing so much the spirit calms as rum and true religion;" but talked over the probability of reaching Cape Pillar on such a boisterous evening. By this time it was nearly dark, and nothing was to be seen around us but black and gathering clouds, from which a vivid gleam of lightening occasionally shot, making the gloom still more awful; "That night, a child might understand, the Deil had business on his hand."' The quotation is from *Tam O'Shanter*, ll. 77–8, by Robert Burns.

[2] Later named Cape Phillip.

[3] [Note in original] 'true bearing'.

[4] Macdouall, *Narrative*, pp. 100–101: '... in passing too near the rocks ... the ship struck three successive times, grating harshly against them, and heeling over fearfully; each shock made sad defalcations in the glass and crockery, both in the officers and ship's company. At this crisis, nothing could exceed the surprising activity of the sick, particularly old Baptiste, the black cook ... who jumping out of his hammock, made directly for the main hatch ... I with some others, rushed up on deck to see what was the matter. The men were all abroad, scarcely knowing which rope to lay hold of; but the danger had been momentary; the foresail having been kept on, had dragged her completely over the rocks, and in a few minutes afterwards we were safely anchored.'

Lieutenant, I proceeded – accompanied by the Master[1] (early in the morning of the 1st. of February) to cross the Straits, and examine the South Coast,[2] and make ourselves acquainted with the anchorages it afforded between such a point as the Beagle might be expected to fetch on her next attempt, and Cape Pillar its western most limit.

February the 1st.

On the day of our departure the weather was favourable, and we crossed over & gained the south coast with ease; but during the remainder of our cruize we had a constant heavy gale from the WNW, with thick weather & incessant, drenching, rain: the consequent discomfort in an open boat will be readily enough conceived. Throughout that interval – five days – we were all constantly wet to the skin; repeatedly in doubling the various head lands, we were obliged – after hours of efforts to pull the boat ahead against the violent squalls and cross turbulent sea that opposed us – to desist for a time and seek rest and shelter in any little cove that chanced to be at hand. The nature of the coast was not such as to invite us much to go on shore; for it was either high steep rocks; or a narrow beach composed of knobs of granite attrited to roundness and slipperiness by the action of the tides, and skirted, almost at high water mark, by a scarcely penetrable jungle – Every where an utter solitude – Fortunately wherever we landed we found good fresh water and wood for fuel; but [p. 23] of the latter we could not always avail ourselves from its being so thoroughly wet. The beach sometimes afforded muscles and limpits; Geese and ducks were seen in considerable numbers, but their rank fishy flavour rendered them uneatable: not a trace of a quadruped of any kind.

However, we accomplished our purpose. We made ourselves acquainted with some anchorages (of which as they were subsequently visited in the ship, I shall say nothing here) and in the forenoon of the 5th. reached the westernmost harbour which the south coast presents, situated 2½ miles to the eastward of Cape Pillar.[3]

February the 5th.

Here we fell in with a party of the Natives: They good naturedly supplied us with fire and quick fuel for cooking our dinner, and, at our invitation, partook of our meal with us. I visited their dwellings, and bestowed upon them some of the knives, scissars, beads &c with which the Admiralty had been pleased to cause the ship to be furnished as presents for conciliating the good will of the natives of the countries we might visit: I likewise distributed several of the brass "Adventure & Beagle" medals which you had had struck as tokens of this Expedition. Our meeting, however, gave rise to no incident worth relating; so I shall reserve for a subsequent page, the scanty notices respecting these people which our few opportunities enabled us to collect on this and other occasions.

---

[1] 'together with Mr Jones, volunteer, and twelve men.': Macdouall, *Narrative*, p. 102.
[2] [Note in original] 'That of the Tiera del Fuego'.
[3] This is Puerto Misericordia, lat. 52°47′S, long. 74°35′W, known to the English as 'The Harbour of Mercy', see also pp. 211, n. 1.

About two in the afternoon – having sounded and made ourselves acquainted with this harbour – we proceeded to cross the Straits and rejoin the Beagle in Tamar Bay: But we found this was an undertaking which the violent squalls that continually assailed us, and a confused breaking mountainous sea, rendered our boat altogether inadequate to perform. After running for about an hour – not, most likely, under such circumstances of dire alarm as (in the year 1588) inspired Pedro Sarmiento (one of the earliest explorers of the Straits) and his [p. 24] followers, when they describe themselves as traversing these very parts in their ship, "con grandissimo trabajo y peligro, llamando a Dios Nuestro Señor, y a su benditissima Madre, y a todos los Santos que intercediesen por nosotros," &c;[1] but with so much of risk as served to excite well-founded apprehensions in the mind of a skillful & resolute seaman as is the Master of the Beagle, Mr. Flinn: I concurred in the opinion he expressed to me that the boat would not live if it were attempted further to prosecute the passage across the Straits: so giving up the design until the weather should become less tempestuous, I regained the south coast and took shelter in a little cove about fifteen miles from the harbour whence I had started in the afternoon.

As soon as the boat was secured, we all went ashore; and while one party was employed in constructing a lodging for the night with such materials as our sails masts oars and the thick jungle that skirted the beach, supplied; another was busy cutting wood, and striving to make a fire. This last design, however, proved altogether abortive; for after exhausting the lungs of half a dozen very able bodied seamen in ineffectual efforts, perseveringly pursued for a couple of hours, to kindle a blaze with wet fuel, we were obliged to desist in despair. By this time, too, we had discovered that the Bower we had formed – though it would, no doubt, have answered admirably well for the purpose of (as Pope sings)
"A soft retreat from sudden vernal showers"[2]

---

[1] This comes from Sarmiento de Gamboa, *Viage al Estrecho de Magallanes*, p. 176. Sarmiento's two ships were in the Pacific having sailed from Puerto Bermejo. The *Capitana* had a brigantine, with eight men on board, in tow. At nightfall it came on to blow furiously and in due course the two ships parted company. It was at this stage that Sarmiento wrote: 'Y en la Capitana se iba con grandísimo trabajo y peligro llamando á Dios Nuestro Señor, y á su benditísima Madre, y á los Sanctos que intercediesen por nosotros con Nuestro Señor Jesu-Christo que hubiese misericordia de nosotros.' ('On board the *Capitana* they went in great anxiety and danger, calling on God our Lord, on His most blessed Mother, and all the Saints, that they would intercede for us with our Lord Jesus Christ, so that He might have mercy upon us.': translation, Markham, *Narratives of the Voyages*, p. 99.) The brigantine was in great danger of sinking. At daylight, with great difficulty, she was hauled alongside and the men transferred to the *Capitana*. One was lost; the brigantine was cut adrift; the storm lasted all that day. The next day land was sighted. It was called Santa Ines 'because we sailed from Puerto Bermejo on her day' (ibid., p. 101); they doubled the cape which was named Espiritu Santo, and just inside it found refuge in a bay which was named Bahía Misericordia (Bay of Mercy), 'seeing that our Lord God had saved us from such dangers as we had passed through during the storm' (ibid., p. 102).

[2] Alexander Pope, 1688–1744. *Spring: The First Pastoral, or Damon*, written in 1704, line 97ff.
>
> Now rise, and haste to yonder woodbine bowers
> A soft retreat from sudden vernal showers;
> The turf with rural dainties shall be crowned,
> While opening blooms diffuse their sweets around.
> For see! the gathering flocks to shelter tend,
> And from the Pleiads fruitful showers descend.

was by no means suited to the kind of weather we had the benefit of; we were litterally washed out of it: so we betook ourselves to the boat, and, having with sails masts & oars rigged us in the dark such shelter as we might from the torrents of rain that were falling, in her we lay – wet and miserable enough until morning broke.

February the 6th.
At day light, the wind and sea having considerably subsided, we proceeded to cross the Straits, and in the course of the forenoon rejoined the Beagle in Tamar Bay.[1]

[p. 25]   I found that, during our absence, they had had at the ship just such weather as we had experienced in the boat; although the Beagle lay within a quarter of a mile of the shore, it had been visited, but twice or thrice, throughout that interval; & then only for the purpose of examining the tide pole which I had caused to be erected near the beach.

On this part of the northern shore of the Straits, the flood tide sets to the eastward: it does not exceed half a mile an hour. In Tamar Bay (or Harbour) the time of high water at the full and change of the moon is 3^h. 5^m. PM, and the perpendicular rise & fall of the tide is 5 feet, 2 inches.

February the 7th.
Early this morning, we again weighed & made sail to work up towards the western mouth of the Straits – wind WNW, fresh breezes and squally, with drizzling rain.

After beating up until late in the afternoon, I found that the only anchorage we could have fetched on the southern shore was one which I had examined in my late boat excursion; and did not consider as an eligible one under the present circumstances of wind & weather: so we bore up & sought shelter for the third time in Tamar Bay. We anchored in it at half past six in the evening, in 13 fathoms, sandy bottom, and veered to 80 fathoms of chain cable on the best bower. In the course of the night, the violent squalls from the WNW obliged us to let go a second anchor.

The horrible weather we had all this day, (the 8th.) obliged me to lie fast at my present anchorage.

... 9th.
At four o'Clock the next morning, we weighed the Beagle's anchor for the purpose of proceeding on a third attempt to get to the westward of Tamar Bay. The wind as on

---

[1] Macdouall, *Narrative*, pp. 103–4: 'On her arriving alongside, we soon perceived an alteration in the looks of all, except Captain Stokes and Mr. Flinn, who appeared none the worse for the cruise; but, on looking after our messmate, he presented such a whimsical appearance … for his face naturally bluff, was puffed up and swollen, and speckled of a blue tint, something resembling that of a drowned person. The effect of the weather upon some of the men had been different; their faces looked as long as an eight shilling teaboard. It was useless for us to ask any questions of Mr. Jones until he had attended to the demands of the "Victualling department"; one of us was obliged to cut up the provision for him, his hands being rendered nearly powerless from the effects of the cold.'

the preceding day, dead foul – yet considerably less violent and squally – much rain throughout the day. There were some tolerably long intervals of clear weather; so I thought it a good opportunity to examine the bay to the eastw^d. of a Cape[1] (bearing from Cape Tamar [ ] dist-[p. 26]tant [ ][2]) from which I represented myself as running on the 31st. of January, on account of the thick weather and heavy squalls which then prevailed.

We found here an excellent anchorage and having secured the ship with the best bower in 15 fathoms, and veered to 75 fathoms of chain cable, we lay fast all the following day, for the purpose of determining the position of the Cape, and making plans of the anchorage and adjacent coast.

This is the easternmost of three successive Capes on the northern shore, which, as they are altogether unnoticed, as well as unnamed, in any preceding chart or account of the Straits, I consider as the property of this Expedition; and therefore request to be allowed to name them in succession, (beginning at the easternmost,) Phillip,[3] Parker,[4] King,[5] as a tribute of respect to my Commanding Officer.

We succeeded in determining the latitude of the Cape & its chronometric diff^{ce}. of Longitude, and in making a plan of the anchorage and adjacent coast. To enter into the minute hydrographical details of every anchorage we visited, would swell this Journal to a most bulky size; & it is quite unnecessary, as the Sailing directions,[6] given with the Charts of the Expedition, at a future period, will be sufficiently explicit on that head: So (referring to the plans we formed of this place) I shall here confine myself to stating that, to a Vessel working to the westward through the Straits, this anchorage under Cape Phillip will be found very valuable; from the nature of the northern coast, which here forms deep & extensive bays, the anchorages under the head-lands, as Cape Providence, Cape Tamar, Cape Phillip &c, are much easier for a stranger to recognize than those on the opposite coast del [p. 27] Fuego, besides the wind hangs somewhat to the northward of west in general, a better start will be obtained[7] will therefore be obtained on the northern shore. Here as in every anchorage in the Straits good freshwater and fuel are easily procurable – & nothing more save a few wild berries, wild celery, muscles & limpits: the brant-goose abounds here, but its naseous fishy taste renders it uneatable – No inhabitants, no quadrupeds.

---

[1] [Note in original] 'Note. This Cape is altogether unnoticed in any preceding Chart of the Straits.'
[2] Spaces left blank in original. On 31 January bearing given from Cape Tamar N15°W distant 11 miles, i.e., Cape Phillip.
[3] Cabo Phillip, lat. 52°45'S, long. 73°55'W.
[4] Cabo Parker, lat. 52°43'S, long. 74°11'W.
[5] Cabo King, lat. 52°24'S, long. 74°40'W.
[6] King, *Sailing Directions*, p. 79: 'Under the lee (the N.E.) of CAPE PHILLIP is SHOLL's BAY, in which the Beagle anchored in 1827. Of this place Captain Stokes writes: "We found, there, an excellent anchorage in 15 fathoms. It is valuable for vessels working through the strait to the westward, inasmuch as, from the discontinuous nature of the northern shore, (which here is formed into deep bays,) this place will be much more easily recognized than the anchorages on the opposite coast; besides the winds hang here, in general, somewhat to the northward of west, hence a better starting-place for the westward is obtained. Here, as in every anchorage on the strait, water and fuel are easily procured; but nothing more, unless we except the wild berries, (*Berberis, sp.*) celery, muscles, and limpets; the wild goose abounds here, but its nauseous, filthy taste, renders it uneatable. No inhabitants, no quadrupeds."' *

* Stokes' MSS.
[7] 'will be obtained' should, presumably, have been deleted.

February the 11th.

We weighed early in the morning & beat to the westward along the southern shore, the wind for the most part a moderate breeze from the WNW, with occasional squalls. a good deal of rain. At 6 PM we anchored in a bay on the south shore, to which the name of Valentine's Bay[1] has been given, 21 fathoms water, sandy bottom. Here we lay all the 12th. and 13th. – employing ourselves during the short intervals of dry weather, in taking angles, with the theodolite, and in forming plans of the bay of the adjacent coast, & in sounding. The details will find their proper place in the sailing directions.[2]

On the morning of the 14th. – the weather having in some slight degree improved, we weighed at 5 o'clock, and commenced our beating to windward – Wind, fresh breezes & squally, from the WNW – much rain.

By sunset Cape Pillar bore from us SW½S, distant only three miles. That Cape is the northwesternmost limit of the Straits of Magellan – the land immediately about the cape are pointed rocks having very much the appearance of the Needles off the Isle of Wight:[3] to the eastward of these the coast is about 1600 feet high, a craggy barren mass – all granite, with viens of quartz running in a NNW direction. the Coast all along is exceedingly bold – no bottom with 100 fathoms of line – there are no dangers – there are some patches of sea-weed here and there – but they grow on rocks at such a depth as would not bring any ship up.

Had our sole object have been to effect the passage of the Straits, that might have been accomplished this night. [p. 28] A few tacks more would have enabled us to double Cape Pillar, and have placed us in the Pacific Ocean; but as one of the objects to which you had directed my attention was the fixing astronomically of Cape Pillar and some of the adjacent points; that could be best effected by my taking up some anchorage in its neighbourhood, whence I could land with my instruments either on the Cape itself, or (if that should be impracticable on account of the surf) on some place near it.

---

[1] Bahía Valentina, lat. 52°56′S, long. 74°15′W. No doubt named for Saint Valentine on whose day they left the bay. A plan was produced of this bay – see Fig. 19.

[2] King, *Sailing Directions*, p. 81: 'At its [Point Felix] western side is VALENTINE HARBOUR, in which the Beagle anchored, of which there is no written description in Captain Stokes' Journal: the plan [Fig. 19], however, will shew the nature of the anchorage, which seems to be commodious and secure, and of easy approach. On hauling round the island, there are some islets half a mile off, which must be avoided, but otherwise there seems to be no dangers.

The anchorage, as a stopping place, is from 20 to 26 fathoms, sand, at nearly a quarter of a mile from either shore: a more sheltered situation may be obtained to the south-west.

The latitude of the mount (marked in the plan) is 52°55′05″[S]. and lon. 74°15′[W]. Variation of the compass 24°10′[E].'

[3] Cabo Pilar is described by Wood, 'This Cape *Disado W.S.W.* from you, makes much like the *Needles* going into the Isle of *Wight*, but higher and not of that Colour.': Hacke, *Collection of Original Voyages*, p. 91.

Voyage of Discovery (p. 43) says, 'The famous Cape Pillar, remarkable for its position on the south side of the west entrance into the Strait of Magellan, and for its elevation over the water, is still more so [the north coast is described in the preceding paragraph as 'the fragments of a ruined world torn to pieces by violent earthquakes'], for two peaks which rise on its summit, both inclining a little to the northwest. That on the east, which is the highest, is connected with a hill from which the cape itself projects; the peak on the west rises up like a great tower from a base on the edge of the water, on the west of the cape: from the general resemblance of that mass of rock to a rude pillar, when seen from the west, the cape probable received its name.'

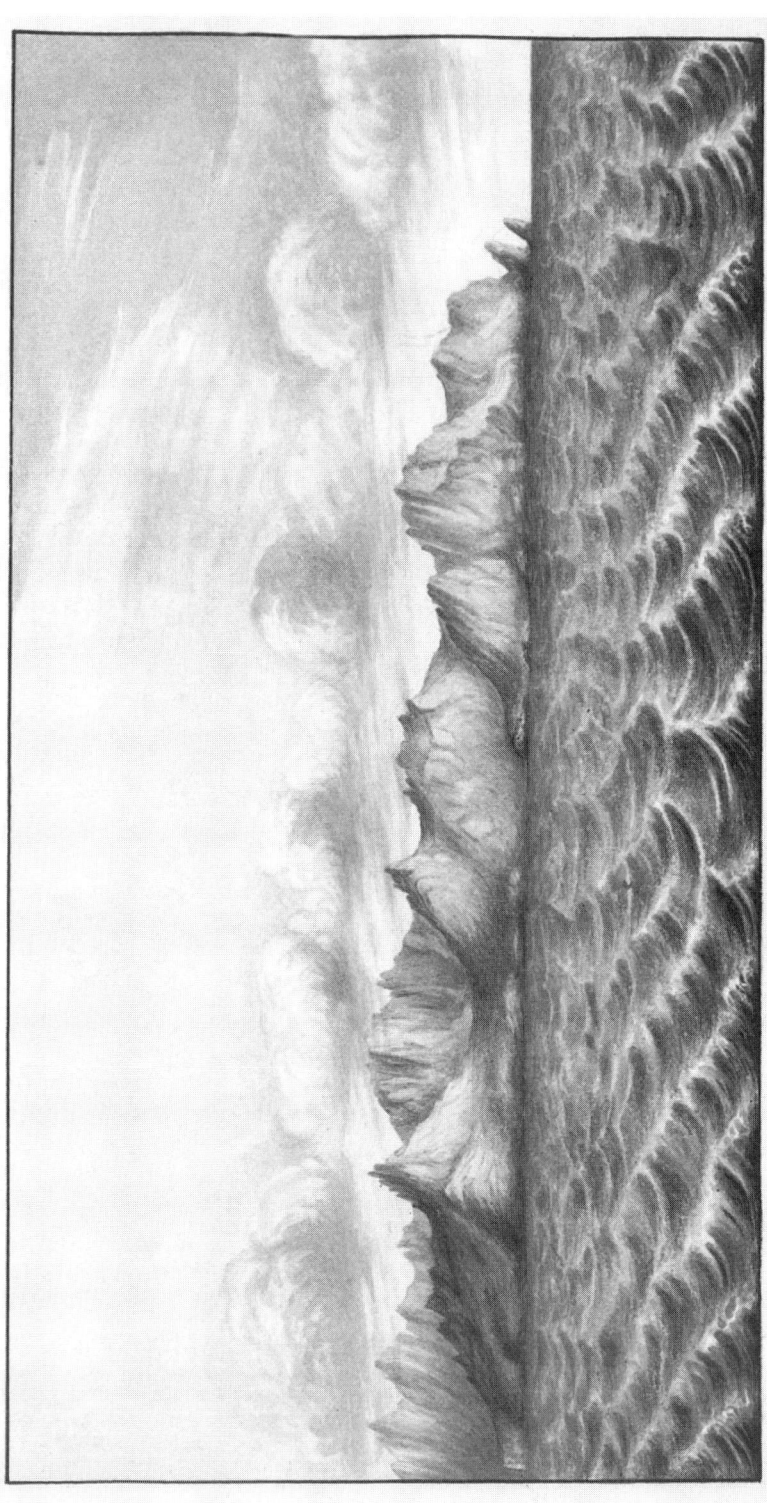

Cape Cortado   *Cape Pillar*

Figure 24: Cape Pillar bearing SW^{b}S. Cape Cortado SE½S and Peak over Separation Harbour also called The Harbour of Mercy S^{b}W, as seen by HMS *Beagle*, Pringle Stokes Commander. Drawn by Lieutenant R. H. Sholl. Provided by the UK Hydrographic Office.

Accordingly I made for the last (the westernmost or nearest to Cape Pillar) anchorage on the south coast; the same which I had visited on the 5th. of the month when on our boat excursion in search of Harbours.

The local knowledge acquired on that occasion was now found of special advantage: the wind as we closed the shore drew right out of the Harbour, so that in consequence of the number of tacks we had to make, it was long dark ere we reached it: our previous visit enabled us, by the soundings, and indistinct view of the land, to take up a very good birth. We came to in 15 fathoms, oazy bottom; and moored ship half a cable (chain bowers) each way, an open hawse to the westward; the low point, which forms the westernmost point of the Bay, bearing SW ¼ of a mile –

Harassing and tedious as had been the Beagle's progress through the Straits of Magellan to the westward of Cape Forward [*sic*]; yet the perusal of the Journals of Byron, Wallis & Carteret; as well as that of Bougainville (the last Navigator who had effected the passage of the Straits in a square-rigged vessel until the Beagle accomplished it,) will prove that we performed it in less time and with less difficulty than that of the recorded attempts of our predecessors. To beat up from Port Famine to the western mouth of the Straits, took Byron, in 1764, forty two days; Wallis in 1766, eighty two days; Carteret in the same year eighty four days; Bougainville in 1768, forty days. [p. 29]

At the end of this Journal will be found an abstract of a very copious register of the wind and weather, kept on board: it will point out the prevalence of the westerly winds, and the incessant rain we experienced almost throughout the period we have been to the Westward of Cape Forward [*sic*]: With respect to the rain, I must here take occasion to say, that though I served a season through the rains on the Western Coast of Africa,[1] I did not experience, on the Coast of Guinea, rain so heavy & so unremitting as during the five weeks I have been in the Western part of the Straits of Magellan. This rainy weather has been in the highest degree prejudicial to the health of my ship's company; for the necessity of keeping all hands constantly on deck[2] caused a most distressing exposure; and the effects were but too manifest by the state of my sick list, rarely fewer than ten being in it, and that from a ship's company of fifty eight only: their complaints were pulmonary, or visceral derangements.

The Harbour of Mercy[3] forms an excellent shelter and starting place for a Vessel working through the Straits. It is distant less than three miles from Cape Pillar the southwesternmost limit of the Fuegian side; so that if a Vessel be foiled in her attempt to clear the western mouth of the Straits, she can find a refuge without the necessity of running back far. Its position is sufficiently indicated by the two remarkable Capes, Cape Pillar, & Cape Cortado[4] (see the annexed drawing) [Fig. 24] and by three islands which lie off the western point of the Harbour: these islands and the westernmost point may be approached within a ¼ of a mile: there are some rocky patches, with seaweed

---

[1] During his time in HMS *Iphigenia* and HMS *Snapper*, 1822, see p. 152.

[2] [Note in original] 'while beating to windward'.

[3] See p. 210, n. 3. A plan of the Harbour was produced by Lieutenant Skyring in 1828: UKHO South America Folio 2, G 236.

[4] Cabo Cortado, lat. 52°51'S, long. 74°25'W, 7½ miles ESE from Puerto Misericordia.

growing on them, but nothing to bring up a ship. Good fresh water & fuel in abundance.[1] [p. 30]

The Natives of the Straits of Magellan

On the shores of the Straits of Magellan are found two distinct races of men: the famous Patagonians respecting whose stature the reports of Voyagers have been so various, and those wandering tribes to whom the appellations of "Indians" and "Fuegians" have been given. The Patagonians are found in the Straits on that part of the northern shore which lies to the eastward of Cape Negro:[2] the Indians (or Fuegians) are met with, in small numbers and at wide intervals, throughout the extensive tracts comprehended between Cape Negro & Cape Victory[3] on the one hand, and the whole range of the Coast del Fuego. As this Journal refers to a period during which we were in that portion which the Indians alone inhabit, it is of them only that I shall endeavour to give some account in the following slender notices – all that the few opportunities enjoyed by my Officers and myself enabled us to collect.

Personal Appearance
As might be expected from the unkindly climate in which they dwell, the personal appearance of these Indians exhibits, in male or female, no indications of activity or strength, nor aught of beauty. A singular and very striking mutual resemblance reigns among them. The average standard of height is five feet, five; the habit of body, in general, spare; the limbs badly twined & of flaccid muscle;[4] the hair of the head, black, straight, & coarse; the beard, whiskers &c, naturally exceedingly scanty, carefully plucked out; the eyes dark and of a moderate size; nose inclining to be prominent, with dilated nostrils; mouth large, the underlip thick; the teeth small & regular, but of a bad colour: the countenance dull and void of expression: Their complexion is a dirty copper colour.[5]

Clothing
To protect them from the rigors of these inclement regions, their clothing is miserably ill suited. Their only covering, in all seasons, is the skin of the seal or the sea otter,

---

[1] King, *Sailing Directions*, p. 83: 'At three miles and a half from the west point of Skyring Harbour is the east head of the HARBOUR OF MERCY, (Puerto de la Misericordia of Sarmiento,* Separation Harbour of Wallis and Carteret,**), one of the best anchorages of the western part of the strait, and being only four miles within Cape Pillar, is very conveniently placed for a ship to anchor at to await a favourable opportunity for leaving the strait. The plan will be a sufficient guide; for there is no danger in entering. The depth is moderate, 12 to 14 fathoms, and the holding ground excellent, being a black clay. A ship may select her position; but the one off the first bight round the point being equally well sheltered, and much more convenient for many purposes, is the best berth.
  * Sarmiento [de Gambóa, *Viage al Estrecho de Magellanes*], p. 182.
  ** Chart of the Strait of Magalhaens in Hawkesworth [*Account of the Voyages*], vol. I.'
[2] Cabo Negro, lat. 52°57'S, long. 70°48'W.
[3] Cabo Victoria, lat. 52°17'S, long. 74°55'W.
[4] Macdouall describes one native he met as 'about five foot six inches in height and exceedingly robust and broad-chested, but had altogether a most miserable appearance.': Macdouall, *Narrative*, p. 109.
[5] [Note in original] 'The annexed drawing [Fig. 22] is the portrait of an <u>individual</u>, one of the Men who visited the Beagle in the Harbour of Mercy.'

thrown over the shoulders, hairy side outermost, in the mode of a mantle; the two upper corners are tied together across the [p. 31] breast with a thong, a girdle of thong secures it round the waist, and the skirts are brought forward so as somewhat to cover those parts which modesty bids conceal.[1] Such is the garb of all – old & young, male & female.[2]

Toilette
The furniture of the toilette is quite in keeping with this simplicity of dress. Their comb is the jaw bone of the porpoise; and they anoint their hair with seal, whale, or porpoise blubber: for removing the beard, whiskers &c. they employ a very primitive kind of tweezers, namely, two muscle shells. They daub their bodies – after no particular mode or pattern – with a red earth, like the ruddle[3] used in England for marking sheep. The women and children wear necklaces formed of a small kind of pretty shell (the Turbo of conchology)[4] neatly attached by a plaiting of the fine fibres of the intestines of the seal.

Food
The tracts they inhabit are altogether destitute of four-footed animals; and they have neither domesticated the geese and ducks that abound there, nor do they use them as an article of food in their wild state: Of tillage of any kind they are utterly ignorant; the only vegetable productions they eat are a few wild berries (arbutus & berberis) and a kind of sea-weed. The staple of their food is the muscle, the limpit, and the sea-egg (Echinus sp.); but they occasionally regale on the porpoise, the sea-otter, and the seal, for we often found in their deserted dwellings bones of these animals which had under- [p. 32] gone the action of fire. Their only mode of cooking their food is by broiling it on the live embers. Former Voyagers have noticed the avidity with which they swallow offal of the most disgusting nature – decomposing seal skins, rancid seal & whale blubber &c. We had nothing of that kind with which to test their appetites. When on board my ship they ate and drank greedily of whatever was given to them, salt beef, salt

---

[1] [Note in original] 'Note. In this last particular, however, they are guided, I take it, not by the promptings of modesty, but by the desire to screen a sensitive organ from injury while making their way through the thick underwood that commonly surrounds their dwellings: for while they sat amongst us about the fires we sometimes kindled on shore for the purpose of cooking our provisions, this never well arranged screen was suffered to slip altogether aside, and they delighted in cherishing those parts and giving them the full benefit of all the warmth which the fire could impart.'

[2] [Note in original] 'Note. We fell in with a party towards the eastern part of the Straits, who (from their intercourse with the Patagonians) had been taught somewhat to improve their raiment. In the drawing of the female figure will be seen the more seemly fashion of the improved mantle' [Fig. 23].

[3] A red variety of ochre.

[4] King, *Narrative*, I, p. 557 states of *Margarita Violacea* 'of this shell the Indians make their necklaces; it is found adhering to the leaves of the *Fucus giganteus*, [*Macrocystis pyrifera*] and is the principal food of the Steamer or Racehorse duck.' The most recent generic name for this shell is *Margarella*.

Sir Hans Sloane had a collection of shells from Estrecho de Magallanes in which he describes 'a small black *Trochus*, which being Strung by the Natives, on Fish Guts, or Nerves and Worn as Bracelets and Necklaces, come to an extraordinary fine Colour, even beyond that of the finest Oriental Pearl.': Sloane *Natural History*, II, introduction, p. ii, part of the caption under Figure xi, from MacGregor, *Sir Hans Sloane*, p. 105.

It is not possible to be certain the shells indicated by Stokes are the same as those identified by King. It could be that they are *Margarella violacea* or possibly *Diloma nigerrima* which also occurs in the area. The only positive statement that can be made is that they appear to be of the family *Trochidae*.

pork, Donkin's preserved meats, bread, pudding, tea, coffee, wine or brandy – nothing came amiss – One little instance occurred, however, to shew that they prefered what was of a fat and greasy nature. As they were being conveyed on shore, a lump of tallow used for arming the lead for sounding was presented to them; this was received with particular delight; it was scrupulously divided among the party, and being placed in their baskets with some of the miscellaneous articles of food enumerated above, was reserved as a delicacy to be eaten last pour la bonne bouche.

Intercourse of the Sexes
Of the nature of their domestic ties – whether the women are in common among the tribe (as some have supposed) – or whether any thing like the matrimonial union obtains among them – we had not sufficient opportunities to form a judgement. Nothing could exceed the jealous care to secrete the women from our advances, that was manifest by the Natives at the harbour of Mercy: and there appeared to be a corresponding reserve on the part of the females; for though we were repeatedly on shore in all directions, they never once threw themselves in our way – although they must have been tempted by the liberality with which we gave away our knives, scissars, beads &c, and might easily have eluded the vigilance of their men. In other parts, too, when our boats approached their canoes almost every woman landed and took to the jungle for concealment. Thus far my own observation; but it should seem it was my fortune to see only correct Indian families: for I learnt that when my boat was away towards the eastern parts of the Straits, so far from any [p. 33] manifestation of jealousy in the one Sex or reserve in the other, the men were found allant en avant as regarded the readiness with which their women abandoned themselves to the embraces of my seamen. On one occasion, an old woman, who appeared to be the mother and chief of the tribe, conducted the bargains of prostitution, and very warmly rebuked the younger females whenever she conceived they evinced aught of coyness or reluctancy.

Children
Their manner towards their children is affectionate and caressing. I have often witnessed the tenderness with which every member of the family endeavoured to dispel the alarm & quiet the outcries which our presence used at first to occasion, & the pleasure they manifested when we bestowed any trifle upon the children. It should seem, too, that they recognize their right of property in various articles of possession, & consult their little whims & wishes respecting the disposal of the same. As I lay alongside one of their canoes in my boat, bargaining for different articles – spears, bows & arrows, baskets &c by way of curiosities – I took a fancy to a dog that lay at the feet of one of the women, & offered a price for it;- one of my seamen supposing the bargain concluded, laid hands on the animal to haul it into the boat, upon which the woman set up a most dismal yell: so bidding him desist, I increased my offers – for some time without effect; but at last they became so magnificent – namely a red cap, three knives, a pair of scissars & an assortment of beads – that after an ineffectual endeavour to get me to take any *two* other dogs instead – the woman called out of the jungle into which he had fled at our approach, a little boy of eight years of age who turned out to be the owner of the one in question: the goods I offered were shewn to him; and all, the men as well as the women, urged him to sell it: the little urchin, however, would not consent. I bought of him his necklace; & what he received in

exchange for it was put away in his own little basket apart – I observed several other instances of children conducting bargains most completely on their own account. [p. 34]

Dwellings

To their dwellings have been given, in different books of voyages, the names of huts, wigwams &c; but, as regards their construction; old Sir John Narborough's term for them will convey the best idea to an english reader, he calls them Arbours.[1] An Arbour, then, is formed of about a couple of dozen boughs of the birch tree, the larger ends of which are pointed, & stuck into the ground around a circular, or elliptical, space about 10 feet diameter at the greatest; the upper ends are brought together so as to form a hollow intertexture of a very regular hemispherical,[2] or semi-ellipsoidal figure, the intersections of the boughs being secured with tyers of grass: over this is thrown a thatching of grass and of seal & sea-otter skins: a hole is left at the lee side as a door, & another at the top as a vent for the smoke. In the centre is the fire place, and around it on the bare earth, the family stow themselves as they may. Their only household goods are two or three large shells (of the kind called in conchology concholepas, per.)[3] which they use as drinking cups; a water tight bucket made of the bark of the birch; and a basket or two, woven of grass, the work of the women – in which they collect the shell fish along the shore. They kindle their fire by the collision of two pieces of mundic;[4] using as tinder, very fine scrapings of the boughs of the berberis bush. To cut through, or split up, a large piece of timber is, with the kind of implements they possess, a very laborious & tedious process: so when they can find no billets of wood of a convenient size for fuel, the end of the large trunk of a drift tree is hauled into the hut, & placed on the fire place, and the outer part gradually dragged in as the interior portion is burned away, until the whole is consumed. Over their fires they are constantly cowering – stirring out of their huts as seldom as possible: hence, when they are seen abroad, instead of – as might be expected from their scanty clothing and inclement climate – a hardy Savage fit to brave every vicissitude of the weather; you see a miserable creature shivering to every breeze. I never met people who appeared so sensitive to cold as these poor Indians. Their [p. 35] dwellings are found sometimes seven or eight together only a few feet apart, sometimes a solitary one is seen, many miles remote from even the trace of another.

Canoes

They migrate from point to point and from one side of the Straits to the other, by means of canoes formed of the bark of the birch tree – the bark used in the construction of these canoes is much broader than any that could be stript from the trees that are found near the coast. The usual length of a canoe is from 14 to 16 feet: the pieces of bark of which it is constructed (three in number, one forming the bottom, the other two, the sides) are sewed together by thongs of seal skin, or strong shreds of the birch

---

[1] Narborough, *Several Late Voyages*, pp. 74 and 119.
[2] [Note in original] 'according to the figure of the ground plan.'
[3] [Note in original] 'concholepas peruviana.' The name *Concholepas peruvianus* has now been replaced by *Concholepas concholepas*.
[4] Iron pyrites.

bark: the timbers are the pliant boughs of the birch: along the bottom is a platform of clay which serves at once for ballast for the vessel, and as a hearth for the fire which, as in their dwellings, is kept constantly burning, as long as they are on board.[1] They make use of very rudely formed paddles. Even in smooth water these canoes would not pass a man of war's jolly boat.[2]

Arms
Their arms are bows and arrows, & spears; the bow is of a hard and elastic wood, of which I could never discover along the coast the parent tree; it is smooth and takes a good polish; the usual length of the bow is three feet and a half, the string being formed of the twisted fibres of the intestines of the seal: the arrow is about two feet long; the shaft is the tough and straight bough of the berberis bush, finely polished by scraping and rubbing; the arrow is feathered at the top; and pointed with a piece of flint, heart-shaped, & exceedingly sharp. Their spear heads are of bone, barbed, and well pointed; they are attached to shafts commonly about 10 feet long. they throw their spears with considerable force and precision, – they are, I think, less dexterous with their bows and arrows.[3] [p. 36]

Language
Their language is in the pronunciation exceedingly harsh & guttural; of its idiom we had not time to acquire a knowledge: Indeed I don't think we brought away with us more than two words of which the significations were correctly ascertained, namely "sheroo", a ship, boat or canoe; and "petit", a child: of this last word (singular enough) not only is the pronunciation strictly French, but (what is still more surprising) that pronunciation varies with every change of gender & number precisely as it does in French. This is one of those philological coincidences for which it were in vain satisfactorily to attempt to account. They have a wonderful facility in imitating the combinations of sounds of strange languages: let a sentence of a dozen of words be distinctly pronounced, & they will instantly repeat it with the utmost precision.[4]

As has been often remarked of Savages, these people never manifested any tokens of thankfulness for what we gave them: what was offered, they clutched at, as if doubtful of its being with-held; and when obtained, they hastily concealed it, as if fearful of its being reclaimed.

I often tried to find out whether they affected any particular colour; and for that purpose held out four strings of beads, black, blue, white & red; they clutched at them

---

[1] Macdouall records that the Fuegians were building a canoe during their visit, so that this must be from direct observation: Macdouall, *Narrative*, p. 120.

[2] These canoes were described by Sir Richard Hawkins and are also described in far greater detail in the account of Córdoba's expedition which also mentions canoes constructed of planks at the western end of the Strait – see Appendix 3.

[3] Córdoba's boat party were presented with a number of ducks by the Fuegians, all of which, he noted, had been wounded in the head, indicating their dexterity with their bows and arrows: *Apéndice a la Relacion*, p. 59.

[4] E.g. Macdouall, *Narrative*, p. 113: 'the younger one [of the two Fuegians], taking his station at the door of the wig-wam (as if to guard the children) cried out, "D—n your eyes," an expression he had picked up amongst us, and of which he was perfect master.'

all in their usual manner; and I found my pantomime just about as unintelligible as my idiom when I tried to make them understand that I wished them to indicate a preference.

Barter
With the exception of the trifles that are procured from them by way of curiosities, they have no articles for barter, except a few seal and sea otter skins: but I learn[1] that even these can be procured from them, only in such inconsiderable number as renders it an altogether insignificant supply towards completing the cargo of a Vessel. [p. 37]

Religion
As to their notions with respect to subjects of a divine nature, I am altogether in the dark: I never witnessed among them the semblance of a religious rite or superstitious practise. [p. 38]

February the 19th.

The incessant rain and thick weather that prevail here, had prevented my completing, until today, my observations for fixing astronomically the position of the <u>island</u> (situated just outside of the Harbour of Mercy)[2] which I intended to make the southern end of the base line for the triangulation of the Islands of Direction; that called Westminster Hall, & the adjacent coast. I determined its Latitude and chronometric difference of Longitude;[3] and the points mentioned above being distinctly seen for a considerable time during the latter part of the afternoon, we were enabled to ascertain with great accuracy their true bearings by several long series of celestial azimuths.[4]

... 20th.

Having completed the business on the southern side, I weighed on the following morning and beat to windward for the purpose of taking up an anchorage somewhere on the other side of the Straits, near which I might land with my instruments and fix the <u>northern</u> end of the base.

The wind through the day was chiefly from WNW, a moderate breeze, the weather clear and fine.

At 2 PM, we doubled <u>Cape Pillar</u>, <u>the southwesternmost limit of the Straits of Magellan</u>.

---

[1] [Note in the original] 'From the Master and Crew of a sealing Vessel.' This may have been Mathew Brisbane and the crew of the *Prince of Saxe-Cobourg* who were rescued on the return passage – see pp. 230–33.

[2] There is a group of five islands to the north of the entrance to Puerto Misericordia, three parallel to the coast and two smaller islets to seaward of the central island. The observations were taken on the central island adjacent to the coast, shown on Skyring's plan (UKHO South America folio 2, G 236) as Observation Islet.

[3] King, *Sailing Directions*, pp. 83–4: 'The observations for latitude and longitude were made upon the largest of the Observation Islets, the summit of which was found to be in lat. 52°44′57″[S], and lon. 74°35′31″[W]; the variation is 23°48′[E].'

[4] The true bearing of a celestial object (probably the sun in this case) can be calculated from its altitude, declination and the observer's latitude, so that using this azimuth it is possible to obtain an accurate bearing of any object visible to the observer.

At 6 PM we anchored among a perfect archipelago of islands, which skirts the northern shore of the Straits to the westward of the Island called Westminster Hall.[1] To this assemblage of islands, islets, rocks (below & above water) has been given the name of the Scilly Islands.[2] Most of the larger islands have <u>each</u> a patch of sand running off the eastern end of it; on which will be found anchorage in from 20 to 12 fathoms, tolerably good holding ground, distant, at the furthest, a quarter of a mile from the island to which it is attached; but the depth of water beyond that distance increases suddenly to 60, 70, & 100 fathoms with a rocky bottom: and with the prevailing <u>westerly</u> winds, a ship would draw against her anchor <u>down</u> this steep descent.

For any of the ordinary purposes of naviga-[p. 39]tion this extensive and very thickly-sown group, need only be pointed out in order to caution a Ship against approaching them. The number and closeness of the rocks both below & above water, would render it most hazardous navigation for any <u>square</u> rigged vessel: nothing but the particular duty on which I was sent would have led me among them. On the Chart of the Western part of the Straits (which I send herewith) <u>I have marked a dotted line, within which a square rigged vessel should not venture.</u> Sealing Vessels, as they are generally <u>fore and aft</u> rigged, may work in this archipelago with comparatively little risk; and as the rocks are frequented by the fur seal in vast numbers, a couple of handy vessels might seal in these parts for a season or two with advantage.[3]

---

[1] Named Westminster Island by Narbrough (Narborough, *Several Late Voyages*, p. 77) – 'this island which I called *Westminster Island*, is a high rocky island shewing like *Westminster Hall*.' It is now part of Islas (Grupo) Westminster, lat. 52°36′S, long. 74°25′W. The group includes Isla Westminster Hall, Islas Lawyers, Isla Cóndor, Isla Stavely, and several other islets and rocks. The group to the westward of Islas Westminster is now called Grupo Narborough.

[2] Van Speilbergen's Journal for May 1615 – 'These islets, lying in the South Sea at the end of the Magellan Strait, we gave the name of Sorlinges, because they are not unlike the Sorlinges* outside the English Channel.'

'*Sorlingues, the French name for the Scilly Islands.' Villiers, *East and West Indian Mirror*, p. 50.

It may be that this is the origin of Stokes's remark on the name, although from van Speilbergen's chart (Fig. 3) it would appear that he gave the name to the group now known as Grupo Evangelistas – Narbrough's Isles of Direction. Martin de Uriarte, the pilot who accompanied Comendador Loaysa into the Strait in January, 1526, states 'On the N.E. coast there are many bays and indications of ports as far as the Cape of *San Alfonso* [Cabo Victoria] which is at the outlet of the strait opposite Cape *Deseado*. Here the outlet is 5 leagues across, and between the last island and Cape *San Alfonso* there are five islands, one large which, with the four smaller ones, looks like the Berlings [Ilha da Berlenga, lat. 39°28′N, long. 9°32′W, off the west coast of Portugal]. They are almost in mid channel.' Markham, *Early Spanish Voyages*, p. 101. These would also appear to be Grupo Evangelistas (described in the Admiralty Sailing Directions as 'A group of four rugged and barren islets and some detached rocks'), since neither Grupo Narborough nor Islas Westminster could be described as being in mid-channel.

[3] King, *Sailing Directions*, p. 83: 'The easternmost island is WESTMINSTER HALL, a high rocky island; and there are two or three other conspicuous points such as the CUPOLA [Islote La Cúpula, lat. 52°32′S, long. 74°37′W] and OBSERVATION MOUNT, that might be noticed. The Beagle ran in amongst the breakers, and anchored near the latter, for the purpose of ascertaining its position, and obtaining bearings for the survey.'

February 21st.

This morning I landed on one of the larger islands (with the requisite instruments)[1] accompanied by the Assistant Surveyor, and having ascended an eminence (which I called on the chart Observation Mount) I determined its latitude & relative chronometric difference of longitude; this fixed the <u>northern</u> end of the base (see preceding page). It was a beautifully clear day; the Isles of Direction, and every point of importance on the adjacent coast were distinctly seen during several hours; & we were enabled to determine their true bearings by long series of celestial azimuths, as we had done at the <u>southern</u> end: thus the triangulation was completed.[2]

I had still the <u>northwesternmost</u> limit of the Straits – called Cape Victory – to fix, and I wished likewise to ascertain whether there were any anchorages in its neighbourhood. [p. 40]

February the 22nd.

Accordingly we weighed early in the morning, and having, not without some difficulty, extricated ourselves from the labyrinth of the Scilly Archipelago; we commenced beating to the westward.

About noon, the Isles of Direction were seen from the deck, bearing SW by S distant 3½ leagues.

The violent squalls and heavy sea, together with the thick weather, which arose in the course of the Afternoon, precluded any chance of my being able to effect my purpose this day: & it was not advisable to remain out here under sail during the night, in as much as, judging from the appearance of the weather, we were liable to be driven so far to the southward as would cost several days to regain: and I certainly had no desire to revisit the Scilly archipelago for an anchorage; so about 6 PM I ran back into the Straits, and anchored in the Harbour of Mercy.

... 23rd.

The next morning the weather being very fine, and the wind and sea moderate, we left the Harbour, and beat towards the Isles of Direction.

---

[1] [Note in original] 'See page 8' [p. 187].

[2] [Note in original] 'Note. It will be seen from the Latitudes of the two ends of the base – Observation Island on the south Coast, and Observation Mount on the northern – and from the meridian distances of each – which, together with the true bearings of the principal of the Isles of the Direction – the southernmost one, which from its figure is called the Sugar Loaf [see p. 225, n. 1] – are subjoined – that a well conditioned triangle was formed.

Observation Island	Lat^{de}	52°44'40"	South	True bearing of the Southernmost Isle of Direction N39°4'40"W
	Long^{de}	3°37'55"	West of Port Famine	
Observation Mount	Lat^{de}	52°28'58"	South	True bearing of the Southernmost Isle of Direction N76°4'34"W'
	Long^{de}	3°34'42"	West of Port Famine	

It should be noted that, while the latitudes agree with those given in the Geographical Position Book (Appendix 1), there are minor differences in the longitudes.'

By sunset we were within a mile and a half of the southernmost of those Isles (called from its conical figure the Sugar-Loaf);[1] and the weather having every appearance of continuing fine, I resolved to pass this night in the Pacific Ocean, with the intention of availing myself, on the following day, of the opportunity of fixing the position of Cape Victory: so I shortened sail; and tacked off and on the Isles of Direction, during the night.

... 24th.

The day broke with the weather in every respect favourable; beautifully clear, the wind moderate, the water tolerably smooth: So manning one of my ten oared boats, & taking a couple of days provisions in the event of an unseen detention, I left the Beagle sounding about the Isles of Direction, & proceeded to effect a landing on Cape Victory. Vast numbers of the black whale were seen around us. We pulled along several miles of these rocky & barren shores threading the labyrinth of islets & rocks that fringe them; all of which were completely covered with the fur seal and the brant-goose, [p. 41] apparently in very friendly joint occupancy.

After a pull of six hours, we landed at 11 AM, upon Cape Victory, the northwesternmost limit of the Straits of Magellan. Here with sextant, artificial Horizon, and chronometer – (certainly the first – most likely they will prove the last – instruments of that kind ever placed on these remote and dreary coasts), I determined the Latitude and Longitude, & took several angles in verification of our former positions. From an eminence of 800 feet, immediately above the Cape, a commanding view was obtained of the adjacent coast and a boundless one of the Pacific Ocean. Our observations here enabled us to rectify some material errors respecting Latitude & Longitude, as well as some respecting the trend of the land to the northward of Cape Victory, as will be evident on comparing our Chart with any of earlier date.

The productions, (animal and vegetable) of this part, were in no wise different from those we had hitherto found in our progress through the Straits. Of the geological structure of the Coasts, I have sent you specimens collected on the various points on which I landed, here as well as elsewhere.

About 8 o'clock in the evening, we rejoined the Beagle off the Isles of Directions; and after passing a second night in the Pacific, commenced, at daylight, our return to the eastward to rejoin you at Port Famine.

February the 25th.

I confess I felt no ordinary satisfaction that it had been my good fortune to effect, without disaster, the passage of these famous Straits, and to have fulfilled the objects of the, not unarduous, service on which you had been pleased to order me.

The Isles of Direction, or (as they have been called by some Spanish Navigator) the Evangelists,[2] are four rocky isles, situated just without the Straits western mouth; they

---

[1] Islote Pan de Azúcar, lat. 52°25'S, long. 75°04'W.

[2] These islands are shown on the chart of Bartolomè and Gonzalo Nodal as Los Evangelistas (1621), Markham, *Early Spanish Voyages*, between pp. 188 and 189. It is presumed that they named them, although this is not actually stated (see Fig. 25).

Figure 25: Evangelists bearing WNW as seen by HMS *Beagle*, Pringle Stokes Commander. Called by English navigators Isles of Direction. Provided by the UK Hydrographic Office.

are barren rocks, of which the southernmost is very remarkable from its conical figure; they all abound with seal, and the approach to them is bold; to land on them is difficult, except on the <u>eastern</u> side of the <u>westernmost</u>. They form the best possible land fall [p. 42] for a ship desirous to effect the passage of the Straits from the <u>westw$^{\underline{d}}$</u>. as they place her at once in the fair way. They should all be passed to the <u>southward</u>, in order to avoid the foul & broken ground along the northern shore (see the Beagle's chart). In tolerably clear weather, they may be seen from a ship's deck at the distance of five or six leagues.[1]

In passing Cape Pillar, I landed at the base of it, with my instruments, and determined the Latitude and chronometric difference of Longitude.[2]

When within a league & a half to the westward of Cape Pillar, a current, setting to the southward nearly at the rate of two knots an hour, was felt. As we <u>re-entered</u> the Straits, the wind was light, and the ship was set rapidly towards the dangerous rocks called the Apostles, when a commanding breeze fortunately sprung up.

By sunset we had got abreast of the Harbour of Mercy, and the wind being extremely light, we towed the ship in with the boats, and (half past ten) anchored in it in 15 fathoms, sand & oaze.

February the 26th.

At nine o'clock the next morning, we made sail for Tuesday's Bay, three leagues to the westward of the Harbour of Mercy: and, at 3 PM, anchored in it in 21 fathoms rocky bottom: we made a plan of this bay & sounded it; Of all the anchorages in the Straits of Magellan, Tuesday's Bay[3] is the most easily recognisable, on account of the figure of the Cape[4] which forms the western extreme of the Bay, it is as perpendicular as the wall of a house, and this upright structure, which it presents both from the eastward & the westward, distinguishes it most completely from every other head-land

---

[1] King, *Sailing Directions*, p. 84: "'THE EVANGELISTS, as they were named by the early Spanish navigators, but THE ISLES OF DIRECTION by Narborough, from their forming a capital leading-mark for the western mouth of the strait, are a group of rocky islets, consisting of four principal ones, and some detached rocks and breakers. The islands are very rugged and barren, and suited only to afford a resting-place or breading-haunt of seals and oceanic birds. There is landing on one of the islands, and anchorage round them, if necessary. The largest and highest may be seen in tolerably clear weather, from a brig's deck, at the distance of seven or eight leagues.* The southernmost, from its shape called the Sugar Loaf, is in latitude 52°24′18″ and longitude 75°02′56″. From the Sugar Loaf, the extremity of Cape Pillar [this is an error and should read 'The Sugar Loaf, from the extremity of Cape Pillar …'. It is corrected in later editions.] bears N.38°,W. twenty three miles and a half, and from Cape Victory, according to Captain Stokes's survey S.42° W 11 miles." Stokes' MSS.

*We saw them twenty-two miles off from the Adventure's deck. P.P.K.'

[2] King, *Sailing Directions*, p. 84: 'Three miles to the westward of the largest Observation Islet, is Cape Pillar, upon which Captain Stokes landed, on 25th February, 1827, but not without considerable difficulty, owing to the great swell that then, and indeed always, prevails near it. Here he observed the latitude. Captain Fitz Roy also landed in a cove under the cape in 1829, with his instruments, to obtain bearings from its summit; but the difficulty of the ascent was so great that he did not risk the destruction of them.

The extremity of Cape Pillar is in lat 52°42′53″, and longitude 74°39′31″, and Cape Victory in 52°16′10″ and 74°50′55″. These points form the western entrance of the strait.'

[3] Bahía Tuesday, lat. 52°51′S, long. 74°27′W. The bay was named by Narbrough, Hacke, *Collection of Original Voyages*, p. 92. No separate plan appears to have been rendered and so the work must have been subsumed in the general chart.

[4] Cabo Cortado. See p. 216, n. 4.

in the Straits. See the drawing facing page [   ][1] in which the Cape (called by the Spaniards) Cabo Cortado, is represented together with the adjoining coast to the westw^d. as far as its extremity <u>Cape Pillar</u>. To find out Tuesday's Bay the Cape is a sufficient leading mark: It affords a very [p. 43] good anchorage as regards the prevalent winds; its extreme points lie NNW and SSE, ½ a mile apart; the holding is tolerably good, but a breeze from the eastward would render it an exceedingly anxious anchorage, in consequence of the heavy sea that it would throw in, and the little distance for drift should an anchor start, or a cable part. Water may be got here, and fuel cut; its neighbourhood, however, affords no spar large enough even for the topgallant yard of a ship of 300 tons. No inhabitants, nor a trace of a quadruped; a few brant-geese, and a good many logger-headed ducks, which, from the splashing they made as they propelled themselves along the surface of the water, our seamen called, <u>steam-boats</u>;[2] a few <u>penguins</u> were likewise seen: but the rank flavour of every bird we shot here rendered it uneatable.[3]

February the 27th.

The next day we weighed and crossed the Straits for the purpose of examining whether any anchorage under the Cape[4] to which I had given (see page [   ])[5] the name of Cape Parker.

We passed within a mile of the rock or island called <u>Westminster Hall</u> by old Narborough from some fancied resemblance:[6] I certainly must say that the resemblance did not strike me, altho' I saw the rock in various points of view. It is about a mile long, utterly barren, and of bold approach; both coming from the eastward and the westward, in making in two high peaks, of which the easternmost is the higher.

A violent, cross, breaking, mountainous sea prevailed, such as I have rarely witnessed.

Under Cape Parker we certainly did find an anchorage, for we brought up in 22 f^ms. rocky bottom; but it was such a place as I would recommend to every <u>square-rigged</u> vessel most carefully to avoid. The violent squalls that prevail here render it difficult in the extreme to get into the [p. 44] bay, and the soundings are so irregular that where we anchored, (altho' we certainly did not chuse the worst birth) our anchor was in 22 fathoms on a <u>rocky bottom</u>, and the rocks above water within a couple of cable lengths astern. So long as the usual westerly winds blow, no damage will arise so long as the cable is good. but the slightest shift to the southward of SW, would throw in a sea

---

[1] Page number blank in original. See Fig. 24.

[2] Steam-driven vessels at this time were propelled by paddle wheels. The Royal Navy had its first steam tug, the *Lightening*, built at Devonport in 1823, and its first demonstration of the propeller in 1837 while the celebrated tug-of-war between the *Rattler* and the *Alecto* did not take place until 1845. The beating of the ducks' wings on the surface of the water as they raced along looked not unlike the double wake of paddle wheels in action.

[3] King, *Sailing Directions*, p. 82: 'There is plenty of wood and water in Truxillo Bay, but nobody will visit it in preference to TUESDAY BAY, or, rather, the more convenient anchorage of TUESDAY COVE, situated three-quarters of a mile south of Cape Cortado. The anchorage is in 12 to 14 fathoms. Tuesday Bay is larger, and, therefore, more exposed to the squalls; but for a ship, perhaps, might be more convenient.'

[4] [Note in original] 'on the north Shore.'

[5] Space left blank in original. The reference is to page 26 of the journal [p. 213].

[6] See p. 223, n. 1.

which, together with the indifferent nature of the holding ground, would render it an anchorage of extreme anxiety.

I staid here all the next day, and determined the latitude and longitude of the Cape, the variation of the compass, and laid down the adjacent Coast. The nature of the coast like all that I have hitherto described, to the westward of Cape Froward, rocky and barren, an utter solitude, uncheered alike by man or any other animal, save a few water fowl, geese, shags, and penguins.

March the 1st.

Next morning we quitted this dismal & unsafe place; and prosecuting our way to the eastward for the purpose of rejoining you at Port Famine, ran down to Cape Upright, under which we anchored at 5.30 PM. in 22 f^{ms}.

The anchorage under Cape Upright is one of the best that the Straits affords: It is perfectly sheltered from the prevailing winds by the high land about the Cape, & it is situated so far up the Straits, that very little sea would be caused by any wind. It is easily recognisible by the black rock which lies off the eastern extreme of the Cape, and to take up a birth in the bay no other directions are necessary than that the closer you get immediately under the pitch of the Cape, the better will be the holding ground.

Here is good fresh water in abundance and fuel: spars of the birch, large enough to furnish us with top masts or topsail yards, might have been cut, close to the shore, had we needed them. A few deserted dwellings proved that the wandering tribes of Indians (or Fuegians) sometimes [p. 45] made this their temporary abode.[1]

Having passed the time of stay in forming a plan of the anchorage,[2] we weighed on the 4th. of March to prosecute our return – the westerly winds which had been our fierce and unremitting antagonists during the last five weeks, now befriending our progress to our destination Port Famine.

We passed along the coast rapidly to the eastward, accomplishing in one hour's run the progress which, in our way to the westward, it had required a laborious day's beat to effect: and by 5 in the afternoon had reached a bay on the northern shore,[3] into which, as the weather looked extremely threatening, I stood for the purpose of securing an anchorage for the night, when an incident occurred which turned our attention and exertions into another channel than the surveying and exploring service into which they had hither to been directed. A Boat under a lug sail was observed making towards

---

[1] King, *Sailing Directions*, pp. 76–7: 'Of UPRIGHT BAY we know little. The Adelaide rode out a gale from the eastward with her stern in the surf of the beach, and the Beagle anchored under the east side of the cape, at about half a mile S.W. of the rocky islet, and, for shelter from the westerly winds, found it to be very good. Of this Captain Stokes says:– "We anchored at a cable's length off a small patch of light-coloured shingle beach, situated at the west side of the bay, in 22 fathoms, sandy bottom. The anchorage, though affording excellent shelter from the prevailing winds, is bad with a southerly one; for the steepness of the bottom requiring a vessel to anchor close to the shore, sufficient scope is not left for veering cable. There is a plan of the bay in Hawkesworth [*Account of the Voyages*, I, to front Commodore Byron's Voyage] from Byron's account, who anchored in the southern part of the bay, perhaps under the lee of the islands to the S.E. of the cape. ..." Stokes' MSS.'

[2] No plan of this anchorage appears to have been rendered. The work was presumably incorporated in the general chart, Fig. 26.

[3] [Note in original] '3 leagues to the westward of Cape Notch.' This is presumably Caleta Marión, the bay surveyed on 25 January.

us from the southern coast. I fired several guns to ensure her marking our position ere we became shut in by the land, and shortly after I anchored had the satisfaction to receive her alongside. She proved to be a whale boat, having in her the 2nd. mate and five men belonging to the Sealing Vessel Prince of Saxe Cobourg, Mathew Brisbane,[1] Master, of London. They informed me that, their vessel had, on the 16th. of December last, been wrecked in a Harbour called Fury Harbour[2] situated on the Coast of Tierra del Fuego, very near the <u>southern</u> entrance of the supposed Barbara Channel;[3] that the Master and the Crew had all escaped from the wreck; and that having rigged them with their sails and spars such accommodation as they might on these dreary and desolate shores, awaited the arrival of any Vessel which chance might bring to their relief – After a lapse of five weeks (in the course of which one of the party died) the Men became impatient; and the Master consented to the <u>Chief</u> mate and six of the crew taking one of the boats for the purpose of making their way to Rio Negro on the East [p. 46] Coast of Patagonia; another boat (the one with which we had fallen in) was despatched through the Straits of Magellan to endeavour to find some Vessel taking seal off the Isles of Direction. As we had so recently passed two days & two nights near those Isles, our knowledge that no Vessel was in their neighbourhood, together with the bad weather we had experienced, afforded us abundant reason to congratulate these poor fellows on their good luck in having fallen in with the Beagle. The second mate presented me with a letter addressed to the "Captain of any Vessel", by the Master, (who, together with seven of the Crew still remained at the wreck) couched in very urgent terms for relief; and he informed me further that the ship-wrecked party had much to fear from the hostility of the natives; some of their canoes having attacked the boat with bow and arrows, and she had only escaped through her superior pulling.

This circumstance determined me not to lose a moment in hastening to their relief.

March the 4th.

Accordingly on the following morning we ran down to Port Gallant, situated on the northern shore of the Straits immediately opposite the northern entrance of the supposed Barbara Channel: and having despatched to you a boat[4] with a letter to dispel any alarm that might be excited in your mind by our protracted absence,

---

[1] Brisbane had commanded the *Beaufoy* on James Weddell's voyage towards the South Pole in the *Jane* (1822–24) and was subsequently murdered at Port Louis, Berkeley Sound in the Falkland Islands, 26 August, 1833 where he was the Superintendent of the settlement. (The settlement was originally called Saint Louis and had been established by Bougainville, see p. 17.) Some details of his career are given in King and Fitz-Roy, *Narrative*, II, pp. 327–35 and A. E. G. Jones, *Polar Portraits*, Whitby, 1992, pp. 111–14; the murder in some detail in Thomas Helsby's manuscript, printed in part in the *Falkland Islands Journal*, 1968, pp. 21–31. See also James Weddell, *A Voyage towards the South Pole &c.*, London, 1825.

[2] Bahía Furia, lat. 54°27′S, long. 72°17′W.

[3] Canal Bárbara, lat. 54°10′S, long. 72°10′W. Sarmiento mentions channels to the south in this vicinity, February, 1580 (Markham, *Narratives of the Voyages*, pp. 121–3), and a channel south to the ocean was also mentioned by Van Speilbergen, April 1615 (Villiers, *East and West Indian Mirror*, pp. 42 and 46). The use of the passage was first reported by Frézier, *Voyage to the South Sea*, pp. 285–7; see p. 169.

[4] 'The whale-boat, manned by Lieutenant Sholl and four men, victualled for twenty-one days': Macdouall, *Narrative*, p. 130.

... 5th.

I left the Beagle moored in Port Gallant in charge of the first Lieutenant, and proceeded, with two ten oared boats,[1] one under my immediate command, and the other under that of my Master, to the relief of our ship wrecked countrymen.[2]

We crossed the Straits, and after bivouacking one night in the boats, succeeded in threading all the intricacies of the various channels which here intersect the Tierra del Fuego; and in the afternoon of the second day of our departure, we reached (after a labori-[p. 47]ous pull along the Coast of the Southern Ocean) Fury Harbour, where we were received by our ship-wrecked countrymen with every mark of joy and thankfulness. We were of course prepared to find our arrival welcome; but it should seem that an incident had occurred to render it still more so: they had been visited by a party of the natives who, tho' they behaved themselves on that occasion peaceably enough, might, from the opportunity they had had of seeing the smallness of the party, and the valuable spoil on the beach, be expected soon to repeat their visit in greater force & in less friendly guise. And indeed so strong was this impression that when our boats were first seen by a man who was looking out on a Hill top, the alarm was given of "the Indians", "the Indians", and each sought to put himself in the best possible state of defence: And it was not until we came pretty close that their apprehension of attack was turned into joy for the relief.

We found the Vessel on the rocks, bilged,[3] an utter wreck; and the Master and the crew exceedingly anxious to get away; so having embarked them all in our boats with as much of their property as they could stow, succeeded (after bivouacking another night in our boats, and a long pull of eighty miles,) in rejoining the Beagle in Cape Gallant Harbour on the 5th. day of our absence from her.[4]

---

[1] [Note in original] 'victualled for a fortnight.'

[2] Macdouall, *Narrative*, p. 130: 'Captain Stokes, the master and Mr. Kirke, midshipman, with the six shipwrecked seamen and some of our own crew, proceeded in the launch and cutter to the relief of Captain Brisbane, at Fury Harbour, about seventy miles distant from Port Gallant.'

[3] The bilge is that part of the hull of a vessel on which she would rest if laid on the ground. A vessel is said to be 'bilged' when this part is fractured: Burney, *Dictionary*, p. 39.

[4] Macdouall, *Narrative*, pp. 130–33: 'Mr. Kirke related to me a few particulars of their passage. About midway in the Barbara Channel, they encountered a great many Indians in their canoes, who endeavoured to keep up with the launch and cutter, and as the boats neared any of the natives occupying the rocks and headlands (which in a great many places were thronged by them) they, as the boats passed underneath, set up a halloo, and discharged their arrows and spears, which, however, fell very greatly short of the mark; but how great was the surprise of Captain Stokes and the rest of the officers, to behold some of these naked savages running along the beach, holding in their hands the identical boarding-pikes which had been missing from Port Gallant on our return [see p. 191, n. 1] … As the party were returning, they landed where a great many Indians had collected, most of whom were painted or daubed over the face and body, red and white, … however they were very friendly, and a good many lances and bows and arrows were obtained from them in exchange for beads, knives &c … Belonging to our ship's crew there was a black man, who had gone in the cutter, and he no sooner landed among the natives than they all gathered round him, astonished at his black face, and uttering strange sounds, pointed at him with their fingers, and kept touching his face and pulling his woolly head, laughing loudly, and indulging in many extravagant gestures, as if delighted at his sable appearance.'

Subsequently, on passage to Montevideo, Captain Brisbane related an incident at Fury Harbour to Macdouall when Captain Brisbane had jumped from the wreck of his ship and landed in a native boat with four men in it, without the knowledge of any of his people. The Fuegians capsized the boat to escape. Macdouall considered 'it substantiates, in some degree, what I have said about their *kindness* when they happen to have the superiority, it places beyond doubt their *friendly* intention towards Captain Brisbane.':

March the 10th.

This morning we weighed and made sail for Port Famine; which place we reached at 5 PM, and having informed you, in answer to your telegraphic signal, that all the ship wrecked men were safe, we dropt our Anchor, and you were good enough to greet us with three cheers.

<u>Pringle Stokes.</u>
Commander, R.N.

Straits of Magellan,
March 1827.

Macdouall, 1833, pp. 175–6. It is also worth noting that Brisbane had visited the Fuegeans with James Weddell in 1823 (*Jane* and *Beaufoy*) and again on his own in 1825 (*Beaufoy*) and found the Fuegians friendly on both occasions: Weddell, *Voyage*, pp. 146–92 and 317–24.

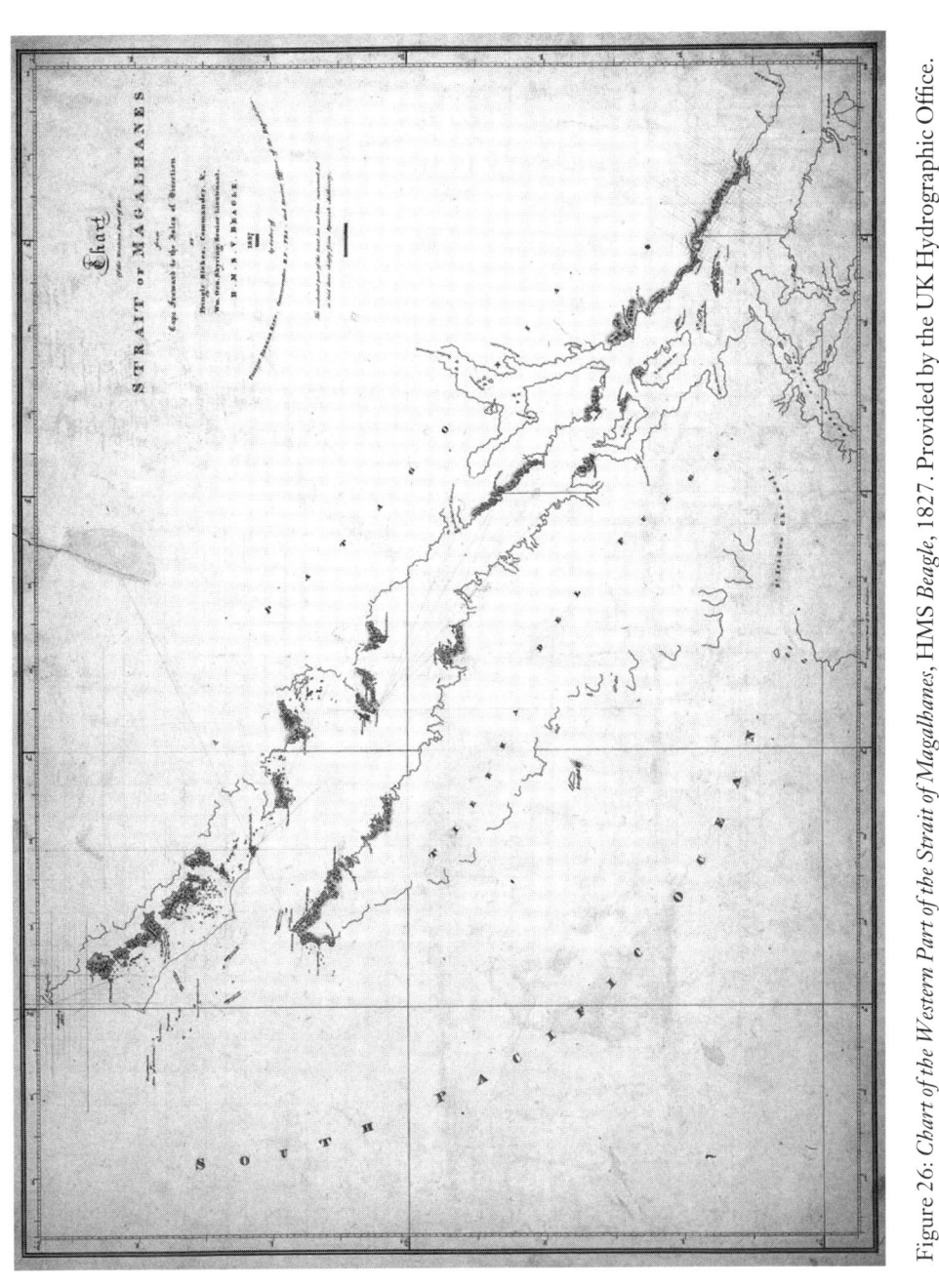

Figure 26: *Chart of the Western Part of the Strait of Magalhanes*, HMS *Beagle*, 1827. Provided by the UK Hydrographic Office.

ON THE PASSAGE ROUND CAPE HORN OR THE STRAITS OF MAGELLAN

If I be required, after this ample detail of our passage – to state my opinion respecting the Straits of Magellan as a channel for ships from the Atlantic into the Pacific Ocean, I shall say that except in cases which I shall specify, the route by Cape Horn is to be preferred: and this opinion I have founded on an attentive perusal of the accounts of our older Navigators, as well as my own experience. It is my belief (and that of every one of my Officers) that no <u>square-rigged</u> Vessel except the Beagle, or one of her class, a ten gun brig, bark-rigged, would have effected the passage under similar adverse circumstances of wind and weather. A westerly wind, strong and squally, reigns throughout the western division of the straits with the unvarying constancy of a <u>trade</u>: current[1] setting always to the eastw^d. at the rate of two knots an hour:[2] rain as heavy and unremitting as ever I have experienced, during the rainy season on the coast of Guinea; Thick misty weather; such is the weather in the summer, what the winter produces we have yet to learn. In point of handiness, and general sailing qualities, the Beagle is the very next to a <u>fore and aft</u> rigged Vessel; yet we have more than once worked in these Straits through twelve successive hours, during which we have made more than thirty boards, without realizing a league to windward. There is another circumstance which contributes greatly to the difficulties of this navigation: the narrowness of the Straits, together with the prevalent bad weather, renders it necessary that an anchorage for the night should always be secured; yet how correctly soever the coast may be laid down in the charts, the general aspect of the coast is one of such unvaried rugged barrenness that for the most part, it is alike impossible to convey or to acquire, by any written description, such leading landmarks as shall enable a Vessel without being subjected more than once (unless she have the rare blessing of a fair wind) to the vexation of running back, after a laborious day's beat, to the anchorage whence she had started in the morning. Local knowledge, gained on the spot, is the only thing likely to conduce – except easterly winds – to render to <u>any</u> description of Vessels the passage of the Straits of Magellan [preferable] to that round Cape Horn. [p. 49] In short, the passage from the Atlantic into the Pacific. may, by <u>fore & aft</u> rigged merchant vessels, be successfully attempted through the Straits; but such as [are] <u>square</u> rigged, will, if they would avoid danger, probable disaster, and certain loss of time, adhere to the route round Cape Horn.

Let me illustrate my ideas by imagining a case: The Commander in Chief on the South America Station orders me to convey, in the <u>Beagle</u>, or any <u>vessel of her class</u>, or,

---

[1] This word is apparently written 'raicurrent' in the original as if the writer had started the next phrase – 'rain as heavy – …', realized his mistake and then failed to cross out the initial three letters.

[2] See p. 182, n. 4.

à-fortiori, a <u>fore and aft</u> rigged Vessel, despatches from Rio de Janeiro, or the Rio de la Plata, to Valparaiso; leaving to me the choice of route; I should, having now the requisite local knowledge, go by way of the Straits of Magellan; should he order me to convey them in a <u>ship sloop</u> or any <u>larger Vessel</u>, I should decline the Straits and go round the Horn.

Pringle Stokes
Commander

H.M.S. Beagle
Straits of Magellan March 1827

    The following abstract of my meteorological Journal will convey at a glance the state of the weather, while we were in the western part of the Straits.

P.S.

An Abstract of the Meteorological Journal of HMS *Beagle*, Pringle Stokes, Commander; in the Western parts of the Straits of Magellan

Month	Day	Thermometer				Barometer			Prevailing Winds	Prevailing Weather
		Air			Water	Maximum	Minimum	Mean		
		Maximum	Minimum	Mean						
January the	15.th	° 52.5	° 47.0	° 50.1	° 48.0	Inches 29.79	Inches 29.58	Inches 29.68	From WNW, via West, to SW	First & latter parts of the day light Airs; middle moderate and fine.
,,	16.	55.0	48.0	52.3	47.8	29.84	29.61	29.79	From NW, via West, to SW	Fresh breezes & squally with rain; the latter part Light airs & calms.
,,	17.	51.0	47.0	49.4	47.3	29.60	29.46	29.53	From NWbyW, via West, to WSW	Strong breezes & heavy squalls, with drizzling rain.
,,	18.	51.0	47.0	48.6	47.1	29.65	29.48	29.55	From SWbyW, via West, to WNW	Fresh breezes & cloudy, with a good deal of rain – very hazy –
,,	19.	45.5	43.0	44.4	47.0	29.50	29.44	29.48	From West, via SW to South	ditto
,,	20.	49.0	42.0	46.2	47.5	29.72	29.44	29.61	From WbyN, via South, to ESE	Light breezes and cloudy; fine pleasant weather – very clear.
,,	21st	51.0	42.0	48.1	48.3	29.72	29.58	29.64	From SE via East to NE	ditto
,,	22d	59.0	47.0	51.1	49.0	29.76	29.58	29.64	ESE, calm; and Easterly	ditto
,,	23rd	52.0	47.0	48.9	48.8	29.76	29.47	29.61	Easterly, calm, and WSW	Light breezes in the fore part of the day; the remainder, Strong breezes.
,,	24th	49.0	46.0	47.7	48.3	29.85	29.48	29/68	From WbyN, via West, to WbyS	Fresh breezes with very heavy squalls, rain & thick weather

Month	Day	Thermometer					Barometer			Prevailing Winds	Prevailing Weather
		Air				Water	Maximum	Minimum	Mean		
		Maximum	Minimum	Mean							
,,	25.	49.5	47.0	48.3		48.8	29.98	29.87	29.91	From NW, via West, to WNW	ditto, incessant drenching rain
,,	26.	52.0	49.0	50.1		50.1	29.97	29.70	29.83	From NW, via West to WSW	ditto ditto
,,	27.	56.0	48.0	50.8		50.8	29.97	29.72	29.86	From WNW, via West, to WSW	ditto ditto
,,	28.	53.0	50.0	51.9		51.0	29.72	29.38	29.59	ditto	ditto ditto
,,	29.	52.0	47.0	49.8		48.8	29.56	29.32	29.41	From SWbyW, via West, to WbyN	ditto ditto
,,	30.	54.0	46.0	49.5		49.0	29.54	29.32	29.47	From WNW, via West, to WbyN	ditto ditto
,,	31st	51.0	48.0	50.6		48.5	29.33	29.26	29.28	NWbyW	Strong breezes & heavy squalls – very thick weather, rain at times & hail
February the 1st		50.0	47.5	48.7		48.9	29.73	29.25	29.47	From WNW, via West, to WSW	fresh breezes with heavy squalls in the morning; remainder fine.
,,	2nd	52.0	46.0	49.2		48.4	29.70	29.62	29.69	From WbN, via West, to WbS	Strong breezes & very hard squalls, thick weather & incessant rain.
,,	3rd	55.5	50.0	52.1		49.0	29.58	29.39	29.54	West	ditto ditto
,,	4th	50.0	47.5	48.8		49.0	29.76	29.50	29.53	ditto	ditto ditto
,,	5.	52.0	48.0	50.3		48.8	29.70	29.60	29.69	WbyN	ditto ditto

Month	Day	Thermometer				Barometer			Prevailing Winds	Prevailing Weather
		Air			Water	Maximum	Minimum	Mean		
		Maximum	Minimum	Mean						
,,	6.	52.0	48.0	49.6	48.6	29.69	29.57	29.65	WNW	ditto ditto
,,	7.	50.0	48.0	49.4	48.9	29.53	29.53	29.39	ditto	ditto ditto
,,	8.	56.0	50.0	54.2	48.0	29.44	29.44	29.47	ditto	Fresh breezes & squally, misty weather & drizzling rain.
,,	9.	55.0	50.0	53.8	47.8	29.51	29.33	29.39	From NWbW, via West, to WbyS	ditto ditto
,,	10.	54.0	50.0	52.3	47.3	29.51	29.39	29.47	West	ditto ditto
,,	11.	49.0	47.5	48.3	47.1	29.67	29.50	29.60	From WNW, via West, to WbS	ditto heavy rain
,,	12.	51.0	48.0	49.8	47.0	29.69	29.63	29.66	Westerly	ditto ditto
,,	13.	52.0	49.0	50.2	47.5	29.69	29.51	29.62	ditto	ditto ditto
,,	14.	52.5	48.0	50.1	48.3	29.88	29.70	29.82	ditto	ditto ditto
,,	15.	53.0	44.0	49.3	49.2	29.70	29.29	29.46	From NW, via West, to WNW	ditto ditto
,,	16.	45.0	43.5	44.7	49.0	29.69	29.46	29.58	From WNW, via West, to WbS	Fresh breezes, with very heavy & sudden gusts, with hail & sleet.
,,	17.	52.0	49.0	49.6	48.5	29.82	29.68	29.77	West	Fresh breezes & squally, with misty weather & incessant drenching rain.
,,	18.	52.0	48.0	50.0	48.9	29.77	29.62	29.67	ditto	ditto ditto

Month	Day	Thermometer				Barometer			Prevailing Winds	Prevailing Weather
		Air Maximum	Air Minimum	Air Mean	Water	Maximum	Minimum	Mean		
,,	19.	52.0	47.0	50.4	49.0	30.15	29.75	29.93	WbyN	Fresh breezes & squally with rain; the afternoon Lt Breezes & fine weather.
,,	20.	55.0	50.0	52.2	49.3	30.18	30.14	30.15	From NW, via West, to WSW	Fresh breezes and cloudy, much of the day moderate & fine.
,,	21st	58.0	53.0	56.4	50.0	30.13	29.85	30.02	From WNW, via West, to WbS	Moderate and fine.
,,	22nd	55.0	53.0	54.7	50.2	29.83	29.62	29.67	From WNW, via West, to West	Fresh breezes and cloudy, with squalls and heavy rain.
,,	23rd	52.0	50.0	51.3	51.5	30.0	29.64	29.84	Westerly	Moderate and cloudy with rain.
,,	24th	55.5	51.0	53.7	51.8	30.29	30.02	30.18	ditto	Moderate breezes & fine clear weather.
,,	25.	54.0	51.0	53.1	50.5	30.27	29.78	30.05	From NEbN, via East, to ENE	First part Modte & fine; latter part, Light airs and calms.
,,	26.	55.0	52.0	53.6	50.0	29.89	29.79	29.83	From NW, via West, to WSW	Moderate and cloudy, during the latter much rain.
,,	27.	52.0	46.0	48.9	49.5	29.78	29.65	29.71	West	Fresh breezes and squally with rain.
,,	28.	47.0	43.0	45.4	49.0	29.71	29.60	29.64	ditto	ditto weather heavy rain and thick mist
March the	1st	45.0	44.0	44.6	49.5	29.93	29.71	29.86	From WNW, via West, to South	ditto ditto ditto

Month	Day	Thermometer					Barometer			Prevailing Winds	Prevailing Weather
		Air			Water	Maximum	Minimum	Mean		
		Maximum	Minimum	Mean						
,,	2nd	46.0	43.0	44.8	49.0	29.93	29.87	29.89	From WbS, via West, to SW	ditto ditto ditto
,,	3d	46.5	45.0	45.8	48.5	30.17	30.00	30.10	ditto	Moderate and cloudy with rain and thick weather.
,,	4th	49.0	43.5	46.1	48.8	30.17	30.05	30.11	ditto	ditto ditto ditto
,,	5.	55.0	45.5	52.4	49.0	29.99	29.41	29.69	NW, via West, to WSW	Moderate and fine.
,,	6.	57.0	45.0	49.7	49.0	29.69	29.49	29.51	West to NW	Fresh breezes and squally with rain.
,,	7.	46.5	43.0	45.1	48.8	29.43	29.31	29.36	WSW to WNW	Moderate and fine.
,,	8.	55.0	44.0	49.2	48.8	29.75	29.65	29.71	West	Calm and cloudy with heavy rain.
,,	9.	54.0	48.0	50.5	48.5	29.62	29.25	29.43	ditto	ditto ditto

[signed] Pringle Stokes.
Commander, R.N.

APPENDIX 1

Geographical Position Book[1]
Observations in the Straits of Magellan – 1827

Latitude of Several Stations in the Western Part of the Straits.

	Latitude	How deduced	Observer	
Cape Gallant	53°42′11″S	Circum Merid. Alt de.		
Point York	53°32′35″S	Merid. Alt de.		
Cape Tamar	52°55′06″S	Circum Merid Alt		
Cape Philip [*sic*] (Sholls Bay)	52°44′05″S	Double Alt de.	Capt Pringle Stokes	The sun was the observed object. A sextant of 8 inches radius by Jones and a mercurial horizon were the instruments employed.
Cape Felix	52°56′12″S	⎫		
Cape Caves	52°53′19″S	⎬ Merid. Alt de.		
Tuesday Bay (West Pt.)	52°50′04″S	⎬		
Cape Parker (Station near)	52°41′49″S	⎭		
Observation Mt, (Scilly Islands)	52°28′58″S	⎫ Circum Merid. Alt de.		
Misericordia Harbour (Obs. Island)	52°44′40″S	⎭		
	52°44′57″S	Capt Fitzroy Cir M Alt		
Cape Victory	52°16′10″S	Merid. Alt de.		

---

[1] UKHO Geographical Position Book No 80. Degree, minute and second symbols have been added for clarity where these were not included in the original.

Following is an extract, omitting all the calculation.

Longitude
Chronometric Determination of the difference of Longitude between Port Famine and several Stations in the Western Part of the Straits.

1827
Mean difference of Longitude by Chronometer between Port Famine and:
Station near Cape Gallant   1°05′5.″452
             Cape Tamar   2°50′0.″6
             Cape Phillip   2°54′18.″7 [Corrected in pencil] 73°48′19.″9
Obs. Island Harbour of Mercy   3°37′5.″67 [Corrected in red ink by Phillip P. King] 3°40′30″

Mean Diff between Harbour of Mercy
& Obs Mt, Scilly Islands   0° 3′13.″33
Port Famine → Harbour of Mercy   3°37′ [figures crossed out in red ink]
Cape Victory
Harbour of Mercy → Cape Victory   0°15′24.″15
                                   3°37′ [figures crossed out in red ink]

Cape Parker
Harbour of Mercy → Cape Parker   0°28′20.″7
                                   3°37′ [figures crossed out in red ink]

Island off Cape Upright
Harbour of Mercy and Cape Upright 1°2′14.″0
                                   3°37′ [figures crossed out in red ink]

[Page at end of calculations in red ink signed Phillip P. King contains following]
Rate recalculated by Fitz Roy Gave H of Mercy W of Port Famine 3°40′30″
            Lat 52°44′57″

Variation observed in 1827
Station near Cape Tamar   23°24′24″ E   Jan 28
Harbour of Mercy   23°48′21″ E   Feb 19
Obs Mt Scilly Islands   25° 9′26″ E   Feb 21
Mean of 4 (5 at Cape Tamar) each a mean of a number of individual Observations with Theodolite.

APPENDIX 2

Comparison of Geographical Positions

Name	Captain King's final value		Modern value	
	Latitude South	Longitude West	Latitude South	Longitude West
Cape Virgins	52°18'35"	68°17'46".2	52°20'	68°21'
		68°16'55"		
Port Famine				
Obs. Position	53°38'11".7	70°54'01".2	53°37'51"	70°55'36"
Pt St Anna	53°37'55"	70°51'19"	53°38'03"	70°54'51"
Cape Froward				
Summit of Morro	53°53'43"	71°14'31"	53°54'06"	71°17'54"
Cape Gallant				
Extremity	53°42'11"	71°59'01"	53°42'21"	72°01'48"
Port Gallant				
Wigwam Point	53°41'42".7	71°56'56".6	53°41'30"	71°59'24"
Point York				
Extremity	53°32'35".2		53°34'39"	72°17'12"
Cape Tamar				
Obs. Island	52°55'06"	73°44'01".8	52°55'36"	73°45'54"
Extremity of Cape	52°55'30"	73°44'26"	52°56'00"	73°46'30"
Cape Phillip				
Sholls Bay	52°44'05"	73°48'19".9	52°44'18"	73°54'00"
Summit of Cape	52°44'20"	73°53'01"	52°44'48"	73°55'06"
Point Felix				
Station E side	52°56'31".4			
Extremity	52°56'00"	74°09'01"	52°56'24"	74°08'36"
Harbour of Mercy				
Obs. Island	52°44'57"*	74°35'31".2	52°46'00"	74°35'30"
Observation Mount				
Scilly Islands	52°28'58"	74°32'17".9	52°29'36"	74°34'30"
Westminster Hall				
Eastern summit	52°37'18"	74°20'26"	52°37'	74°22'
Cape Pillar				
Extremity	52°42'40"	74°39'31"	52°43'45"	74°40'30"
	52°42'53"	74°37'41"		
Cape Victory	52°16'10"	74°50'55".3	52°17'15"	74°55'12"

*Value taken from King, *Narrative*, I, p. 484. See also note at end of Mean Difference of Longitude, Appendix 1, p. 243 above.

Captain King's final values have been taken from the final section of his Observation Book, 1830, (UKHO Geographical Data Book No. 30). which contains Chronometer errors and the working for determining positions together with lists of positions. The above are an extract only. Modern values have been taken for Cape Virgins from the Admiralty Sailing Directions, the remainder from Admiralty charts. When comparing results it must be recognized that it is not always possible to be sure the modern point chosen is exactly the same place as that where the observations were taken. E.g. no summits are shown on Admiralty Chart 1282 for Isla Westminster Hall.

APPENDIX 3

Canoes

In *Observations of Sir Richard Hawkins Knight, in his voyage into the South Sea. Anno Domini 1593*, Hawkins describes the native canoes – 'we saw a Cannoa made fast under a Rocke with a Wyth, most artificially made with the rindes of Trees, and sewed together with the finnes of Whales; at both ends sharpe, and turning up, with a greene bough in either end, and ribbes for strengthening it.'[1]

The construction of these canoes is described in such detail in the account of Córdoba's expedition[2] that one is tempted to think he, or one of his people, had watched a canoe being built. He says:

> Their boats or canoes are made of the bark of the tree producing resin, whose greatest thickness does not exceed one inch. They are formed of three pieces; one in the middle, forming the bottom and the keel, and the other two pieces the sides. Their patience and application are admirable in stripping the bark of those trees, having for the purpose no other instrument than a flint somewhat shaped and sharpened, with which they make an incision round the trunk at each end, and then another lengthways, to join them; afterwards with vast patience and management, they strip off the whole bark of the tree in one entire piece, of the proper length for the intended canoe, which, in some, including the bend of the middle-piece, which forms the stem and stern as well as the keel and bottom, is from 30 to 32 feet; and the true length of this frail boat, when finished, is from 24 to 26 feet; the greatest breadth four feet, and the depth from two to three feet.
>
> In order to make this bark acquire the proper curvature and shape, and loose what it has on the tree, they lay it on the ground, with the inside downwards, and on each end place a heap of stones, leaving it so for two or three days, in which time it dries and becomes fit for being employed. Then they place the side-pieces almost perpendicular to that in the middle, joining them together with seams of dry rushes, and caulking or filling up the interstices with dried grass and clay or mud, as much as possible to prevent the entrance of the water. To give some strength and resistance to the sides, they lay across the canoe pieces of wood resembling pipe-staves, one by the side of another, all along the length of the canoe, giving it the shape of a semiellipsoid, and make the ledge or gunwale of two strong poles, well joined together at each end, and in them are fixed the ends of the cross-pieces, which serve as ribs or timbers; the whole being tied and sowed together with rushes; placing also crosswise from time to time, some pieces of wood, that answer the purpose of benches or thwarts to sit on.
>
> When the canoe is in this state, they line almost the whole of the inside with pieces of the same bark, about one foot broad, laid across, and having the ends made fast in the

---

[1] Purchas, *Hakluytus Posthumus*, IV, p. 1387, and reprinted ibid., 1906, XVII, p. 117; and Williamson, *Observations*, pp. 82–3.
[2] *Relacion*, pp. 343–6, translated in *Voyage of Discovery*, pp. 99–101.

gunwale on each side. In order to give these pieces the requisite bend, they heat them by the fire, and, when they are half dried, apply them to their proper situations. Besides this, they form a kind of floor from the fourth part from stem and stern, placing it about half a foot from the bottom, leaving an opening in the middle, to throw out the water. This sort of floor consists of boards laid lengthwise over others placed crosswise; and, as well as all the rest of the canoe, is covered with bark.
Such is the construction of their boats …

Many of these canoes are capable of containing nine or ten Indians; they are moved along with a sort of paddles, and rowing them is the ordinary employment of the women. When they enter on a long voyage, which is always either in a calm or with a fair wind, they set up a pole as a mast in the bow of the canoe, and across it another like a yard, having fixed to it the skin of a seal, and keep the lower parts of it steady with their hands: and this scanty sail relieves them from the fatigue of rowing. In the middle of the canoe are some stones, with heaps of shells and sand, on which sort of hearth they make their fire keeping it up with branches and sticks.

On the second voyage, when the boat party was in the vicinity of Punta Echeñique (lat. 53°02′S, long. 73°51′W), towards the western end of the strait, they were visited by a group of Fuegians and their boats were described.[1]

The construction of their canoes displays some superiority of talent over the others in the Strait; since they are not made of frail bark and badly put together like the ones of these [other people] but of planks joined together by means of a half inch thick cord with a caulking which seems to be made of grasses and a particular clay so adhesive and tough that it does not let the water in: each side consists of two quite strong planks, which are skilfully given the necessary curvature to decrease the beam regularly towards the stern and bow, which has its maximum width at the water line: the keel is a thick plank, long and narrow, fixed to the side plank in the same way as they are fixed to each other: the thwarts and floor timbers are the same as in the canoes of the other natives, although somewhat stronger; even though their shape did not seem so suitable for moving at speed, they at least have the advantage of greater stability and strength, and also of not running the risk of sinking by letting in water like the others. They rowed in the same way as Europeans and their oars were of a well proportioned length, by which it can be seen that they knew the merit [of long oars] over the paddles that the other natives use, and the need of a certain length [of oar] to make the most of their strength: their only use of a paddle was as a rudder, just as the oar is used by Europeans particularly in rough seas. It is to be believed that living on the coast where the seas and winds are usually stronger than in the rest of the Strait, has required these men to improve the construction of their canoes; since progress in human ingenuity has always followed the law of necessity.

[1] *Apéndice a la Relacion*, pp. 59–60, *Voyage of Discovery*, p. 38), translated for me by Mr T. N. Snow.

APPENDIX 4

Names in Estrecho de Magallanes given by Pedro Sarmiento de Gambóa still in use

Captain King made a very real effort to retain the names given by his predecessors. He says of Sarmiento 'Any name given by this excellent old navigator is too classical to be omitted.' (King, *Narrative*, I, p. 145.)

The page references given below are to Sarmiento de Gambóa, *Viage al Estrecho de Magallanes*.

Abra, Isla, p. 206, 53°22'S, 73°04'W, in entrance to Canal Abra.
Agua-dulce, Caleta de, p. 217, now Caleta Agua Dulce, 53°55'S, 71°45'W.
Alguilgua, Ensenada, p. 203, now Bahía Alguilgua, 53°04'S, 73°46'W, name collected by Sarmiento from native Fuegians.
Anegada, Punta, p. 273, 52°28'S, 69°25'W.
Angosto, Puerto, p. 204, 53°13'S, 73°22'W.
Barranca, [Punta] p. 273, 52°33'S, 69°42'W.
Baxa, Punta, p. 273, 52°35'S, 69°36'W.
Boqueron Punta del, p. 223, now Cabo Boquerón, 53°29'S, 70°12'W.
Campana, Bahía de la, p. 213, now Bahía Bell, 53°55'S, 71°48'W. Sarmiento named the bay for Campana de Roldan, a hill named in Magellan's expedition and shown on Admiralty chart 554 as Roldan's Bell until 1964 (53°58'S, 71°45'W). The name appears to have been anglicized by King and retained in that form by the Chilean authorities.
Delgada, Punta, p. 272, 52°27'S, 69°33'W.
Gente-Grande, Punta de, p. 243, name retained in Bahía Gente Grande, 52°57'S, 70°12'W.
Lomas, Ensenada de, p. 275, now Bahía Lomas, 52°32'S, 69°00'W.
Lomas, Morro de, p. 221, name preserved in Bahía Lomas, 53°50'S, 70°40'W.
Madalena, Canal-de-la-, p. 220, now Seno Magdalena, 54°05'S, 70°57'W.
Madalena, Isla, p. 254, now Isla Magdalena, 52°55'S, 70°35'W.
Misericordia, Puerto de la, p. 182, now Puerto Misericordia, 52°47'S, 74°35'W.
Mucha-nieve, Ensenada de, pp. 206–7, now Seno de las Nieves, 53°31'S, 72°42'W.
Playa-Parda, p. 206, now Caleta Playa Parda, 53°19'S, 73°01'W and Estero Playa Parda 53°19'S, 73°02'W.
Puchachailgua, Ensenada, p. 203, now Estero Puchachailgua, 53°02'S, 73°51'W, name collected by Sarmiento from native Fuegians.
San-Felipe, Ensenada de, p. 270, now Bahía Felipe, 52°44'S, 69°55'W.
San-Gabriel, Canal-de-, p. 220, now Canal Gabriel, 54°12'S, 70°40'W.

San-Gregorio, Bahía de, p. 260, now Bahía Gregorio, 52°36′S, 70°06′W.

San-Gregorio, Punta de, p. 251, (referred to as Cabo de San-Gregorio at p. 259) now Cabo Gregorio, 52°39′S, 70°33′W.

San-Ildefonso, Punta de (Sanctelifonso, Punta de), p. 203, now Cabo Upright, 53°05′S, 73°36′W, name retained in Bahía Ildefonso close west of the cape.

San-Isidro, Punta de, p. 257, now Punta San Isidro, 52°44′S, 70°07′W.

San-Juan, Rio de, p. 223, where Sarmiento took possession of the land, pp. 230, 241, and then named Rio de San-Juan-de-la-Posesion, p. 241, now Puerto San Juan de la Posesión, 53°38′S, 70°56′W. The bay was called by Cavendish Port Famine and it is also known as Puerto del Hambre.

San-Pedro, Bahía de, p. 217, now Seno Pedro, 53°56′S, 71°37′W.

San-Simon, Bahía de, p. 213, now Bahía Simón, 53°51′S, 72°00′W.

San-Valentin, Punta de, p. 223, now Cabo Valentín, 53°34′S, 70°31′W.

San-Vicente, Cabo de, p. 249, now Cabo San Vicente, 52°47′S, 70°26′W.

Sancti-Isidro, Punta de, p. 220, now Cabo San Isidro, 53°47′S, 70°58′W.

Santa-Agueda, Punta y Morro de, p. 218, now Cabo Froward with Morro Santa Agueda, 53°54′S, 71°18′W, above it.

Santa-Ana, Punta de, p. 222, now Punta Santa Ana, 53°38′S, 70°55′W.

Santa-Catalina, Ensenada de, p. 255, now Bahía Catalina, 53°06′S, 70°51′W.

Santa-Ines, Isla de, p. 180, now Isla Santa Inés, 53°40′S, 73°00′W; Isla Desolación, 53°05′S, 74°00′W, was the island named by Sarmiento but the name has been transferred.

Santa-Marta, Isla de, p. 254, now Isla Marta, 52°51′S, 70°35′W.

Santa-Mónica, Puerto de, p. 202, now Caleta Santa Mónica, 53°02′S, 73°52′W.

Santa-Susana, Ancon de, p. 256, now Caleta Susana, 52°39′S, 70°20′W.

Santiago, Ensenada de, p. 271, now Bahía Santiago, 52°33′S, 69°52′W.

Tinquichisgua, Punta de, p. 213, now Punta Tinquichisgua, 53°52′S, 71°51′W, name collected by Sarmiento from native Fuegians.

Voces, Playa de-las, p. 222, now Bahía Voces, 53°41′S, 70°57′W.

Xaultegua, Ensenada, pp. 203–4, now Golfo Xaultegua, 53°09′S, 73°05′W, name collected by Sarmiento from native Fuegians.

Whenever possible Sarmiento consulted the native Fuegians and used their names for topographic features. Considerably more than those given above are mentioned in his account, but these are the only ones which are in use today.

BIBLIOGRAPHY

Apéndice a la Relacion del viage al Magallanes de la Fragata de guerra Santa María de la Cabeza que contiene el de Paquebotes Santa Casilda [under the command of A. de Córdoba] *y Santa Eulalia* [under F. de Miera] *para completar el reconocimiento del Estrecho en los años de 1788 y 1789* [by Josef de Vargas y Ponce], Madrid, 1793.

Beaglehole, J. C., *The Journals of Captain James Cook on his Voyages of Discovery, I, The Voyage of the Endeavour 1768–1771*, Cambridge, 1955.

Blunt, Wilfred, and Raphael, Sandra, *The Illustrated Herbal*, London, 1979.

Bougainville, L.-A. de, *Voyage autour du monde par la frégate du roi Boudeuse et la flûte l'Étoile de 1766 à 1769*, Paris, 1771.

Bougainville, Lewis de, *A Voyage Round the World &c.*, transl. by John Reinhold Forster, London, 1772.

Bradley, Peter T., 'Narborough's Don Carlos', *Mariner's Mirror*, 72, 1986, pp. 465–75.

Bulkeley, John and Cummins, John, *A Voyage to the South Seas by His Majesty's Ship the Wager in the years 1740–1741*, [first edition, 1743], 2nd edn, Philadelphia, 1757; 3rd edn, London, 1927.

Burney, James, *A Chronological History of the Discoveries in the South Sea or Pacific Ocean*, 5 vols, London, 1803–17 (reprinted, Amsterdam, 1967).

Burney, William, *A New and Universal Dictionary of the Marine &c.*, London, 1830.

Byron, The Hon. John, *The Narrative of the Honorable John Byron containing an Account of the Great Distresses suffered ... on the coasts of Patagonia &c.*, London, 1768.

Carrington, Hugh, ed., *The Discovery of Tahiti &c.*, Hakluyt Society, 2nd ser. 98, London, 1948.

Colledge, J. J., *Ships of the Royal Navy*, London, 1987.

Cortesão, A. and Teixeira da Mota, A., *Portugaliae monumenta cartographica*, 5 vols, Lisbon, 1960 (reprinted Lisbon 1987).

Dawson, L. S., *Memoirs of Hydrography*, Eastbourne, 1885 (reprinted London, 1969).

Desmond, Ray, *Dictionary of British and Irish Botanists and Horticulturists*, London, 1977.

Drake, Sir Francis, *The World Encompassed*, London, 1628 (reprinted Cleveland, Ohio, 1966).

Dunmore, John, ed., *The Pacific Journal of Louis-Antoine de Bougainville, 1767–1768*, Hakluyt Society, 3rd ser. 9, London, 2002.

Dyer, Florence E., *The Life of Admiral Sir John Narbrough*, London, 1931.

Falconer, William, *An Universal Dictionary of the Marine*, London, 1769.

Frézier, Amédée François, *A Voyage to the South Sea and along the Coasts of Chili and Peru in the years 1712, 1713, and 1714*, for Jonah Bowyer, London, 1717

Froger, François, *Relation d'un voyage fait en 1695, 1696 et 1697 aux côtes d'Afrique, détroit de Magellan, Brezil, Cayenne et Isles Antilles par une escadre commandée par M. de Gennes*, Paris, 1698.

Froger, François, *Relation of a Voyage made in the Years 1695, 1696 and 1697 on the Coasts of Africa, Streights of Magellan &c.*, London, 1698.
Gallagher, Robert E., ed., *Byron's Journal of his Circumnavigation, 1764–1766*, Hakluyt Society, 2nd ser. 122, Cambridge, 1964.
Hacke, William, *A Collection of Original Voyages*, London, 1699.
Hakluyt, Richard, *The Principall Navigations Voiages and Discoveries of the English Nation*, 2 vols, London, 1589 (facsimile, Hakluyt Society, Cambridge, 1965).
Hakluyt, Richard, *The Principal Navigations, Voiages, Traffiques and Discoveries of the English Nation &c.*, 3 vols, London, 1598–1600.
Hawkesworth, John, *An Account of the Voyages undertaken by Order of His Present Majesty for Making Discoveries in the Southern Hemisphere, &c.*, I, London, 1773.
Hooker, Joseph Dalton, *The Botany of the Antarctic Voyage of H.M. Discovery Ships Erebus and Terror in the Years 1839–1843*, 6 vols, London, 1844–60, I, 1847 (reprinted 1963).
Jacobs, Els M. *In Pursuit of Pepper and Tea. The Story of the Dutch East India Company*, Zutphen, 1991.
King, Captain P. P., *Sailing Directions for the Coasts of Eastern and Western Patagonia from Port St. Elena on the East Side, to Cape Tres Montes on the West Side, including Strait of Magalhens and the Sea Coast of Tierra del Fuego*, London, 1832.
King, Captain P. P., and Fitz-Roy, Captain R., *Narrative of the Surveying Voyages of His Majesty's Ships Adventure and Beagle*, 3 vols, London, 1839.
Macdouall, John, *Narrative of a Voyage to Patagonia and Terra del Fuego through the Straits of Magellan in H.M.S. Adventure and Beagle in 1826 and 1827*, London, 1833.
MacGregor, A., ed., *Sir Hans Sloane: Collector, Scientist, Antiquary, Founding Father of the British Museum*, British Museum, 1994.
Markham, Sir Clements, KCB, ed., *Early Spanish Voyages to the Strait of Magellan*, Hakluyt Society, 2nd ser. 28, London, 1911.
Markham, Clements R., *A Life of John Davis the Navigator 1550–1605*, 2nd edn, London, 1891.
Markham, Clements R., ed., *Narratives of the Voyages of Pedro Sarmiento de Gambóa to the Straits of Magellan*, Hakluyt Society, 1st ser. 91, London, 1895.
Mercator, Gerardus, *Gerardi Mercatoris Atlas sive cosmographicae meditationes de fabrica mundi et fabricati figura &c. auctus et illustratus a Iudoco Hondio &c.*, Amsterdam, 1606.
Mercator, Gerardus, *Historia Mundi or Mercator's Atlas containing his Cosmographicall Descriptions of the Fabricke and Figure of the world lately rectified in divers places as also beautified and enlarged ... by Iudocus Hondy. Englished by W. S. Generosus, ...*, 2nd edn, London, 1637.
Moore, David M., *Flora of Tierra del Fuego*, Oswestry, 1983.
Narborough [Narbrough], Sir John, *An Account of Several Late Voyages and Discoveries, &c.*, London, 1694 (reprinted Amsterdam, 1969).
Narbrough, Sir John, *An Account of Several Late Voyages and Discoveries, &c.*, 2nd edn, London, 1711.
Navy List, Admiralty Office, London, various dates, 1805 to 1833.
Noort, Olivier van, *Description du penible voyage faict en tour de l'Univers ou Globe Terrestre, par Sr. Olivier du Norte d'Utrecht*, Amsterdam, 1602.
Nuttall, Z., ed., *New Light on Drake*, Hakluyt Society, 2nd ser. 34, London, 1914.

Pigafetta, Antonio, *Magellan's Voyage. A Narrative Account of the First Circumnavigation*, trans. and ed. R. A. Skelton, New Haven and London, 1969.
Purchas, Samuel, *Purchas his Pilgrimage &c.*, London, 1613.
Purchas, Samuel, *Hakluytus Posthumus or Purchas His Pilgrimes*, London, 1625 (reprinted Hakluyt Society, 20 vols, Glasgow 1905–7).
Quinn, David B., ed., *The Last Voyage of Thomas Cavendish 1591–1592*, Chicago, 1975.
Relacion del último viage al Estrecho de Magallanes de la Fragata de S. M. Santa María de la Cabeza en los años de 1785 y 1786 [under the command of A. de Córdoba, by Josef de Vargas y Ponce], Madrid, 1788.
Sarmiento de Gambóa, Pedro de, *Viage al Estrecho de Magallanes &c.*, Madrid, 1768.
Schouten, William Cornelison, *The Relation of a Wonderfull Voiage made by William Cornelison Schouten of Horne*, London, 1619 (reprinted Cleveland, Ohio, 1966).
Smyth, Admiral W. H., *The Sailor's Word-Book*, London, 1867 (reprinted London, 1991).
South America Pilot, Volume II, HMSO, London, 1993.
Stanley, Lord, of Alderley, ed., *The First Voyage round the World*, Hakluyt Society, 1st ser. 52, London, 1874.
Taillemite, Étienne, *Bougainville et ses compagnons autour du monde 1766–1769*, 2 vols, Paris, 1977.
Taylor, E. G. R., 'The Dawn of Navigation', *Journal of the Institute of Navigation*, 1948, pp. 283–9.
Taylor, E. G. R., *The Mathematical Practitioners of Hanoverian England*, Cambridge, 1966.
Taylor, E. G. R., 'More Light on Drake', *Mariner's Mirror*, 16, 1930, pp. 134–51.
Taylor, E. G. R., ed., *The Troublesome Voyage of Captain Edward Fenton 1582–1583*, Hakluyt Society, 2nd ser. 113, Cambridge, 1959 (1957).
Villiers, J. A. J. de, ed., *The East and West Indian Mirror, being an Account of Joris van Speilbergen's Voyage round the World (1614–1617)*, Hakluyt Society, 2nd ser. 18, Cambridge, 1906.
A Voyage of Discovery to the Strait of Magellan ... undertaken by ... Admiral Don A. de Cordoba ... translated from the Spanish [of Josef de Vargas y Ponce] (New Voyages and Travels, 2, Sir Richard Phillips), London, n.d. [1819].
Wallis, Helen, ed., *Carteret's Voyage Round the World, 1766–1769*, 2 vols, Hakluyt Society, 2nd ser. 124–5, Cambridge, 1965.
Walter, R., *Anson's Voyage Round the World*, ed. G. S. Laird Clowes, London, 1928.
Weddell, James, *A Voyage towards the South Pole &c.*, 2nd edn, London, 1827.
Williamson, James A., *The Observations of Sir Richard Hawkins*, London 1933.
Woodville, William, M.D., *Medical Botany*, 4 vols, 2nd edn, London, 1810.

PART 3

JOURNAL KEPT BY MIDSHIPMAN JOSEPH HENRY KAY DURING THE VOYAGE OF HMS *CHANTICLEER*, 1828–1831

Edited by
ANN SAVOURS and ANITA McCONNELL

CONTENTS

List of Illustrations and Maps	256
Acknowledgements	258
INTRODUCTION	261
1. Description of the Manuscript	261
2. The Voyage and its Commander	261
2.1. Midshipman Joseph Henry Kay: the Writer and his Journal	267
2.2. Some of Kay's Shipmates aboard HMS *Chanticleer*	269
3. Scientific Aspects of the Voyage of HMS *Chanticleer*	271
3.1 Pendulum Measurements	271
3.2 Instruments for Terrestrial Magnetism, Astronomy, Surveying and Meteorology	275
THE EDITED TEXT OF JOSEPH HENRY KAY'S JOURNAL	279
Bibliography	325
Index	397

LIST OF ILLUSTRATIONS AND MAPS

Figure 1: Photograph of the fine monument in the parish church of St Anne, Woodplumpton, near Preston, 'Sacred to the Memory of Henry Foster, R.N., F.R.S. ... erected by several of his companions and friends, as a memorial of the high esteem they entertained for his character, and of the deep regret they felt for his untimely death'. By kind permission of the Vicar and with the help of Mr Peter Sheppard, churchwarden, who kindly took the photograph. — 266

Figure 2: Pendulum apparatus used during the voyage of HMS *Chanticleer* in the collection of the Science Museum, South Kensington (Inv. 1939–388). Reproduced by permission of the Science Museum, Science and Society Picture Library. — 273

Figure 3: Watercolour, presumably by E. N. Kendall, of the interior of the portable pendulum house, depicting an observer and Kater's invariable pendulum, probably at Monte Video in August 1828. The house was constructed of panels arranged to form a room of about twelve feet square and fourteen or fifteen high, with eight windows. Reproduced by permission of the Science Museum, South Kensington, Science and Society Picture Library. — 273

Figure 4: First page of the 'Private Journal' kept by Midshipman Joseph Henry Kay. Reproduced by permission of the Scott Polar Research Institute, Cambridge. — 218

Figure 5: 'Ascension, Red Hill, S.E. by E.', showing HMS *Chanticleer* in the foreground, drawn by E. N. Kendall and published in W. H. B. Webster, *Narrative of a Voyage to the Southern Atlantic Ocean in the years 1828, 29, 30, performed in H.M. Sloop Chanticleer, Captain Henry Foster, FRS*, 2 vols, London, 1834, I, facing p. 385. — 280

Figure 6: 'Pendulum station at Port Cook, Staten Island', drawn by E. N. Kendall and published in Webster, *Narrative of a Voyage*, I, facing p. 99. — 299

Figure 7: 'Neptune's Bellows', at the entrance to Port Foster, Deception Island, South Shetland Islands. Taken in 1927, probably during the *Discovery* (Oceanographic) Expedition, 1925–27. Reproduced by permission of the Southampton Oceanography Centre. Photo 5160. — 307

Figure 8: 'Port Foster' (inner basin of the island), with whale factory ship. Photograph taken in 1927 at Deception Island, South Shetland Islands, probably during the *Discovery* (Oceanographic) Expedition, 1925–27. Reproduced by permission of the Southampton Oceanography Centre. Photo 5162.   308

Figure 9: Another view of Port Foster from ashore, also with vessel. Photograph taken in 1927 at Deception Island, South Shetland Islands, probably during the *Discovery* (Oceanographic) Expedition, 1925–27. Reproduced by permission of the Southampton Oceanography Centre. Photo 5163.   309

Figure 10: 'Pendulum Cove, Deception Island (South Shetland)', drawn by E. N. Kendall and published in Webster, *Narrative of a Voyage*, I, facing p. 147.   310

Figure 11: Watercolour by E. N. Kendall entitled 'Sea-Leopards at Deception Island New South Shetland... January 7th, 1828' [*sic.* 1829]. The border is stamped 'Hydrog: Office, 9 Jy. 31', together with its seal of a foul anchor. Reproduced by permission of the Science Museum, South Kensington, Science and Society Picture Library.   312

Figure 12: Watercolour by E. N. Kendall of 'Sea Leopards at the Island of Deception', dated 9 January 1829, in a volume labelled on the outer cover, 'H.M.S. Chanticleer Com^r Henry Foster 1828. Remarks and observations', in the archives of the United Kingdom Hydrographic Office, reference OD39. Reproduced by kind permission.   312

Figure 13: 'St. Martin's Cove near Cape Horn', drawn by E. N. Kendall and published in Webster, *Narrative of a Voyage*, I, facing p. 177.   320

Map 1: Map of the Atlantic Ocean showing the 'Track of H.M.S. Chanticleer, Capt. Henry Foster, F.R.S., &c. In 1828–30.' Printed in Webster, *Narrative of a Voyage*, I, facing p. 1, in editors' copy.   259

Map 2: 'Deception Island New South Shetland by Lieut. E. N. Kendall, 1829', published in the *Journal of the Royal Geographical Society*, 1, 1831 (London, 1833). Reproduced by permission of the Syndics of Cambridge University Library.   314

ACKNOWLEDGEMENTS

The text of this edition of the journal of J. H. Kay, 1828–31 (SPRI MS 894) is reproduced by kind permission of the Scott Polar Research Institute, University of Cambridge. The editors warmly acknowledge assistance given by the library of the Royal Society; the United Kingdom Hydrographic Office, Taunton; the Map Curator of the Royal Geographical Society, Mr Francis Herbert; Cambridge University library; the Royal Astronomical Society; the Public Record Office, Kew (now incorporated into The National Archives); the Science Museum; and the National Maritime Museum, especially Miss B. Tomlinson; Captain Richard Campbell, RN, for suggesting numerous apposite amendments and additions throughout; Captain Michael Barritt, RN, and Captain Campbell for identifying the sea birds; Dr E. Charles Nelson and Dr Wolfram Lobin for elucidating two plant names; and R. K. Headland for his generous assistance with the manuscript. The award to one of the editors of an HSBC Small Research Grant by the Royal Geographical Society in 1998 has helped the compilation of this edition in various ways.

They also thank Mr Peter Sheppard of Woodplumpton, near Preston, for his kind welcome to the church and Mrs Jean Sheppard for transcribing the memorial to Henry Foster; Mr Brian Vale, CBE, MPhil, for many enlightening comments regarding the *Chanticleer* and South America; the late Professor G. E. Fogg, FRS; Dr E. Charles Nelson; Rear Admiral G. S. Ritchie, CB, DSC; the late Rear Admiral M. J. Ross, CB, DSC; Dr G. Hattersley-Smith; Philip Robson; Mrs Karen MacLellan, Mrs Jane Lovelock and especially Mrs Lorraine Mackey for typing or checking the manuscript; and Mrs. E. Follett for her hospitality. In addition we are greatly indebted to Professors Michael Brennan and Will Ryan, Joint Honorary Series Editors of the Hakluyt Society, for their careful work on the draft.

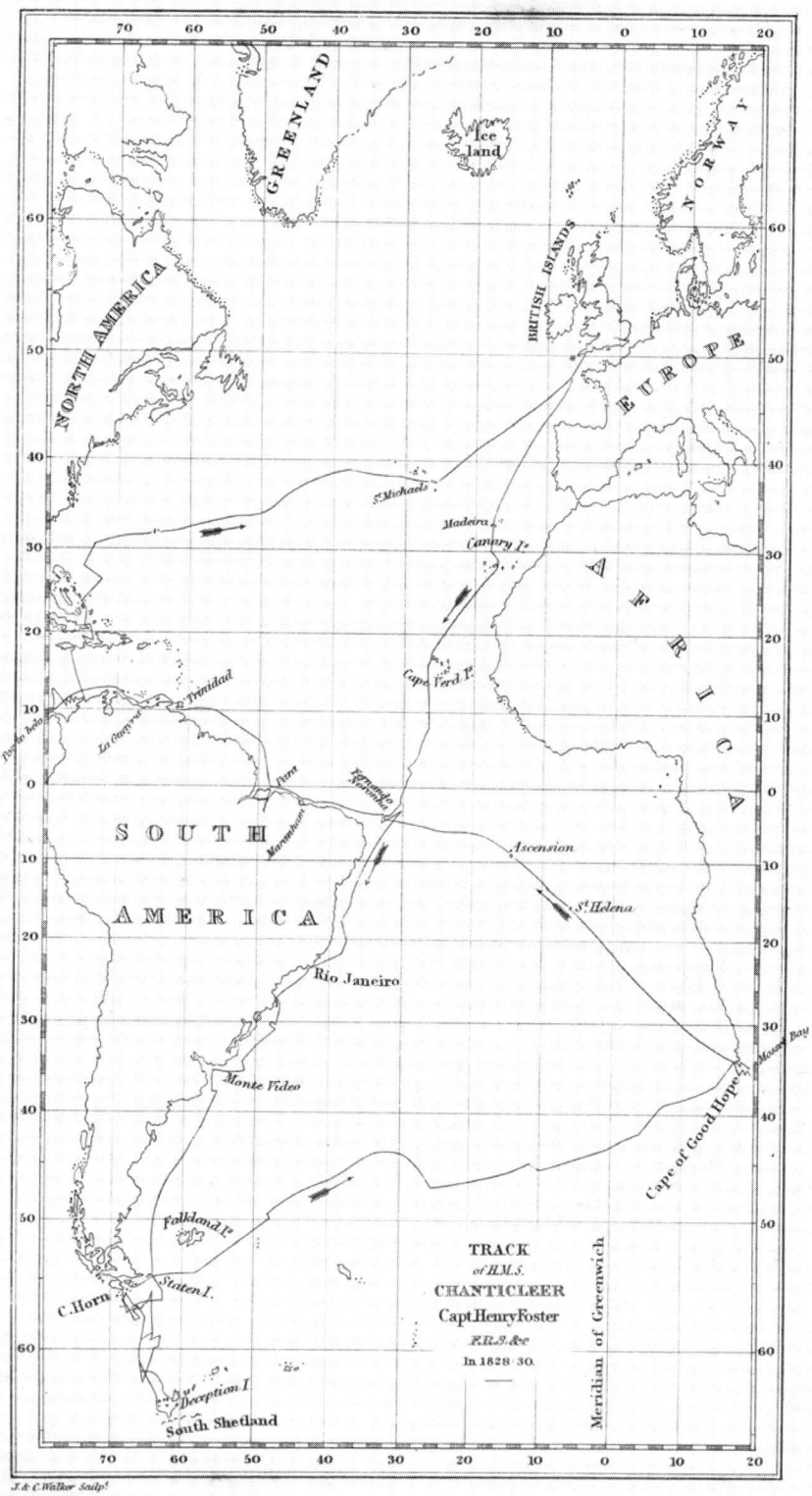

Map 1: Map of the Atlantic Ocean showing the 'Track of H.M.S. Chanticleer, Capt. Henry Foster, F.R.S., &c. In 1828–30.' Printed in Webster, *Narrative of a Voyage*, I, facing p. 1, in editors' copy.

INTRODUCTION

1. Description of the Manuscript

Handwritten in ink, Kay's foolscap private journal, covering the period 21 April 1828 to 23 May 1829, was purchased for £130 by the Friends of the Scott Polar Research Institute, Cambridge, at Sotheby's auction in London on 29 May 1970, lot 526. It is now a slim volume, bound by Cockerell, with the manuscript accession number ms 894. A draft typed transcript was made by the late Mr Alec Laurie, probably in the early 1970s. The vendor was Mrs G. Hatch.

2. The Voyage and its Commander

In the following pages can be found the lively private journal of fourteen-year-old Joseph Henry Kay, kept during the voyage of HMS *Chanticleer* to the South Atlantic, Cape Horn and Deception Island. Sent on the recommendation of the Royal Society, the expedition's main object was to take a series of pendulum observations in order to establish the shape of the earth. Only one narrative was ever published, written by the ship's surgeon, W. H. B. Webster. Kay's journal adds much detail to Webster's account, since he was steadily engaged in the scientific work, often carried out in whirling snow or driving rain.

The scientific voyage of HMS *Chanticleer*, 1828–31, was the first command of Commander Henry Foster, RN, FRS (1796–1831). He was the son of the Reverend Henry Foster and his wife Alice (formerly Croft) of Great Eccleston, Lancashire. He was baptized on 4 September 1796 at Copp Chapel (now Copp Church), an outstation of St Michael's on Wyre.[1] Foster entered the Royal Navy as a First Class Volunteer in 1812. In HMS *Conway* (Captain: Basil Hall) as Master's Mate 1820–23, Foster had made pendulum and other observations in South America, followed by similar scientific and survey work aboard HMS *Griper* (Commander: Douglas Clavering), in Spitsbergen and East Greenland, 1823–4. He served next in HMS *Hecla*, 1824–5, on Sir William Edward Parry's third Arctic voyage in search of the North West Passage. During Parry's fourth voyage, also in the *Hecla*, 1827, the attempt was made by Parry and Commander James Clark Ross to sledge to the North Pole with two ship's boats. Foster remained behind to make surveys of *Hecla*'s anchorage on the north coast of Spitsbergen and of Hinlopenstretet (Hinlopen Strait) as far south as the islets named after him.

The esteem in which Foster was held is shown by his election as a Fellow of the Royal Society in May 1824. His first proposer, Captain Basil Hall, later described him

---

[1] *Registers of St. Michael on Wyre.*

as a 'master spirit',[1] and he was cited as 'a gentleman well versed in practical Astronomy and especially in those branches relating to Nautical and Hydrographical Science'.[2] This distinction was crowned by the decision of the Council of the Royal Society on 15 November 1827 to give him the Copley Medal, its highest honour, 'for his magnetic and other observations made during the Arctic expedition to Port Bowen' during Parry's third voyage.[3] Concerning Foster's award, Basil Hall related, that:

> He thought only of the service, and proceeding straight to the Admiralty, he shewed the medal and declared modestly but firmly to their Lordships that he considered the honour only nominally bestowed upon himself, but essentially conferred upon the naval profession at large. The generous manly appeal could not fail to make its due impression and within the same hour his commission as commander was signed; his appointment to a ship ordered and a voyage of scientific research planned out for him.[4]

The ship in question, HMS *Chanticleer*, had been built in 1807–8 as a ten-gun brig of some 234 tons. She was re-rigged as a barque, and carried only two guns. The Royal Society convened a special committee, which drew up a final Report on 31 January 1828. In it the Committee began by:

> congratulating the country and the scientific world in general on the liberal and enlightened views which have actuated his Royal Highness the Lord High Admiral in directing the outfitting of an enterprise destined solely and simply for the promotion of scientific research, and the extension of the bounds of human knowledge … .[5]

It went on to declare that of all the subjects worthy of attention the most important was the determination of the true figure of the Earth 'and the law of the variation of gravity in different points of its surface'. The committee strongly recommended that the determination of the length of the seconds pendulum 'at a great number of stations, in all parts of the globe' should be the principal object of the expedition.[6] A line of observing stations should be set up in what the committee termed 'Equatorial, Middle and Southern' latitudes, from 10°N to 'New South Shetland' (off the Antarctic Peninsula). Of even greater concern, as regards this 'most important of all the southern stations' would be land of a 'still higher southern latitude' and 'a departure from a beaten track' might be justified in view of the importance of observations in those latitudes partly to counterbalance those made in the far North.

Of other important objects, the exact determination of the geographical positions of 'remarkable points' and 'investigations connected with the theory and application of magnetism' were given pride of place. As a matter of great concern to navigation, both Captain Parry, Hydrographer of the Navy, and James Horsburgh, Hydrographer to the East India Company, had submitted a list of ports whose longitudes could usefully be determined.

---

[1] Hall, *Fragments of Voyages*, I, p. 21. Quoted in Jackson, *Tales of Woodplumpton*, p. 47; photograph of the death mask of Henry Foster appears following p. 74.
[2] *London*, Royal Society, Election Certificate EC/1824/06.
[3] London, Royal Society, Minutes of the Council of the Royal Society, 15 November 1827.
[4] Hall, *Fragments of Voyages*, p. 22.
[5] London, Royal Society, Committee Minute Books, CMB.1, pp. 216–36. Report of the Committee on which the foregoing voyage was ordered. Printed in Webster, *Narrative of a Voyage*, II, pp. 369–82.
[6] Ibid.

Astronomical observations in southern latitudes should include the frequently recurring comet Encke, while various meteorological observations should be made. The Committee would have recommended the appointment of one or more competent naturalists, had not the small dimensions of the *Chanticleer* and 'arrangements connected with the naval service' precluded this. In fact it was the surgeon W. H. B. Webster who was directed to collect and preserve specimens in zoology, mineralogy and geology. The committee of the Royal Society had recommended a round-the-world voyage. In the event, their lordships directed that, in view of the time it would take to complete the entire programme, the voyage be divided in two and that the sloop *Chanticleer* be largely confined to the South Atlantic Ocean, including Staten Island (Isla de los Estados off the south-east tip of Tierra del Fuego), Cape Horn and Deception Island, one of the South Shetlands, which had been discovered by Captain William Smith in the merchant brig *Williams*,[1] a decade earlier.

Chanticleer was helped to find a secure anchorage at Staten Island by a chance meeting with Captain Alexander S. Palmer of the Stonington, Connecticut, sealer *Penguin*, brother of the better known Nathaniel B. Palmer, after whom a stretch of the Antarctic Peninsula has been named. The surgeon, Dr Webster, described him as a 'kind and good-hearted man', who brought a basket of albatross eggs as a present.[2] Both the Palmers were American sealers, but in the years 1829 to 1831, their vessels, the brig *Annawan* and the schooner *Penguin*, together with the brig *Seraph*, Captain Benjamin Pendleton, formed a private sealing and exploring expedition to the Antarctic regions. It was the first such venture from the United States and it carried Dr James Eights, whose reports proved of considerable scientific importance.[3]

During her outward passage, *Chanticleer* had called at Madeira, the Cape Verde Islands, Rio de Janeiro and Monte Video, being challenged at times off the coast of South America by warships of the Brazilian Navy, commanded by former British officers, engaged in conflict with Argentina.[4] With her hull further strengthened for ice navigation at Staten Island with the oak she must have been carrying, *Chanticleer* sailed in between the South Shetland islands avoiding many icebergs before thick fog came down. Commander Foster and Lieutenant Edward N. Kendall, who had travelled with Sir John Franklin in the Arctic, landed at Cape Possession, Hoseason Island and deposited a document written in Latin in a copper cylinder on 7 January 1829, claiming it for the British Crown.

Deception Island is the flooded crater of a volcano, by no means dormant. To reach the inner harbour, now Port Foster, a vessel enters through a narrow passage, known as Neptune's Bellows. Once within, *Chanticleer* was eventually secured in a small cove, known as Pendulum Cove, where the observatories were erected, beneath the encircling cliffs of the crater, which rose to about 1,800 feet. Numerous fumaroles could be seen from the top of one of the hills above the *Chanticleer* and most of the

---

[1] See Campbell (ed.), *Discovery of the South Shetland Islands*.

[2] Webster, *Narrative of a Voyage*, I, pp. 98–9.

[3] Bertrand, *Americans in Antarctica*, chapter 9, pp. 144–58.

[4] For background to this, see Graham and Humphreys (eds), *Navy and South America*. See also the publications of Brian Vale, including *Independence or Death*; *War betwixt Englishmen*; and *A Frigate of King George*.

island consisted of layers of ash and ice. Its inhabitants were sea leopards, sea birds and thousands of penguins. Its fur seals had been decimated by American and British sealers earlier in the decade.

Two months were spent at Deception Island, from early January to 8 March 1829. Lieut. Kendall described their stay:

> It was, however, cheerless work. The fogs were so frequent that, for the first ten days, we saw neither sun nor star; and it was withal so raw and cold, that I do not recollect having suffered more at any time in the arctic regions, even at the lowest range of the thermometer. When to these discomforts is added, that the short allowance, to which we had been reduced, barely formed sufficient for a healthy man's breakfast, you may judge whether what we have accomplished has not been *à la force*. I assure you we were perfectly ravenous, so much so, that on a moderate calculation, upwards of seven thousand penguins were eaten during our stay, which nothing but the most absolute necessity could have induced us to touch; and even portions of the sea-leopard, fried in their own blubber, were accounted palatable food.
>
> Notwithstanding this, we did our work; and being allowed a boat and four men, I surveyed the island, sleeping at night on the cindery beach, with no other covering than a canvas tent.[1]

In his book the surgeon Dr Webster contrasted Deception Island with Staten Island in a more literary style:

> A more dreary or more cheerless scene cannot be imagined than that which Deception Island of Shetland presented: the wild and solitary woods of Staten Island, which we had just left, lonely and uninviting as they appeared to us, were desirable to this. There the visiter [*sic*], although far removed from the busy scenes of life, and destitute of the social comforts of civilized beings, finds vegetation flourishing; and in the animated face of nature there is much to gladden his heart and to employ his mind in its solitary glens; but here all is joyless and comfortless, huge masses of cinders and ashes lie strewed about, which imagination converts into the refuse of Vulcan's forge. No vestige of vegetation relieves the eye, tired of contemplating ashes and lava, from which it can find no other relief than snow: instead of grand and beautiful rocks towering above each other, and overhanging the water in magnificent precipices of awful height, their summits covered with Nature's richest mantle a beautiful foliage, we had here hills of black dust and ashes topped with snow, and enormous icebergs buried beneath immense loads of volcanic matter: instead of the variety of birds of elegant plumage which adorn other happier regions, hosts of penguins here strut about with stupid mien, harmless and happy in their dreary abode as they are unsuspicious of harm from man.[2]

Lieutenant Kendall found the wreck of a large vessel, sundry relics of sealing gangs and the well-preserved body of a sailor in a rough coffin. A maximum and minimum thermometer was deposited, which was found in 1842 by a United States sealer, William H. Smiley. This registered a lowest temperature of –4.9° Fahrenheit, maximum not recorded.[3] Kendall concluded his account:

> We took the hint of the freezing over of the cove, and effected our retreat with much difficulty and severe labour, from the fury of the gales, whose violent gusts had before

---

[1] Kendall, 'Island of Deception', pp. 65–6. His chart of the island accompanies the text (Map 2, p. 314).
[2] Webster, *Narrative of a Voyage*, I, pp. 148–9.
[3] Great Britain, *Antarctic Pilot*, p. 170.

blown down all our tents, and broken many of the instruments. We quitted it amidst the acclamations of thousands of penguins, who croaked a most discordant chorus; and indeed it was a day of rejoicing to us also.[1]

Some wet, windy and hungry weeks were spent observing and surveying near Cape Horn, relieved by the arrival of HMS *Adventure*, Captain Philip Parker King, on Good Friday 17 April 1829, a rendezvous agreed earlier between the two captains at Monte Video, during *Chanticleer*'s outward passage. Supplies of food, including more of Donkin's excellent preserved meat in tins, and the company of their fellow seafarers raised the spirits of the 'Chantys' during their sojourn off a coast 'as broken, desolate, remote and sea-battered ... as the world has to offer', in the words of a modern hydrographer.[2] According to Captain King of the *Adventure*, before the two vessels parted in May 1829, Commander Foster told him of a 'presentiment, which he could not shake off, that he should not survive the voyage.'[3]

The *Chanticleer* departed on 24 May 1829 for the Cape and arrived there on 16 July, blown by following winds across the South Atlantic. Fearon Fallows, Astronomer at the Cape Observatory, much enjoyed Foster's company, finding it 'a treat in this quarter of the Globe to be visited by a scientific man'.[4] Departing on 13 December 1829 from the Cape, after an extended stay owing to Foster's illness, *Chanticleer* sailed north, calling at St Helena, Fernando Noronha, Trinidad and Chagres off the Isthmus of Panama, where there were more pendulum experiments made. It was while endeavouring to measure the difference in longitude between Panama and Chagres by timing rockets fired at each point simultaneously, that Commander Foster was drowned when he fell from a canoe into the River Chagres. On 5 February 1831, after the successful conclusion of their labours across the Isthmus of Darien, the canoe carrying Foster, two young officers of the *Chanticleer* and his servant was descending the River Chagres from Cruces towards the sea, where soon they expected to board the *Chanticleer* and sail for the Azores and home. In the late afternoon, Commander Foster was accidentally precipitated into the river. He was seen to sink and rise no more, to the consternation and horror of his three companions. Unable to find his body, the canoe continued downstream and brought the news to Lieut. Austin, who sent a search party; the body was eventually found caught between the branches of an old fallen tree, and surrounded by turkey buzzards which had already eaten parts of the corpse. Local Indians had stolen Foster's watch and notebook. He was buried beside the river in a grave dug, from the want of a spade, with a loose thwart and an axe. He was only thirty-five years old and his death, as a most talented scientific naval officer, was very greatly lamented. 'What a loss to England and to Science', wrote Robert FitzRoy on hearing of Foster's 'melancholy end'.[5] Lieutenant Horatio T. Austin replaced Foster as commanding officer. The very fine memorial in the parish church of Woodplumpton near Preston, where his grieving father was Rector, can still be seen. It was described by Dr Webster as consisting 'of an Urn, from which the British flag hangs in negligent

---

[1] Kendall, 'Island of Deception', p. 66.
[2] Ritchie, *The Admiralty Chart*, p. 204.
[3] FitzRoy (ed), *Narrative of the Surveying Voyages*, I, p. 205.
[4] Fearon Fallows, quoted in Warner, *Royal Observatory*, p. 197.
[5] Taunton, Hydrographic Office, archives, Letters-in, Box F301–476, No. F452): Robert FitzRoy to Sir Francis Beaufort, Hydrographer of the Navy, from Stratton Street, London, 10 May 1831.

Figure 1: Photograph of the fine monument in the parish church of St Anne, Woodplumpton, near Preston, 'Sacred to the Memory of Henry Foster, R.N., F.R.S. ... erected by several of his companions and friends, as a memorial of the high esteem they entertained for his character, and of the deep regret they felt for his untimely death'. By kind permission of the Vicar and with the help of Mr Peter Sheppard, churchwarden, who kindly took the photograph.

SACRED TO THE MEMORY OF
HENRY FOSTER, R.N. F.R.S.
DISTINGUISHED AS WELL FOR SUPERIORITY OF INTELLECT AS URBANITY OF
MANNERS, BY A ZEALOUS AND FIRM DISCHARGE OF DUTY
HE GAINED THE CONFIDENCE AND REGARD OF HIS BROTHER OFFICERS,
AND BY A SUCCESSFUL PURSUIT OF KNOWLEDGE
ATTRACTED THE NOTICE OF MEN OF SCIENCE
FOR HIS PHILOSOPHICAL EXPERIMENTS MADE IN THE ARCTIC REGIONS
THE COPLEY MEDAL OF THE ROYAL SOCIETY
WAS PRESENTED TO HIM ON THE 30TH NOVEMBER 1827;
WHEN THE LORD HIGH ADMIRAL OF ENGLAND WITH AN ALACRITY
HONOURABLE TO HIMSELF AND TO THE SUBJECT OF HIS PATRONAGE
INSTANTLY PROMOTED HIM TO THE RANK OF COMMANDER.
IN THE YEAR FOLLOWING HE SAILED ON A VOYAGE OF SCIENTIFIC RESEARCH,
HE HAD COMPLETED HIS ASTRONOMICAL OBSERVATIONS AT PANAMA,

AND ALL THINGS HAD PROSPERED IN HIS HAND; WHEN, PROCEEDING TO HIS SHIP AND
ANTICIPATING A SPEEDY RETURN TO HIS NATIVE SHORE, HE FELL FROM A CANOE
AND IN A MOMENT WAS LOST TO HIS COUNTRY AND HIS FRIENDS.
HIS BODY SHROUDED IN THE BRITISH FLAG WAS INTERRED NEAR TO THE FATAL SPOT
ON THE BANK OF THE RIVER GHAGRES IN THE GULPH OF MEXICO
ON THE 8TH OF FEBRUARY 1831, AND IN THE 34TH YEAR OF HIS AGE.

THIS MONUMENT WAS ERECTED BY SEVERAL OF HIS COMPANIONS AND FRIENDS
AS A MEMORIAL OF THE HIGH ESTEEM THEY ENTERTAINED FOR HIS CHARACTER,
AND OF THE DEEP REGRET THEY FELT FOR HIS UNTIMELY DEATH.

HE WAS THE SON OF THE REV: HENRY FOSTER, INCUMBENT OF THIS CHAPELRY.[1]

folds, and against which a sailor is leaning in the attitude of grief. An anchor and quadrant and a few nautical and scientific instruments are also introduced.' A lengthy inscription, reproduced above, summarized Foster's life and scientific achievements.

At least three place-names in British Antarctic Territory commemorate him: Port Foster, Deception Island; Cape Foster, James Ross Island; and Mount Foster, Smith Island. Foster's death mask is in Preston Museum. Through the efforts of the Hydrographer of the Navy, a pension of £75 per annum was to be paid to Foster's two unmarried sisters.[2]

Three years after *Chanticleer*'s return, an extensive *Report on the Pendulum Experiments made by the late Captain Henry Foster, RN in his Scientific Voyage in the Years 1828–1831 with a View to Determining the Figure of the Earth* was drawn up by Francis Baily, President of the Royal Astronomical Society, and published at the expense of the Admiralty in 1834 as Vol. 7 of the *Memoirs of the Royal Astronomical Society*. Because of Foster's death, the narrative of the voyage was written by the surgeon, Dr W. H. B. Webster.[3] The second volume of this work contains a lengthy appendix summarizing the scientific and survey work. It includes reports by Commander Foster and Lieutenant Kendall, as well as Webster's own 'Observations on the natural productions of the places visited by the *Chanticleer*'.

2.1. Midshipman Joseph Henry Kay: the Writer and his Journal

Joseph Henry Kay was born in London on 6 November 1814, the second son of Joseph Kay, the architect (1775–1847) and his wife Sarah Henrietta, *née* Porden. Through her he was a nephew of Sir John Franklin, the naval officer and Arctic explorer, by the latter's first marriage.

Kay entered the Navy on 18 December 1827 as a First Class Volunteer, being appointed to the *Chanticleer* that same month. According to a footnote in Dr Webster's

---

[1] The Editors are most grateful to Jean Sheppard for this exact transcription which corrects several errors included in other printed versions. The 'River Ghagres' should read 'River Chagres'. The date of Foster's death was 5 February, followed by his burial on the 8th. See Webster, *Narrative of a Voyage*, I, pp. iv–v.

[2] Taunton, Hydrographic Office, archives, Out-letter-book, No. 3, 1830–32, 27 May 1831: Sir Francis Beaufort to the Reverend Henry Foster.

[3] Webster, *Narrative of a Voyage*, 2 vols.

narrative, Kay 'was of much service to Captain Foster in his scientific operations', being also 'spoken of in high terms by Captain Sir John Franklin, with whom he has since served in His Majesty's ship Rainbow.'[1] He was one of the small party accompanying Captain Foster in the canoe descending the Chagres on the day he was drowned. Kay passed for Lieutenant on 6 April 1839, soon afterwards serving in HMS *Terror*, Captain Francis Rawdon Moira Crozier, during her outward passage to Van Diemen's Land (Tasmania). This was the first leg of the celebrated Antarctic circumnavigation of *Erebus* and *Terror*, 1839–43, commanded by that most experienced of Arctic officers, Captain James Clark Ross. On arrival in Hobart, Kay was appointed by Ross to direct the newly established Rossbank magnetic observatory, which he did from 1840 to 1853, finding himself tied to an endless and sedentary routine of observations and calculations, far away from the professional duties of a naval officer. At first full of zeal, despite the monotony and the difficulties of running an official station so remote from its headquarters in the British Isles, Kay came to resent his exile and his lack of promotion. The data which he and his assistants sent to Britain, were of the utmost importance in understanding the phenomenon of terrestrial magnetism, because of Hobart's relative proximity to the South Magnetic Pole, itself of great relevance to navigation. Rossbank was one of a network of British colonial observatories forming part of Carl Friedrich Gauss's global *Magnetische Verein*.[2]

Kay was elected a Fellow of the Royal Society on 26 February 1846 and was a founding member of the Royal Society of Tasmania, for whose publications he wrote the articles listed below. He was appointed Commander on 23 August 1849 and Retired Captain under order-in-Council on 9 July 1864. After leaving Tasmania, Kay held various posts in the public service of the State of Victoria, and at his death was Clerk to the Executive Council. He died from diabetes on 17 July 1875 at South Yarra, Melbourne, and was buried in St Kilda cemetery.

The young Midshipman's lively and carefree 'Private Journal' covers only the first half of the voyage of HMS *Chanticleer*, but his account, now published for the first time, forms a useful addition to the literature of hydrographic surveying and scientific work in the nineteenth century, supplementing the more discursive two-volume narrative by the surgeon, based on Webster's own journal of the voyage. Interesting to the modern reader are Kay's concern for the slaves still in thrall at Rio de Janeiro and his regret for the necessary killing of many hundreds of penguins for food at Deception Island.

The following articles by Joseph Henry Kay were published by the Royal Society of Van Diemen's Land (Tasmania):

1. 'On the Aneroid Barometer', *Papers and Proceedings of the Royal Society of Van Diemen's Land*, 1, 1849–50 (pub. 1851), pp. 83–87. [Explains the principles upon which 'this elegant little philosophical instrument of very modern invention' was constructed; tests of its powers on Mount Wellington; comparison with a mercurial barometer at the Royal Observatory, Hobart Town.]

---

[1] Webster, *Narrative of a Voyage*, II, p. 192.
[2] On the magnetic observatory and Kay's tenure as Director, see Savours and McConnell, 'Rossbank Observatory', pp. 527–64, and by the same authors 'Return to Rossbank', pp. 49–58.

2. 'Meteorological Tables for the Years 1847 and 1848; made at the Royal Observatory, Hobart Town. Lat. 42°52′S., long. 9h.50m.E.', ibid., pp. 144–53.

3. 'Meteorological Tables for the Year 1849; Royal Observatory, Hobart Town …', ibid., pp. 255–7.

4. 'Observations made for Determining the Geographical Position of the Magnetic Observatory at Hobart Town, Van Diemen's Land', *Report, Papers and Proceedings of the Royal Society of Van Diemen's Land*, 1852–4, pp. 264–87. ['By Commander Kay, RN, FRS, etc.' in connexion with the Trigonometrical Survey 'now in course of operation in this Island'.]

5. 'Meteorological Tables, Royal Observatory, Hobart Town', ibid., pp. 292–307. ['By Commander Kay, RN, FRS, for years 1850–52, with means for 1841 to 1848'. The article refers to the publication by the British Government of two volumes 'of the magnetical and meteorological observations made at Hobart Town under my direction' mainly devoted to investigation of horary, diurnal and annual variations of the magnetic elements, with their peculiar changes. This great object having been attained, 'the connection of the British Government with the Observatory in Tasmania will cease in April 1853'. The third volume of the Hobart Town observations 'will discuss the peculiarities of the climate of Hobart Town, as exhibited in the extensive series of meteorological observations which have been made'.]

2.2. Some of Kay's Shipmates aboard HMS *Chanticleer*

Lieut. Horatio Thomas Austin. First Lieutenant, was the son of Captain Sir Thomas Hardy's boatswain, who had lost an arm at the Battle of the Nile, when Nelson defeated the French fleet in 1788. Horatio Austin took part in Captain William Edward Parry's North West Passage expedition of 1824–5, serving in HMS *Fury*, Commander Henry Parkyns Hoppner, which vessel left her timbers on an icy Arctic shore. Captain Austin was much later Commodore of the Arctic squadron searching for Sir John Franklin in 1850–51. A member of the American squadron, also searching for Franklin at the time, described him as 'the jolliest old Englishman ever seen'.[1]

Austin was an enthusiast for steam and commanded the navy's first steam vessel, *Salamander*, in 1832. Sir Clements Markham described him as a 'very experienced officer, who based his decisions on wide knowledge', admirable 'for managing the internal economy of a ship and organising the details of an expedition'. He was also 'genial and warm-hearted'.[2] Kay's journal illustrates how effectively Austin could deal with a crisis, while Webster's narrative shows how he took command on Foster's greatly lamented death. During operations in 1840 off the coast of Syria, he distinguished himself and was awarded the CB. He rose from sheer ability to become Captain Superintendent of Deptford Dockyard during the Crimean War and Admiral Superintendent of Malta Dockyard in 1863. He was knighted (KCB) and died in 1865.

---

[1] Quoted in Savours, *North West Passage*, p. 202.
[2] London, Royal Geographical Society Archives, CRM/40: Sir Clements Markham, 'Story of my Service'.

Lieut. Edward Nicholas Kendall. Second Lieutenant and Surveyor. He was born in October 1800, his father being a Captain in the Royal Navy. He was educated at the Royal Naval College, Portsmouth, entering the Navy on 26 October 1814. He served as midshipman in a number of vessels, including the *Erne*, wrecked off the Cape Verde Islands in 1819. He worked for three years surveying the North Sea and also served in HMS *Griper*, Captain George Francis Lyon, as assistant surveyor, during an unsuccessful and hazardous Arctic voyage in 1824. One of his drawings was engraved for Lyon's *Brief Narrative*, published in 1825. He was then appointed with the rank of Admiralty Mate as assistant surveyor during Franklin's second overland expedition of 1825–7, accompanying Dr John Richardson along the north coast of America eastwards from the delta of the Mackenzie River. He made a number of paintings and drawings, some reproduced in Franklin's *Narrative*. He was promoted Lieutenant on 30 April 1827. He was transferred from the *Chanticleer* to HMS *Hecla*, surveying the west coast of Africa in 1830. That same year he was sent by the Colonial Office to ascertain the boundary between New Brunswick and Maine, also mapping the former province in 1831. He then married Mary Anne Kay, Sir John Franklin's niece. Although recommended for promotion, he remained a lieutenant and saw no more naval service. He died a superintendent in the Peninsular and Oriental Steam Navigation Company on 12 February 1845.[1] The fine plates in the first volume of Dr Webster's narrative of the voyage of HMS *Chanticleer* were engraved from Kendall's drawings. A number of his watercolours are held by the Science Museum, London. A watercolour sketch of 'Sea Leopards on Deception Island', similar to one in the Science Museum, is in the Hydrographic Office, Taunton (OD 39). Kendall's original surveys of both Staten and Deception Islands are in the archives of the Hydrographic Office, Taunton.

Dr William Henry Bayley Webster. Born in 1793, he studied medicine and surgery at St Bartholomew's and St George's Hospitals, London. He became a Licentiate of the Society of Apothecaries in April 1816, having entered the Navy as an Assistant Surgeon in August 1815. Webster is said to have corresponded with the Admiralty about a method of preserving meat. His only period at sea appears to have been as surgeon of HMS *Chanticleer*, where he acted also as naturalist. Because of the death of Commander Henry Foster towards the end of the voyage, it fell to Webster to write the two-volume narrative of the expedition, published by Bentley in 1835. He was not employed again in the Navy and spent the rest of his life in Ipswich, practising as a doctor there for forty years. He published and lectured locally on meteorology, and died on 24 November 1875.[2] The Assistant Surgeon, *Peter Conolan MD*, had trained in Edinburgh. He died in 1862.[3]

Richard Collinson. The son of a clergyman in the north of England, Richard Collinson entered the Navy at the age of twelve in 1823. He served in HMS *Cambridge*, Captain T. J. Maling, on the Pacific Station, 1823–27. Through the influence of Admiral Sir

---

[1] Biographical details largely summarized from *Sir John Franklin's Journals and Correspondence*, ed. Davis, pp. xxxi–xxxiii.
[2] Summarized from Jones, 'W. H. B. Webster', pp. 143–5.
[3] Biographical index in the Wellcome Library for the History and Understanding of Medicine, London.

Byam Martin and on Maling's recommendation, he secured a berth as a midshipman aboard HMS *Chanticleer*. From 1831 to 1833 he served in HMS *Aetna*, Captain Edward Belcher, during a surveying cruise mainly off the west coast of Africa and in the Mediterranean. He next joined the steamers *Salamander* and *Medea*, under Captain Horatio Austin. He was promoted lieutenant in April 1835, having proved his worth by then as a hydrographer. He sailed in HMS *Sulphur*, on a surveying expedition to the Pacific, as third lieutenant and assistant surveyor, returning to England in November 1839. He did survey work in Chinese waters 1840–46. Collinson is best known as Captain of HMS *Enterprise*, which sailed in 1850 with HMS *Investigator*, Captain Robert McClure, to search for the lost Franklin expedition of 1845–8 from the Pacific end of the North West Passage. The ships were separated and the *Enterprise* spent several long and dreary years in isolation in what is now the Canadian Arctic archipelago. On his return to England in 1855, Collinson gained little credit for a remarkable voyage, in contrast to McClure who was hailed as the discoverer of the North West Passage. Active as a Fellow of the Royal Geographical Society, and a member of the Hakluyt Society and the Royal United Services Institution, it was through his work for Trinity House that he regained his reputation, being knighted in 1875, after promotion to Rear Admiral in 1862. He died in September 1883.[1]

3. Scientific Aspects of the Voyage of HMS *Chanticleer*

3.1. Pendulum Measurements

A freely-swinging pendulum – that is, one not regulated by a clock – beats at a frequency determined principally by its length, but modified by the local force of gravity. Depending on its location on the globe, a pendulum of about 40 inches (102 cm) in length has a one-second beat; one of about 20 inches has a half-second beat, and so *pro rata*. If two of these factors are known, it is possible to derive the third. Thus, knowing the exact length of a pendulum, and its beat, the local force of gravity may be calculated. The discovery that the length of a pendulum beating seconds differed between Paris, close to 49°N and Cayenne (French Guiana), at nearly 5°N, was first remarked on by the astronomer Jean Richer when he was sent there in 1672 by the Académie royale des sciences to observe the more southerly stars than could be seen from Paris and to consider the length of the seconds pendulum. During the year that he was in Cayenne, Richer found that his clock lost time, and he had to shorten its pendulum by 1¼ Paris lines (0.0226 cm) in order to keep it beating seconds.[2] When it was realized that this change in rate arose from a lower force of gravity at Cayenne, the way was open for the use of a pendulum for investigating the force of gravity in different parts of the world, and thus – it was hoped – to ascertain the true form of the earth.[3]

---

[1] Summarized from the biographical memoir following the text of his *Journal of HMS 'Enterprise'*, pp. 436–520.

[2] Richer, 'Observations astronomiques', pp. 231–326. The Paris inch was divided into 12 lines.

[3] The earth is an oblate spheroid, i.e., it is flattened at the poles, with its equatorial radius being greater than the polar radius. As the attraction due to gravity decreases with distance, it follows that gravity is less in equatorial regions than it is towards the poles.

Pendulum experiments were regularly included in the instructions given to commanders of British scientific voyages, notably those into Arctic regions, starting with that of Captain Constantine John Phipps in 1773.[1] The scientist, Captain Henry Kater, Royal Artillery, FRS, devoted much of his time to the design and testing of pendulums for both relative and absolute measurements within the British Isles.[2] His invariable pendulum for relative measurements, some five feet long overall, consisted of a brass bar with a fixed knife-edge near one end and a heavy weight at the other, ending in a long tail. The pendulum was set up on agate planes, the apparatus suspended in a stout frame erected in front of a pendulum clock. It was set swinging, and the times when the pendulum tail coincided with a white disc on the swinging tail of the clock pendulum were noted. The observed values were then corrected for temperature, friction, expansion of metals, and other lesser effects.

Foster had considerable experience with invariable pendulum experiments, first with Captain Basil Hall in *Conway*, in the Pacific,[3] then in the Arctic, with Edward Sabine in *Griper*, Captain Clavering, and with Captain Parry in *Hecla*,[4] work for which the Royal Society awarded him its Copley medal in 1827.[5] Promoted Commander, he was appointed to *Chanticleer*, and on 13 December 1827 he wrote to Thomas Young, Secretary to the Board of Longitude, listing the astronomical, magnetic and meteorological instruments, in the ownership of the Admiralty, the Board and the Royal Society, which he wished to take.[6]

Pendulums of two different forms were issued to *Chanticleer*: one pair, the property of the Royal Society, made by Thomas Jones and bearing the numbers 10 and 11,[7] were of the conventional Kater pattern; the second pair was provided by the retired commercial mathematician and amateur astronomer Francis Baily and made by Troughton & Simms to his own design.[8] They consisted of straight bars with knife-edges but no lower weight, one bar being made of copper, the other of iron, to test the response of metals of differing density to the force of gravity. Each pendulum was packed in its own fitted box, together with its agate planes, thermometers, and other tools and fittings. The manoeuvering from ship to shore of these heavy boxes with their precious contents, often in icy, wet and windy conditions was particularly difficult.

In inhabited regions a suitable building might be found ashore, otherwise Foster had to land and assemble the observatory house, which had been prefabricated in frame, with double walls and roof. Erected with the long axis at right angles to the meridian and partitioned into one room 10 by 10 feet, and one 10 by 5 feet, the large room could

---

[1] Phipps, *Voyage towards the North Pole*, p. 13; Savours, "'A Very Interesting Point in Geography'", 1984, pp. 402–28.

[2] Kater, 'Account of Experiments', 1818, pp. 33–102; 'Account of Experiments', 1819, pp. 337–508.

[3] Hall, 'Letter', pp. 211–88.

[4] Foster, 'Account of Experiments', pp. 1–70.

[5] Exceptionally, the Society's Council decided to award a second Copley medal that year: Royal Society Council Minutes, 15 November 1827.

[6] Cambridge, University Library, Royal Greenwich Observatory (hereafter RGO), 14/49, f. 58^{r-v} and f. 60^{r-v}.

[7] Now in the National Museum of Science and Industry (hereafter NMSI), Inv. 1914–586 and 587.

[8] Their construction is described in Baily, 'Report on Experiments', pp. 1–378. The surviving copper pendulum is now NMSI Inv. 1939–388.

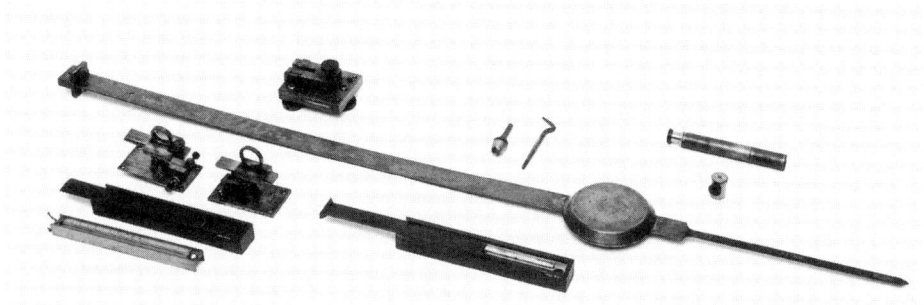

Figure 2: Pendulum apparatus used during the voyage of HMS *Chanticleer* in the collection of the Science Museum, South Kensington (Inv. 1939–388). Reproduced by permission of the Science Museum, Science and Society Picture Library.

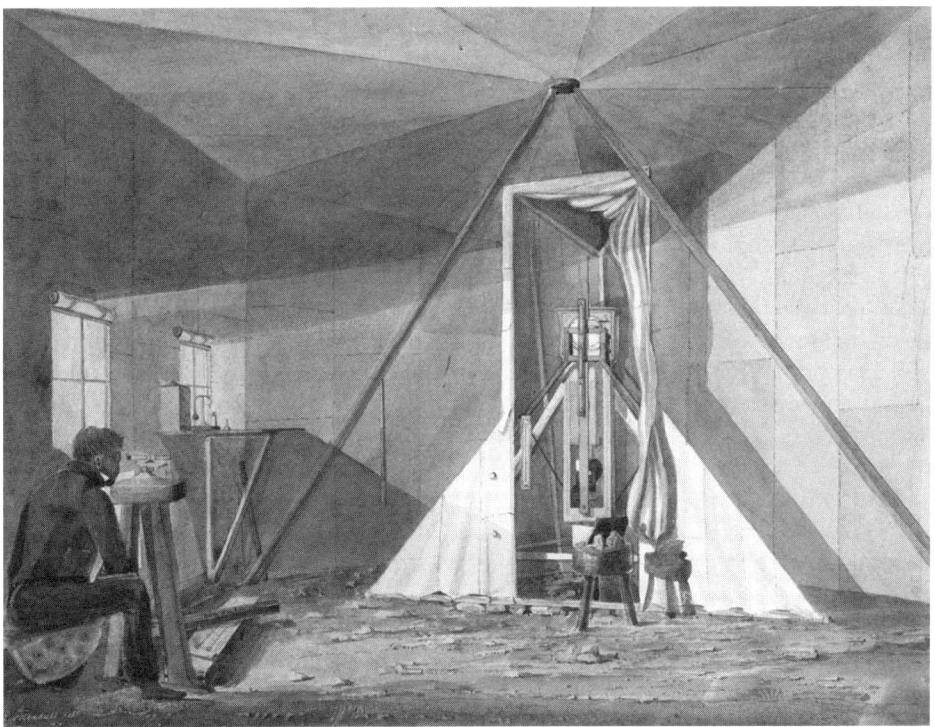

Figure 3: Watercolour, presumably by E. N. Kendall, of the interior of the portable pendulum house, depicting an observer and Kater's invariable pendulum, probably at Monte Video in August 1828. The house was constructed of panels arranged to form a room of about twelve feet square and fourteen or fifteen high, with eight windows. Reproduced by permission of the Science Museum, South Kensington, Science and Society Picture Library.

be lined with thick woollen 'fearnought' cloth, its floor boarded, and a stove installed. In the small room the 30-inch transit telescope, for which astronomical observations were used to provide the time, was firmly set up on a large stone, capping a cask filled with sand, viewing the sky through an 18-inch slot cut in walls and roof.

The Royal Society provided two clocks, with gridiron pendulums and wooden stands; the Kater and Baily pendulums had their own iron stands. Foster followed Kater's practice of determining intervals between coincidences, that is, from the disappearance of the white disc on the clock pendulum behind the pendulum tail-piece, and its reappearance. When pendulum observations were to begin, the large room was made as secure from draughts and temperature variation as possible. The outside door was blocked, the only entrance being through the slit from the small room, this entrance being guarded by canvas and fearnought screens on both sides. The floor boards were lifted and a solid foundation was prepared of sand, on which large flat stones were set, such that the supports for clock and pendulum stood on their individual stones. Lead weights of 40 to 50 pounds were hung on the back of these supports, to ensure their stability when the pendulums were swinging. Coincidences were observed through a small telescope with a diaphragm, set up 9¾ feet from the pendulum, in a shielded outside porch. The stove was found to vary the temperature excessively, so to maintain the observation room at 50° F, the same temperature that the pendululms had been tested in at Greenwich, it was found necessary to move the stove outside the partition and to warm the observing room by leading the smoke pipe through it. A wrap of fearnought lined with racoon skins enveloped the pendulum apparatus, apart from the portion to be observed. The free pendulum was set in motion by drawing it aside with a piece of twine secured to one leg of the stand until its tail point was about 1°2′ along the arc, releasing it just before the clock pendulum reached its highest ascent on that side. This allowed five coincidences to be timed before the swing became too small.

Baily's Report informs us that Foster's practice was to record on his proformas the height above sea level, readings from barometer, hygrometer, and each of three thermometers fixed level with the top, centre and base of the pendulums. He noted the compass bearing of the plane of swing and the time of high tide. For each set of observations he noted the first three and last three coincidences.

Foster arrived at the Cape of Good Hope on 16 July 1829, while Fearon Fallows, astronomer at the Cape Observatory, was in the middle of his own winter pendulum observations. As Fallows wrote to John Barrow, Second Secretary of the Admiralty, he had not been informed of the *Chanticleer*'s voyage, with its scientific programme, but he immediately ceased his own operations in order to allow Foster to set up at the same location, the centre room of the observatory, and also to make use of an adjacent small temporary observatory.[1]

The Kater pendulums were swung in London before and after the voyage,[2] and in twelve places around the Atlantic; the iron pendulum was swung at six places, the copper pendulum at eight. In total 1,017 sets of observations were carried out, on

---

[1] Warner, *Royal Observatory*, pp. 194–203.

[2] At Henry Browne's house in Portland Place (where Kater had also swung his pendulums), between 29 January and 24 March 1828, then at Greenwich Observatory, between 24 and 28 February 1828. On return, Baily swung the pendulums in his own house in Tavistock Square, between July and August, in October, and again in January 1832.

which Foster spent 2,710 hours. During the voyage there was no time for Foster to deduce any results from his experiments. When *Chanticleer* returned to England the pendulums and the proformas were handed to Baily, who undertook another set of observations at his own house. The computations were done by William Richardson at Greenwich to Baily's instructions.

In summary, Baily found that the brass pendulum results differed from those of the copper pendulum, this divergence being greatest at the South Shetlands and least at Maranham,[1] near the Equator. The iron pendulum likewise differed, though to a lesser extent. This was something which Foster had foreseen; writing from Table Bay on 16 July 1829 he had commented on the iron-rich hornblende rock at Cape Horn which might have affected the iron pendulum, had weather conditions allowed the iron and copper pendulums to be swung.[2] The force of gravity also seemed greater in islands remote from the continents, such as Ascension, St Helena and Fernando de Noronha.[3] Baily then consulted all the previous records available to him; they confirmed Foster's results, and clearly showed local anomalies which defeated all attempts to deduce the true figure of the Earth by isolated pendulum experiments. Baily was aware that most mid-ocean islands were composed of dense volcanic rocks but he had supposed that their effect would be nullified by the surrounding sea.

3.2. Instruments for Terrestrial Magnetism, Astronomy, Surveying and Meteorology

At any point on the earth's surface a freely-suspended magnetic needle takes up a position with respect to three axes: geographic north, geographic east and the vertical. The force or intensity of the earth's magnetic field can be measured in terms of its vertical and horizontal components, also by means of a suspended magnetic needle, Hansteen's apparatus being in use at this time. These values of direction and force vary from place to place, and over time, both short-term (within the day) and long-term (over a period of years). Edmund Halley's famous chart, published in 1701, showing misleadingly confident lines of equal variation (isogonics) in the Atlantic had been drawn from relatively few observations, as had the chart showing lines of equal dip (isoclinics) across the Atlantic and Indian Oceans, prepared by Johan Wilke in 1768. In the early nineteenth century, charts of equal intensity (isodynamics) were published for various regions. As more and better observations were made, the extent to which the values changed or migrated with time became more apparent. But there was still considerable debate and disagreement regarding magnetic values in high latitudes, although it seemed that the north and south magnetic poles were not opposite one another.

For this reason, magnetic observations were a fundamental part of scientific work on Arctic expeditions. In polar regions the horizontal magnetic force was at its weakest, and the compass needle moved sluggishly in response to any change of course. Vertical

---

[1] Now San Luis de Maranhão.
[2] The National Archives, Public Record Office, Foster: letter 5 ADM 1/1817, F67.
[3] Nineteenth-century English accounts often mis-spelt this island, discovered by and named after Fernando de Noronha.

force was at its strongest, with the dip needle attracted into the near-vertical. At the magnetic equator, a line through the tropics whose precise path was still uncertain, the horizontal needle was strongly attracted towards the north magnetic pole (also as yet unlocated) while the vertical force was weakest, so that the dip needle lay horizontal. Variation could be measured at sea, by comparing the direction of the ship's compass needle with that of the sun (or other heavenly body) rising or setting, but there was as yet no dip circle or magnetometer able to function with any degree of accuracy on board a moving ship. The practice was generally to observe from coastal stations and mid-ocean islands. It was not then appreciated that the mid-Atlantic islands were composed largely of basaltic rocks, which distorted the magnetic readings.

In 1824–5 during Parry's third Arctic expedition, Henry Foster and Captain Edward Parry had made copious observations of variation, dip and intensity measurements, and these were written up in a lengthy series of reports published as Part IV of the *Philosophical Transactions*, 116, 1826. Using a variety of forms of needle, they found hourly changes in both variation and force, but were unable to detect such changes in dip values.

Magnetic measurements were therefore an integral part of *Chanticleer*'s scientific programme and Foster made preparations accordingly. In March he had receipted George Dollond for the delivery of a dip circle with ancillary apparatus for measuring force, 'a diurnal variation instrument on a new plan, in mahogany and ivory', a 2½ ft transit instrument complete, a small Hansteen's magnetometer, a 46-inch telescope 'completely repaired', 'a magnetic apparatus in mahogany, for general experiments, on a plan by Captain Kater'.[1] Magnetic apparatus was also included on Foster's list of 13 December 1827[2] and it is unclear whether the diurnal variation instrument requested was a standard instrument in addition to that 'on a new plan' received from Dollond. It seems, however, that he took with him one 6-inch and one 12-inch dipping needle, four Kater compasses, one or two variation compasses and one or two Hansteen's magnetometers, plus various magnets and ancillary fittings.

For determining longitudes, for sundry surveying tasks, and for the normal needs of the vessel, Foster requested 30 chronometers, a 46-inch achromatic telescope with micrometers, mounted on an equatorial stand, three simple telescopes, an astronomical circle and vertical collimator, a repeating circle, a reflecting circle, a large theodolite, two dip sectors, two sextants and six artificial horizons, a chain, a level and staffs, sundry drawing instruments, a compound microscope, and Nautical Almanacs from 1828 to 1831. The meteorological list included six Newman's mountain barometers, two marine barometers, thirty-six common thermometers, eight max-min thermometers, four Marcet's water bottles, hydrometers, hygrometers, two actinometers and atmospheric electricity apparatus.

Magnetic observations of variation and intensity during the voyage were made hourly under Foster's supervision at Cape Horn 'by young gentlemen of the Chanticleer ... under circumstances of great discomfort to them'.[3]

---

[1] RGO 14/49, f. 57ʳ.
[2] RGO 14/49 f. 58ʳ⁻ᵛ.
[3] The National Archives, Public Record Office (PRO) ADM 1/1817 F67: Captains' letters, Foster to Admiralty, Table Bay, 16 July 1829.

Not much is said otherwise, regarding the success or otherwise of magnetic observations during the voyage. On *Chanticleer*'s return, some, or perhaps all, of the observations were passed to Samuel Hunter Christie FRS, professor at Woolwich Academy. Hunter was particularly interested in the daily shifts in the magnetic force, which he believed to arise from electrical currents generated by unequal solar heating at the surface. He reported to the British Association for the Advancement of Science that 'These valuable observations have been placed in my hands by His Royal Highness the President, and the Council of the Royal Society, and I intend, when I have sufficient leisure, to compare them, and likewise those to which I have already referred in the northern hemisphere [ie, those of Midshipman Robert Hood and Lieut. Henry Foster] with the diurnal deviations that would result at the corresponding places on the earth's surface, on the supposition that such electric currents as I have supposed are excited on contrary sides of the equator, in consequence of different parts of the earth's surface becoming successively the places of greatest heat, during its revolution upon its axis'.[1] But nothing was published and the records themselves have not been found. The only survival is a manuscript book in the Royal Society labelled 'Fair copy of Magnetic experiments and meteorological observations, Book 2. made by C. Williams, mate and lieutenant, and R. Collinson, A. Hodskin, and J. H. Kay, midshipmen',[2] these last three being presumably among the young gentlemen mentioned above. In his exhaustive survey, 'Report on the variations of the magnetic intensity observed at different points of the earth's surface', Edward Sabine included results from British and foreign voyages and expeditions, from that of E.-P.-E. Rossel in 1791–4[3] to that of James Clark Ross in 1836 with, sadly, no mention whatsoever of Foster.[4]

---

[1] Christie, 'Terrestrial Magnetic Force', pp. 106–30, 114–15.
[2] Royal Society MA 159.
[3] Rossel had taken command following the death of D'Entrecasteaux during the Pacific search for La Pérouse.
[4] Sabine, *Report*, pp. 1–85.

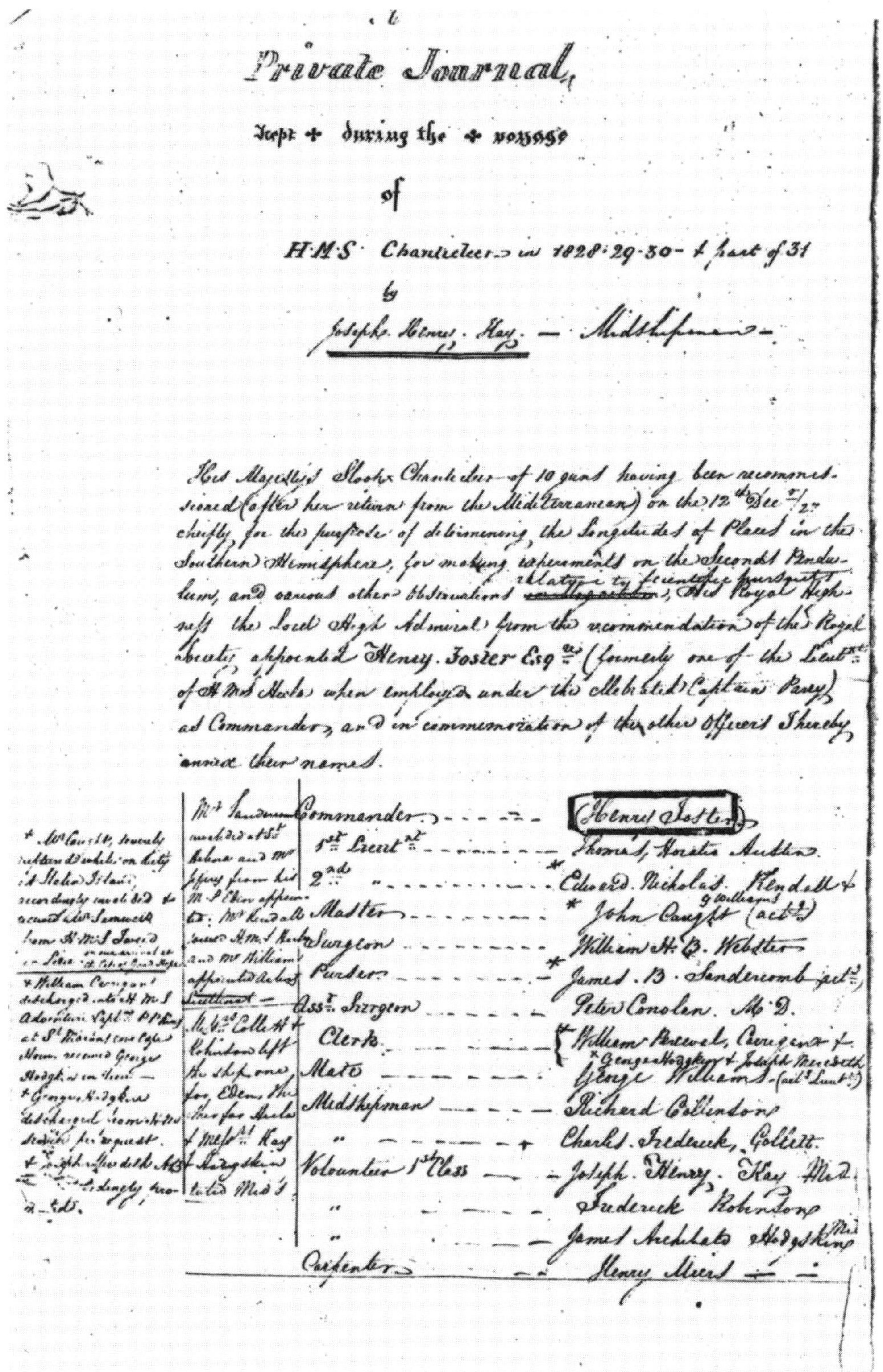

Private Journal,

kept during the voyage

of

H.M.S. Chanticleer in 1828·29·30· & part of 31

by

Joseph. Henry. Kay. — Midshipman —

His Majesty's Sloop Chanticleer of 10 guns having been recommissioned (after her return from the Mediterranean) on the 12th Decr /27 chiefly for the purpose of determining the Longitudes of Places in the Southern Hemisphere, for making experiments on the Second Pendulum, and various other observations relative to scientific pursuits, His Royal Highness the Lord High Admiral, from the recommendation of the Royal Society, appointed Henry Foster Esqr. (formerly one of the Lieuts. of H.M.S. Hecla when employed under the Celebrated Captain Parry) as Commander, and in commemoration of the other Officers I hereby annex their names.

Commander	—	Henry Foster
1st Lieut	—	Thomas Horatio Austin
2nd "	—	Edward Nicholas Kendall
Master	—	John Caught (actg) (William)
Surgeon	—	William H. B. Webster
Purser	—	James. B. Sandercomb actg
Assr. Surgeon	—	Peter Conolen. M.D.
Clerk	—	William Brewel, Reugen & George Haggsky & Joseph Meredith
Mate	—	George Williams (actg Lieut.)
Midshipman	—	Richard Collinson
"	—	Charles Frederick Gollett
Volunteer 1st Class	—	Joseph. Henry. Kay Mid
"	—	Frederick Robinson
"	—	James Archibald Hodgskin Mid
Carpenter	—	Henry Meers

Figure 4: First page of the 'Private Journal' kept by Midshipman Joseph Henry Kay. Reproduced by permission of the Scott Polar Research Institute, Cambridge.

THE JOURNAL

A
Private Journal
Kept during the voyage
of
HMS Chanticleer in 1828, 29, 30 – and part of 31
by
Joseph Henry Kay, Midshipman

His Majesty's Sloop Chanticleer of 10 guns[1] having been recommissioned (after her return from the Mediterranean)[2] on the 12th December, 1827, chiefly for the purpose of determining the longitudes of places in the Southern Hemisphere, for making experiments on the Seconds Pendulum, and various other observations relative to scientific pursuits, His Royal Highness the Lord High Admiral[3] from the recommendation of the Royal Society appointed Henry Foster Esqre (formerly one of the Lieut^{nts} of HMS Hecla when employed under the celebrated Captain Parry) as Commander, and in commemoration of the other Officers, hereby annex their names.

Commander	..	Henry Foster
1st Lieutenant	..	Thomas Horatio Austin
2nd Lieutenant	..	Edward Nicholas Kendall* and G. Williams
Master	..	John Caught (actg.)*
Surgeon	..	William H.B. Webster
Purser	..	James B. Sandercomb* (actg.)
Asst. Surgeon	..	Peter Conolan M.D.

*Mr. Caught, severely ruptured while on duty at Staten Island accordingly invalided and received Mr. Samwick from HMS Tweed in Lieu on our arrival at the Cape of Good Hope.

* William Carrigan discharged into HMS Adventure, Captn PP King at St. Martin's Cove Cape Horn, received George Hodgkin in Lieu

* George Hodgkin discharged from HMS [illegible] per request and Joseph Meredith A.B. [accor]dingly promoted.

---

[1] Built by Daniel List of East Cowes in 1807–8, *Chanticleer* was one of the numerous class of ten-gun brigs (known initially as coffin brigs). After service in the Mediterranean and South Atlantic, she became a Coastguard Watch vessel in 1830, being renamed CGWV.5 in 1863. She was broken up at Sheerness in August 1871: Lyon and Winfield, *Sail and Steam*, p. 75.

[2] During the years 1824 to 1827, *Chanticleer* formed part of the Mediterranean Squadron. Her Surgeon, William Black LRCSE, wrote a narrative of her cruises during the Greek War of Independence (from the Ottoman Empire) published Edinburgh, 1900. As with the title on the pictorial cover, *Cruises in the Mediterranean of H.M.S. Chanticleer during the Greek War of 1824–26*, *Chanticleer* is pictured in gold. She is also shown at the Piraeus, in an engraving facing p. 190.

[3] The office of Lord High Admiral was revived in 1827 especially for Prince William Henry, Duke of Clarence (1765–1837), the third son of King George III, who had served in the Royal Navy from 1779 and who became known on his accession to the throne as the 'Sailor King', William IV. He was Lord High Admiral for just over one year, May 1827 to June 1828.

Figure 5: 'Ascension, Red Hill, S.E. by E.', showing HMS *Chanticleer* in the foreground, drawn by E. N. Kendall and published in W. H. B. Webster, *Narrative of a Voyage to the Southern Atlantic Ocean in the years 1828, 29, 30, performed in H.M. Sloop Chanticleer, Captain Henry Foster, FRS*, 2 vols, London, 1834, I, facing p. 385.

Clerk	..	William Perceval Carrigan*, and George Hodgkin*, and Joseph Meredith
Mate	..	George Williams (actg. Lieu^t)
Midshipman	..	Richard Collinson
Midshipman	..	Charles Frederick Collett
Volunteer 1st Class		Joseph Henry Kay. Mid. Frederick Robinson James Archibald Hodgskin. Mid.
Carpenter	..	Henry Miers

* Mr. Sandercomb invalided at St. Helena and Mr. Jeffery from his M.S. Eden appointed. Mr. Kendall joined HMS Hecla and Mr. Williams appointed Acting Lieutenant.

Messrs Collett and Robinson left the ship one for Eden, the other for Hecla and Messrs Kay and Hodgskin rated Mid's.

The whole complement of the Chanticleer is 56, viz. 18 AB, 15 Officers, 15 Petty Officers, 6 Private Marines, a Sergeant and a Corporal. She was rigged, fitted and manned under the direction of the 1st Lieut^{nt}., Mr. Austin, whose indefatigable pride and attention in the fitting of her for the mutual convenience and comfort of the Officers and Men cannot be too much admired. As the Captain and 2nd Lieut^{nt} were in London chiefly, the whole duty of preparing the Ship was of course involved [*sic*] on Mr. A. We were most handsomely supplied by Government with Instruments for the different scientific observations intended to be made, and in fact a ship could not be more completely fitted up for a hazardous scientific or long cruize than we were. As it was deemed likely that we should have a great deal of ice to encounter the fore and main channels were filled up with stout oak so as to secure the Chain Plates and Lower Deadeyes,[1] and as we were of a Reduced complement she was Barque rigged and carried a Boom foresail. We were also supplied with Donkins preserved meats,[2] and Pickles and had a stove or range on the Lower Deck of Mr. Frasers construction[3] which entirely dispensed with that great nuisance on board a ship, viz. Smokes. As a Midshipman's Chest on board a 10 gun brig. is too cumbersome and moreover having no convenient place to keep them in but the hold, they were supplied with Drawers in lieu which are not only more convenient but look well on the Lower deck. And as the whole number of guns would have lumbered her exceedingly, and not being a fighting ship, we only took two 18lb. Carronades. All these little contrivances tended greatly to our comfort, and last though not least we were supplied with the Patent Illuminators[4] for the purpose of more ventilation throughout the Vessel.

Having now I believe entered into all the minutiae of her equipment I will proceed by saying that after a period of 131 days from the time of her first being commissioned, on the 21st of April, favored by a light breeze from the N.W. we slipped from our Hulk, the Prothee lying in Portsmouth Harbour and went out under single reefed topsails, courses and jib and at 3.30 came too in Spithead in 5fm furled Sails. HM Ships Asia and Blonde went out to Spithead the same day for the Mediterranean. It blew very fresh most of the time we were laying and the same night we dropped the BB [Best bower anchor] under foot but weighed it again in the morning.

Tuesday 22nd [April]. We had this day the accustomed visit of Adml Sir Robert Stopford who went all over the ship and mustered the Ships Company; he appeared and expressed himself pleased with her arrangements and said we had a crew of fine stout young men. He declined the usual ceremony of manning yards and saluting.

Thursday. 24th. This day the Pay clerks came on board and the 2 months advance was paid to the Ships Company, and by an order from the Captain, every man was obliged to provide himself before going to 'blue water' with a new blue cloth jacket and trowsers, a blue Guernsey frock with the Device of a Cock on it, and a few other articles that would ensure their allways appearing in a neat and clean uniform when

---

[1] The shrouds, part of the standing rigging supporting the masts, were secured at their lower ends by deadeyes, round flat wooden blocks, through which a lanyard was laced to another deadeye secured to the channels by the chain-plate, the lower end of which was bolted to the ship's side.

[2] The first preserves were made in glass jars, but it was soon realised that the new process of tinplate rolling would be preferable to produce less breakable containers. John Hall's patent was bought by Bryan Donkin and the firm of Donkin, Hall and Gamble were making regular supplies for the Royal Navy by 1815. See Laing, 'Introduction of Canned Food', pp. 146–8 and Savours and Deacon, 'Nutritional Aspects', in Watt, *Starving Sailors*, pp. 131–62, 203–5.

[3] There are three patents for James Fraser's stoves: 4210 of 15 January 1818; 4310 of 12 November 1818 and 4706 of 27 September 1822, this last for 'ships' cabooses or hearths and apparatus connected therewith for evaporating and condensing water', together with an oven, steam cookers and hot plates. Patent 4706 was granted to John Dowell Moxon, merchant and shipowner of Liverpool, with James Fraser, engineer and coppersmith of Long Acre, Westminster, as second patentee.

[4] According to L. G. Carr Laughton, illuminators were glass bullseyes: 'HMS Victory', pp. 173–211. Bullseyes were pieces of thick glass inserted in the deck, port-lids and scuttle-hatches, to allow light below decks and these were constructed to open like a modern port-hole.

occasion required it.[1] The pay day is a day generally dreaded on board a man of war on account of the drunken and noisome scene that usually follows but through good management our men were very sober indeed.

26th. A party of shipwrights came on board and fitted the Patent Illuminators in the ships decks for better ventilation and the only inconvenience attending them is that they are seldom or ever water-tight but a little wet occasionally is certainly preferable to being stifled in hot weather. A moderate breeze from the S.W. and proving fine in the evening we fidded topgallant masts[2] and crossed the yards and made all preparations for weighing next morning.

27th. Light breezes and fine. 10.30 Weighed and made Sail. Wind Light and variable. However we dropped down as far as Cowes where we came too at 4 P.M. in 5 f^{ms}. HMS Blonde got down as far as Yarmouth roads that day. On the following day the breeze again springing up from the S.W. we weighed, and attempted backing and filling through the Needles in which we succeeded at 2.30 p.m. when we made all sail for Falmouth.

30th. Saw the land occasionally.

31st. 8.30 P.M. Saw the Eddystone Light. 10. Lost sight of d°.

1st May. Made Sail to gain the anchorage, and at 7.30 shortened sail and came too in 10 f^{ms}. In a very outer anchorage.
Pendennis castle)   N.25.W
St. Anthony's pt.)   N.75.E
There was a heavy swell running from the S.Eastward and the same day we completed our water by the tank lighter which supplies vessels. Falmouth is the principal resort for the Packets and there is a Commodore of the Red there with the Astrea frigate as flag ship. It is not usually visited by men of war I believe except in cases of necessity. The principal anchorage is in the Carrick roads but at the best it is an open place, and with the wind from the Southward there is a heavy rolling sea.

2nd May. In the afternoon we weighed, but owing to the swell in weighing the anchor we carried it away 18 inches from the crown leaving the flukes and buoy rope when we were obliged to bear up for Carrick roads where we anchored in the evening. As it was necessary to have another anchor we received one of the packets anchors from the Astrea that evening. The town of Falmouth appears large and well built but owing to not having landed I can give no description. There are 2 castles, one called St. Mawes and the other Pendennis.

3rd. We weighed again this morning and succeeded in getting to sea and at 1.30 P.M. I took a last farewell of dear England for 3 years I believe at least, however I do not say

---

[1] Since uniform for naval ratings was not introduced until 1857, this stipulation is unusual. Sea officers, on the other hand, had been ordered from 13 April 1748 to wear the uniform of their rank: Jarrett, *British Naval Dress*, p. 30.

[2] Fidded topgallant masts: hoisted the topgallant masts and secured them by inserting a fid (a bar of wood or iron).

with any feeling of regret as my prospects were as encouraging as could possibly be ... Our passage was very beautifully made by the little Chantie in 8 days having a fine stiff breeze across the Bay of Biscay with a pretty good sea. On the 4th we passed a man of war brig which we supposed to be the Onyx but owing to the haziness of the weather we could not distinguish her number. The Porpoises were in great numbers swimming about in all directions but we had not the luck to catch any. On the same afternoon it looked rather 'dirty' and as we did not like to risk our boats, which were hanging to the quarters, we got them in on the Skids and took in 3rd reefs.

11th. This morning at 6 A.M. saw Porto Santo bearing S.S.W.; by noon we were running along it and then altered course to S.W. by S for Madeira. There appear to be a few trees on the island scattered about in various directions and a few fisherman's huts. The beach is sandy and hills rather high. At 6 P.M. we came too in Funchal roads but owing to having missed the bank the anchor could find no bottom. We soon picked it up again and came too in 35 f^{ms}.
Brazen Head N. 75. E
Loo Rock N.5.W

The road is open to all winds from the Southward and frequently violent gusts of wind come down the Valley and endanger vessels laying at a short scope. It is usual to lay at single anchor and veer away quite ready for slipping. Water is procured in the town and shipping are supplied by boats full. To the westward is the Loo rock on which there is a castle and the principal landing place is under the lee of it, but the shore boats land on the sandy beach directly under the town but with the wind in I fancy there is a rolling surf. The town, was then in the possession of the Portuguese, is situated at the bottom of the valley of Funchal and is of good size. There are several churches into which we entered, but not one that you would call a fine building. They are the same as most Catholic churches, being ornamented with pictures and long tapers. We were most kindly received by H. Vitch [Veitch], Esq., the English consul resident there, who invited us to go and see his daughters in the country; we set out on horseback from the town at 6 A.M. and after about a 3 hours hard riding we arrived at his country residence where we breakfasted. The whole country throughout abounds in vineyards and fruit trees of all descriptions, very little wood is to be seen and what there is of no size. The young ladies were rather shy and we could not persuade them to accompany us to a hill called the *'Caral' from whence you look down on the valley beneath of a most beautiful green, with a few huts scattered about. The inhabitants viz. peasantry are apparently very courteous, always touching their hats to you while passing but the instant your back is turned they will pelt you with stones and laugh at you. We caught one or two in the fact of which we sickened them for the time I believe, but to return to my subject – after we left the Caral we proceeded into the town again and honored Mr. Paine's, an English pastry cook, with our company to 'tiffins' where we dressed and proceeded in full tog to Mr. Vitch's house and dined with him. He pressed us sadly to stay all night but not being able to we left for the ship where we arrived in safety at 9 P.M. after having spent a most delightful day and very much fatigued. The Horses are dear but pretty good and I was much astonished to see the men, or perhaps owners of them, accompany us all the way walking by the side of us when we went moderately and catching hold of the horses tails when we galloped singing out 'Poco a Poco' which means gently. They went the whole way with us occasionally requiring 'refresheners' of wine which was easily procured all along the road and it was really charity to walk part of the way for their sakes. The ship being watered and Chronometers rated we made preparations for departing, which we did at 4.30 P.M. on the 17th May 1828, for Hidalgo point, Teneriffe. The weather was fine and favoured with a good breeze we made the Peak, topping its lofty head above the clouds on the 18th. From the distance we were off it, the very Peak was the only land visible and it was a beautiful sight to see it appearing at such an immense height through the clouds. It is supposed to be elevated 12.300 feet above the level of the Sea.

* rather 'Cural' – a very good account of it is given in Daniells Book on meteorology when he visited it.[1]

20th. 10.30 A.M. came too in Santa Cruz bay in 22 f^{ms}. with the Church bearing N.64.W and directly after our arrival a Ship and a Schooner were seen in the offing which proved on closing to be the Hecla Cap^{tn}. Boteler for the Coast of Africa and her tender the Albatross. It is rather remarkable that the Hecla left Spithead almost the same day as we did but owing to contrary winds could not get in to Madeira, while we on the other hand had been at Madeira in a week and still got in Teneriffe before her. Ships of war generally anchor about half-a-mile off the northernmost fort from 15 to 18 fathoms. The best anchoring place for Merchant ships is between the middle of the town and one of the forts about a mile to the Northward of it in 6, 7 or 8 f^{ms}.; it is exposed to the N.E. and S.E. winds. The town is on the whole well built I think and in it is a fine square called the 'Plaza Real' where our troops met after the attempted siege of the town by Captn. Troubridge, and on the beach is the fort at which the Hero Nelson lost his arm. There is one church in the town, a fine building. The town produces excellent wine and is very reasonable. There is a fine pier built for landing on and is one of the principal ornaments and promenades of the town. Fruit is not to be obtained except in the mornings and it does not grow in the vicinity of the town but at Liguna, a small village about 5 miles out. It is brought down in the mornings to market and is not sold during the day on account of the great heat. I intended to ride out there but horses were so intolerably dear that I gave up the attempt. Sharks are very numerous in the bay but fish are few; the whole country in the vicinity of the town is exceedingly barren and sandy. Ice is procured from the top of the Peak which is very acceptable after broiling all day in the Sun. A few small craft laying in the bay.

21st. P.M. weighed and made sail for S$^{t.}$ Antonio, one of the Cape De Verd[e] islands. On the afternoon of the 22nd the Peak of Teneriffe[2] was in sight bearing N.38.E.

May 28th. P.M. saw the island of S$^{t.}$ Antonio and being close in and night coming on we shortened sail and hove too off the island – at midnight tacked to close with the land again and at 8 A.M. the Captain and 2nd Lieut$^{nt.}$ left to make observations. The only description I can give is such as I have heard different people say of the island. It is

---

[1] Daniell, *Meteorological Essays*. At Daniell's request, information on Madeira, which included a visit to Curral (p. 309) had been provided by Edward Sabine, who visited it during his 1822 voyage.

[2] Pico de Teide, 3,715m (12,188 feet).

remarkable for 2 high hills, one of which is called the Sugar loaf and is elevated 7,400 feet above the level of the sea. The island is woody on the S.E. side but where we were there is none. There is also a town there and refreshments are easily obtained. It is the Northwesternmost of the Cape De Verde islands about 9 leagues in length, 4 in breadth. At 5 P.M. the Captain returned having effected a tolerable landing and obtained sights for time and latitude. He brought a few specimens of rock and some fish which an old hermit gave him. The rock was a species of granite and some flowers of the Asclepius venosa.[1] We then braced up on [the] starboard tack and filled for the small island of St. Paul intending to pass in sight of it to obtain the Latitude of it. On the 17th [June] at 5 A.M. saw the Island of St. Paul S.59.E, 3 or 4 leagues; it was only just visible from the deck and appeared nothing but a few rocks without verdure at all and only the habitation of wild birds.

June 17th. 9 P.M. Crossed the Equinoctial Line and were hailed by Neptune who enquired what Green horns we had on board, where from, where bound, how long out and concluded by saying he should call on us tomorrow at 9 o'clock A.M.

18th. According to promise Neptune this morning came on board, preceded by fifes and drums and was received by the officers in full Uniform on the Quarter Deck. Soon afterwards the operation of shaving commenced and was carried on in a ludicrous manner by the barber and party, the Sable God giving directions as to the treatment, of the Green horns which was generally accomplished by daubing their face over with a mixture of oil, paint, tar, grease etc., or any filth that could be accumulated, or if you did not choose to undergo the operation a promise of grog or money would mitigate it to a good ducking in a sail filled with water for the purpose. Myself with the rest of the youngsters got off with the latter according to the Captain's wish but the Green 'Men' were all lathered, shaved and ducked. One poor Black fellow had white paint daubed over his face by way of making it show. But the cream of the joke was to come for ours was the strongest party, and after we had quietly submitted to their pleasure we thought it high time to have a little fun ourselves. Accordingly we fisted Lord and Lady Neptune[2] neck and crop and soused them well in the same sail which a few minutes before they had the full command of. It's needless to add that the whole of his party shared the same fate and were highly incensed at it which was great satisfaction to us. This terminated the days fun which was peculiarly favourable for it, and by dinner time the ship was the same as if nothing had occurred, all well cleaned and decks swabbed down. This same afternoon we saw a Brig on the Weather bow standing to the N.E. and we were in hopes of sending letters, but she did not close.

19th. We are just now got into the Latitude of Squalls, Rain and Variables, and a great deal of rain has fallen accompanied by lightning. The squalls are only perceptible by a little white cloud which rises in the direction the wind is coming from, and sail should be shortened in time to prevent accidents which might and often do occur from

---

[1] Possibly *Sarcostemma daltoni* or *Calotropis procera*, the first a leafless succulent growing on dry cliffs on the island and the second a shrub found in coastal areas. Asclepias are milkweeds or silkweeds (Asclepiadaceae). This identification kindly made by Dr E. Charles Nelson and Dr Wolfram Lobin.

[2] 'Classically speaking Neptune & Amphitrite', inserted in the margin.

negligence. I find flannels of great service in keeping myself from chillyness during my watch in heavy rain and the best plan is to go without shoes or stockings when it does rain. We are closing Fernando Noronha very fast and expect to see it tomorrow.

20th. 2.50 P.M. The welcome news of 'Land ahead' has just been received and re-echoed from aloft, and bears NW^{b}W.¾W. – 20 miles. It is a remarkable island and very readily known by a high rocky pyramid very barren and rugged off which at a considerable distance lies a sunken Rock extremely dangerous to approach. At 6.20 came too in 17 f$^{ms.}$ off the town, furled sails, and were boarded by the Governor of the island, incognito, with whom we soon made friends. Coming on squally we struck top gallant masts and yards and made all snug.

21st. Continued heavy rain. Employed watering ship which is very miserable work indeed. I see no chance of its moderating and it is very close below indeed.

22nd. The rain has now ceased and has turned out a most beautiful morning. We sent the Ships company on shore for a walk this afternoon and a party of mids[hipmen] went on a shooting excursion to the South end of the island. Nothing but Boobies[1] were to be found and the swell was too great to admit of landing. We returned at 6 P.M. where [when] we passed the Governor dining on board with the Captain. The men returned the same evening and as may be guessed not all sober. The Governor was dressed in the Brazilian uniform and after an agreable evening departed at 9 P.M. The principal produce of the island are fine cattle, sheep, poultry, melons, corn in a small proportion and cocoa nuts. The Cattle are in general small but very good meat, as also the sheep. The island is in fact a receptacle for Portuguese exiles, guarded by a strong garrison and all the little points in the island are defended by small forts. They appear to be in very good order and from what I could see the Governor is just the man for such a command, punishing with great severity and keeping them in that sort of Discipline that they are afraid to mutiny or rise up any disturbance that might tend to that. The general character of the island is fertile except on the Pyramid which is bare rock. Fruit is only grown for the consumption of the officers in charge of the place and were a ship to lie any time there, they could not supply them with fresh provisions, I fancy, as the produce of the island is not much more than the daily consumption of the troops. With regard to water in a dry season, it is a scarce article, I believe, and is procured at a well near the Governor's house. The surf on the beach is too high to admit of landing without swamping your boat, and the best method of watering is by breakers floated in and out by the surf. It is a great pity that some sort of a pier or landing place is not erected to facilitate the getting on shore as at present it is difficult to land without a wetting and swimming is not agreeable to every one barring the quantity of sharks which abound. The Island is about 7 miles in length and 2 in breadth. The road [anchorage] is on the North side near the N.E. end of the island and the Anchorage is tolerably good in from 9 to 13 f$^{ms.}$ water about a mile from the shore

---

[1] This would appear to be the Brown Booby, *Sula leucogaster*, which breeds on oceanic islands not far from the Equator, including Fernando Noronha. It is related to the Gannet of the British Isles. The common name derives from the Spanish 'bobo', a dunce, which was given by sailors because the birds were so easy to catch.

and the Pyramid about S.S.W. to S.W. It is exposed to Northerly and Westerly winds, which are said to prevail here from December to April, at other times they are Easterly. About the full of the Moon heavy rollers set in from the N.W. I have observed which at present my poor head cannot comprehend. However before I leave the Atlantic I hope to discover some clue to the reason why. It could not be worse if a heavy gale of wind from that quarter had been blowing a few days previous. [*margin*: I have since heard and seen it is the same at all islands in the Atlantic, for instance St. Helena, Ascension, etc.]

The Island abounds in Woods and Swamps full of gnats and Musquitoes, very uncomfortable and dangerous to walk through. Anything in the shape of Specimens of Natural History are not easily procured. Having completed our water we sailed for Cape Frio on the coast of South America on the 26th June at 6 P.M. with 3 sheep and 2 fine stout bullocks on board. It blew a fresh breeze from the N.E. which soon carried us far enough from this lonely island, to be again visited by us on our return from the southward in 30.

8th July. Our passage from the 26th to this time was extremely favourable when by this time we were closing a bank called the Abrolhos shoals. As their extent is unknown it was our object to discover if possible, (as far as lay in our route) their extent, the depth of water on them, and all the various particulars respecting such a dangerous obstacle to our trade, and merchant vessels. We tacked about in hopes of getting at some satisfactory evidence as to their extent, but the shoalest water we arrived at was 13 fathoms over a bottom of coralline and sand.

On the 10th the water gradually increased till we could get no bottom at 100 fathoms and we pursued our course, but just before Noon we perceived a strange sail bearing down on us, at which no notice was taken of at first, till however she got so close that she fired a gun at us and hoisted Brazilian colours. She continued hovering about us till 1 P.M. when she hauled her wind and stood off to repair some damage in her Jib. It was evident she did [not] know what to make of us as our rig was different from the commonality of men of war brigs. At 2.30 she again bore down on us, when we prepared for her and wore to close her at which she again hauled her wind and stood off on the starboard tack. Making Bold we stood after her but she soon outstripped us in sailing and by Sunset she was out of sight. We accordingly resumed our course keeping a look out during the night, when at 11 P.M. we again observed a light astern, piped to quarters, but it disappearing, called the watch. Thus ended the exploits of the day, a day always to be renowned in the Annals of the Chanticleer for) having without firing a shot, beat off and prevented the great effusion of blood that must necessarily have followed such an unequal contest. } quiz

July 1828 13th. At 6 A.M. we saw land which proved to be Cape Frio bearing W. by S., and by 1 P.M. we were close enough for the Captain to land in his boat and take the necessary observations for the Chronometric differences of Meridians. While hove to off the island a Brazilian Corvette called the 'Marie Isabel' passed under our stern and hailed us. She kept her guns to bear on us as she gradually hauled up round us, and had we been any thing else but a British man of war it would have looked rather awful, but that alone was quite sufficient to ensure our safety. She was commanded by an

Englishman, Capt[n]. Grenville[67] and I am sorry to notice it is too frequently the case that our countrymen, not succeeding according to their expectations in our service, join that of a foreign nation for the sake of a little advance in pay and should the war which now exists in those Nations become finished, most likely they will be reduced, and not being again received into ours perhaps become destitute.[68] The appearance of the Cape from the ship was to my mind extremely unsatisfactory, the Coasts being completely Iron bound and landing extremely Difficult, but the interior of the Country appeared fertile and well wooded. Night coming on and the Captain not having returned it was considered necessary to commence burning blue lights and firing muskets, as it was probable he might miss the ship. However he returned at 11 P.M. and we stood off the land. It appeared he had found landing extremely bad and he had a long pull off.

14th. At Daylight we tacked to close the land and at 8 a.m. the Captain left to complete the observations begun yesterday. The coast has not risen in my estimation since yesterday. At 4 P.M. the Captain having returned we departed for Rio De Janeiro, about 90 miles distant.

15th. 9 a.m. We are now standing into the Magnificent Harbour of Rio Janeiro, and have just spoken HMS Cadmus, Sir T. B. Thompson,[69] Son of the Late Treasurer to Greenwich Hospital. Our rate of sailing is about equal, however I like Chantie best.

On entering a Foreign Port, to those on board the ship who have not visited it before there is always a delightful feeling of curiosity and uncertainty, which recalls to our memory the Juvenile Emotions with which we have read any little trifling book when young. The place itself generally (not always) comes up to the vivid promise which a lively imagination holds out, nor is this imagination abated by the repeated sight of new objects, on the contrary, I think each new place seems more curious than the last, for as our sphere of observation is enlarged, our curiosity becomes more impatient, though at the same time I think, more easily satisfied. The world indeed in every place is so crowded, with new and varied objects, that no one can hope, even by the most acute observation, to be perfectly acquainted with the details of every scene, and it is by this variety that we are kept in hopes and the curiosity enlivened, by the almost certainty of finding something new. At first our pleasure springs out of ignorance, in time we derive it from knowledge.

Having I think sufficiently digressed from my Detail, I will proceed by stating that after having been buffeted about by calms, light winds, and variables, for a period of 6

---

[67] During the war between Brazil and Argentina of 1825–8, Brazil came to rely heavily on British naval officers and seamen recruited into her service. Captain 'Grenville' must have been John Pascoe Grenfell (1800–69) commanding the Brazilian man-of-war *Maria Isabela*, in which vessel he captured the privateer *Peruano* off Cape Frio in July 1828: Vale, *Independence or Death*, p. 191.

[68] For the attitude of officers of the Royal Navy's South America Squadron to their countrymen serving under the Brazilian flag, see ibid., pp. 186–7. Kay was wrong to worry about the fate of these officers. They all had contracts, which enabled them either to remain in the Brazilian navy or retire to Great Britain on Brazilian half-pay for life. Brian Vale, personal communication, 30 June 2000.

[69] According to O'Byrne, *Naval Biographical Dictionary*, p. 1175, and the Navy List, December 1829, this was Commander Sir Thomas Raikes Trigge Thompson, only son of Vice Admiral Sir Thomas Boulden Thompson, Bart. He was employed again in South America in the *Talbot*, 1842–47.

hours, at 3 P.M. we succeeded in getting the Sea Breeze which carried us safely in to the Magnificent Harbour of Rio De Janeiro, a Harbour which I suppose is not equalled for splendour of scenery (and I think I may say safe anchorage) throughout the world. I am too poor a writer to attempt describing it in the terms it deserves but having undertaken this task I cannot omit such a most pre-eminent part. On your right hand on entering is a a large fort built by the Portuguese, mounting 3 Ramparts of heavy guns one above the other, and in other respects a well fortified looking place from which they hail you as you pass to know who you are. On the otherside is the far famed Sugar Loaf (a remarkable rock) with only a very few signs of verdure on it which are at the top. [*margin:* Its estimated height is [   ] feet]. Proceeding onward you perceive a small island with a fort on it, where they send all thieves, robbers, etc. and are confined underground below the level of the sea, so close surrounding them. From that the town soon bursts upon your view, which is certainly the worst scene in the harbour. On your left again you will perceive a lofty hill called the 'Cockervado' [Corcovado], and immediately afterwards you will pass another small island (on which the French had formerly a settlement) called the 'Ville Gagnon'. The best anchoring place is with the flag staff on the Ville Gagnon on with the summit of the Sugar Loaf hill. To speak of the harbour in regard to anchorage it cannot be more briefly stated than to say at once it is a good one. Heavy gusts of wind blow occasionally during the summer months from the S.E. to which a ship might drive were her ground tackle not good but then the water is smooth and it is not so much the wind as the sea which does mischief to a ship. Regular land and sea breezes blow during most part of the year, and seldom if ever so strong as to render them inconvenient or troublesome. The atmosphere is naturally hot during the months of December, January, February etc. but (more especially as the Southern are found to be comparatively colder than the Northern Latitudes) it enjoys a sort of winter very delightful to them during our Summer months in England. It appears as if all the sublime and beautiful features of Nature had been called into action to enrich the scenery of Rio Janeiro. The rich and luxuriant descriptions of former travellers, it must be acknowledged, are not overdrawn for I think it would indeed be difficult to exceed the truth in pourtraying [*sic*] the grandeur of such a scene, its richly wooded banks rising like an amphitheatre on either side, studded with villages and country seats, added to the distant view of lofty and picturesque mountains form a scene more easily conceived than described. The town is to the Westw^d of the harbour near which is a tolerable citadel and several minor forts where a tolerable good look out is kept and a vessels approach signified to the town. The principal building is the Palace situated in the square on your right on landing, very inferior when compared to some of our Magnificent Edifices in England. The streets are straight and narrow, and in most parts forming right angles to one another, and are denominated according to the different trades and manufactures carried on in them, for instance, Gold Street, Silver Street, etc. In one part of the town a fine market is kept with a great abundance of tropical fruits. The town is supplied with water by means of a large aqueduct carried across the vallies [*sic*] and supported by a double tier of arches. The Numerous Slaves running about the Town in all parts and Directions mostly loaded, do not convey to the eye of a stranger feelings the most agreeable, and reminds one how happy a thing it is that the abominable System of Slavery is abolished, in England at least.[1] Numerous

---

[1] Slavery was not abolished in Brazil until May 1888.

insects are flying about in all directions and on the Braganza side of the harbour oranges and limes are growing in all directions. Soon after our arrival HMS Ganges, Rear Adml. Sir Robert Otway,[1] came in to harbour from a cruize.

21st. HMS Blossom, Captain F. Beechey, arrived this evening from her enterprising undertaking and it is with true regret I hear of her near approach to Captn. Franklin's party without knowledge of it.[2] Went a few miles up the country this evening in company with the surgeon and was very much amused with the surrounding scenery and beautiful fertility that abounds. We slept at a friend's house and I was very much tormented with the Musquitoes to which as yet I was a stranger. On returning I visited a large quarry of gneiss which is very plentiful in the suburbs of the town and was much astonished to find great quantities of garnet abounding in it, but in very small particles. I brought 2 or 3 of the best specimens of it away. We had a large squadron of French and Brazilian ships in the harbour,[3] and 2 or 3 Americans. The ship being ready for sea, and Captn. Foster having finished his observations, on the 28th at 6 a.m. we weighed for the beautiful island of St. Catherines[4] on the coast of Brazil and it being calm, received great assistance in getting out from the boats belonging to the ships of the Squadron.

August 1st. Nothing particular occurred in our passage to this place except the same unwearied round of exercising in making and shortening sail, mustering at quarters etc. etc. very uninteresting to the reader, and the detail of which (to any person who has ever heard of or seen a ship) cannot fail to be stale news. At noon we made the land and soon afterwards were boarded by the Captain of a Brazilian man of war brig, another Englishman.[5] The harbour is a great length and defended by several forts with an anchorage in good tough mud in about 6 or 8 fathoms. A very strong current runs with the ebb tide and it requires a look [out] to be kept on the ship to keep her clear of her anchor. We lay some distance from the small town of St. Catherine's and consequently I could not get an opportunity of visiting it, but the part of the island on which we landed was of the most luxurious vegetation. You would be almost dazzled with the splendour of the orange and lemon trees, sinking apparently under their weight.

---

[1] Sir Robert Waller Otway, 1772–1846, was Commander-in-Chief on the South America Station, 1826–9, during which period he was presented with the Brazilian Order of the Southern Cross. During his career afloat, he had taken part in some one hundred war-time engagements: O'Byrne, *Naval Biographical Dictionary*, pp. 841–4.

[2] Captain Frederick William Beechey (1796–1856) was on his homeward voyage from the North Pacific (Bering Strait) where he had been waiting in vain for the emergence from the east of Captain John Franklin and party during Franklin's second overland expedition along the north coast of America, 1825–27. The voyage of HMS *Blossom*, some 73,000 miles in length, which lasted for well over three years, resulted in the discovery of a number of Pacific islands and the charting and exploration of the north-west coast of Russian America (now Alaska). Both outward and homeward passages were via Cape Horn. Beechey's two-volume *Narrative of a Voyage to the Pacific and Beering's Strait … in the years 1825, 26, 27, 28*, was published by Henry Colburn and Richard Bentley in both quarto and octavo editions in 1831.

[3] The huge French squadron had arrived in Rio on 12 July to force the Brazilians to pay compensation for the seven French merchantmen arrested for attempting to break the blockade of the River Plate. Brian Vale, personal communication 20 June 2000.

[4] Santa Catarina, a large island, at roughly 47°W 27°S and present name for this state of Brazil.

[5] Probably Commander William James Inglis, recruited in 1823, Captain of the *Caboclo*. Brian Vale, personal communication 20 June 2000.

They are to be obtained merely for the trouble of gathering them of which I assure you we took ample advantage. There are a very few inhabitants on the island, but remarkably courteous and civil. Fish is in great plenty off the island of Arvaredo[1] and some fine fishing boats are kept for the purpose. Wood is to be cut in great plenty and also Broom stuff.[2] It blows fresh occasionally from all quarters and accompanied by Dirty weather. Water is in great abundance on all sides but the landing is not at all good on the sandy beaches which abound, on account of the rolling surf. Insects of all kinds were in great abundance and all the flowers of a tropical climate have a fine opportunity here of displaying their greatest splendour. Of venomous reptiles I cannot say anything as I had no opportunity of seeing any but I should imagine from the vicinity of the island to the main land and the many opportunities which might occur for conveying them thither that they were likely to be in great plenty.

August 7th. At 9 A.M. this morning we got under weigh and made sail for Monte Video with a fresh breeze from the S.E. which very soon carried us out of sight of the vast continent of South America. When entering the mouth of the River Plate or Rio de la Plata we had a fresh breeze from the N.E. with a very heavy swell which made her roll most excessively.

August 14th. At 4 this afternoon we made the small island of Lobos which is always a sure guide to a ship going up the river. The course by the World[3] from Lobos to Flores island is W.4°.30'N., distance 52 miles. Seals are to be found in great plenty I believe on this island. Having a good breeze we soon enough left it on our Starboard Quarter and made the best of our way hoping to get in to morrow.

August 15th. 8 a.m. we have just made the land to the N.W.d and can now plainly distinguish it to be Flores island – by 4 P.M. the mount was visible towering above the low land which abounds here. We continued under easy sail and by 9.40 P.M. we dropped our anchor in the outer roads not liking to run further in on account of the numerous Ships in all directions.

August 16th. At Daylight we found the Sapphire 28 [guns] Captain Dundas,[4] the U.S. corvette Boston, and a Brazilian Squadron lying here. Saluted the Brazilian governor with 13 Guns by Proxy through the Sapphire which was returned by an equal number. It continued to blow fresh with rain all the day and we therefore made no attempt to run inside.

---

[1] Ilha do Avoredo.

[2] This shrub was used to burn off the weed and barnacles grown on a ship's bottom: Kemp (ed.), *Oxford Companion*, p. 112.

[3] This would appear to be a reference to a chart in an atlas or folio of the World, from which the bearing and distance to Isla de Flores from Isla de Lobos could be taken. Captain Richard Campbell, RN, personal communication, 15 August 2005.

[4] Captain Henry Dundas had been appointed to HMS *Sapphire* on the Jamaica and South America Stations on 20 December 1826: O'Byrne, *Naval Biographical Dictionary*, p. 312.

August 17th. It continued to blow fresh all this day with a rolling sea right in the wind being from the S.E.

August 18th. The wind moderated greatly this morning, and in the afternoon we attempted to purchase our anchor[1] which as luck would have it we did without any accident although a heavy swell was running at the time. We came to on the Northern side of the open roadstead of Monte Video which I think is a very poor harbour for ships, should the wind hang from the Southward for any time, it being then open to the whole swell of an immense river at least 120 miles in breadth in that part and where (on account of the shallowness of the water and the Numerous banks which occur), a very heavy sea rises in a very short space of time. The only advantage which the roadstead derives from having the wind from the Southward is that it causes the water to rise about 2 feet, or probably more.

With the wind from the Northward the harbour of Monte Video is very well protected, and if the wind is of any strength it causes the water to run out leaving generally about 11 or 10½ foot water only and therefore ships drawing more than that lay aground in the soft mud very securely. With the 'Pamperos' which so frequently occur during the Summer months the ships are sheltered greatly by the mount itself and receive no damage except what may occur from rain, thunder and lightning. For the General trade to the River Plate I should certainly recommend ships not drawing more than 13 feet and under as the average depth of the river will not admit larger ships than that. There are 2 or 3 small islands in the Harbour on one of which the Pendulum experiments of the Chanticleer were carried on called Rat island.[2] It being the first place at which the Pendulum had been landed since our leaving England great anxiety was felt as to the accomplishment of this most important part of the voyage and I am happy to say that the experiments were concluded without the slightest accident occurring.

While the Captain was employed in making the necessary observations, Lieut[nt]. Austin was daily employed in completing the refit of the vessel in readiness for a cruize off Cape Horn. The little barque was very much loaded indeed being obliged to take 9 months provisions in and it was feared she was too heavy for the rough and dangerous sea we knew we should meet off Cape Horn, and it will hereafter be seen whether or not our conjectures were right. Great civility was experienced from the different English merchants in the town by all the officers of the Chanticleer which together with the good shooting in the neighbourhood of the mount afforded a very tolerable recreation. Partridges were numerous, as were also peroquites,[3] teal and several other birds of most luxuriant plumage. On one occasion I saw 2 tiger cats[4] and I am informed that tigers were plentiful in the Neighbourhood of the town. No scenery of any note is

---

[1] Purchase our anchor, i.e., weigh our anchor.

[2] So named on early Admiralty charts; that of 1884 marks 'obsy' (observatory) but this could be a later building. The Spanish place-name Isla de Ratas was later superseded by Isla de Libertad.

[3] Parakeets.

[4] Probably either *Felis colocolo*, the Pampas Cat, or *Felis geoffroyi*, Geoffrey's Cat. The former inhabits a variety of habitats, such as open grassland, forest, mountainous areas and swamps, while the latter inhabits scrubby woodland, open bush and the pampas grasslands. Possible identification courtesy of the Mammal Group, Natural History Museum, South Kensington.

to be seen on account of the great want of trees and throughout you have nothing but a flat uninteresting country to view.

The principal commerce of Monte Video is in hides and tallow, but at this present time our merchant ships are greatly detained by the continual unsettled state of affairs here and in fact many of them have been stopped from proceeding up to Buenos Ayres by the Brasilian Squadrons in the river.[1] They go on in such a curious manner that there is no divining their final determination. 2 prizes have already been taken from the Buenos Ayreans, and escorted in here by a grand fleet, mustered on the occasion, consisting of all the schooners and brigs in the place.

The best anchorage for any sloop of war or frigate drawing more water than about 14 or 15 feet is with Point Braba[2] by compass $W^bN\ ½N$

Cathedral " NE^bN
Mount " NW^bN

in about 4 fathoms with the harbour quite open and about 2 miles from the town.

Provisions of all sorts are very easily obtained at Monte Video but the Beef generally speaking is not good. Fruit is very scarce and can hardly be obtained of any sort at any rate, most of it being brought round from Rio Janeiro. Water is to be obtained in several places at the foot of the mountain from small springs running from the top. Fish is not plentiful, neither is it worthy of much notice, nothing of any consequence being caught there. A great quantity of flour is imported from America, as also wines from Portugal.

The town of Monte Video is built on a projecting point, that and the Mount itself on an opposite point forming the Small and not over secure inner anchorage already mentioned. The town itself had nothing particularly inviting in its appearance, it being built irregularly and the only building worthy of notice is the Cathedral which has rather a curious appearance from the circumstance of its being tiled or covered with china. A Wooden pier is erected at the landing place which when the swell sets in is much sheltered by the projecting point on which the town is built. Few of the English merchants have houses of their own, most of them living in lodging houses and boarding at a sort of table d'hote kept by an American, which was a great place of resort for all the officers of the different ships in the Bay. After dinner we used generally to walk or ride a little way outside the town. It was considered very dangerous to go outside the Barriers on account of the then unsettled state of the country, in fact some of the Sapphire's officers had been seized one day while outside and carried up to the Patriot Camp from whence of course they were immediately released when they were found to be British Officers. I really think had they not been released that the Sapphire could have almost battered the town to pieces and then have started for the interior for their rescue.[3]

---

[1] For three years from 1825 to 1828, the Empire of Brazil was at war with the United Provinces of the River Plate (Republic of Argentina), fighting for the control of the north bank of the River Plate, after the dissolution of the Spanish Empire: see Vale, *War betwixt Englishmen*.

[2] Presumably Punta Brava.

[3] Kay here shows his ignorance of the situation ashore: in fact the interior was dominated by the rebel Uruguayans and their Argentine allies, while Montevideo was held by the Brazilians. The idea of bombarding the town as a lesson, and then rescuing the officers from the interior would not therefore have worked! Brian Vale, personal communication, 30 June 2000.

Friday Oct. 3rd/28 This morning arrived a corvette from Buenos Ayres which had Commodore Brown[1] on board, and is come down with treaties of peace I believe. The old commodore was most graciously received on landing. We were employed variously preparing for sea.

Oct. 4th Employed getting the launch in and stowing the quarter boats on the skids – also getting ready for sea.

Sunday Oct. 5th Employed stowing the observatories away. At 3 P.M. every thing being on board we weighed and ran out to clear the different banks in the river before night. Having just cleared the English Bank running with the wind and a heavy sea abeam, the little barque staggering under all canvass and so loaded with provisions and lumber as to lose all her liveliness, she shipped a tremendous 'Green Sea' which knocked several men down and completely floated the Upper Deck. It steadied her for a moment or so, and had another one followed I will not take on me to say what might have been the consequence. Great doubts arose in consequence as to the safety of our proceeding in such a situation but somehow or other on she did go and from that very circumstance that day was named by the officers 'Eventful Sunday'. We were much favored in our passage down until we reached the latitude of 52° and began really to hope we might accomplish our voyage to Staten island without any of those dangers which we had prognosticated.

Octr. 16th This evening we are shortening sail by degrees and there seems every chance of a very heavy gale.

Oct. 17th We are now laying to under a close reefed m[ain] topsail with a tremendous gale from the Westw^d and it is really fearfully magnificent to see the little Barque riding at one time triumphantly o'er the curling foam and at another buried in the immense abyss, each successive wave seeming as if it would overwhelm her with its irresistible force. The grandeur of the scene is beyond description, and completely outsteps any idea that I could have imagined. The Chanticleer has the decided name of a good 'sea boat' and I am sure well she deserves it, for if any thing could ever appear lively on the water it is her. We are deserted by all our tropical visitors and the pintado's[2] alone, (that most elegant little bird) appear to be the only living thing delighting in the fury of the elements. He sits during the heaviest parts of the squalls under the lee of the ship in the hollow of one of the waves and rides as triumphantly as the little Chanticleer herself. Occasionally a mother Carey's chicken is seen.[3]

The heavy gales so much noted off Cape Horn generally begin with a light breeze from N.N.E. gradually increasing in strength and veering round to N.W. or westerly

---

[1] Commodore William Brown (1776–1857), an Irishman, had led the Argentine blockade of Monte Video, so ending Spanish rule on that river in 1814. After some years in obscurity as a result of a disastrous privateering expedition, he was appointed with the rank of Commodore, later Admiral, to command the Navy of the United Provinces against that of the Empire of Brazil. Peace was finally ratified on 4 October 1828: see Vale, *War betwixt Englishmen*.
[2] Pintado Petrel or Cape Pigeon (*Daption capense*).
[3] Wilson's Cape Petrel or Storm-Petrel (*Oceanites oceanicus*).

as the strength of the Gale increases. Its duration is commonly about 36 hours and generally after about 24 it begins to moderate its violence. During its strength, the sea that it raises is most tremendous and from experience it has been found that a 'Storm – or monkey – reefed main topsail' is by far the best sail to lay to under as it enlivens the vessel when in the trough of the sea and is not so likely to be becalmed by the deep trough in a small vessel.

October 18th 28. At daylight this morning we saw Staten Island bearing South. It was blowing most violently from W.NW. veering occasionally a point or two each way, and there did not appear the slightest chance of any thing like its moderating. In the evening we wore [ship] to prevent the possibility of our drifting in with the land during the night. Sadness was on every countenance at the idea of being actually in sight of our Destination. It was the first time I ever was fairly blown off the land in my life and never shall I forget that one night.

The Gale continued with unabated violence occasionally moderating a little and then again increasing in violence until the morning of the 21st during which time all the ship's head and weather waist netting had been carried away by the violence of the sea, it all but breaking on board of her several times. On the evening of the 20th it fell light and gradually dwindled away to a calm and on the 21st we were enabled to make sail towards the island once more. On the 24th we were off Cape St. Johns[1] and by the 25th had worked up to Deadmans island[2] where we anchored at 10 a.m. in 20 f^{ms}.

25th October. This evening we made up a party to land on the island called Deadmans island situated off the entrance of North Hatchett Harbour.[3] On landing we were much amused to notice the Penguins jumping off the rocks in all directions at our approach. Kelp Geese[4] and a sort of duck were also numerous. Walking was difficult on account of the long tussac grass that the island was covered with. We made 3 or 4 attempts to shoot some of the Penguins but could never succeed owing to their quickness of sight, as instantly that the flash was made in the pan of your fowling piece they were down fathoms below the surface of the water.

27th October. This morning I set out in company with the first Lieut^{nt}. (and a Captain Palmer[5] who had come out with the Captain the day he went in to examine New Years Harbour[6] having the command of a small schooner on a sealing excursion) to examine

---

[1] Cabo San Juan.
[2] Presumably renamed Isla Observatorio.
[3] Now Puerto Cook.
[4] Probably *Cloëphaga hybrida hybrida*.
[5] Captain Alexander Palmer and Captain Henry Foster became friends, according to John R. Spears's biography of Captain Nathaniel Brown Palmer. As a sealer, Alexander Palmer knew Staten Island and the South Shetlands well. He kindly advised the British commander as to the coasts, the islands and safe havens. A letter of acknowledgement dated 28 October 1828 concerning Palmer's piloting of *Chanticleer* to a safe mooring in North Port Hatchet, Staten Island, is quoted in the biography: Spears, *Captain Nathaniel Brown Palmer*, p. 83.
[6] Puerto Año Nuevo.

Deadmans island. We landed about 10 a.m. and proceeded to the N.E. point of the island. Our way lay chiefly through long tussac grass, under the roots of which the Penguins burrow (if I may use the term) like Rabbits and there deposit their eggs during the breeding season. The Female stays till the eggs are hatched, the Male not landing except occasionally to bring his mate some food. We were unfortunately too late to procure any eggs as we were late in the season and they were all addled. On any attempt being made to take the eggs from the female they instantly show fight, and it is the hardest thing possible to kill one of them being as tenacious of life as a cat itself. I have seen one of our men actually leaning his whole weight on one of their breasts and yet not kill it.

Penguins are generally classed in 3 distinct species and known by the names of the 'King Penguin', 'Jackass Penguin' and 'Macaroni Penguin'[1] The King Penguin is not surpassed in pride by the Peacock although in Plumage I think he is certainly inferior. During the time of moulting they repel each other with disgust, as if they were ashamed of being seen one by the other in their ragged coats, but immediately they arrive at their full plumage, they assemble and any one who is not complete is not allowed to enter their community. They stand erect and walk with a waddling gait frequently looking down their front in order to discover any 'faux pas' in their 'toilette'. From December to February they pair and lay their eggs and I believe the period of the female remaining on shore is usually about 4 or 5 months, but the short stay of the Chanticleer at each place prevented our witnessing any thing definite as to the times of pairing or hatching, such information as we have got being generally assumed on someone else's authority or knowledge of the circumstances. The two other classes are denominated by the Macaroni, on account of a tuft of yellow and black feathers which he carries on his head and his peculiar dandy and foppish appearance, and the Jackass, from the noise he makes resembling that animal – the latter Sir John Narborough[2] compares 'to little Children with their white aprons on', it having a snow white breast and black head. We proceeded to the N.E. end of the island without any difficulty having shot several shags and Penguins by way of trial for the table – and returned on board about 3 P.M.

The island was covered with low bushes about 4 feet high of the 'Arbutus Aculeata' bearing an abundance of red berries very pleasant indeed to the taste.[3] Wild celery was very abundant in all directions but more particularly on the beaches or near the sea.

---

[1] King Penguin: *Aptenodytes patagonicus*; Jackass Penguin (Magellanic Penguin): *Spheniscus magellanicus*; Macaroni Penguin: *Eudyptes chrysolophus*.

[2] Narborough in the *Sweepstakes* sailed through the Straits of Magellan in both directions during his voyage of 1669–71, narrated in his *Account of Several Late Voyages and Discoveries to the South and North towards the Streights of Magellan &c &c.*, London, 1694, reprinted Amsterdam 1969, p. 59.

[3] The surgeon of the *Chanticleer*, Dr W. H. B. Webster, described 'the arbutus with sharp pointed leaf' as the 'pride of these regions', an evergreen of an 'elegant and most pleasing' appearance, resembling a fine myrtle, bearing 'white, cup-shaped flowers, followed by a profusion of fine red shining berries, which ornament the tree throughout the winter'. Mixed with a few raisins, these were made into tarts. See Webster, *Narrative of a Voyage*, II, pp. 293–4. This name cannot be traced in botanical literature. The most likely plant is *Gaultheria mucronata* (formerly *Pernettya mucronata*, syn. *Arbutus mucronata*, prickly heath). The editors are grateful to Dr E. Charles Nelson for this identification (personal communcation, 4 October 2005). See Moore, *Flora of Tierra del Fuego*, p. 121–2.

28th. At 9 this morning we weighed and ran into Hatchetts Harbour under the pilotage of Captain Palmer of the American sealing schooner Penguin of Stonington [Connecticut], who kindly came out with Captain Foster in his gig to pilot the Chanticleer to her destination. The only danger to be avoided in going into Hatchetts harbour is a small rock on the left hand side of the entrance about ¼ mile from the shore which is generally perfectly visible as the sea at times breaks on it with great violence. Giving this a tolerably clear berth you will have from 22 to 19 fms. at about mid-channel when soon after a small island will appear on the Larb^d Hand called High water island about 1 cable length from the Left hand shore off this island about 8 fms. and then increasing. You may still run on in mid-channel and will have from 17 to 19 f^{ms}. sandy bottom when the head of the harbour will appear and then you may choose your berth as close to the shore as you like. The Chanticleer anchored off a small run of water in the corner of the harbour in 14 fms. sandy bottom. Having once fairly got into Hatchetts Harbour it is one of the most secure and snug little places imaginable, the direction of it laying N.E. and S.W., that part or the mouth of it which is open to the N.E. being sheltered from any thing like a sea rising by a curve like this ∫ in the middle of it and at the other end of it which is open or exposed to the S.W. there is not room for any thing like even a swell to rise and you lay quite close to the shore. The shore and the water to about ½ cables length is filled and covered with thick kelp which is very annoying to you when pulling in a boat as it entangles the oars and prevents your proceeding in a great measure. Lofty hills surround you on all sides and completely line the harbour right up and down except at the S.W. end on which is a small Level neck of land of a ¼ mile long connecting the harbour we were laying in to another one on the other side of the island. These 2 harbours were called by the names of North and South Hatchetts but we distinguished them by the names of Port Cook and Port Vancouver in honor of the two great Navigators of that name.[1] This neck of land is called the Haul over from the circumstance of the Sealers at former periods having frequently hauled their boats over there to prevent the necessity of taking them round the island to the other side when employed on sealing excursions at the South side of the island, and is I believe without exaggeration almost the only level piece of land in Staten Isl^d.

Staten Island in general
Staten Island is deeply indented with Bays and Harbours all round but more particularly on the Northern side than the Southern. Its greatest length extends from East to West and the Easternmost extremity is called Cape S^t. John off which is a strong tide rip and must not be approached too close. Off Port Hatchett are several small islands known by the names of the Deadman's islands and off the mouth of the New Years Harbour (a harbour more to the Eastw^d.) are several more. The Western extremity is called Cape S^t. Bartholomew,[2] and is seen when going through the Straits of Le Maire, it being very lofty land.

---

[1] Puerto Cook and Puerto Vancouver and many other place-names derived from their British originals appear on the Admiralty chart of Isla de los Estados.
[2] Cabo San Bartolomé.

The Whole of Staten island is covered with immense high hills, bare at the summit but clothed downwards to their very bases with the ('Fagus antarctica')[1] intermixed occasionally with that beautiful evergreen the Arbutus aculeata (but on account of its not growing to that height is not perceptible till you enter the woods) and the Winters Bark[2] – (Wintera Aromatica). The Beech of Staten Island attains the height of about 50 or 60 feet and in some instances may be large enough for many purposes but generally speaking, it is too small to be made much use of. It is watered by three or four Large freshwater lakes which empty themselves into the numerous harbours being themselves replenished from the different hills surrounding them which must be, I should think, always saturated with moisture, the summits of them except in very fine weather being covered with dense mists and most frequently Snow.

Productions of Staten Island
The Shores of Staten island abound in wild fowl, such as Kelp geese[3] (so called from its feeding on the Kelp and fish and which also gives it a disagreeable flavour), another kind of Goose called the Brent Goose[4] and which is a very fine Bird, Loggerhead Ducks called the Steamer duck[5] and numerous other smaller birds. The Loggerhead duck is called by the Sealers the Steam boat on account of its peculiar manner of propelling itself when chased along the surface of the water, it having no wings but short pinions which as I before said it works on the surface of the water like the paddles of a steam vessel and splashing the water up, accompanied with that noise, very much resembles it in motion. It proceeds in this manner at an immensely quick rate and I do not know whether it would go faster with wings than in this novel method of proceeding. These birds before mentioned deposit their eggs in the hollows of the rocky shores almost inaccessible to human beings and then breed and bring up their young without the slightest disturbance. Several of the sandy beaches in the island afforded a great quantity of clams being a large species of the [   ] of which we took ample advantage. These were to be got buried in the sand at Low-water and more especially in Port Cook than any other place. Limpets (Patelloe) abounded on all the rocks but any other kind of shell fish or shells were not to be obtained. Wild celery was plentiful in all parts of the island as also land cresses both of which we used in great quantities as an antiscorbutic. No animals of any kind except Otters are on the island – no person having I suppose thought the island worth bestowing any on, but I should imagine that many kinds would thrive well. The General character of the island is marshy and from the great thickness of the woods the sun seldom penetrates many parts of the island.

---

[1] Now *Nothofagus antarctica*.
[2] Winter's Bark. See Journal of Commander Pringle Stokes in this volume, p.185, n. 1.
[3] Kelp Goose, *Cloëphaga hybrida hybrida*.
[4] Brent Goose: there are no Brent Geese in the South, but on the Falkland Islands 'Brent' is a vernacular term for the Ruddy-headed Goose, *Cloëphaga rubidiceps*, which is also found on the mainland.
[5] The Magellanic Flightless Steamer Duck, *Tachyeres pteneres*.

Figure 6: 'Pendulum station at Port Cook, Staten Island', drawn by E. N. Kendall and published in Webster, *Narrative of a Voyage*, I, facing p. 99.

General Winds and Weather

The prevalent winds of Staten Island may be generally estimated at this, and from all information I could collect they appear to me much the same as the usual winds experienced off Cape Horn during the different months, the distance between the two being so comparatively small that any wind that would affect the one would act the same on the other. They prevail from the Northward with tolerably fine weather from about November to February, when from that during March, April, May, and beginning of June, the Westerly gales are particularly tempestuous and the sea raised by them is mountainous. Yet even although during these above mentioned months such winds and weather are said to prevail, in our passage down to Staten island in November we experienced a tremendous Westerly gale and therefore the best rule to lay down is when your winds begin light from the Northward and gradually veer round to the Westward you may expect a severe gale. June is a fine month and the S.W. and N.W. winds prevail again from that until November. The Gales frequently blow in squalls, having moderate intervals of fine w^r [weather] and during the squalls the spray of the sea is carried along at a tremendous rate more resembling snow drift than any thing else – etc. etc. etc. etc.

Account of Observations Carried on

The pendulum and instruments were landed in the [ ] corner of the Harbour in a place cleared away among the Beech trees for the purpose. The square house for the Pendulum was erected on a firm bed of greywacke covered over with a large tent made out of the stud^g. sails and other small sails belonging to the ship and strengthened by guys fastened to trees around. Marquees and Tents were erected for carrying on the different observa^{ns}. in Magnetism under the superindendence of [ ] and consisted of the Hourly Horizontal intensity of the Magnetic Needle, the Diurnal Variation of the Magnetic Needle with an instrument constructed by Dollond and the variation of a needle by a compass well graduated of rather a superior construction and under the influence of a magnet.

A series of observa^{ns} on the Hourly oscillations of the Barometer, the temperature of the Air and Sea water every 2 hours as well as the range of temperature in 24 hours as indicated by a Sixes thermometer[1] were also carried on by one of the Mid's. Four Midshipmen were appointed to make the experiments on Magnetism Day and Night without intermission, being for that purpose in what seamen call watch and watch.[2] The series lasted one month, that time being supposed sufficient to show every change of the moon. The Pendulum experiments were conducted solely by Captain Foster nobody being allowed even to get a look at it except per the most special favor and he had his own tent or Marquee pitched near the transit house for the purpose. In a small octagon house, which moved round in a circle with rollers, was placed the transit Telescope of a most beautiful kind and in another house, but rather larger, was the Astronomical Circle for ascertaining the latitude. These two latter were also for Captn.F's own use, no one ever touching them but himself.

While these obs^{ns} were carrying on on shore the 1st Lieut^{nt}. generally employed the men in exercising at the Musquet or else sent parties away to procure such Birds and Fish as were to be got. I ought to have mentioned that on our first arrival the Captain had thought it proper to place the officers and Ships Company on 2/3 allowance of all species of provisions, which in a cold climate where the stomach is tenacious of hunger was not the most agreeable thing in the world, and therefore as I before said, to drive 'ennui' from the sailors, parties were made up to fish or shoot or some picked celery for the daily consumption as vegetables which was regularly boiled with the salt meat. One of the ship's cutters with 4 men under the command of the Second lieut^{nt}. was despatched to make a complete survey of the Eastern end of the island, having done that to run in to Port Vancouver, leave the boat there and walk across the Haulover. He left the ship on the 30th of October and arrived at Port Vancouver on the 7th November having surveyed all the bays and harbours to the Eastward of Port Cook using the site of the Observatory at Port Cook as the first meridian, and having doubled the most Eastern cape called Cape St. John off which as I before said, such strong tide rips exist. He then left Port Vancouver on the 16th to survey to the Westward and see if possible the land of Terra del Fuego over the Straits of Le Maire. He returned from this cruize in good health on the 2nd December.

It may appear odd my having in some parts introduced dates and individual circumstances as they happened, but when a ship is laying in harbour for 6 weeks or 2 months at a stretch I find it becomes stupidly tedious to put down single occurrences every day as they happen and that my log book would answer quite well to my readers for that matter. I have therefore endeavoured to sum up all our occupations and employments in one mass, as necessarily in detailing the duty of a ship so much

---

[1] Six's (or Sixes) Thermometer is a mercury max.-min. thermometer, still today a popular way of registering extremes of temperature in the absence of an observer.
[2] Watch and watch: normally four hours on duty, and four hours off, although it may be that the period was different, see pp. 310–11 below, entry for 16 January 1829.

repetition and the same unwearied round goes on that it renders a journal at once uninteresting and annoying to peruse. Shut up as we are in an uninhabited place there is little for me to speak of except the productions of the place or the general character of it. There are no manners or customs of the inhabitants to treat of, no buildings to describe, no pic-nic or party of pleasure to relate and I am almost inclined to leave off writing. Besides, another drawback to writing a journal is the want of a general knowledge of Geology, and all the other parts connected with Natural History which it cannot be expected I am master of. I have most rigidly refrained from introducing into it any of the results of observations made during the voyage and being in the first place against the wishes of the Admiralty and in the next I could easily perceive that it was not wished that they should be generally known before the Admiralty themselves published them for the general information of the Public.* etc. etc. etc. etc.

*From Private Thoughts and Feelings. J. H. Kay.

During our stay the schooner sent her boats away for seals and returned with 20 or 30 each time which I believe was not considered good luck. I heard them say that of late years the seals were not near so numerous as formerly. Seal hunting is attended with some little difficulty and they will frequently, if on a rock overhanging or above you, let themselves fall on you and when you attack them they will frequently turn and show fight. They delight to bask in the sun on a rock and frequently herds of 40 or 50 are met with, with one as a sentry ready to give notice of a stranger's or intruder's approach. The men accustomed to go after them always go to leeward of them as their scent is so extremely good that they can smell any human being a long distance off and immediately that is the case they are off in the water instantly. They are killed by a good blow on the nose and die immediately. The seal has been so often described by other more able people than myself I shall not attempt to say any thing about it, and as I had hardly any opportunities of seeing them, my information must be chiefly gathered from other books or people's authority and not from my own actual observation.

Dec. 20th. The observations onshore were completed today and all the instruments embarked, the ship unmoored and every thing got ready for sea. The 2nd Lieutnt Mr. Kendall left the ship this evening at moon rise to deposit a cylinder with documents at Deadman's island for Captn. P. P. King of HMS Adventure according to previous arrangements made between him and Captain Foster. It stated that he had found it necessary to reduce us to 2/3 allowance of provisions and was then just starting for South Shetland calling at Cape Horn per via, and expected to be at Wigwam Cove, Cape Horn again in April 1829 at which time it was hoped the Adventure would come bringing us some provisions to recruit our scanty stock. He returned early the next morning having succeeded in depositing the cylinder on the most prominent part of the island and erected a flag staff to attract the Adventure's attention.

Latitude of the observatory   S Variation of the Compass $21°40'$E
Longitude   "   "   "   W Dip of the Needle $58°23'$S

Dec. 21st, 1828. We commenced working out of the harbour this afternoon at 1 o'clock but the wind blowing right in and what there was being very scant, were obliged to commence warping out by the kedge and continued kedging[1] without intermission all the night. We had by daylight on the 22nd so far succeeded as to be able to make sail to clear the land and then stretched out on our course to Cape Horn as fast as ever H.M. Sloop could carry us.

Dec.r 25th Xmas Day. We had a fine stiff breeze from the Westward and the wind seemed inclined to hang in that Q^r. At noon we hove to and spoke the American Whaler Pacific bound to Nantucket who had seen the Horn yesterday. The officers reefed the foretopsail this afternoon so that the men might not be disturbed it being Christmas day, and at dinner we all thought of 'those far away' and many a prayer was offered for the safety of our relations and friends and success of our arduous voyage with the most heartfelt acknowledgements of gratitude to that Almighty Being for having thus far preserved us.

Dec.r 27th. We made the Horn at daylight this morning and continued running along the land until we hauled up for the splendid bay of St. Francis.[2] Running strait up this bay we at noon bore up for Wigwam cove[3] and anchored. At 3, while we were in the act of shortening in to shift our berth, some of the most violent squalls arose without the least notice rushing down between the high land which formed a valley or kind of funnel as it were for it, and sweeping every thing before it. I happened with 2 or 3 messmates to be in the dingy to windward and by the most Providential escape (which was only effected by keeping the boat end on) happened just to fetch the ship who was laying with her whole broadside exposed to these squalls. Although we had not the smallest piece of canvass set it blew so hard as to heel her, three streak's[4] of her copper, and were obliged to veer to a whole cable, let go another anchor and be all ready for starting which every body expected would have been the case. Even the oldest seamen on board declared that the only thing they ever saw to any way equal it were the 'Tornadoes' on the coast of Africa. They were complete hurricane squalls which lasted themselves about 2 minutes and then for about 2 minutes more it would be a perfect calm, and then return to their former strength carrying the water or spray to the height of 80 or 120 feet. I could see there was just anxiety expressed by those who knew more about it than I did but as for myself I looked with a wild and vacant stare not knowing whether to be afraid or not or whether to take it in earnest or no. Had we missed the ship in the dingy we should undoubtedly have been lost or driven out to sea when the probabilities were against the boats living, and had the ship herself gone from her anchorage we should most likely all have met a watery grave not being able to show canvass to clear the numerous rocks and islands in the Bay. The Squalls continued without intermission until about ½ past 4 when it became beautifully fine and appeared as if nothing had happened.

---

[1] Kedging: To carry a kedge (relatively small) anchor away by boat, drop it at the full extent of its cable and then winch it from the ship, thus hauling the ship up to the anchor, then another anchor is carried forward, the first weighed, and the ship hauled up again, and so on until the required position is reached.
[2] Bahía San Francisco.
[3] Caleta San Martín (St Martin's Bay).
[4] Streak/strake: one breadth of a plank in a ship's hull.

28th Dec. We were employed getting water on board today. Squalls very strong from off the Hills. (The harb^r will be described next time we visit).

29th. Weighed at 1 P.M. Weather threatening. Having cleared the Horn we found as I had anticipated that it was blowing a very tremendous gale, and we only carried sail enough to clear the land. When we deemed ourselves sufficiently clear of the land we hove to. The weather was tremendous and it seemed as if Eolus[1] had summoned all his force this night and centred it in the West^{ly} wind which was then raging with all its greatest fury.

30th. The Gale has not at all abated in its violence and the sea is terrifically awful. Sailors are not generally speaking timid or fearful. They have habituated themselves to danger from their earliest infancy and from having accustomed themselves to view danger in many various forms and shapes they are not cowards in necessity. In no place can men witness God's wonders to a more awful extent than in the deep. The vast expanse of water that he views in so many forms, sometimes rising mountain high when excited or agitated by storms and winds, at another glassy smooth without the slightest ripple, all tend to inspire him with awe and veneration for the Being whom 'seas and winds obey'. Not that I mean to state that Sailors are more religious or more fervently believe in God than any one else, on the contrary, no where perhaps do you meet such varieties of character with regard to religion. Some are profligate and wicked in the extreme, others superstitiously foolish, and others real Christians, but they are all the same in appearance when dangers surround them from, as I said, having habituated themselves to it. Who can doubt but that the religious man feels most confidence in himself?

Dec. 31st/28 the gale is still unabated, and we have now been battened down below for 2 days with only fresh air at intervals. At ½ past 5 a tremendous sea carried away all the ship's head and starboard waist hammock netting. We find our stove of Mr. Fraser's Construction to answer most admirably not having any smoke below which is a great comfort. Our Lat^{de} 57°S. Long^{de} 66°W. This afternoon, a faint hope is entertained that the wind may expire with the Old Year – if not we shall spend a truly miserable and unhappy New Year's Day.

1st. Jan^{ry}. 1829 The wind has greatly moderated this morning and we have been able to make all sail. At noon we saw an American Ship standing to the Southward, apparently bound round the Horn. Lat. Noon 57°40'S. Long. 66°W.

2nd Jan^{ry}. 1829 An Iceberg (the first we had seen) was discovered to day apparently drifting to the Northw^d. We sent a thermo^{er} to the depth of 140 f^{ms}. below the surface of the water to try whether our being in the neighbourhood of ice had any effect on its temperature. It indicated at that depth the surface water being [ ] and air [ ]. This same afternoon we also sounded with 1170 fathoms of line from a boat lowered for the purpose but got no bottom. Lat. 59°S. Long. 65°W.

---

[1] Eolus, Aeolus: king of winds and storms in Greek mythology.

3rd Jan^{ry} Passed several more icebergs. Temp^e. [ ] air [ ] water.

4th Jan^{ry} Made and shortened sail occasionally in the fog to clear the icebergs. Lat. 62°S. Long. 64°W.

5th Jan^{ry} At Daylight this morning we saw Smith's Island. It was a miserably wet and cold morning and I thought I had never seen any place half so miserable as it appeared to be. It was black, barren and summits covered with snow. The wind was light with rain and a heavy swell from the recent gale caused us great uneasiness. In the afternoon a breeze sprung up from N.E. bringing on with it an intolerably severe snow storm. From an order of the Captain's, the Midshipman of the watch had to keep his watch to look out on the forecastle which came severe on the youngsters and I had the first watch. From the pitching and rolling of the vessel it was impossible to walk about on her little forecastle (without running an imminent risk of going overboard) to keep yourself warm and exposed as we were standing it was most piteously cold. My feet were fairly dead and I think I never was so truly miserable for the time, the snow beating down on you while perhaps you were anxiously looking at some iceberg and giving directions how to steer that you might avoid it. In this peculiar navigation it became necessary that great attention should be paid to the steerage of the ship among the numerous icebergs as one accident might have been the means of our having to spend the remainder of our lives there (however short they might have been). The ice was very thick all round us but luckily none was of the 'field' species and with little attention we went along perfectly safe.[1]

6th Jan^{ry}. The Wind is falling light and the snow has discontinued; numerous penguins are swimming about the ship in all directions. This evening we passed through 2 immense icebergs supposed to be 300 feet high and certainly the most splendid ones I have yet seen. They must I should think have been at least 1200 feet in depth as I judged that from the specific Gravity of that kind of blue and solid ice that 3 times its visible height must have been under water. The Supposed Southern Land is seen to the Southward this evening.

7th Jan^{ry}. The wind was quite light this morning and the boats were lowered to clear several icebergs. It was one of the most lovely mornings ever witnessed under the heavens and was a most complete relief after the weather we have had for some days back. Surrounded on all sides by snow-capped hills belonging to the Southern supposed continent, the sun shining on them gave a pleasing glare, to add to which immense quantities of whales were sporting about the ship, these huge monsters of the deep spouting columns of water and occasionally passing so close to us as to be able to completely view them to advantage. This afternoon we had so far neared the land as to be able to land and take possession of it in the name of His Most Gracious Majesty

---

[1] The nineteenth-century authority on the Arctic regions, Dr William Scoresby, defined a *field* as 'a continued sheet of ice, so large, that its boundaries cannot be seen from the summit of a ship's mast'. 'Some of them have been observed near a hundred miles in length, and more than half that breadth, each consisting of a single sheet of ice, having its surface raised in general four or six feet above the level of the water, and its base depressed to the depth of nearly twenty feet beneath': Scoresby, 'Greenland or Polar Ice', p. 264.

which was done by Capt[n]. F. and Mr. Kendall and they deposited a cylinder enclosing accounts of our proceedings written in several languages. They procured some pieces of rock [ ] and several limpet shells which were found adhering to the rocks of which I was fortunate enough to obtain some. Having completed his obs[ns] on its latitude and longitude and erected some prominent beacon to attract attention, in the evening he returned. The point on which he landed was called Cape Possession.[1] As far as the eye could reach land appeared stretching away to the Southward and I believe it is said that it has before been visited for produce but not finding it yield any was abandoned. The sun did not set till ½ past 9 this evening and I could see to read a book at midnight with the greatest ease. After the Captain's return we bore up for Deception Island.

Jan[ry]. 8th. We are navigating among numerous icebergs, shaving close bye some, giving others a wide berth, as circumstances best admit of. Great quantities of penguins are leaping about the low icebergs which are of a most brilliant blue color, many of which are evidently thawing fast away.
Temp. max. min. Water max. min. Bar.
Deception island in sight at noon. No land in sight to the Southward.
Lat. 63.20S Long. 61°.W.

In going along this afternoon with a light breeze and numerous detached pieces of ice all about, we attempted to give one (about half the size of the Brig's hull) a close shave, but having neared it considerably just as it was within 20 yards of us we attempted to sheer the ship to starboard but found she would not answer her helm. She went slap on top of it and by the little way she had on her split it in 2 halves, when the shock which it threw throughout the vessel brought every body on deck all apprehensive that great mischief was done. The Well was immediately sounded and the Stem examined when it was found that the only damage received was a small bruize on the cutwater and we went on as if nothing had happened. It rendered us more cautious for the future and no doubt was a good lesson to us.

Jan[ry] 9th. By this morning we had worked up off the mouth of the Harbour and the Captain left to examine its anchorage. In the evening it fell calm and the tide drifted us rapidly towards a large iceberg, one part of which was overhanging and appeared ready to fall at the slightest touch, but finding we could not avoid coming in contact with it we laid her right alongside of it by the boats, and then got what spars out we could possibly muster. By this means the brig was shoved ahead into clear water and fortunately the overhanging piece was just lofty enough to admit our mastheads underneath it or we should have all inevitably been crushed to pieces. This providential escape again caused me to reflect how Grateful we ought to be for the many narrow escapes in various manners that we have had. Nothing struck me more forcibly than the Comm[g]. Officer's (Mr. Austin) great coolness, and presence of mind in this most imminent danger. He gave all his orders towards the booming the ship off in the most deliberate manner and so fully convinced me of the foolishness when in command of evincing any fear in danger as it only puts every body out and may frequently be the

---

[1] Now identified as Cape Possession on Hoseason Island. Hattersley-Smith, *Place-names*, p. 21.

cause of more mischief arising than actually would otherwise have happened. At ½ past 9 the Captain returned and we sent boats ahead to tow the brig into the basin of Deception island.

Before going in to Deception island it may be as well to attempt to describe or give some account of the first discovery of the [South] Shetland islands. Captains Cook and Furneaux must have narrowly escaped seeing them in their voyage in 1773 and 74 as according to the best accounts they must have passed within 50 or 60 miles of their Eastern extremity and had they seen them we should have had a knowledge of them 50 years ago. In that voyage they penetrated as far as 72° South but a later Navigator, viz. Weddel, went to 74° with a clear sea at the time of returning.[1] The Archipelago called the Shetland islands were first accidentally discovered by a person named Smith in a trading voyage from Monte Video to Valparaiso having gone thus far South to make a fair wind round the Horn to the Westward. He succeeded in landing on them and found seals in abundance but (not liking to risk his cargo) he returned on his voyage merely having fixed their latitude and longitude to enable other navigators to find them. They extend from about 61° to 63° latitude and 54° to 63° longitude and consist I think of about 14 islands in all, Deception island being the most Southern island of all.

The Public are principally indebted to Mr. Weddell and Barnsfield [*sic* Bransfield] for any information they may have received concerning them, they having taken some pains to ascertain their extent and positions.[2] Their produce consists alone of Amphibious animals and from a calculation of Mr. Weddel's it was immense. The quantities of seals taken from these islands during the years 21 and 22 he estimates at 320,000 and by a similar law which in England restrains fishermen in the size of their net they might have yielded annually 100,000 furs for many years to come. But the fact was a system of extermination was practised and whenever a seal reached the beach of any denomination he was immediately killed and by this means in 2 years they became nearly extinct, and the young ones, losing their mothers before old enough to take the water, died. Thus the commercial avarice of some few (principally American) have rendered these islands unproductive, desolate and dreary.[3]

On entering the mouth of this magnificent basin on the left side you are assailed by the unceasing chatter of thousands of Penguins who have fixed their abode or rookery on the left side of the harbour and cause a most abominable stench from non-mentionable circumstances, making their nests by burrowing small circular holes in the soil large enough to admit their body for hatching their eggs. This Rookery covers an extent of about 3 to 400 yards all ways on the side of a hill, a reef of rocks extending directly underneath them across 1 half of the entrance and obliges you to keep the Starb^d. shore close aboard. The Starbd. side is of rocks 380 feet high under which a vessel gets becalmed and very steep to having 16 and 18 fms. close in. The entrance between the reef and shore is about [ ] wide and after that no danger need be

---

[1] See Weddell, *Voyage*.

[2] See Campbell (ed.), *Discovery of the South Shetland Islands*.

[3] Dr Webster commented as follows: 'The harvest of these seas has been so effectually reaped, that not a single fur-seal was seen by us during our visit to the South Shetland group; and although it is but a few years back since countless multitudes covered the shores, the ruthless spirit of barbarism slaughtered young and old alike, so as to destroy the race.' Webster, *Narrative of a Voyage*, II, p. 302.

Figure 7: 'Neptune's Bellows', at the entrance to Port Foster, Deception Island, South Shetland Islands. Taken in 1927, probably during the *Discovery* (Oceanographic) Expedition, 1925–27. Reproduced by permission of the Southampton Oceanography Centre. Photo 5160.

apprehended up to Pendulum Cove, which is situated in the [ ] corner, the place where the Chanticleer was moored, except that the usual precaution of not nearing the shore too much must be attended to. It was now Midnight, solemn midnight, and here was a really fine scene with broad daylight and water as smooth as glass. Here we are in an uninhabited and remote part of the world making our way up a harbour large enough to contain the largest fleet that ever swam, surrounded with lofty snow capped hills, every body happy at the prospect of getting on shore, on our own resources, and nobody to help us, which convinces me that true enjoyment at sea does not consist always in the expectation of what's to come. We have no letters or news to expect, no friends to see or nothing to look forward to except the time at which we shall quit these dreary regions. We anchored off the mouth of the cove at 2 A.M. and next day i.e. the 11th got her moored safe in the Cove called Pendulum in honor of the observations going to be made.

Jan 11th. On one side as per figure was a tolerably level plain extending from the base of some lofty hills to the water's edge founded on a firm foundation of ice covered to the depth of 4 feet with a mixture of ashes, stones and a little earth. Close to the water, smoke or rather steam was issuing in columns and on walking past, your feet and body would become sensibly warmer. By digging a small hole in the ground it would be immediately filled with hot water at the temp^e. of about [ ] and indeed would (had we possessed any) easily have boiled an egg. It may well be called Deception island, such an island of contrarieties is hardly credible, its shores reeking with steam within 2 yards of sea water at the temp^e. of [ ]; its hills emitting smoke and evidently volcanic; while at the same time immense masses of never melting ice are surrounding and clothing their very summits.

12th Jan^{ry}. Preparations were begun very early this morning (for commencing the observations as soon as possible) by erecting the necessary observatories etc, etc.

Figure 8: 'Port Foster' (inner basin of the island), with whale factory ship. Photograph taken in 1927 at Deception Island, South Shetland Islands, probably during the *Discovery* (Oceanographic) Expedition, 1925–27. Reproduced by permission of the Southampton Oceanography Centre. Photo 5162.

Figure 9: Another view of Port Foster from ashore, also with vessel. Photograph taken in 1927 at Deception Island, South Shetland Islands, probably during the *Discovery* (Oceanographic) Expedition, 1925–27. Reproduced by permission of the Southampton Oceanography Centre. Photo 5163.

Strong squalls with sleet and snow are frequent off the hills. Temp. Water [ ] air [ ].

13th. As yesterday, and securing the Ship head and stern.

14th. Mids and all their trapperies were landed to day to commence magnetizing as soon as they could. It was confoundedly cold sleeping under no other shelter than a canvass marquee and all the blankets I could muster were put in requisition, besides monkey jackets etc, etc. The tent at night was 34° but I managed to bring inside my bed up to blood heat I think.

15th. The same obs.ns are carrying on here as at Staten Island and the Pendulum is erected in the Square house brought out from England for the purpose well cased and covered with sails for protection. I really expect to see Captain, Mids, instruments, tents and all go flying over the adjacent hills some of these days, the squalls are so very strong and sudden. Temp. max. min. water.

Figure 10: 'Pendulum Cove, Deception Island (South Shetland)', drawn by E. N. Kendall and published in Webster, *Narrative of a Voyage*, I, facing p. 147.

16th. A party was landed from the ship this morning to dig a well for watering. The runs of snow water are plentiful all around, but what is more curious, during the short time we have been here, they have entirely altered the appearance of the plain on which we are now living, from the soil being extremely loose and therefore easily washed away by them. We are also apprehensive of a frost coming on as several indications of one have been made and then water is troublesome to procure. A great quantity of penguins are continually landing right under our tent and the poor simple things are so innocent that they suffer any one to approach them quite close looking at you as significantly as a human being and making as many bows and cringes as the best bred dandy. They even stand and suffer themselves to be knocked down with a stick which is frequently practised by every one and are much eaten by the seamen. I will myself be candid enough to say that I have frequently been the death of many of them but not through a wanton cruelty or delight of killing them. They are much akin to carrion in appearance and I dare say taste like it, but I acquit myself and every one else of cruelty; necessity and hunger obliging us to sacrifice them as we do, for what will not a hungry man accomplish to have the cravings of his appetite appeased. Consider there are 56 of us, and 2/3 allowance is not much in a cold climate. (I mention this, because at future periods of our stay here I have occasion to relate several other massacres).

The crew are employed generally as the weather admits. If it is fine, they have a run on shore, play at leap frog, some procure penguins for the daily consumption, others exercise firing at a target, others get water for the ships use, and as we have lower yards and topmasts struck there is not much to do. It may be as well to describe our employments on shore at the observatory. In the first place the obs.^{ns} on the Intensity of the Needle occupy us 25 minutes out of each hour day and night; the diurnal variation takes about 5 minutes more every hour and as there [are] only 4 of us we are in watch and watch day and night also, it requiring 2 to take the obs.^{ns}. At night instead of dividing it into 3 watches of 4 hours each = 12 hrs., we divide it into 2 of 6 hours, when those who have the first watch stay up till 2 A.M. the next morning, and then are relieved by the other 2 until 8 A.M. It is extremely tedious work and so very monotonous and stupid counting the vibrations of a needle for ½ an hour at a time. I have been thus minute in describing our employments, for the mere satisfaction of looking over it at some future period, and just to show to those who may chance to peruse my journal that we were not exactly idle. I think the daily consumption of penguins cannot be less than 10 or 15 a day, each on an average yielding 4 lbs. of solid matter.

27 Jan[ry]. The launch was sent this morning to one of the penguin Rookeries at the mouth of the harbour to procure some for salting as Sea stock. I went down in her to have a look at the island for mere curiosity and on the way shot several Albatrosses (Diomedea) and Eglets, Nellys[1] etc. whose craws I was very much astonished to find filled with shrimps which plainly showed that the water was not so unproductive as I had supposed. We landed amidst myriads of penguins all drawn up in battle array and our work of execution soon began, disdaining to shoot them, but using a small clubbed stick, some killing, while others loaded the boat. We massacred 500 thinking that would be enough for the present.

28th. We have been very fortunate indeed in our weather for (barring the cold) we have had it beautiful. Smoke is issuing from the top of the high snow capped mount behind us but can only be seen when it is quite clear. We fancy sometimes (and I believe it is reality) that subterraneous rumbling noises are heard as if proceeding from the very entrails of the adjacent hills; indeed it is such a strange place altogether that I can almost believe anything and the contrasts so opposite that nothing comes amiss.

31st Jan. To day they took one of the ship's 18 lb carronades to the head of the bay for the purpose of measuring an extensive base line for the survey of the basin which is going to be made by Mr. Kendall. It can be done by sound, which according to the best authority flies 1142 feet in a second, and therefore by two people in different places noting the times of the flash and report, a very correct distance may be estimated. Temp. max. min. Water. Barom[r]. (Memo. I find the bread is not near enough for us.)

2nd February. Mr. Kendall left in one of the Cutters to commence his survey of the basin and a party was despatched to fire the Gun for measuring his base line. The interval between the flash and report at the tent by the mean of several obs[ns] was 20 beats of a watch beating 10 times in 4 seconds therefore the distance between the tent and gun will be 9136 feet or 1 mile and ½.

They found within a few yards of the Gun a sea leopard[2] an amphibious animal particularly peculiar to these islands, and of which a small account may not be uninteresting. He was lying on the beach within a few yards of the water apparently asleep, and made very little resistance when struck, appearing to be most unwieldly animal on land, having nothing to propel himself forward with but his two flippers.

---

[1] Southern Giant Petrel, *Macronectes giganteus*.
[2] Sea leopard: Leopard seal (*Hydrurga leptonyx*).

Figure 11: Watercolour by E. N. Kendall entitled 'Sea-Leopards at Deception Island New South Shetland . . . January 7th, 1828' [*sic.* 1829]. The border is stamped 'Hydrog: Office, 9 Jy. 31', together with its seal of a foul anchor. Reproduced by permission of the Science Museum, South Kensington, Science and Society Picture Library.

Figure 12: Watercolour by E. N. Kendall of 'Sea Leopards at the Island of Deception', dated 9 January 1829, in a volume labelled on the outer cover, 'H.M.S. Chanticleer Com^r Henry Foster 1828. Remarks and observations', in the archives of the United Kingdom Hydrographic Office, reference OD39. Reproduced by kind permission.

The head was small in proportion to the rest of his body, which was spotted like the quadruped from which it derives its name.

Febr^y. 8th. I was much surprised on rising this morning after the 8 to 2 Watch to find ice had formed in this cove of a quarter of an inch thick during the night. The thermometer was as low as 24° and sea water congeals at 2° generally. I ascended one of the hills today to see whether ice had formed outside, and found great difficulty in ascending, owing to the looseness of the soil not allowing your feet to have any hold, besides the hill itself was very little out of a perpendicular, and had a fine view of the adjacent islands, but more particularly of Livingstons island whose very shores were encompassed with massive icebergs. The sea appeared to be much clearer of icebergs and ice than when we came in which I ascribe to the Northerly winds so prevalent of late having driven them all to the Southward. Indeed hardly any are to be seen. The temp^e in the hot springs was 150° to day and the beach was smoking greatly all along.

Feby. 9th. From the great havocs that our men have made among the penguins on this part of the harbour, the supply is become but scanty and therefore a cutter was despatched to day to replenish our present use stock. Quantities of birds of the following species are flying about our tents all day long. First, the Silvery Gull (Larus Argentea)[1] that really most elegant bird uttering a harsh scream but its plumage of a silvery white. The Eglet, the Nelly, so called by Sailors, but in reality the grey peterel, Penguins, a most delicate species of Tern with red beaks,[2] and a kind of white pigeon with a Parrots bill which latter is only found near the mouth of the Harbour.[3] The Nelly has such a predilection for blubber or any sort of oily food that I have frequently seen them so gorged as not to be able to fly. This unaccountable taste renders their flesh unpalatable, nor are their eggs so good as those birds who are not so voracious. The Eglet something resembles a young eagle in appearance and is a most voracious, ravenous bird, of a dark brown colour. I have also frequently seen them contending one with the other, who shall have the honor or pleasure of picking the entrails out of a Penguin just killed and yet warm left there perhaps by some one gone after some more. Indeed if a Sea leopard or Seal was killed and his carcase left on the beach, these gentlemen assisted by Mr. Nelly would sufficiently dissect him in a few hours. They remind me much of the Turkey buzzards in South America which are like Scavengers in London.

On the 10th Mr. Kendall returned from his survey of the interior basin and on the 12th left again to survey its exterior. Our weather was fine and wind moderate.

Max. air min. temp. water

13th Feby. The wind is blowing strong from the N.E. to day, and I am very anxious to see Mr. Kendall back. The cinders and dust are blown along like whirlwinds more than anything I ever saw and it is utterly impossible to face them as they come sweeping over

---

[1] Silvery gull: probably the Dolphin Gull, *Larus scoresbii*, whose distinctive call is noted in R. C. Murphy's classic work, *Oceanic Birds of South America* …, New York, 1936.

[2] Probably the Antarctic Tern, *Sterna vittata*.

[3] Probably the pale-faced sheathbill, *Chionis alba*, white, with a heavy bill, which behaves rather like a chicken.

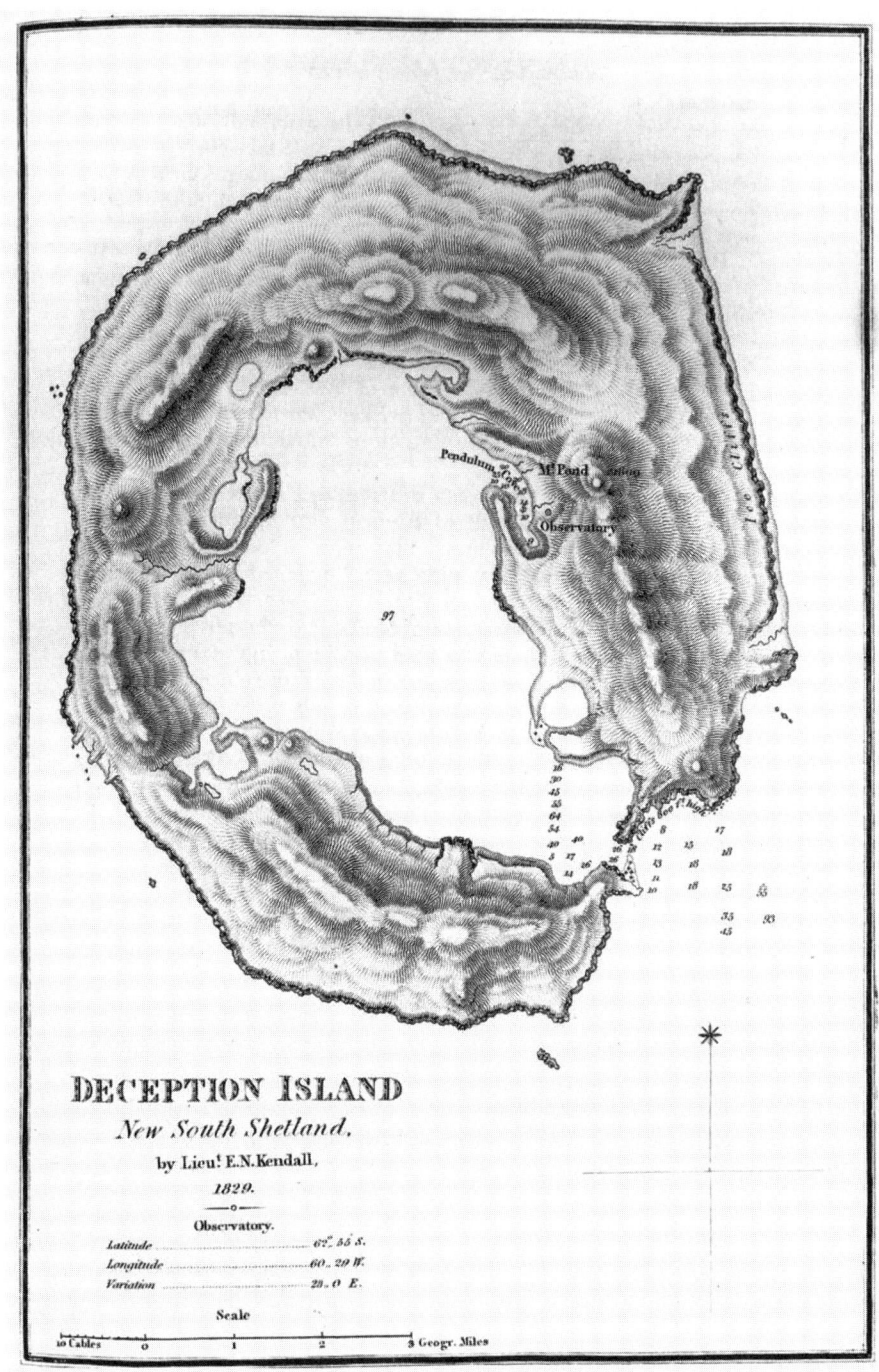

Map 2: 'Deception Island New South Shetland by Lieut. E. N. Kendall, 1829', published in the *Journal of the Royal Geographical Society*, 1, 1831 (London, 1833). Reproduced by permission of the Syndics of Cambridge University Library.

the Plain entirely covering everything with a dingy brown. Our Poor Canvass Marquee has quaked for its life several times and I really expect some day to find myself and messmates with nothing but the heavens for a covering. One of the Instrument Tents has been already blown down and the instrument damaged. Our provisions are so very scanty that we cannot afford to drink tea of an evening and must now content ourselves with fasting from 12 i.e. dinner time to 8 a.m. the next morning and therefore go supperless to bed.

14th. Mr. Kendall returned this evening having finished his survey of the island and reports dirty weather. Parties are employed all day long with the Purser, skinning and salting the Penguins for Sea Stock of which I believe 700 are going to be prepared.

During the 18th and 19th the wind was fresh from the N.E. with snow and sleet which made it to us in the tents very uncomfortable, indeed (except when duty forced us out) we very seldom quitted our blanket bags but lay on our beds talking to each other or reading as circumstances were. I had previously heard how accommodating and useful the blanket bags were with Captn. Franklin and his party and therefore determined on adopting them. In truth mine could not be called a bag, but bags, for I had 3 blankets sewn one within the other and used to 'turn in ' during the day 'all standing' except my shoes.

On the evening of the 19th the wind shifted suddenly to complete hurricane squalls from the S.E. and continued without intermission all night. Just when 'the iron tongue of Midnight had tolled Twelve' as I was going out with my watchmate Collinson, a squall took the Intensity Tent and as a Sailor said Dous'd it in a twinkling. I ran to save the Chronor and while doing that the same squall took all the others successively leaving none but our own and the Captain's Marquee standing. It was all the work of a moment and I half began to fear whether some such another might not take me into the water which at this season would have been far from pleasant. Having secured the instruments between us and carried them up to the Captain in his tent which was the only one we thought would stand owing to its superior fastenings, I ran down to hail the Ship and while waiting for assistance might well have repeated the lines of King Lear when he says in the tempest –

> Blow wind and crack your cheeks! rage! blow!
> Yon cataracts and hurricanoes spout
> Till you have doused our tents, drowned our persons …
> Etc. etc. etc. etc. etc.

The lines are applicable with slight alteration. I could not stand head to wind and the dust that the wind carried along with it was so cutting and sharp that my face soon became lacerated dreadfully and the snow beating down soon covered the ground to the thickness of several inches. Assistance having arrived in due time, the Pendulum house and 2 Marquees were secured with extra fastenings and at 3 a.m. I turned in to my blanket bag expecting to be blown out and therefore all ready for a start. Hungry, cold, and wet, I forgot all in a sweet sleep.

20th Feb^ry. Gale unabated and dreadfully cold and having no obs^ns to make we lay in bed till dinner time, not being able to get any breakfast as the snow had blocked up our little fire place and we could not get the fire to light. At noon it moderated and I ventured out of my bag, like a Mole from his hole, to see how the land lay, and succeeded in getting on board to dinner by dint of great industry, for the dingy, about 12 foot long, the boat that was appropriated for our use had been hauled up last night several yards from high water mark, in a place of great safety as we thought, but somehow or other (by the strangest circumstances I ever saw or heard of) the wind had fairly lifted her and blown her right across the cove, a distance of not less than 30 feet by land and ¼ mile by water. The same night our gig, a 22 foot boat hauled up on the beach to repair, was blown over and over. I think I do not exaggerate when I say that hundreds and hundreds of penguins have come up during the night and taken shelter in the vicinity of our tents from the inclemency of the weather, their black shining backs speckling the white snow all around. In the evening the wind gradually dwindled away to a calm and at night all around was deathlike silence so delightful, contrasted with the boisterous scene we had just encountered. It was quite time to be off for our series of observa^ns had been twice broken into and several instruments damaged among which was a valuable Repeating Circle[1] entirely rendered useless by a squall of wind having blown down the house in which it was contained.

26th Feby 1829 Myself and Collett ascended St. Georges Mount[2] today for the purpose of placing a flag on its summit which was evidently composed of one immense iceberg. We commenced the ascent at 8 over a rugged path never before visited by human beings, composed of cinders and ice intersected with streams of snow water which had began to thaw and rendered the walking extremely troublesome and by ½ past 9 had reached a more level part of the hill. In a few minutes more we reached the summit on which as we had conjectured an immense iceberg was formed covering an area of I should imagine 60 or 80 feet. All around the ground was perforated with holes from whence issued hot steam with a loud hissing noise and at such a temperature that we could not bear our hands near them. The Ground was also cracked in several places as if it had been done by the heat. Unfortunately just as we attained the summit a thick fog came gradually stealing over us and completely intercepted our view of the adjacent islands or no doubt it would have been fine as we were now at the height of [ ] feet above the level of the sea. Having stayed some minutes in hopes the fog would clear, and sufficient to warm ourselves, we commenced the ascent of the iceberg [glacier], grappling our way up through the loose snow with poles and advancing in the proportion of 2 foot ahead and 1 astern. Every step we were up to our knees and in about 10 minutes or so we had arrived at the part directly overhanging the observatory on which we stuck the flag and gave 3 cheers as we were now about as far from all human intercourse as we well could be. Had we possessed some grog we might have drunk His Majesty's health but as it was I took a mouthful of snow instead. We were

---

[1] The repeating circle is mentioned in the list of surveying instruments in Cambridge University Library (ref. RGO 14/49, f. 60^(r–v)).

[2] Today's Mount Pond, which appears on Kendall's chart of Deception Island, was called after John Pond (1767–1836), Astronomer Royal (1811–35). It is the highest point on Deception Island (540m. 1,772 feet). The point fixed by Kendall (and presumably the one climbed by Kay and Collett) and shown on his chart is the lower summit. Hattersley-Smith, *Place-names*, part 2.

favored by a partial clear just as we were so loyally drinking and were much gratified in discovering our tents just under us, looking like specks on the black ground, and being then quite content I seated myself down and soon arrived at its base, altho' the after part of my pantaloons suffered considerably by the expedition with which I descended. By 12.30 we reached the bottom and it having cleared, our flag was seen waving above the icy mass, quite proudly. The various observations which constituted the Chanticleers visit to these islands having been completed, the Survey by Lieut^{nt}. Kendall also being finished, great preparations were made for our starting from these dreary Regions to those of a more civilised and agreeable part of the world, where we might at least hear some news or some intelligence of our friends and country. The weather was fine and sea apparently very free from ice of which we had great reason to be most thankful, for with a gale of wind among this loose drifting ice our situation would have been most perilous. The island of Deception as far as regards description may be summed up in a very few words. It is a place of strange contrarities, it is agreeable, it is disagreeable, its scenery is beautiful in one sense, and horribly ugly in another, it is cold, it is hot, it is different from any other part of the world and yet in many aspects much the same, in short it is volcanic; perhaps as our men were heard to say 'it is the last place that Nature made', and if volcanoes are the habitations of demons and goblins, they may live there in perfect peace and quietness. The only vegetable production that was found by any of us was a little moss – most likely produced by dampness in the caves and holes where it was found. As for any other the island does not produce it and we could not therefore make a weekly demand for fresh provisions and vegetables.

March 1st. In one of our Rambles to day we found a pile of stones evidently put together artificially, and as our curiosity was naturally much excited we agreed on opening the place to see what it contained. After having cleared away the surface for 4 or 5 feet we discovered a coffin, rudely constructed and containing the body of a man in a red woollen shirt, in a high state of preservation, the features being very little distorted by time or decay, which evidently shews the general state of the atmosphere in these islands. I was sorry afterwards that we had thus abruptly disturbed it, but it was natural for one to be curious in such a place as this and a few yards further off there were evident symptoms of a house having been erected for the purpose of boiling the seal oil which was once procured here in great abundance. Several Hoops and Staves of Casks were found which fully convinced us that they must have been Americans from the marks on them.

2nd March. To shew the changeable nature of the soil or cinder of this island, a very good example was this day presented by the circumstance of one of our guns which had been landed for some experiments before mentioned, being totally missing, when sent for. It was well known that nobody was here to take it away nor was it probable that the wind could have blown it into the water, being well secured, (as was thought) and besides much too heavy for such an excursion. On digging to the depth of 3 feet below the surface all round its supposed position, it was at length found completely imbedded, and standing in exactly the same position as when last left, which can only be accounted for by the continual shifting off the cinders, which had at length covered it entirely. Indeed parts of the island have actually undergone 3 or 4 changes even during the short time we've been here. The Harbour or Cove that we are laying in has in some parts got shallower and who knows but in the space of a few years, the whole of this splendid basin which the island contains may be filled up also, new harbours formed, and the face and character of the island entirely changed. It appears likely enough to me!

March 4th. We attempted getting out to day and were actually going along with a fine fair breeze when all of a sudden the wind took us right aback and drove Chanticleer ashore on the Eastern bank. It was fortunately formed of that soft cindery stuff and moreover had 10 or 12 fathoms water close along side of it, so she lay as snug as possible until we underran the end of one of our warps which had been slipped in the morning and by 8, hove her into her former berth without the least damage. The wind was too baffling to make sail, and we after breakfast commenced warping out, which was continued without intermission until 5 P.M., when we again anchored for the night.

5th March. We got at our work early again this morning but the wind shifted and blew so strong that we were now on a dead lee shore and there seemed but little chance of saving the ship, except by getting the launch out and laying another anchor out to windward. Warp her we could not, for the bottom was so bad that nothing but a bower anchor would hold. By 3 the Launch was got out and we got the B.B. Anchor and cable into her, and laid it out in hopes of it holding her for the night; at Midnight, it blew in complete hurricanes with heavy snow.

6th March. The ship was most dangerously situated all night but the anchor held and she kept afloat. It having somewhat moderated towards noon we warped her further off shore, and made snug again for the night still indulged with the faint hope of rescuing ourselves from that most perilous place. Should the ice form in the harbour and block the entrance up it would be difficult to say what would become of us; we have no provisions to winter it out with; no fuel nor no prospect of getting any. Our situation will soon become desperate if we do not succeed in getting out.

7th March. The attempts at Warping out were made again to day, with as little prospects of success as ever. The wind still hanging strong and violently to the S.W. It really looks as if we should be obliged to stay here. At Midnight however the wind suddenly changed to the N.N.E. and of course our expectation is high and are waiting most patiently for daylight, for fear we should lose this merciful opportunity.

8th. Sunday. By 7 a.m. we were in the narrow entrance, when the wind took us right aback threatning the vessel with instant destruction on the neighbouring reef. The boats were instantly lowered but by the same good fortune which has so often attended us in all our adventures, the sails were again filled with a favoring air and she shot out triumphantly clear of danger. Such a Providential Escape could hardly be anticipated and I am sure the Chanticleer contained no one on board who was not glowing with thankfulness for all our hair breadth escapes. The Ship's head was then turned North or so and a course was shaped for our running between Smith's Island and [Snow Island].

March 8th. The night was dreary and dark and almost blowing a gale, but we were obliged to keep a press of canvass to extricate ourselves from the narrow channel in which we were running. Of course we expected nothing less than to meet a great quantity of ice, and a most vigilant look out was kept in consequence, as the sea ran tremendously high and a 'rencontre' with a piece of ice would have been extremely dangerous. Fortunately we saw none and the wind still holding, we cleared all.

In the Evening of the 10th the wind gradually increased and by the 11th at 8 a.m. we were brought under a close reefed main topsail with an increasing Sea. The Ship was so uneasy with the cross breaking sea that it was necessary to start [empty] some of our water to lighten her, and we rode it out till the next day when it moderated. We had nothing but a succession of heavy gales and strong Westerly Winds all the way up which together with a foul wind made it a most uncomfortable and dangerous passage. Had we not been in such a superlative vessel as the little Chanticleer we might have fared worse but as it was we made the Horn (that celebrated termination of the immense continent of South America so much formerly dreaded by Navigators, indeed Lord Anson's Sufferings were sufficient to make it a dread to all, without the dreadful relations of various storms and tempests which almost everyone has heard.

On the 24th March we kept in sight of it all night and the next day entered the splendid bay of St Francis (which is formed by the Hermite Islands)[1] and anchored in S^t. Martin's or Wigwam Cove by 5 P.M. It may not be generally known that Cape Horn itself is an island belonging to a set called the Hermite Islands off the extreme end of the Continent of South America and is a bold headland, round which you may run tolerably close taking care of 2 half tide rocks that are off it with generally a leading wind to take you up the bay of S^t. Francis. I believe from all authentic accounts, that this is w[h]ere the Wager, one of Lord Anson's Squadron was lost, she having run into the bay during a thick haze without seeing the land on either side of her.[2] There is no danger but what shews itself and the Cove may easily be found by a small island laying off its mouth which we called Chanticleer island.[3]

25th March. The anchor was hardly gone before a canoe full of Fuegian Indians came off hallooing and making all kinds of gestures and grimaces, going round and round the ship, minutely inspecting all parts of her exterior, but not approaching sufficiently close for communication. They appeared to be an extremely wild and savage set as far as regards feature, destitute of the most common clothing although the thermometer was then down at 36°, and their canoes were the most rudely constructed that have ever been described by even Capt^{ns} Parry or Franklin, being about 20 foot long I should imagine.

26th March. Our friends again visited us this morning and were persuaded to venture alongside by dint of signs and wonders, when some few particulars of them were collected. They consisted of 2 women, 2 men and 1 child, all naked and the women

---

[1] *Islas Hermite*
[2] HMS *Wager* was not lost near Cape Horn, but on the rocks of the small island later named after her in the Golfo de Peñas, south of Valdivia on the coast of Chile.
[3] Isla Chanticleer can still be found in the Bahía San Francisco on the Admiralty chart.

Figure 13: 'St. Martin's Cove near Cape Horn', drawn by E. N. Kendall and published in Webster, *Narrative of a Voyage*, I, facing p. 177.

extremely bulky and loosely built. The men were I should say of rather diminutive stature with extremely long hair and copper complexion appearing to take great delight in what was shewn them and minutely and attentively examining every thing which appeared of interest to them. A needle and thread was given to the women and they really understood the use of it remarkably well, but above all appeared highly delighted with the manner of threading it. The Child was triced up on the Woman's back and did not give her much trouble apparently. They unfortunately staied such a little time with us that more of their manners could not be collected as I only saw them once, and then only for a short time.[128]

17th April. After an immensity of labour, (which could only be conceived by those who were spectators and assistants) the miserable, wet, cold, place was prepared at the

---

[1] Dr Webster (*Narrative of a Voyage*, I, pp. 172–3 and pp. 175–85) provides a fuller account than Kay of the Yahgan Indians, whose habits he described as 'simple and inoffensive', with a demeanour 'modest and becoming' (p. 181), their riches consisting of 'a canoe, a dog, and their fishing tackle'. He describes the construction of their canoes and gives their length as no more than 9 feet long (pp. 183–4). One of their semi-circular huts is illustrated facing p. 177 in Lieut. E. N. Kendall's charming view of 'St. Martin's Cove near Cape Horn' (Fig. 13). More than fifty years after the *Chanticleer*'s visit, a monumental study of the Yahgan people inhabiting the islands south of Tierra del Fuego was published in 1891 as Tome VII (Anthropologie, Ethnographie) of the results of the French Mission scientifique de Cap Horn, 1882–3. Illustrated with many fine photographic plates, including handicrafts, weapons, a boat and the people themselves, the volume resulted from the field work of the surgeon, Paul-Daniel-Jules Hyadès and Dr J. Deniker of the Bibliothèque du Museum d'Histoire Naturelle. The authors had the advantage of the vast knowledge accumulated over the years at his Christian Mission at Ushuaia of the Reverend Thomas Bridges during the years spent by the expedition based at Bahía Orange, Isla Hoste. Webster's two-volume narrative of the voyage of the *Chanticleer* is listed in the bibliography of this work – pp. 393–402. For a resumé in English, see Barr, 'French Expedition', pp. 101–18.

head of the harbour for our observations but before a sufficiently solid and staple place could be made, it was necessary to dry and clear away, to drive staples and stakes, to cut dykes for draining the water off; in fact to change the whole face of the place, for such a collection of muck and marsh could not be found again as was overcome by the Chanti boys at this place. The deposit of years of rain was entirely saturating the ground, and as I say all this had to be done besides building a pier of spars for landing on, as the surf was so heavy that we could not land the Pendulums without. We laboured under a great disadvantage for our men were badly off for shoes, and really as for myself I hadn't a pair to my feet. The rain fell incessantly and the intervals of fine weather were succeeded with sleet and hail, snow drift from the tops of the hills, and squalls of wind hurricane like rushing with overwhelming impetuosity through the channel formed by the adjacent hills. However every thing was to be overcome by the zeal and activity that was displayed and it was by the 9th that we got on shore to commence our observations on magnetism.

12th April. I ought to have mentioned that the Indians left just after our arrival apparently much frightened at some Rockets we 'd fired as signals to a boat that was away late with the Captain; for they fled in the woods, and did not come back again for 1 day when they immediately got into their canoes and went out of the harbour. As I have given the details of our observations before it would be useless for me to again relate them, but most can easily imagine our situation when we were living in a canvass marquee in a climate where it rained incessantly. We had to visit the observation tents every hour during the night and day and really, (had it not been so uncomfortable in this cold and eternally wet climate) it would have been laughable to see us go sprawling half a dozen times among the mud and snow before we could reach the tent. When other duties did not occupy us we used to cut wood for keeping our fire in, and commenced building a good stout house of trees to windward of our tent well wattled with bushes. Another amusement (or rather necessity) was go into the boat among the kelp and catch fish for our dinner, of which sometimes several dozen might be caught in a morning, a hint which we received from the Indians who used to catch them by enticing them to the top of the water with a small piece of seal's fat or anything and then as dexterously quick as lightning put out their hands and seize hold of them. The usual employment of the men was catching fish when the weather permitted which was indeed seldom. There was no level spot of ground there 3 yards in circumference on which any one could walk, and the celery was again picked and eaten as before. The Musquet exercise, singing or dancing or any thing was invented to keep them in health, extra fire was kept on the lower deck, and every invention for amusement was called forth in hopes of keeping the scurvy and 'ennui' away from them. On shore we were always wet and worse than all had no shoes. It was really a most abominable place.

13th April 1829. The 2/3 allowance of provisions was again commenced today and all contrivances were in vogue to obtain something to eat. The berries of the Arbutus Aculeata were much sought for although few were to be got and I used to go into the woods after dinner, and search for them by way of dessert. Parties were appointed to cut wood and make charcoal, some to fish, some to pick celery and berries and all had an employment. My gun was a constant companion, in search of something for the 'pot' for these were hungry times, and although he had suffered severely in 2 or 3 late excursions (in one of which I broke the stock) yet by dint of cobling I got him to 'gee' very well. I remember well with what exulting glee I used to go down to the rock close bye when a kelp goose or duck was reported in sight, in hopes of getting him for dinner, creeping along ready to let fly at the first intimation of a start.

15th April. There did not seem to be the slightest chance of having any fine weather, and the celestial observations requisite for the Pendulum experiments w[e]re sadly interupted.[1] The hurricane squalls of wind have been exceedingly strong and our poor tent was blown about our ears threatning to level it with the ground every instant.

17th April. This was a most glorious day in the Annals of the Chanticleer, for at noon a square rigged vessel hove in sight round the point, and signals, boats, hawsers and anchors, were instantly alive to help our long wanted and wished for relief, HMS Adventure, into the harbour.[2] The most sudden revolution in affairs took place consequently (almost as much as any change in prices during a famine would excite in England) and plenty to eat was I'm sure thought of by all, barring all the news we had to devour and con over. Our gratitude to the generosity of Captain King was naturally great for his coming down with provisions for us so strictly to the request left at Staten Island and all except the observations (which nothing ought to disarrange) was left aside to enjoy ourselves after our weeks of trouble.

18th. Little parties of joviality and friendship were soon formed with each other, and the disinterested kindness with which every Chanticleer was treated by the Adventures will long be remembered with feelings of gratitude and thankfulness. The Mids, with Mids, Lieut^{nts}. With Lieut^{nts}. and Captain with Captain, all fell together as if we/they had known each other for years. Every thing they had was offered. The Mids supplied the Mids with part of their own stock and appeared offended if we did not take it, even sent them all and wanted to see us every day. 'Pendulum town' as our observatory was quizzingly called was frequently a place of great glee, and our canvass walls almost re-echoed the shouts and songs of mirth with which they were frequently filled. Even the badness of the weather appeared more reconciled; the miserable cold nights passed off well. The whole face of the country appeared almost as if more agreeable, and we might almost have been sorry when our time came for parting.

25th. The same round went on more pleasant than ever until the end of the month, and Cape Horn never contained 2 more agreeable ships than H.M.Ships Adventure and Chanticleer. The weather was still however very unfavourable for all celestial observations and we were kept here much longer in the faint hope of it still mending. The General character of Cape Horn and the adjacent islands was much as that I have described at Staten Island, exceedingly fertile and clothed with evergreen woods of the

---

[1] Celestial observations: transit observations of stars are needed to establish the exact local time when making pendulum and magnetic observations.

[2] Commanded by Captain Phillip Parker King, FRS who had already surveyed much of the coast of Australia from 1817 to 1821, as a result of which he wrote a *Narrative of the Survey of the Intertropical and Western Coasts of Australia* (1827) and compiled an atlas. In HMS *Adventure* from 1826 to 1830, in company with HMS *Beagle*, Commander Pringle Stokes, 1826–8 and Commander Robert FitzRoy, 1828–30, he was engaged in surveying Tierra del Fuego and other southern shores of South America.

'Fagus antarctica' among which quantities of the beautiful little bush, 'Arbutus aculeata' and various other shrubs were prominent. Among the marshes and by the sides of the streams of water, a kind of rush grew of which baskets exceedingly strong were made by the Fuegians, and appeared to be highly useful in various ways, and with the view of trying to get it propagated in England, many of the seeds were collected by Mr Webster, our surgeon.

The Shores of the Hermite Islands are extremely rugged and ironbound and mostly difficult of landing. St Francis bay is formed between the different islands and a heavy swell generally sets into it, from its being open to the whole swell of the Southern Atlantic, but nevertheless contains two safe harbours for any vessel, whose approach are so easy that on making the Horn, steer for its western side, when the opening of the bay will soon be distinguished. You may safely run along the Western shore, and Wigwam, or S^t. Martin's Cove will soon be distinguished by a small island laying off its entrance. Round it at a proper distance and anchor when the South Head or point of land shuts in Cape Horn. The heavy squalls which I have described require a vessel to be well secured but in other respects the anchorage is safe. The second Anchorage is Maxwells harbour,[1] a little further up but it was not visited. The island called the Horn[2] is a very prominent one and generally makes at sea like a continuation of the whole Continent. It is easily distinguished from Cape Spencer by its remarkable appearance, it forming the East^n point of S^t. Francis bay and Cape Spencer[3] the Western. The prevalent winds are those before described, and no doubt but the Month of March, the time of the Westerly winds, is the worst time possible to attempt rounding the Horn but that it has been described in blacker terms than just by some people I think is most probable, while others who have been successful, speak vice-versa.

1st May. Another Month came, and with it as little hopes of fine weather as ever. The climate appeared utterly hopeless of change, for it was bleak and wet, and our situation was getting worse than ever for the snow was frequent, and still more frequent the thaws, which made our little Pendulum town abominably bad. It was just that miserable muggy weather that makes every one 'half dead and alive' accompanied with rain and sleet, the ground too wet for exercise and our poor tents soaked all day long, with an accumulation of thawing snow and mud in all our footpaths, through which every hour the 'Redoutable observers of Magnetism' had to wade will or no. Our friends the Adventures still remained with us and kept us alive and merry and I had frequent marks of kindness from Captn. King of whom I had had a previous knowledge, and whose great kindness to my feet and stomach will long be remembered by my grateful memory.

The meteorological Register kept on shore gave for the max. temp during the month of April min [ ] mean [ ] The max Bar^r. was [ ] and min [ ]. Deposit of rain in a pluviameter was [ ] and evaporation [ ] From these Results it would appear as a question whether is the Southern Hemisphere −10 degrees comparatively colder than the Northern, as it is generally supposed to be, or what proofs can those who

---

[1] Puerto Maxwell.
[2] Isla Hornos.
[3] Cabo Spencer.

assert it, bring to support their argument. Take for supposition, parts of the Island of Newfoundland, or Equal latitudes in both Hemispheres and compare the quantities of ice found in the Northern at the Lat^{des} of 55, 56, 57, 58 and all the way up to 85 or as far as our Polar navigators have reached. Is there any ice at Cape Horn or its vicinity in Lat^{de}. 56°? I query much or should we not have seen symptoms. In the Chanticleer we did not see any till we reached 61° and that only a drift berg, and Weddell was to 74° with a then clear sea at the time of his bearing up. Captain Cook does not mention ice off Cape Horn and from all accounts to the Lat^{de}. of 60°. the Southern Atlantic is free from ice. The Southern land is clothed with ice I admit, but so is the Northern. Whether then at the same Latitude (if ice is a symptom of cold) is apparently coldest? They may say the Northern ice is blown down from Higher Latitudes, but on many parts of Newfoundland and North America in that latitude it is regularly formed, and found and seen forming. As long as these undubitable proofs remain I shall not admit of this Contrariety in the System of Nature, as I have been an eye witness of one half and have excellent authority for the other. I would like to know the reasons which set this absurd notion afloat!

22nd May. The Last and parting meeting was held this night and each one felt partially sad at parting, for now we had become entirely associated together. Mirth, Songs and Glee went [on] till a late hour each hoping to meet in another ship or at least at some future period. The Pendulum and all the Instruments were embarked also in the afternoon and 'Chanti' prepared for a long run across the Atlantic to the other vast Continent, which perhaps might have sounded awful but we were so full of getting news and intelligence from home that it was thought lightly of. At 11 we left the Adventure under a heavy salute of cheers and goodbyes answering in Return with all our Stentorian lungs, shouting like so many Young Bulls. There is in my idea a pleasing feeling and such heartfelt expressions of mutual good will to each other, and now under these circumstances in a dark and stilly night, 'ere slumber's chain had bound us' it was particularly cheering and enlivening. We had all met under peculiarity of circumstances, theirs was the pleasure of affording assistance to fellow creatures, ours the gratitude of receiving such assistance, and we parted in the same happy manner perhaps never to meet again. That fortune might favour them was not silently wished from the people on board HBM Sloop Chanticleer.

23rd May. We left our anchorage early this morning under a strong double salute of cheers in passing HMS *Adventure*, we being bound for the Cape, she for Valparaiso. Having cleared the land the voyage began.

(End of the Southern Cruze)

BIBLIOGRAPHY

The Antarctic Pilot. Fourth edition, Taunton, 1974.
Baily, F., 'A Report on Experiments Made by the Late Captain Henry Foster, R.N., with the Pendulums Taken out by Him, in his Scientific Voyage, in the Years 1828–31', *Memoirs of the Royal Astronomical Society*, 7, 1834, pp. 1–378.
Barr, W., 'The French Expedition of the First International Polar Year to Cabo de Hornos', *Inter-Nord* (Paris), 19, 1990, pp. 101–18
Beechey, F. W., *Narrative of a Voyage to the Pacific and Beering's Strait ... in the Years 1825, 26, 27, 28*, 2 vols, London, 1831.
Bertrand, K. J., *Americans in Antarctica, 1775–1948*, New York, 1971.
Black, W., *Cruises in the Mediterranean of H.M.S. Chanticleer During the Greek War of 1824–26*, Edinburgh, 1900.
Campbell, R. J., ed., *The Discovery of the South Shetland Islands: the Voyages of the Brig Williams 1819–1820 ... and the Journal of Midshipman C. W. Poynter*, Hakluyt Society, 3rd series, 4, London, 2000.
Christie, S. H., 'The Direction of the Terrestrial Magnetic Force', *Report of the British Association for the Advancement of Science, 1833*, London, 1834, pp. 106–30.
Collinson, R., *Journal of H.M.S. 'Enterprise' on the Expedition in Search of Sir John Franklin's Ships by Behring Strait 1850–55*, London, 1889.
Daniell, J. F., *Meteorological Essays and Observations*, London, 1823.
FitzRoy, R. (ed.), *Narrative of the Surveying Voyages of His Majesty's Ships* Adventure *and* Beagle *... in Three Volumes*, London, 1839. Reprinted New York, 1966.
Foster, Henry, 'Account of Experiments Made with an Invariable Pendulum at the Royal Observatory, Greenwich, and at Port Bowen, on the Eastern Side of Prince Regent's Inlet, under Captain W. E. Parry, 1823–4', *Philosophical Transactions of the Royal Society*, 116, pt. 4, 1826, pp. 1–70.
Sir John Franklin's Journals and Correspondence: the Second Arctic Land Expedition 1825–27, ed. Richard C. Davis, Champlain Society, Toronto, 1998.
Graham, G. S. and Humphreys, R. A., eds, *The Navy and South America 1807–1823: Correspondence of the Commanders-in-Chief on the South American Station*, London, 1962.
Hall, B., 'Letter from Captain Basil Hall to Captain Kater, communicating the details of experiments made by him and Mr. Henry Foster, with an invariable pendulum; in London; at the Galapagos Islands in the Pacific; near the Equator; at San Blas de California on the N.W. coast of Mexico; and at Rio de Janeiro in Brazil. With an Appendix containing the second series of experiments in London, on the return', *Philosophical Transactions of the Royal Society*, 113, 1823, pp. 211–88.
Hall, B., *Fragments of Voyages and Travels*, 2nd series, 2 vols, Edinburgh, 1832.
Hattersley-Smith, G., *Place-names in the British Antarctic Territory*, Cambridge, 1991.

Hyadès, P. and Deniker, J., *Anthropologie, Ethnographie*, Paris, 1891. (*Mission scientifique du Cap Horn*, Tome VII.)

Jackson, George, *Tales of Woodplumpton and Kirkham*, Preston, privately printed, n.d. [c. 1975].

Jarrett, D., *British Naval Dress*. London, 1960.

Jones, A. G. E., 'Dr. W. H. B. Webster, 1793–1875: Antarctic Scientist', *Polar Record*, 17, 1974, pp. 143–5.

Kater, H., 'An Account of Experiments for Determining the Length of the Pendulum Vibrating Seconds in the Latitude of London', *Philosophical Transactions of the Royal Society*, 108, 1818, pp. 33–102.

Kater, H., 'An Account of Experiments to Determine the Variation in the Length of the Pendulum Vibrating Seconds, at the Principal Stations of the Trigonometrical Survey of Great Britain', *Philosophical Transactions of the Royal Society*, 109, 1819, pp. 337–508.

Kemp, Peter (ed.), *Oxford Companion to Ships and the Sea*, London, 1979.

Kendall, E. N., 'Account of the Island of Deception ...', *Journal of the Royal Geographical Society*, 1, 1831, pp. 65–6, chart.

King, P. P., *Narrative of the Survey of the Intertropical and Western Coasts of Australia*, London, 1827.

Laing, E. A. M., 'The Introduction of Canned Food into the Royal Navy, 1811–1852', *Mariner's Mirror*, 5, 1964, pp. 146–8

Laughton, L. G. C., 'H.M.S. Victory: report to the Victory Technical Committee of a Search Among the Admiralty records', *Mariner's Mirror*, 10, 1924, pp. 173–211.

Lyon, D. and Winfield, R., *The Sail and Steam Navy List*, London, 2004.

Moore, David M., *Flora of Tierra del Fuego*, Nelson, England, 1983.

O'Byrne, W. R., *Naval Biographical Dictionary*, London, 1849.

Phipps, Constantine John, *A Voyage Towards the North Pole Undertaken by His Majesty's Command, 1773*, London, 1774.

The Registers of St Michael on Wyre, 1707–1765 and of Copp Chapel, 1728–1837, transcribed by Albert J. Clayton, Lancashire Parish Register Society, Newport, 1998.

Richer, J., 'Observations astronomiques et physiques faites en l'isle de Caïenne', *Mémoires de l'Académie Royale des Sciences*, 7, 1679, pp. 231–326

Ritchie, G. S., *The Admiralty Chart: British Naval Hydrography in the Nineteenth Century*, New Edition, Edinburgh, 1995.

Sabine, E., *Report of the British Association for the Advancement of Science, 1837*, London, 1838, pp. 1–85.

Savours, A., *The Search for the North West Passage*, London, 1999.

Savours, A., '"A Very Interesting Point in Geography": the Phipps Expedition Towards the North Pole', *Arctic*, 37, 1984, pp. 402–28

Savours, Ann and Deacon, Margaret, 'Nutritional Aspects of the British Arctic (Nares) Expedition of 1875–76 and its Predecessors', in James Watt et al., eds, *Starving Sailors*, London, 1981, pp. 131–62, 203–5.

Savours, A. and McConnell, A., 'The History of the Rossbank Observatory, Tasmania', *Annals of Science*, 39, 1982, pp. 527–64.

Savours, A. and McConnell, A., 'Return to Rossbank: Magnetism and Meteorology at Hobart in Theory and Practice, 1840–54', in J. Kenworthy and M. Walker, eds, *Colonial Observatories and Observations*, Durham University, 1997, pp. 49–58.

Scoresby, W. jr., 'On the Greenland or Polar Ice', *Memoirs of the Wernerian Society*, 2, 1815. Reprinted Whitby, 1980, pp. 328–36.

Spears, J. R., *Captain Nathaniel Brown Palmer*, New York, 1922.

Vale, B., *Independence or Death: British Sailors and Brazilian Independence, 1822–1825*, London, 1995.

Vale, B., *A War Betwixt Englishmen: Brazil Against Argentina on the River Plate 1825–1830*, London, 2000.

Vale, B., *A Frigate of King George: Life and Duty on a British Man-of-War, 1807–1829*, London, 2001.

Warner, B., *Royal Observatory, Cape of Good Hope 1820–1831, the Founding of a Colonial Observatory*, Dordrecht and London, 1995.

Webster, W. H. B., *Narrative of a Voyage to the Southern Atlantic Ocean in the years 1828, 29, 30, performed in H.M. Sloop Chanticleer, Captain Henry Foster, FRS*, 2 vols, London, 1834.

Weddell, J., *A Voyage Towards the South Pole*, 2nd edn, London, 1827.

PART 4

'A DANGEROUS AND TOILSOME JOURNEY'. JACOB WAINWRIGHT'S DIARY OF THE TRANSPORTATION OF DR LIVINGSTONE'S BODY TO THE COAST, 4 MAY 1873–18 FEBRUARY 1874

Translated from the German and edited by
ROY BRIDGES

CONTENTS

List of Illustrations and Map	332
Acknowledgements	333
INTRODUCTION	335
1. The Nature and Importance of the Text and Comparable Accounts	335
2. The History of the Text	339
3. The Character and Career of Jacob Wainwright	343
4. East and Central Africa in 1873–1874	348
5. The European Position in East and Central Africa in 1873–1874	352
6. How Good was Jacob Wainwright's English?	355
7. Notes on the Translation	356
8. Establishing the Route	358
THE EDITED TEXT OF JACOB WAINWRIGHT'S 'DIARY'	361
Bibliography	382
Index	400

LIST OF ILLUSTRATIONS AND MAP

Figure 1: Jacob Wainwright from a photograph taken in 1874. Reproduced by permission of the National Library of Scotland. — 336

Figure 2: Jacob Wainwright on board the SS *Malwa* with Livingstone's coffin and trunks. Reproduced by permission of the National Trust for Scotland. — 344

Figure 3: Wainwright follows Livingstone's coffin as it is landed at Southampton. Reproduced from the *Illustrated London News*. — 345

Figure 4: Wainwright as a pall-bearer at Livingstone's funeral in Westminster Abbey. Reproduced from the *Illustrated London News*. — 347

Figure 5: Carving by C d'O. Pilkington Jackson, symbolically representing the journey to the coast. Reproduced by permission of the National Trust for Scotland. — 354

Figure 6: A specimen of Wainwright's handwriting from a letter to the Revd Joseph Moore. Reproduced by permission of the National Library of Scotland. — 357

Figure 7: The mupundu or mbura tree beneath which Livingstone's heart was buried. Photographed about twenty years later. Reproduced by permission of the Royal Geographical Society. — 364

Figure 8: The carving made on the mbura tree by Wainwright. Reproduced by permission of the Royal Geographical Society. — 366

Figure 9: Wainwright's letter to Captain Cameron. Reproduced from Cameron, *Across Africa*. — 374

Map 1: Showing Wainwright's route and the principal peoples encountered; with an inset showing Lake Bangweulu as depicted in Waller's *Last Journals of Livingstone*. — 362

ACKNOWLEDGEMENTS

I am grateful to the Council of the Hakluyt Society for agreeing to publish this edition of Wainwright's 'Diary' which has been overlooked, at least in the English-speaking world, for over 130 years. In particular, I would like to acknowledge the advice and help of the Series Editors, Professors Will Ryan and Michael Brennan. Professor Brennan has had the major responsibility for this volume and I have benefited from his patient guidance. One of the Society's Vice-presidents, Dr John Bockstoce, drew the attention of Professor Brennan and myself to the fact that two letters of Wainwright's had been offered for sale at auction. Recognizing the importance of the letters in connection with its existing collection of materials relating to David Livingstone, the National Library of Scotland secured the letters and a *carte de visite* of Wainwright's. All this was achieved through the good offices of Dr Iain Brown, Principal Curator of Manuscripts at the Library, who, together with his colleagues, afforded me help in securing copies of the materials. I acknowledge with gratitude the permission extended to me by the National Library to quote from as well as to reproduce as Figs 1 and 6 in this edition part of one letter together with the *carte de visite* photograph of Wainwright. At the Royal Geographical Society, I received kind help from Ms Pauline Hubner of the Society's Picture Library and I acknowledge the permission of the Society to reproduce Figs 7 and 8 of which it holds the copyright. At the Scottish National Memorial to David Livingstone at Blantyre, Ms Karen Carruthers was interested and helpful and I acknowledge the permission of the National Trust for Scotland who now manage the Memorial, to reproduce Figs 2 and 5. The map showing Wainwright's route was professionally drawn for me by Mrs Julie Snook who most patiently and efficiently interpreted my wishes.

The cost of reproducing the illustrations which accompany this edition of Wainwright's 'Diary' has been met in large part through grants from the Scottish Institute of Missionary Studies, whose Director, Professor Andrew Walls, was personally generous with his help, and from Brian Bridges Esq. of Townsend House, Ullingswick, Herefordshire.

As the version of Wainwright's text nearest to the original was a translation into German, its retranslation into English was a matter of considerable importance and I have to acknowledge my great debt in this respect to Mr Frank Gatter, sometime member of Hamburg University, who aided me many years ago by producing a preliminary draft. More recently, Dr André Reibig has given me the benefit of his advice and been very helpful over certain technical points. Neither of these scholars, however, is responsible for any mistakes or misapprehensions; as is explained more fully below, I take full responsibility for the translation offered here.

The Introduction and notes to the text will make apparent my academic debts. First, I must acknowledge the work of Professor François Bontinck who has not only pioneered studies of the Africans associated with European expeditions in Africa in the nineteenth century but also produced his own version in French of the Wainwright text. Although I have ventured to disagree with him over certain minor points, Professor Bontinck's work has been of immense value to me. The late Donald Simpson, who was Librarian of the Royal Commonwealth Society, was the first scholar in Britain to make serious studies of Africans associated with exploration; he drew my attention to many important sources of information. My colleague and friend Dr Jeffrey Stone has not only provided me with invaluable advice on maps but also guided me in many matters connected with the history and geography of the parts of present-day Zambia through which Wainwright and his companions passed. I also acknowledge the help of Mr Kenneth Chilles of Inverurie, Mrs Jackie Brown of the Map Control Room at St Mary's in the University of Aberdeen and of the staff of the University's Library.

My more general academic debts will appear from the notes and Bibliography. I believe that the pioneer work which historians carried out in the 1960s and 1970s on the history of Africa was a remarkable development. In this connection, as far as this edition is concerned, it will be seen that I owe much to the work of Professor Andrew Roberts, formerly of the School of Oriental and African Studies in the University of London; it is difficult to see how his work, especially on the Bemba, can ever be surpassed. The rather different studies of nineteenth-century East African developments carried out by Professor Norman Bennett, formerly of Boston University, are also outstanding.

Perhaps my most important academic acknowledgement should be to Jacob Wainwright himself, and the leaders of his party, James Chuma and Susi Amoda. They it was who made it possible for the records of the last parts of David Livingstone's travels to be saved for future study as well as, in Wainwright's case, producing the rather remarkable text which is the subject of this short study. I believe this to be the first Hakluyt Society text by an African.

Finally, I am grateful as ever for the support, counsel and encouragement of Jill, my wife.

Roy Bridges
Newmachar, Aberdeenshire

INTRODUCTION

1. The Nature and Importance of the Text and Comparable Accounts

'A dangerous and toilsome journey'[1] was the phrase the young freed slave Jacob Wainwright used to describe the expedition which is the subject of this study. After the death of Dr Livingstone, on or about 1 May 1873[2] at Ilala in what is now northern Zambia, his leading followers eviscerated and roughly embalmed his body. They resolved, with the consent of the whole party of some fifty-six porters and escorts, to carry the corpse approximately one thousand miles back to the coast and Zanzibar. This was an act both of devotion and of prudence – devotion because they did admire Livingstone despite his vagaries, and prudence because they knew they must be able to prove that they had not deserted the great explorer.[3] The venture took ten months. Jacob Wainwright's diary is the only direct first-hand record of this journey and he brought it to Britain with him when he accompanied Livingstone's coffin on the voyage from Zanzibar and attended the funeral in Westminster Abbey as a pall-bearer. Unfortunately, the original diary has disappeared but a German translation was made in 1874 and it is this version which is retranslated and printed here.

In total, five accounts of Livingstone's last days and death exist. The best known and the one hitherto regarded as standard is that provided by the undoubted leaders of the group of followers, David Abdallah Susi[4] and James Chuma.[5] Livingstone's old friend James Young arranged and paid for them to be brought to Britain in May 1874, a month after Wainwright, and they were interviewed at length by Horace Waller who later became the editor of Livingstone's *Last Journals*. As a result, Waller produced the account of the journey to the coast which occupies the last chapter.[6] Although this is reasonably systematic in its list of places, no dates are provided. This version of the

---

[1] National Library of Scotland, [hereafter NLS] Acc. 12444, Wainwright to Moore, 23 May 1874. Joseph Moore (1816–97) was a London Missionary Society [hereafter LMS] missionary who trained with Livingstone and remained his friend. Moore served in Tahiti and was later pastor of a church in Congleton for forty years. Clendennen and Cunningham, *Catalogue*, p. 174.

[2] The precise date is uncertain but Wainwright believed it to be 4 May 1873. See also n. 14, below.

[3] One group who had been recruited from Johanna in the Comoros in 1866 deserted and claimed Livingstone was dead to justify their reappearance at the coast; their leader was imprisoned in Johanna. Coupland, *Last Journey*, p. 66.

[4] Susi (c. 1831–91) first entered Livingstone's service in 1861 and was with him almost constantly in the periods 1861–4 and 1866–73. See Bontinck, 'Voyageurs africaines; XVII, David Abdallah Susi'.

[5] Chuma (c. 1850–82), originally a freed slave attached to the Universities' Mission to Central Africa [hereafter UMCA] expedition to the Shire Valley, came into Livingstone's service in 1864. He spoke and could write reasonable English but presumably did not consider his own skills equal to those of Wainwright. Simpson, *Companions*, pp. 56, 64, 67–9, 160–61, 193.

[6] *Last Journals*, II, pp. 319–46.

Figure 1: Jacob Wainwright from a photograph taken in 1874. Reproduced by permission of the National Library of Scotland.

venture can be checked and elaborated to a certain extent by reference to Waller's original notes of the interviews;[1] these tend to suggest that Waller, perhaps understandably enough, was more interested in Livingstone's last days and death than the journey to the coast. Indeed, the last stages of the journey from Unyanyembe to Bagomoyo are not described at all.

A second orally-transmitted account of the journey was the work of Carus Farrar[2] who, like Wainwright, joined Livingstone in 1872. Farrar told his story to a Church Missionary Society missionary in India on 9 September 1874.[3] His description of the journey back to the coast with the body, more concerned with how he was recruited at Nasik, is a very short summary with many of his words obviously misheard by his interviewer.[4] However, unlike either Wainwright or Waller, he does mention the fact that Livingstone's body was received at Bagomoyo by the French Holy Ghost Fathers who made a proper coffin and held a service.[5] Farrar also talked of splits between Muslims and Christians in the party.[6]

A third account is that dictated by Matthew Wellington[7] some thirty-seven years after the events.[8] His treatment of the journey amounts to no more than half a dozen sentences but, interestingly, he refers to Wainwright as the 'leader' of the six Nasik boys recruited in 1872. He says Livingstone died on a Saturday night which means, if his memory was sound, that Wainwright was correct in giving the date as 4 May and Waller wrong to insist on 1 May.[9] A fourth testimony, again oral, was made by Majwara who was actually with Livingstone in the hut when he died but he did not describe the journey back to the coast with the body.[10]

Jacob Wainwright's diary, then, is the fifth account but is arguably the most important simply because it is the only one of the five which has the journey to the coast as its main subject and which is the result of a written record clearly kept during that journey. Since Chuma and Susi told Wainwright to add an entry to Livingstone's

---

[1] Oxford, Bodleian Library MS Afr S 16/4. Notes made by Horace Waller from the testimony of Chuma and Susi concerning travels with Dr Livingstone, 1865–74.

[2] Farrar (d. 1906) was another Yao freed slave. Simpson, *Companions*, p. 193.

[3] The text has been printed twice. Thomas, 'Carus Farrar's Narrative', pp. 115–28. In this *Uganda Journal* article, Thomas provides a short introduction and an epilogue comparing the account with what can be found in *Last Journals*. Even so assiduous a researcher as Mr Thomas apparently knew nothing of Wainwright's 'Diary'. The second printing, 'Carras Farar's Story ...', is in *The Zambia [Northern Rhodesia] Journal*, VI, 1965, pp. 95–9. The spelling of both of Farar's names varies between these two versions.

[4] E.g., 'Pipa' for 'Fipa' and the European name 'Moffat' for 'Murphy'. The latter confusion is understandable in that Livingstone's nephew, Robert Moffat, had been with Cameron's party but died in May 1873. Murphy survived. Cameron, *Across Africa*, I, p. 71.

[5] Wainwright stops making entries as Bagomoyo is reached but Waller's failure to mention the contribution of the Roman Catholic community must have been the result of narrow sectarianism and nationalistic chauvinism.

[6] He also says that Chuma and Susi were both Muslims while with Livingstone, although later they clearly were not.

[7] Wellington (c. 1847–1935): another Yao freed and educated at Nasik. Simpson, *Companions*, p. 198.

[8] The account, 'Account of the Life of Matthew Wellington in his Own Words' was printed with Farrar's in *Zambia [Northern Rhodesia] Journal*, 6, 1965, pp. 99–102.

[9] Wainwright stated his belief about the date on his arrival in Zanzibar according to Frederic Holmwood of the British Agency there. *Proceedings of the Royal Geographical Society* [hereafter *Proc. RGS*], 18, 1873–4, p. 245. See also Bontinck, 'Diaire', p. 407.

[10] 'Majwara's Account' recorded by Frederic Holmwood, 12 Mar 1874, *Proc. RGS*, 18, 1873–4, pp. 244–6.

notebook constituting an inventory of their leader's possessions and to cut the inscription on the tree where Livingstone's heart was buried,[1] it seems not altogether improbable that they also suggested the desirability of his keeping a written record of their proceedings. Whether this is the case or not, clearly Wainwright did not begin until Livingstone had died so that the first two paragraphs of the account are a brief general account of what had happened between August 1872 and May 1873. There is no internal evidence to indicate whether, once the record was fairly begun, Wainwright made immediate notes and later 'wrote them up'. Nor is it possible to say why, at certain points, he chose to insert more discursive accounts of people or places. A phrase in the general account of Usagara might be held to indicate that Wainwright sought to interview informants who could give him relevant information. Perhaps he had observed Livingstone following this practice. Most of the day-to-day records are in the past tense, but this is normal if one is writing about each day in the evening, for example. Whatever the obscurities over precisely how Wainwright set about his task, it is for the great part contemporary with the party's actual adventures.[2] As such, it is evidence not only of the means by which Livingstone's body and records were preserved and taken back to Zanzibar but also a travel journal in its own right. W. S. Price, Wainwright's former Superintendent at Nasik and his mentor wrote:

> It contains observations of the natural features of the countries through which they passed, and some notices of the customs of the different tribes of people with whom they came into contact.[3]

More particularly, one may add, the route taken around the western side of Lake Bangweulu had not hitherto been followed by any traveller who also kept a record and so the 'Diary' is original in this respect. The party also followed a previously unrecorded route between the southern end of Lake Tanganyika and Unyanyembe. Unfortunately, for some reason which we shall now never know, Wainwright for a time abandoned his day-to-day diary entries when the party had crossed the Lufubu River into the land of the Lungu people near the southern end of Lake Tanganyika. Equally puzzling is his insertion at this point of a general account of the Bisa people which would have been much more relevant earlier on in the record. There follows only the most general account of the area between the lake end and Unyanyembe. This is a pity because no traveller recording the event directly had ever before been through the main part of the territory of the Fipa people. Even so, some interesting things are said. It is noticeable that Wainwright's day-to-day entries also break off at the points where the story of the expedition concerns dealings with Europeans – that is, at Unyanyembe where they met Verney Lovett Cameron[4] who tried to persuade them to bury the body there and then, and at Bagomoyo where they encountered the Catholic fathers and soon British representatives in Zanzibar. Whether this was because Wainwright was self-conscious, fearful of creating suspicion with his note taking, or

---

[1] *Last Journals*, II, pp. 311, 317.
[2] In other words, it is 'raw' record. See Bridges, 'Nineteenth-Century East African Travel Records', pp. 180–81, for a discussion on the forms travellers' evidence can take.
[3] Letter to *The Times*, 22 April 1874.
[4] Cameron is equally reticent about Wainwright and the other members of the party in his book. Cameron, *Across Africa*, I, pp. 165–9.

simply assuming that he must concentrate on 'geographical' matters, it is now impossible to say. Fortunately, the day-to-day entries resume after the encounter with Cameron and are supplemented by general accounts of the Nyamwezi, Gogo and Sagara peoples and their areas.

Although there is now an excellent edition of Wainwright's journal translated from the German into French and edited and annotated by Professor Bontinck,[1] it has never, as far as one knows, been published in English.[2] Why this is the case involves an extremely complicated story and one which is by no means altogether edifying.

2. The History of the Text

Many of those who became involved in the question of what was to happen to Wainwright's journal do not come well out of the story. Their behaviour was affected by simple greed, religious sectarianism, jealousy, literary one-upmanship, and racial or class prejudice. What happened to the text between Wainwright's arrival in Zanzibar with it in February 1874 and his return there in September is a complicated story. Only three historians have given any attention to the matter but none of their accounts covers all the problems. Simpson provides a very short account of what he knows of the fate of the 'Diary' which concludes with the simple but true statement that 'it vanished from the record'.[3] Dorothy Helly, on the other hand, provides a very long analysis of the situation which is principally designed to show why Waller's edition of the *Last Journals* did not make use of Wainwright's evidence.[4] Impressive as her assemblage of the evidence is, it must be noted that Professor Helly was not concerned at all with the content of the journal and clearly had not seen the German translation.[5] François Bontinck's annotated French translation is accompanied by a sensitive and detailed account of what he knows of the journal and Wainwright's life.[6]

What neither Helly nor Bontinck seems to have realized is that two copies of the journal were apparently made whilst Wainwright was in Zanzibar waiting to sail to Britain. One was by an 'enterprising gentleman',[7] possibly Dr James Christie, the surgeon attached to the British Consulate.[8] Whoever he was, he claimed he wanted to transmit the contents to Livingstone's family but actually planned to send it to the Bombay newspapers. He paid Wainwright for the loan with two shillings![9] Whether this transcription exists cannot now be said. The second copy was made by Richard

---

[1] Bontinck, 'Diaire'. Arguably, it is a pity that this version was not published in a major French-language academic journal; it might then have become better known.

[2] It is also, I believe, the first travel record written by an African from south of the Sahara to be published by the Hakluyt Society. Leo Africanus (c. 1494–1552), of course, was a North African if he was African at all. His great work, *The History and Description of Africa*, was published by the Society in 1895.

[3] Simpson, *Companions*, p. 101.

[4] Helly, *Legacy*, pp. 76, 69–73, 113–14.

[5] Ibid., p. 97, n. 21.

[6] Bontinck, 'Diaire', pp. 402–11.

[7] W. S. Price to *The Times*, 22 April 1874.

[8] Helly, *Legacy*, p. 97, n. 22; although Helly links this story to what became the German version.

[9] W. S. Price to *The Times*, 22 April 1874. I have not been able to discover whether anything was published in Bombay. Simpson, *Companions*, p. 101, implies that something did appear but gives no information about it.

Brenner, the Austrian Consul at Aden but visiting Zanzibar in February 1874. He himself was a former explorer who had accompanied Von der Decken. Because of this, he knew some of the porters who had served him and were now with Wainwright's party and they told him about the journal. He must have persuaded Wainwright to let him borrow it and he certainly made a copy which was forthwith sent to Germany's leading geographer, August Petermann (1822–78). Petermann just as quickly made a translation into German accompanied by a preliminary account of how the diary had reached him via Brenner.[1] The German version was put into print with remarkable rapidity; it was ready for distribution by 25 April.[2] What then happened to Brenner's transcription of the original English text is not apparent; perhaps it still exists somewhere in Germany.

Back in Zanzibar, it seems that Acting Consul Prideaux decided that the original diary ought not to be hawked around. He obtained it and then entrusted it to a British merchant, Arthur Laing, who accompanied Wainwright and Livingstone's coffin on the *Malwa* as far as Suez before taking a faster passage home via Marseilles and the railways, apparently retaining the diary. He reported the existence of the diary, together with a favourable report on Wainwright, to the Royal Geographical Society on 13th April.[3] When Wainwright reached England, Laing returned his diary to him in the presence of W. S. Price.[4] The document was soon in other hands again: Wainwright gave it up to the Lay Secretary of the Church Missionary Society, Edward Hutchinson, and it was subsequently to be seen by several other figures in Britain.

One of the many oddities of the situation which was now unfolding was that none of the people who saw or handled Wainwright's diary in Zanzibar or Britain seems to have left a record of its physical appearance. Did Wainwright have with him his own notebooks or perhaps school exercise books or did he borrow one of Livingstone's blank notebooks? Frustratingly, it is impossible to answer these questions.

Edward Hutchinson took control of Wainwright's diary in the course of what was obviously an attempt by the Church Missionary Society to take advantage of the great national interest in Livingstone. It was they who had seized on the fact that Wainwright was with the body and telegraphed that he should be sent home with the coffin at their expense whilst Susi, Chuma and the other followers were being rather overlooked. Not long after the funeral, Wainwright was taken to the Anniversary meeting and, later, other assemblies of the Church Missionary Society; he was clearly being used as part of the ongoing campaign to raise money and support for their missionary cause.[5] After all, Wainwright was their product, as it were, since they ran the Nasik School in India where he had been educated.

The interest of the Church Missionary Society must have been an important factor among those affecting the question of the use which was to be made of Wainwright's remarkable document. In April 1874, everything seemed straightforward. Sir Bartle Frere, the President of the Royal Geographical Society, explained publicly on the 27th

---

[1] 'Tagebuch von Jacob Wainwright'. Brenner's explanation occupies pp. 187–8.
[2] See *Mittheilungen*, 20, 1874, p. 200.
[3] *Proc. RGS*, 18, 1873–4, p. 246.
[4] W. S. Price to *The Times*, 22 April 1874.
[5] Stock, *The History of the Church Missionary Society* [hereafter *History of CMS*], 3, pp. 77–8.

that the great explorer's son, Tom Livingstone, was to edit his father's papers for publication and noted that Wainwright:

> had himself contributed very valuable materials in the shape of a journal ...; and this with the aid of such illustrations which it might receive from his oral communications ... would, it was hoped, enable Mr Livingstone to draw out a connected narrative of the route by which the body of the great traveller was brought home.[1]

Tom Livingstone had himself joined the *Malwa* at Alexandria and had presumably come to know and like Wainwright well enough to favour the kind of co-operation with him that Frere had outlined.

The situation now became complicated because the Church Missionary Society developed plans to publish Wainwright's diary themselves. In fact, Hutchinson was in contact with the great publisher John Murray, and sent the diary to him suggesting that he might like to pay £200 for it. Murray declined the offer. No doubt this was because he had already agreed to publish the materials on the last journey edited by the son, just as he had published Livingstone's earlier travel works; a separate publication of Wainwright would spoil the market. Further than this, Murray advised against using Wainwright's material at all. He had read the journal and concluded it would not much interest readers. If unfortunate for posterity, his attitude was perhaps understandable because Wainwright says comparatively little about the details of Livingstone's last days and death. Moreover, it seems Murray had become aware of the imminent publication of Petermann's German version.

Tom Livingstone had very much wanted to use Wainwright's record and negotiations now ensued which involved himself, the Church Missionary Society, John Murray, Henry Morton Stanley (who rather championed Wainwright's cause) and other members of the Livingstone family. Should Wainwright be regarded as in some sense still in the employ of the heirs of Dr Livingstone or was he, as a product of Nasik, under the control of the Church Missionary Society? However, by 10 June, Tom Livingstone had been reluctantly persuaded that he could do without Wainwright because information was available from Susi and Chuma. The two men had been brought to Britain at the expense of Livingstone's old friend James Young and visited Newstead Abbey where Tom was supposed to be working on his father's notebooks and journals.[2]

The visit of Susi and Chuma to Newstead was in the first part of June 1874. By this time, it was beginning to become apparent that Tom was making rather heavy weather of the confusing task of editing his father's records. There were family tensions no doubt exacerbated by Murray's desire for the rapid production of a book whilst public interest in Livingstone was still high. With some persuasion from him, the family eventually agreed that the editorial task should be taken over from Tom by the Revd Horace Waller.[3]

---

[1] 'Remarks of Sir Bartle Frere, 27 April 1874', *Proc. RGS*, 18, 1873–4, p. 255.

[2] Tom Livingstone worked in Egypt and so was given hospitality by the owner of Newstead, W. F. Webb, who had known Livingstone senior in South Africa and later put him up whilst he prepared his Zambesi book in 1865. See Fraser, *Livingstone and Newstead*, pp. 210–19.

[3] The foregoing two paragraphs are largely based on the very detailed account of how Waller came to be the editor of Livingstone's *Last Journals* in Helly, *Legacy*, pp. 73–91. Helly reports that Tom Livingstone remained bitter and frustrated about the turn of events up to the time of his death in 1876.

Waller was a former member of the Universities' Mission to Central Africa [hereafter UMCA] who had come to know Livingstone at the time of the Zambesi Expedition in the early 1860s and got on well with him. He also knew Chuma, who served the UMCA in that period, and Susi who had been with Livingstone. Waller publicly championed the cause of these two men as the most important of Livingstone's followers as early as May 1874, suggesting that they rather than Wainwright should have been brought to Britain with the coffin.[1] It was Waller who had brought the two men to Newstead in June 1874 to meet Tom Livingstone and it was, in practice, he who began to draw out from them information about Livingstone. When he finally took over the editorship, he relied entirely upon their oral testimony for the account in the *Last Journals* of the journey to the coast with the body.[2] In fact, Wainwright's record of the journey is not mentioned at all.

The train of events between February and June 1874 which has been outlined above makes it possible to begin to understand why what remains the standard account of Livingstone's last expedition, and which has been relied upon by hundreds of later biographers and historians, does not acknowledge Wainwright's role as chronicler of the final stages of the Livingstone saga.

Why, in the end, did Waller choose to ignore the diary? His references to Wainwright in *Last Journals* are not in the least hostile. Indeed his first mention of him is full of praise:

> It was to the intelligence and superior education of Jacob Wainwright ... that we were indebted for the earliest account of the eventful eighteen months during which he was attached to the party.[3]

Although this seems to hint that his information will be useful, it perhaps hints even more strongly that there are later and better sources of information. Clearly, Waller does favour Susi and Chuma but the fact that he does not mention Wainwright's account even as an adjunct to what he learned from them may indicate nothing more than the fact that the 'Diary' was not available to him when he prepared the last chapters of *Last Journals;* the proprietorial attitude of the Church Missionary Society and the unwillingness of Murray to pay for it makes one wonder whether Waller ever actually saw it. Moreover, by August, Wainwright was on his way back to Zanzibar, presumably taking his manuscript with him. Even if circumstances meant Waller did not gain access to the 'Diary', there were reasons enough why Waller would not have exerted himself to get a sight of Wainwright's work. The overwhelming desire on the part of the public was, as he knew, to learn about the last days and death of Livingstone. The testimony of Susi and Chuma was from men who had been closer to their leader than Wainwright, while that oral testimony could more easily be adapted to meet the public curiosity than the very short and simple written account provided by Wainwright. The evidence afforded by Waller's notes of his interviews with Susi and Chuma shows that he constantly reverted to questioning them about Livingstone's end.[4] The account which he gives of the journey to the coast has no dates and needs to be corrected in certain respects. It was important to Waller mainly insofar as it

---

[1] Ibid., p. 73.
[2] *Last Journals*, II, pp. 319–44.
[3] Ibid., p. 230.
[4] Oxford, Bodleian Library, MS Afr S 16/4.

enhanced the image he wanted to cultivate of a dedicated group of followers taking the body of their beloved leader back to his friends. Waller began interviewing Susi and Chuma when they stayed at the Webb family's house, Newstead, in early June 1874.[1] They appear to have been old enough and perceptive enough to realize that they must play the parts expected of them as servants devoted to the memory of their master. When Wainwright arrived at Newstead a week or two later, he created a very different impression on the family according to the daughter who later, as Mrs A. Z. Fraser, described the meeting. Wainwright was not content, for example, to eat with the servants.[2] No doubt Waller was told about this. Fraser hints clearly enough that the Church Missionary Society missionary and mentor of Wainwright, W. S. Price, who was also present but whom she does not actually name, encouraged what was in her eyes Wainwright's unacceptable behaviour.[3] Waller's own position as a high-church Anglican of the UMCA probably disposed him in any case to be suspicious of the low-church evangelicals associated with the Church Missionary Society.[4]

Whatever the precise reasons for Waller's reliance entirely upon Susi and Chuma to tell the story of the journey to the coast, the effect was to consign Wainwright's account to oblivion for the succeeding hundred years or more.

3. The Character and Career of Jacob Wainwright

Jacob Wainwright has not been given a good character in such of the books on exploration and missionary endeavour in Africa which bother to mention him. The impression conveyed is that, although rescued from slavery, given an education and granted the privilege of attending Livingstone's funeral, he became conceited and arrogant.[5] Being 'lionized' in certain missionary circles in Britain, it is suggested, meant he 'received an exaggerated idea of his future prospects' and so began a 'drifting career' which was to end in tragedy.[6] The chief reason for such verdicts is that the most accessible Livingstone study which purports to describe Wainwright at any length is the extremely hostile account of 1913 by Mrs A. Z. Fraser which has already been mentioned.[7] Among all his other failings, we are told, Wainwright was 'ugly'. Although Mrs Fraser did not see the diary, she confidently described the young man's writing skills as 'very elementary'.

Mrs Fraser was the daughter of W. F. Webb, a former associate of Livingstone in South Africa, who gave Livingstone hospitality at Newstead in 1864–65 to write his Zambesi book. Wainwright's visit there with W. S. Price was for part of one day in June 1874 when, as noted above, he took lunch with the family. That someone who was black and clearly of the servant class did not behave in a deferential fashion was

---

[1] Fraser, *Newstead*, pp. 210–19.
[2] Ibid., pp. 220–28.
[3] Ibid, p. 223.
[4] The significance of the division among Anglicans is a point which is not mentioned by Helly in her account of the situation; even Bontinck, 'Diaire', p. 401, seems to see no more than the rivalry of two missionary societies.
[5] See, for example, Debenham, *Way to Ilala*, p. 294.
[6] Simpson, *Companions*, p. 142.
[7] Fraser, *Newstead*, pp. 220–28.

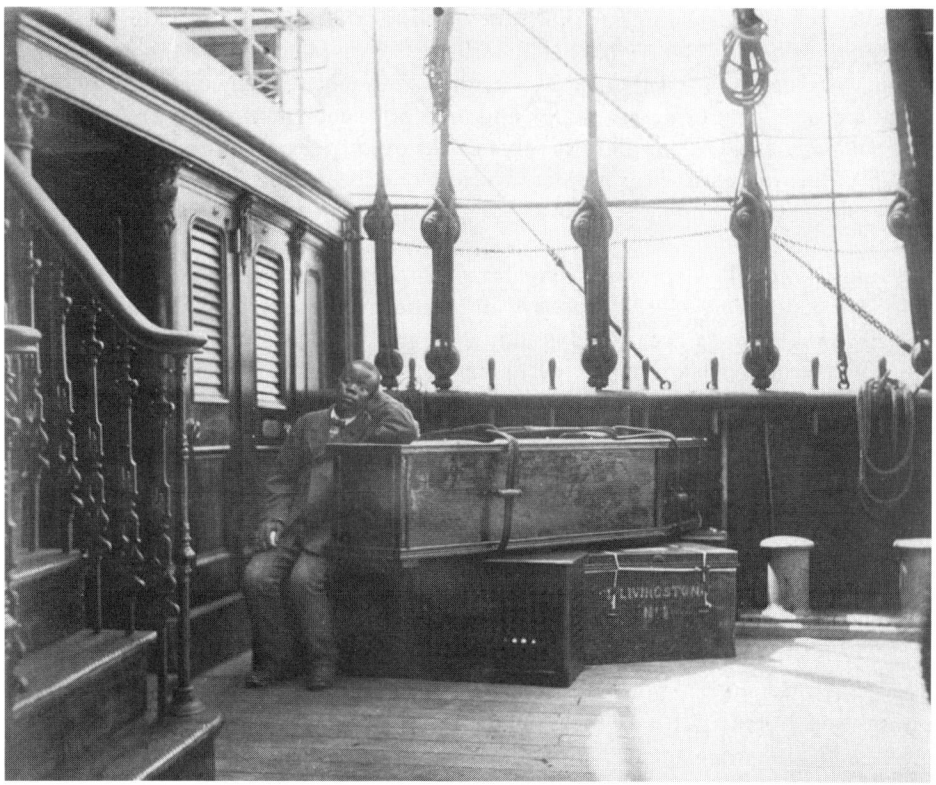

Figure 2: Jacob Wainwright on board the SS *Malwa* with Livingstone's coffin and trunks. Reproduced by permission of the National Trust for Scotland.

obviously seen as an affront by the Webbs and this must have impressed itself deeply upon Mrs Fraser. She was then only a child but made much of the matter when she wrote about it thirty-nine years later. One suspects that Price may have been seen as *declassé* to add to the bad impression.

Although a full study of his career is beyond the scope of this short study, it is worth trying to look a little beyond Mrs Fraser's chapter on Wainwright. One statement she makes which everyone seems to accept was that Wainwright, when with Dr Livingstone, was 'perhaps, then four and twenty'.[1] This would mean he was born in about 1849 or 1850. Yet everyone refers to the recruits for Livingstone from Nasik in 1872 as 'boys'. It is true that 'boy' in colonial contexts can be a demeaning term for a servant of any age but in this case it seems to the writer likely that the term does mean a young male person. The *Daily Telegraph* journalist who described Livingstone's funeral in Westminster Abbey wrote of Wainwright as 'the coloured boy from Nassick School' and then 'the African – a simple-looking quiet honest lad …'.[2] The photograph taken for Wainwright's *carte de visite* in 1874 (see Fig. 1) looks more like that of someone in his late teens than a young man of twenty-five. W. S. Price, who, as

---

[1] Ibid., p. 221.
[2] *Daily Telegraph* report reprinted in Stanley, *How I Found Livingstone*, pp. lxxi–lxxviii.

Figure 3: Wainwright follows Livingstone's coffin as it is landed at Southampton. Reproduced from the *Illustrated London News*.

Superintendent of Nasik, was surely best placed to know, reports only that Wainwright had been 'rescued from slavery some years ago'.[1] Since his age when rescued is not known, the statement is not helpful. Richard Brenner had been told by Wainwright that 'when a small boy he was transported to Kilwa by an Arab' but, again, there is no indication of when this happened.[2] Nevertheless, the strong likelihood is that Wainwright was still a boy when recruited to join Livingstone.

If Wainwright was only, say, eighteen years of age in 1874, his brashness may be at least understandable as a product of immaturity and inexperience. Whatever his age,

---

[1] Price to *The Times*, 22 April 1874.
[2] 'Tagebuch' p. 187.

Wainwright's real problem was that he was almost the first East African *evolué*. Just like the Krio freed slaves in Sierra Leone, he was taught to live, dress and think like a European and then found himself an object of suspicion or hostility among Europeans.[1]

Whether he was more the victim of his situation or had basic character defects, Wainwright's career was in the end tragic. A Yao, and therefore presumably from an area to the east of Lake Malawi in present-day Mozambique, he was sold as a slave, probably by fellow Yao, and was being carried from Kilwa north to Zanzibar when the slave ship was captured by the British anti-slave trade patrol.[2] He was sent to Sharanpur School at Nasik, near Bombay. Set up by the Church Missionary Society in India, the School had become more and more adapted to the needs of Africans freed from slavery by Royal Navy patrols in the Indian Ocean. Given a new name,[3] Wainwright was still at the School when volunteers were sought for service with an expedition organized by the Royal Geographical Society to be led by L. S. Dawson of the Royal Navy which would find Livingstone and take supplies to him. When that expedition collapsed on the arrival at the coast of H. M. Stanley after his famous meeting with Livingstone in October 1871, Wainwright and his companions agreed to join the party of fifty-one porters which Stanley now organized to take supplies to Livingstone.[4] Wainwright's own rather more succinct account takes up the story:

> I was sent by Mr Price who was then my pastor to join the Livingstone Search Expedition, and so we went on and we met and found Livingstone in the Central Africa [*sic*], in the place called Unyanyembe about the month of July 1872 [*sic*], we travelled with him for 10 months, all the time he was suffering from his original desease [*sic*], till we came at the place called Ilala where he died.[5]

In fact it was August when the party joined Livingstone but his records tell little or nothing about the subsequent doings of Wainwright (or of anyone else in the party). Fraser claims Wainwright was a cattle driver.[6] His importance began when he read the burial service from the Anglican Book of Common Prayer as Livingstone's heart was buried and cut the inscription on the tree, the inscribed portion of which can still be seen at the Royal Geographical Society's house in London. (See Figs 7 and 8.) He then began to keep the diary; as noted above, one does not know whether this was his own decision or suggested by Susi and Chuma.

The fate of the diary has been discussed above. Wainwright's own subsequent history is equally problematic. He was described by Price in February 1874 as a 'well-educated, thoroughly reliable, earnest good young man'.[7] He also made a good

---

[1] See Wyse, *Krio*, pp. 33–59. It may be added that Africans, too, found it difficult to accept what one distinguished Nigerian author called 'deluded hybrids'.

[2] This is according to Brenner's letter to Petermann, 12 March 1874, 'Tagebuch', p. 187. If the information is correct, the British cruiser was exceeding the authority given in the 1845 Slave Trade Treaty with Oman since transportation of slaves within the Omani dominions, which would have been held to include Kilwa and Zanzibar, was allowed. But the Royal Navy often made their own rules.

[3] Wainwright was presumably the name of a CMS supporter but I have not yet been able to trace him.

[4] For the highly complicated story of the various relief expeditions organized for Livingstone and the bitter arguments surrounding especially Stanley's involvement, see Bridges, 'Sponsorship and Financing of Livingstone's Last Journey', pp. 98–102.

[5] NLS, MS Acc. 12444, Wainwright to Moore, 23 May 1874.

[6] Fraser, *Newstead*, p. 221.

[7] *Proc. RGS*, 18, 1873–4, p. 182.

Figure 4: Wainwright as a pall-bearer at Livingstone's funeral in Westminster Abbey. Reproduced from the *Illustrated London News*.

impression on Arthur Laing in March. However, Horace Waller wrote to *The Times* on the 31st of that month saying that Susi and Chuma ought to have been allowed to accompany Livingstone's body back to Britain with the implication that Wainwright was not really the appropriate representative from among Livingstone's followers.[1] Nevertheless, Wainwright was a pall-bearer at the great funeral in Westminster Abbey in April,[2] attracted considerable interest and acclaim, was later presented to the Queen at Windsor,[3] and began, as we have seen, to address Church Missionary Society rallies around the country. Perhaps all this did turn his head.

In July Wainwright was staying with Price and his wife at Kessingland near Lowestoft and preparing for work in the new Church Missionary Society station to be set up by Price near Mombasa.[4] Wainwright left England on 18 August and was then

---

[1] Helly, *Livingstone's Legacy*, pp. 113–14.

[2] The *Graphic*, 9, No. 230, 25 April 1874. The artist's attempts to picture Wainwright are terrible. More acceptable are the depictions in *Illustrated London News*, LXIV, No. 1809, 25 Apr 1874. See Figs 3 and 4.

[3] 'Her Majesty presented me a beautiful book'. NLS, Acc 12444, Wainwright to Moore, 23 May 1874. Apparently the Queen had Wainwright's Diary in her possession for some days. Ibid., MS 12604, Hutchinson to Murray, 7 May 1874. Queen Victoria herself slightly misunderstood Wainwright's role. Just before the meeting, she wrote: 'I am going to see Jacob Wainwright, poor Livingstone's black servant who so faithfully nursed him and carried back his remains at the risk of his life.' Victoria to Crown Princess of Germany, 29 April 1874, Fulford, *Darling Child*, p. 137.

[4] NLS, MS Acc 12444, Wainwright to Moore, 10 July 1874.

employed at the existing Church Missionary Society station at Kisulidini near Mombasa as a teacher. When Price arrived to open the new establishment which was named 'Frere Town' and designed to receive freed slaves, Wainwright transferred there and was a 'great help'. Later, something went wrong and he was dismissed in 1876.[1] Returned to Zanzibar, he took what Joseph Thomson obviously considered a menial position as a door porter. Thomson explained Wainwright's habit of pointing out to his English missionary leaders that *they* had not been presented to the Queen as the reason for his downfall.[2] Whatever the truth of the matter, he was re-engaged by the Church Missionary Society to accompany a party to Buganda in 1880. He was spoken of very warmly by the Revd Philip O'Flaherty as his 'schoolmaster, interpreter, personal servant and friend'. Yet he left O'Flaherty to serve Mutesa, the *kabaka* of Buganda as that monarch's adviser, scribe and interpreter.[3] How long this employment continued is not clear but may have been until Mutesa died in 1884. In the late 1880s he became a helper at the London Missionary Society station at Urambo in Unyamwezi, that is, back in the territory he had traversed in late 1873. He died after a scalding accident with a fire and a pot of water in April 1892. Moravian missionaries put up a memorial plaque in 1923[4] but Wainwright had died almost forgotten, and with his great achievement as a recorder of the epic 1873–4 journey altogether forgotten. Moreover, since his death, the damning verdict based on one lunch-time encounter at Newstead has blighted his reputation.

Perhaps the reprinting of his journal will do something to restore Wainwright's reputation. At the very least, the conscientious way in which this boy or very young man attempted to create a record of the epic journey of the survivors deserves to be commended. It would appear that Wainwright conceived it to be his duty to produce a purely factual account. His careful approach meant that the record was compiled in such a way as to tell us little about the writer himself and his feelings. Occasionally one hears of the 'wicked natives' but his judgments on the Africans the party encountered were normally neutral. When they are not, they seem to reflect what one would expect of someone being taught to become an evangelist assistant to the British protestant missionary enterprise. Sometimes, perhaps, he remembers his African roots as when he hears music in Unyamwezi. A rounded account of the career of Wainwright and those of some of the other Nasik products would be a worthy academic enterprise.

4. East and Central Africa in 1873–1874

East and Central Africa were disturbed areas at the time of Livingstone's death. The settlements which Wainwright and his companions visited were stockaded for good reason. At the time, the standard European explanation for disorder was to say that the country was ravaged by wicked slave traders whose depredations the saintly Livingstone had been trying to prevent. Although this stereotype is not without elements of truth, there was a complex mixture of influences bringing about the state of instability.

---

[1] Simpson, *Companions*, pp. 137–8.
[2] Thomson, *Central Lakes*, I, p. 34; Bontinck, 'Diaire', p. 416, somewhat unconvincingly tries to argue that rather than a door porter, he was a dragoman.
[3] Thomas, 'Wainwright in Buganda'. Ironically, he succeeded a UMCA product, Dallington Muftaa.
[4] Simpson, *Companions*, p. 142.

Over a period of about two thousand years, the people known now collectively as the Bantu had settled the whole region which was, for the most part, a difficult environment. In the process, which had not really ended by the middle of the nineteenth century, polities of various kinds had come and gone as they competed for economic advantage or political control. For example, in the region where the porters began their journey, among the Lunda peoples, the empire of the monarch known as the Cazembe had emerged to widespread power and influence by 1800 but was now in decline.[1] This was partly for internal reasons but partly also because of encroachments by traders and predators from the coasts of the continent. The combination of tensions internal to Africa's Bantu societies and upsets caused by influences from outside created the atmosphere of disorder.

The most obvious invasion which was purely African and which could be seen, perhaps, as part of the flux and reflux of the Bantu migrations over the region, was the arrival of groups of Ngoni raiders from the south. One of the consequences of the rise of Zulu power in South Africa in the 1820s was the 'scattering' of peoples, notably, but not only, groups of Ngoni.[2] In 1835, one group under Zwangendaba had crossed the Zambesi not far from Zumbo and moved northwards as a series of marauding groups who lived by plunder and perpetuated themselves by recruiting local young men as warriors. Zwangendaba died in 1835 and his following split up into several groups, often referred to as the Mazitu, one reaching as far north as Lake Victoria. The central area of what is now Tanzania was affected both directly and by emulation: the Nyamwezi warlord Mirambo,[3] about whom Wainwright tells us something, was greatly influenced by Ngoni practices as he recruited the 'ruga-ruga' armies which he used to extend his rule. One group of Ngoni under Mperembe had doubled back south and preyed on or settled in areas to the east of Lake Bangweulu through which Livingstone and his followers had come in 1873.

By the time of the Ngoni incursions, East and Central Africa was already being subjected to ever more powerful external influences. During the previous 250 years or more, the Indian Ocean system of trade had been affecting the region both directly and indirectly. In the medieval period, the East African coastal cities had been part of this system as the evidence of great travellers like Ibn Baṭṭūta attests.[4] The arrival of the Portuguese in the Indian Ocean upset patterns of trade but did not materially lessen the demands for gold, ivory and slaves. After other trading Europeans – the Dutch, the French and the British – arrived in the Indian region, the whole tempo of trade from Eastern Africa, especially to Arabia and the Indian sub-continent, increased, even though there was as yet little direct European contact with the region.[5]

By and large, the pattern that had developed by about 1800 was for African systems of inter-regional trade, for example, in copper from the Cazembe's Empire, or in iron and salt and, later, ivory, from Unyamwezi, to extend to the coast and become

---

[1] Roberts, *Zambia*, pp. 122–3.
[2] Omer-Cooper, *Zulu Aftermath*, pp. 64–85.
[3] Bennett, *Mirambo*, pp. 35–6.
[4] *The Travels of Ibn Baṭṭūta*, II, pp. 373–82.
[5] It is worth noting that the area of the Lunda Cazembe was also affected by the Atlantic system of trade with trade routes developing from the west coast. Slaves and ivory were the major although not the only commodities demanded.

international.¹ Wainwright's own people, the Yao, who live east of Lake Malawi, had taken both slaves and ivory to Mozambique to trade with the Portuguese but now went increasingly to Kilwa, further north.² Here they could link to the currently more powerful system of trade organized by Arabs from Oman who had captured the great Fort Jesus at Mombasa from the Portuguese in 1698. Increasingly backed by Indian finance, the Omani developed a key outpost at Zanzibar.

Prominent among the African groups who had come to engage in this coast-linked trade were, besides the Yao, the Nyamwezi and the Bisa. The latter, living to the north-west of Lake Malawi, traded with the Yao and sometimes got to Kilwa themselves.³ In the end, of much greater importance were the Nyamwezi⁴ who, in the early nineteenth century, had opened up a new route more directly opposite Zanzibar – that from Unyanyembe to the coast which Wainwright was to follow on the last stages of his journey.

To be sure, people from the coast, not least the Portuguese and their allies, had attempted to make direct contact with sources of the supply in the interior for their ivory and slaves. Dr Lacerda's journey to Cazembe in 1798 was the most important of these in the relatively modern period.⁵ Yet this was a failing effort; except along the lower Zambesi, Portuguese power and influence was not apparent at all in the interior. A much more serious challenge to African traders was posed by the Omani Arabs from Zanzibar who began to lead their own trading caravans into the interior from the 1830s.⁶ Encouraged by their sovereign, Seyyid Said, who reigned from 1806 to 1856, they sought ivory and slaves not only for Arabian and Asian markets but also labour for the increasingly important clove plantations established on Zanzibar and Pemba. A development of great importance was the establishment of an Arab depot in the interior. Fundikira, the leader of the major Nyamwezi chiefdom of Unyanyembe, seems to have given them permission in the 1840s or 1850s to establish a station and settlement at what was first known as Kaze but later as Tabora. By the 1870s, Arab and Swahili traders, the most notable of whom included Tippu Tib and Kumba Kumba whom Wainwright refers to, were having massive effects on African life far beyond Tabora. The areas Livingstone had been trying to explore in the last two or three years before his death – Manyuema beyond Tanganyika and the upper Lualaba – became the stamping ground of Tippu Tib. The Omani Arabs were usually accompanied by partners, porters and guards drawn from the people of the East African coast, the Swahili, whose language became the *lingua franca* of the region.⁷

African traders did not simply give in to Arab and Swahili rivals. Many, it is true, were content to be recruited as porters. Every trade item had to be carried by humans because of sleeping sickness affecting draught animals. The Nyamwezi, in particular,

---

¹ Roberts, 'Nyamwezi Trade', pp. 41–5.
² Surprisingly, there is no one good study of the Yao, but see Alpers, *Ivory and Slaves*, chapters 6 and 7.
³ St John, 'Kazembe', pp. 202, 217, 240.
⁴ Roberts, 'Nyamwezi Trade', p. 49.
⁵ *Lacerda's Journey to Cazembe*.
⁶ The founder of the Hakluyt Society, W. D. Cooley, was one of the first to note and record this process. Cooley, 'Geography of Nyassi'.
⁷ For a comprehensive account of the whole process by which Zanzibar and the coast became linked to the interior, see Sheriff, *Slaves, Spices and Ivory*.

were regarded as the best porters in East Africa.[1] But many Nyamwezi groups fought for their position as independent traders and rulers against the Arabs. Fundikira's immediate successor in Unyanyembe, Mnwa Sele, disputed with them until killed in 1865 but the warlord Mirambo is the prime example among the Nyamwezi. In Wainwright's period, the Luapula and Lake Mweru region was much affected by another Nyamwezi group, known as the Yeke, who, led by the warlord-cum-trader Msiri, had migrated from their homeland.[2]

Another means of contesting or at least making something of trading activities whether by Arabs or African groups was to demand presents for the right to pass through a territory. Refusal to pay 'hongo', as it was known, could lead to violence. In any case, the distinction between trading and raiding was often a very ill-defined one – hence the stockaded villages Wainwright describes.

The arrival of European travellers was as yet a minor influence on the overall situation. Nevertheless, exploratory expeditions were organized and equipped like Arab trading caravans and were expected to behave like one of these, not least by paying 'hongo'. To some small extent, European ventures contributed to the loss of stability and order.

One effect of disorder and uncertainty was to encourage the development of new forms of political authority. It has already been noted that Mirambo copied Ngoni methods of aggressive warfare to create his 'empire'. Yet Mirambo was also responding to the external trade influences. His aim was to monopolize or control the supply of ivory from west of Unyanyembe. To realize his ambitions, he became in effect a warlord. Warlords great and small sprang up in other areas, especially after about 1860 when increasing quantities of firearms were brought into East Africa. Many warlords had some claim to traditional authority as 'chiefs' (as did Mirambo) and some had none.

Chiefship is a complicated subject. Traditional European assumptions that a 'chief' is a person wielding unlimited and usually tyrannical power over a discrete group of people in a specific area must be abandoned. Yet, as Wainwright's careful noting of the names of those whom he encountered testifies, who was chief was a matter of importance. The general tendency, although there are many reservations and qualifications to be made, was for chiefs to become more powerful to the extent that they could throw off traditional restraints on their authority or on succession rights. For example, in some basically matrilineal societies like Wainwright's Yao, certain figures sought with considerable success to establish patrilineal inheritance of their power. Secondly, and perhaps even more importantly, there was a tendency for political leaders to claim territorial authority. By and large, traditional African modes of thought and behaviour in a situation where land was relatively abundant but labour was scarce did not need to emphasize property in land; by 1873 that was changing and territorial control was beginning to be insisted upon.[3]

The disorder and uncertainty outlined above finds echoes in what Wainwright says at some points and in the way his party behaved on at least one occasion – using force to

---

[1] Burton, *Lake Regions*, I, p. 337.
[2] Roberts, *Zambia*, pp. 122–3.
[3] For a discussion of the evolution of chiefship in the important case of the Nyamwezi, see Abrahams, *Political Organization*, pp. 31–47.

achieve their ends. However, what is remarkable is that, in the face of all the actual and potential dangers which could have broken up their party or prevented its passage, especially as they were relatively poorly endowed with goods and without overwhelming military authority, they did succeed in reaching the coast at Bagomoyo. No doubt the leadership skills and diplomatic good sense of Susi and Chuma had much to do with this success. Unfortunately, Wainwright's account tells us little of the way the party was organized and conducted but one can guess from some of his comments that leadership was a matter of obtaining consensus. Whether Wainwright himself had any particular influence in the counsels of the party, one does not know.

5. The European Position in East and Central Africa in 1873–1874

In 1798 Dr Francisco Lacerda made his way from the lower Zambesi valley to the area of Lake Mweru and established some contact with the Cazembe's empire but died there.[1] After his death there were further attempts by the Portuguese to maintain a trading link with Cazembe and open up some communication between their settlements on the east and west coasts, notably the sending of two semi-literate traders or 'pombeiros' who spent the years 1806 to 1811 on a protracted journey from Angola to Tete.[2] Then in 1831–2, Captains José Monteiro and Antonio Gamitto made another visit to the Cazembe's kingdom.[3] Trade from Cazembe to the lower Zambesi fell away, as we have seen, because new possibilities of contact with Zanzibar opened up and so these journeys were never followed up. As far as geographical exploration was concerned, it has to be noted that neither the 'pombeiros' nor the two army officers were able to make astronomical observations and produce clear accounts of the configuration of the territories they passed through. In any case, these Portuguese exploits did not become very widely known.[4] To the north and east lay an enormous stretch of territory reaching as far as the confines of Ethiopia for which knowledge of even the most elementary geographical facts concerning its lakes, mountains and rivers did not exist before 1848. In fact, the serious exploration of East and Central Africa by scientific travellers did not properly begin until 1856. Then, in the space of not much more than twenty years, a small group of outstanding travellers, among whom David Livingstone was pre-eminent, managed to produce more basic information than had been accumulated in the previous two thousand years.

Livingstone's discovery of Lake Ngami in 1849 brought him to the notice of geographers. Then his attaining the Zambesi valley from the south was the prelude to his great transcontinental journey of 1852–56. This was made with the aid of, and for the sake of, his Makololo allies. His discoveries suggested that he in turn could help them by opening up a route for European commerce to the centre of the continent via the Zambesi. The British government backed him and the Zambesi Expedition was launched in 1858. Whatever his other failures, including his inability to get to the centre of Africa with steamboats, Livingstone did succeed in putting Lake Malawi and the Shire

---

[1] *Lacerda's Journey to Cazembe*, pp. 55–164.
[2] Ibid., pp. 165–240.
[3] Ibid., pp. 247–57; Roberts, *Zambia*, pp. 111–15.
[4] It was not until 1853 that Gamitto published his interesting journal and an English translation had to wait until 1960. Gamitto, *King Cazembe*, I, p. 23.

Valley on the map in 1859. Meanwhile, further north, a series of expeditions was launched from Zanzibar which took advantage of Arab and African trade routes to reach the lakes and mountains of the interior. The Church Missionary Society missionaries Ludwig Krapf and Johann Rebmann had begun the process in a fairly amateur way when they reported sighting the 'snow mountains', Kilimanjaro and Kenya, in 1848 and 1849 respectively. Their work, plus that of 'armchair geographers' who drew attention to information filtering out from Arab and Swahili traders, and a certain amount of official Indian interest, encouraged the Royal Geographical Society to send Richard Burton and John Hanning Speke on a mission which reached Lake Tanganyika in 1858. Speke alone broke away north to reach the southern end of what he christened Lake Victoria and claimed to be the ultimate feeder of the Nile. The Royal Geographical Society sent Speke back with James Grant to vindicate his claim and Speke alone in July 1862 reached the spot where the Nile does indeed debouch from Lake Victoria. Not everyone believed Speke had found the ultimate source (or even, in some critics' estimation, a source at all). Controversy and speculation were heightened by Samuel Baker's report of his visit to Lake Albert of 1864 and his claim that it stretched far to the south; surely its feeders must be the furthest sources of the Nile. Livingstone's last journey of 1866–73, in which Wainwright was to become involved, was essentially an attempt to find those furthest sources. Livingstone actually killed himself by putting a misplaced faith in the story, which Herodotus relayed, of there existing a hill with four fountains which were the sources of the Nile. Not even Herodotus himself had believed this but Livingstone thought that the fountains were somewhere to the south-west of Lake Tanganyika. Hence his visits to Lake Mweru in November 1867, Lake Bangweulu in July 1868, and protracted investigations of the Lualaba valley west of Lake Tanganyika in 1869–71 before he returned to Ujiji. It was here that Henry Morton Stanley met him in the newspaper 'scoop of the century' in October 1871. Livingstone refused to return with Stanley but waited for the supplies and new personnel, including Wainwright, which reached him in August 1872. He would now attempt to get around the south of Lake Bangweulu, prove there were no sources south of there and then go on to the elusive fountains a little further west. One result was a series of uncharacteristic observational mistakes which gave an exaggerated east–west extent to Lake Bangweulu on the maps based on his data[1] and, the other, far more tragic, his own death.

Verney Lovett Cameron, sent by the Royal Geographical Society to find Livingstone in 1873 and meeting only Wainwright, his companions and the corpse, went on to cross the continent and show beyond reasonable doubt that Livingstone had really been working on the upper waters of the Congo. But it was Stanley, leaping to fame through meeting Livingstone, who, in the years 1874–7, actually demonstrated not only this by following the Congo to the Atlantic but also showed that, for all practical purposes, the source of the Nile was where Speke had said it was. Despite Livingstone's confusions at the end of his life, he and three or four other explorers had achieved an enormous amount; Wainwright's work added some detail and he and his companions brought home Livingstone's results and, it might be said, created even more fame for him and his causes.[2]

---

[1] See below, p. 366, n. 3 and the inset to Map 1 on p. 362.
[2] For a somewhat more extended description of these exploratory activities, see *Times Atlas of World Exploration*, pp. 198–207.

Figure 5: Carving by C. d'O. Pilkington Jackson, symbolically representing the journey to the coast. Reproduced by permission of the National Trust for Scotland.

The explorers had succeeded because they were backed by an increasingly powerful British government and Indian administration and a prosperous geographical society, the Royal Geographical Society, keen to obtain reliable information about the African interior.[1] Much popular attention was attracted in Britain and Europe because of the adventure element in the resulting travel literature. There was, too, the fascinating mystery of who would be able to solve the question which had baffled the Ancient Egyptians, the Greeks, the Romans and the Arab geographers – where are the sources of the Nile? Beyond this, however, interest was generated because serious questions of public policy were involved. Could free trade be extended into Africa? Could the newly-discovered lakes be used by steamboats? Could the Gospel be spread among various African peoples whose existence was now made known to the educated world? What should be done about the slave trade? What, in fact, was to be the relationship between this region and the outside world? The journey which Wainwright chronicled was a small part of the process by which that outside world was to become more closely involved with Africa.

---

[1] Of course, French, Germans, the Egyptian government and King Leopold of Belgium, not to mention the *New York Herald*, were involved to an increasing degree but the basic exploration of East and Central Africa from 1856 to the death of Livingstone was very much a British affair.

The British public, however much or little they might understand about the Nile source question, thought that Livingstone died a martyr to the cause of revealing the horrors of the slave trade and seeking to end it.[1] Consequently his death resulted in the imposition of a new treaty on Zanzibar and increased economic, exploratory, scientific and missionary activity in East Africa on the part of British and increasing numbers of other European organizations. There was much emphasis on what Livingstone had seemed intent upon – opening up routes from coast to interior which could be used by improved means of transport. Yet it is altogether wrong to assume that this increased tempo of activity was part of a colonial drive or even a process which must inevitably lead to European rule. The British government, powerful, committed to ending the slave trade and promoting free trade, was content to exert indirect influence via Zanzibar. To be sure, there were those among Indian officials who would have preferred a rather more active British policy in East Africa but there was in the 1870s little prospect of this. Explorers, missionaries and their allies who wanted to redeem Africa knew that they would have to do it by themselves. They began to try to create what I have ventured to call an 'unofficial empire'.[2]

Insofar as Wainwright became involved in the activities which followed the death of Livingstone, therefore, he was dealing with a group of high-minded officials, scientists, churchmen and their supporters, who wanted to create a new infrastructure for East and Central Africa and a new moral climate, too. High-mindedness did not always translate into toleration and sensible treatment of those who were caught up in the plans as was Wainwright. In the longer run, however, we can see that he did make a contribution to the scientific exploration of East Africa and the extension of Christianity, which processes, for better or worse, were such significant stages in the region's historical development.

6. How Good was Jacob Wainwright's English?

The translation of Wainwright's original text into German and its re-translation here into English makes it extremely difficult to determine just how good his English originally was. Once he returned to East Africa after his schooling at Nasik, he would have been associated with groups of people, notably caravan porters, who used Swahili and possibly not a few of them would have known his own native tongue, the language of the Yao.[3] Presumably Livingstone used Swahili for converse with his porters and near the end, apparently, with Susi.[4] However, it is clear that Chuma had a

---

[1] As he had in the sense that he cared passionately about the slave trade and believed that geographical exploration was an essential first step in the process of bringing Africa into community with the outside world, a process which, he believed, would end the evil.

[2] Bridges, 'Towards the Prelude', pp. 105–6. A more active official policy was forced upon Britain in the 1880s and '90s simply by international rivalry. 'Protectorates' were established which it was hoped would not need much direct supervision. But the fragmented sovereignty which Wainwright experienced was eventually to mean direct rule was seen to be needed.

[3] There is no indication whether he retained this language; presumably much would have depended on what his age was when he was enslaved.

[4] *Last Journals*, II, p. 307.

considerable understanding of English, both spoken and written,[1] and when Waller interviewed him and Susi, he soon found that their English was adequate for his purposes.[2] Insofar as the Nasik-educated young men in the party who joined Livingstone in 1872 tended to stick together as a group, they may well have used English among themselves. Whatever these possibilities, it is clear that, in October 1873, Wainwright was able to write a perfectly clear and coherent letter to the person he and his companions were led to believe was Livingstone's son.[3] The letter has some relatively minor grammatical mistakes and perhaps indicates that Wainwright had trouble over past participles. The letter also exhibits a characteristic which did survive the German translation – a habit of giving alternative words as when he writes of the 'sultan or king'.

The merchant Arthur Laing, who travelled with Wainwright from Zanzibar to Suez, affirmed that 'Jacob Wainwright writes and speaks English very well'.[4] Other pieces of evidence which have recently come to light are Wainwright's two letters to Livingstone's old friend Joseph Moore,[5] which are quoted at other points in this study. The letters to Moore show that Wainwright's English was adequate and clear but by no means entirely correct and idiomatic.[6]

No attempt has been made in presenting the text to try to imitate how Wainwright may have written. Rather, the intention is to produce a reasonably clear version of the information he was trying to convey as far as that is possible through the distortions of two translations. Even this modest aim is not always easily realized, as will be explained below.

It is difficult to avoid the impression that Wainwright must have become aware of how constantly and insistently Livingstone asked 'what river is this?'. Hence his own attempts to record all river names and, often, to indicate where the rivers flowed to. Equally assiduously, he tries to record chiefs' names which, again, tended to be the case with Livingstone. As noted above, the conscientious day-to-day record is sometimes abandoned altogether for reasons which are unclear, but these lacunae are compensated for to a considerable extent by the more general accounts of certain peoples. Altogether, it is remarkable just how sophisticated Wainwright's 'Diary' proves to be.

7. Notes on the Translation

The text of Jacob Wainwright's 'Diary' presented here is a re-translation[7] of August Petermann's translation into German of Richard Brenner's transcription of

---

[1] See Chuma's letter to Waller, quoted in Simpson, *Companions*, p. 100 which includes the information, 'Susi don't know write'.

[2] Fraser, *Newstead*, pp. 213, 215.

[3] See below, p. 374 and Fig. 9. It is worth adding that it was also written in an admirably clear and well-formed handwriting although Mrs Fraser's description of Wainwright's writing as 'a very large childish round hand' is perhaps an alternative way of characterizing it. Fraser, *Newstead*, p. 221.

[4] *Proc. RGS*, 18, p. 246.

[5] NLS, MS Acc 12444.

[6] There is no apparent sign that the English in these letters had been corrected by Price and others with whom Wainwright was associating at this time. See Fig. 6.

[7] I am indebted to Mr Frank Gatter for furnishing me with an initial translation and to Dr André Reibig for making a further check and giving me advice. However, I have ventured to produce my own versions of several passages or individual words and take full responsibility for the whole translation.

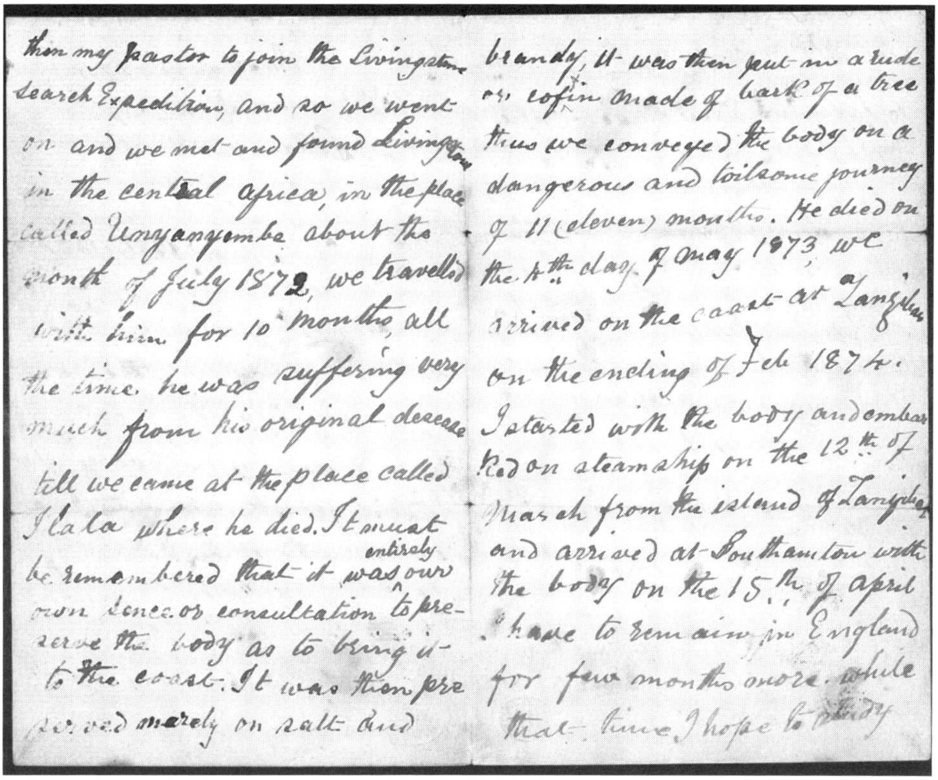
then my pastor to join the Livingstone
search Expedition, and so we went
on and we met and found Livingstone
in the central africa, in the place
called Unyanyembe about the
month of July 1872, we travelled
with him for 10 months, all
the time, he was suffering very
much from his original decease
till we came at the place called
Ilala where he died. It must
be remembered that it was entirely our
own sence or consultation to preserve the body as to bring it
to the coast. It was then pre
served merely on salt and
brandy, it was then put in a rude
or cofin made of bark of a tree
thus we conveyed the body on a
dangerous and toilsome journey
of 11 (eleven) months. He died on
the 4th day of may 1873, we
arrived on the coast at Zanzibar
on the ending of Feb 1874.
I started with the body and embar
ked on steamship on the 12th of
March from the island of Zanzibar
and arrived at Southamton with
the body on the 15th of april
I have to remain in England
for few months more. while
that time I hope to study

Figure 6: A specimen of Wainwright's handwriting from a letter to the Revd Joseph Moore. Reproduced by permission of the National Library of Scotland.

Wainwright's original manuscript. Hence, three processes have intervened between the original and my version. It should be noted that Petermann claims that he had tried to preserve the 'unsophisticated' character of Wainwright's work in his German version.[1] Just what implications this attempt had for the accuracy and fidelity of the translation it is impossible to say.

Even if one accepts Petermann's German version as a reasonably faithful representation of the sense of the original, difficulties remain over certain words and phrases. For example, a frequently recurring term in the German text is the noun 'Wüste' for which the normal and literal translation would be 'desert'. For most readers of English, this word now implies a waterless sandy waste like the Sahara. It would not be entirely appropriate in the context of most of Central and Eastern Africa. Waller, in telling the story of the expedition in the *Last Journals*[2] often uses the word 'jungle' but this, too, is open to objection as being inappropriate. Faced with this difficulty when translating the German text into French, Bontinck normally used

[1] Petermann wrote of Wainwright's notes, 'Wir geben sie in möglichst wörtlicher Übersetzung, um die kunstlose und eigenthümliche Schreibweise des wackeren Negers …', 'Tagebuch', p. 188.

[2] *Last Journals*, II, pp. 319–46.

'brousse', in other words, 'bush' and this is certainly much more apt than 'desert'. Wainwright himself may well have had the Bantu root 'nyika' in mind which, according to one definition, signifies 'open, bare, treeless wilderness [or] open forest with high grass [or] a barren, desolate region'. In the event, I have chosen in most instances to use the term 'wilderness' because this seems closest to 'nyika' and perhaps is the sort of word with which Wainwright would have become familiar in his Bible reading.

There are similar difficulties about using the terms 'village' or 'town' for the stockaded settlements the party moved between. I have used 'village' in most cases. 'Chief', 'King' and 'Sultan' are terms even more fraught with difficulty. All will in one way or another mislead. In this case, I have made no attempt to correct what seems to be the simple translation of the German word given by Petermann. Wainwright's habit of giving two terms, e.g. 'sultan or king', as alternatives adds to the complications. Whether he did this as a sort of literary device for emphasis or because he was himself unsure which English term would be correct, it is difficult to say. However, it does seem to be, as it were, a hallmark of his style.

Punctuation has presented several problems. Possibly, the oddities in Petermann's German version were the result of his attempt to imitate Wainwright's 'lack of sophistication'. The evidence from Wainwright's letters to Joseph Moore and that to 'Livingstone's son' suggests that his punctuation was not very good. I have chosen to punctuate this new version in English in such a way as to make the text comprehensible and readable; there is no attempt to reflect either the German or any presumed original.

8. Establishing the Route

Establishing the precise route taken by Jacob Wainwright and his party is by no means a straightforward procedure. For certain periods, notably that between 31 August and 8 October 1873 when the party were passing through the territories of the Lungu and the Fipa, no details are given of the places visited and one has to resort to other sources and guesswork. Even where Wainwright's record names places, it is not always easy to identify these on modern maps. Where identification does seem reasonably certain, co-ordinates are given in footnotes. On the basis of what can be gleaned, the route shown on the accompanying map has been plotted. Of course, at the scale made necessary by the format of this volume, it is not possible to include any more than a small proportion of the points visited and identified by Wainwright. Nevertheless, the route shown provides at least a general indication of the path followed by the party.

The map is in fact, I believe, the first attempt to depict in full the route taken by the party which brought Livingstone's body out of the interior. There seem to have been only two previous attempts to plot the route on a map and both of these are partial and inaccurate.[1] The map with Waller's *The Last Journals of David Livingstone*[2] shows the

---

[1] It is surprising that Petermann did not attempt a map in the edition of *Mittheilungen* which carried Wainwright's 'Diary'. The periodical was notable for its excellent maps. Perhaps the article was prepared in so short a time that it was impossible to produce a map as well.

[2] In a slip case with Vol. I of *Last Journals*. The scale of the map is 34.3 miles to the inch.

party as travelling clockwise round the wrongly-shaped Lake Bangweulu[1] in a totally misleading way. It is then assumed that the party merely followed Livingstone's own route back to Lake Tanganyika's southern end. From there the route is shown as a straight line through Ufipa until Livingstone's Tabora–Ujiji route was reached at about 6°S. From that (wrong) point, no further details are given at all. The other map in question provides an even more restricted part of the route. It appears in Johnston's biography of Livingstone and tries to show the route to the west of a more correctly-shaped Lake Bangweulu.[2] The result is better than the map in *Last Journals* but still rather unsatisfactory. This map was reproduced to illustrate Bontinck's translation into French and was his only attempt to show the route.[3]

In attempting to establish Wainwright's route, I have consulted modern maps of Zambia and Tanzania at the scales of 1:50,000 and 1:200,000. I have also been able to use some of the excellent official maps produced in Berlin by Dietrich Reimer's Geographische Verlagshandlung when what is now Tanzania was German East Africa. The maps I used were at the scale of 1:300,000. They were drawn and printed in about 1905 and often show the routes of nineteenth-century travellers.

---

[1] See inset to Map 1.
[2] Johnston, *Livingstone*, facing p. 317.
[3] 'Diaire', p. 435. Like Waller, Bontinck seems to have assumed that from Lake Tanganyika the party merely followed Livingstone's former route.

THE TEXT OF WAINWRIGHT'S 'DIARY'

Herr Consul Brenner's introduction is omitted together with remarks presumably by Petermann which occupy about 1,150 words.

It should be noted that, in Bantu languages, the prefixes U, Mu, Bu, and sometimes other variants, indicate places, e.g. (U)nyamwezi, while A, Ba and Wa signify people, e.g. (Wa)nyamwezi or (Ba)Nyamwezi. Such usage naturally occurs in Wainwright's text but in the notes, in order to avoid confusion, the prefixes are as far as possible omitted and the simple root used.

Where possible, the co-ordinates of places mentioned by Wainwright are given in the notes and these positions have been used to establish the route shown on the map.

Dr Livingstone's Death

After Chambezi River,[1] Dr Livingstone's illness gets worse.[2] On the other side of the Chambezi, we reached Manikazi River.[3] The chief of the village where his disease got more and more serious is called Katenkera. From there we went to Kopa, [a] chief on Mitikira River,[4] and from him to Chief Zawamba; the name of the river is Lookulu. From Mazawamba[5] we started carrying him in a hammock.[6] From there we went to Kalongandyofu on the border of Wabisa-Land; the river there is the Lulimala.[7] After crossing Lulimala River we entered into another and different land.[8] The name of the

---

[1] The Chambeshi rises in the highlands between Lakes Malawi and Tanganyika and flows south westwards into Lake Bangweulu. The party crossed it going north to south.

[2] 'An artery gives off a copious stream, and takes away my strength'. *Last Journals*, II, p. 294. Waller's account of Livingstone's last days and death was given a distinct hagiographical slant. For example, Chuma and Susi told him that Livingstone had died 'kneeling on his bed with his head on the pillow' which Waller interpreted as that Livingstone had 'appeared to be engaged in prayer'. Cf. MS Afr S 16/4 and *Last Journals*, II, p. 308. Nevertheless, Waller is substantially accurate. Oral tradition collected later and recorded in RGS MSS File Livingstone, Deposition by District Commissioner, Mpika, 1936, talks of a fair-headed follower who was Livingstone's son and of the 'boys' playing football. At least one author has built upon this an account assuming Livingstone had an illegitimate son. The tradition is confused, amateurishly collected and cannot be accepted at all as having anything to do with Livingstone or Wainwright and his companions.

[3] Munikazi on map in Hughes, *Bangweulu*. 'Rivulets without number. They are so deep as to damp all ardour.' *Last Journals*, II, p. 274.

[4] Luitikila. Hughes, map. Kopa is the title taken by the ruler in this particular chiefdom of the Bisa people. Roberts, *Bemba*, p. 55.

[5] This is possibly a corruption of another Bisa chiefly family name, Mwanasabamba.

[6] See *Last Journals*, II, pp. 299–300. When interviewed at Suez for the *New York Herald*, Wainwright reported that, when ill, Livingstone prayed much and had said, 'Build me a hut to die in'. *The Times*, 30 March 1874. This interview is also reprinted in full in the 1895 edition of Stanley's *How I Found Livingstone*, pp. lviii–lx.

[7] Livingstone, in the very last word he ever wrote, called it the 'Molilamo'. *Last Journals*, II, p. 303.

[8] They were leaving an area then occupied by the Bisa and entering the land of the Lala people. Wainwright here calls it Muilala. On the Lala, who were part of what ethnographers call the 'Bemba cluster' and were noted as iron smelters, see Murdock, *Africa*, p. 294; Roberts, *Bemba*, pp. 91, 101, 153.

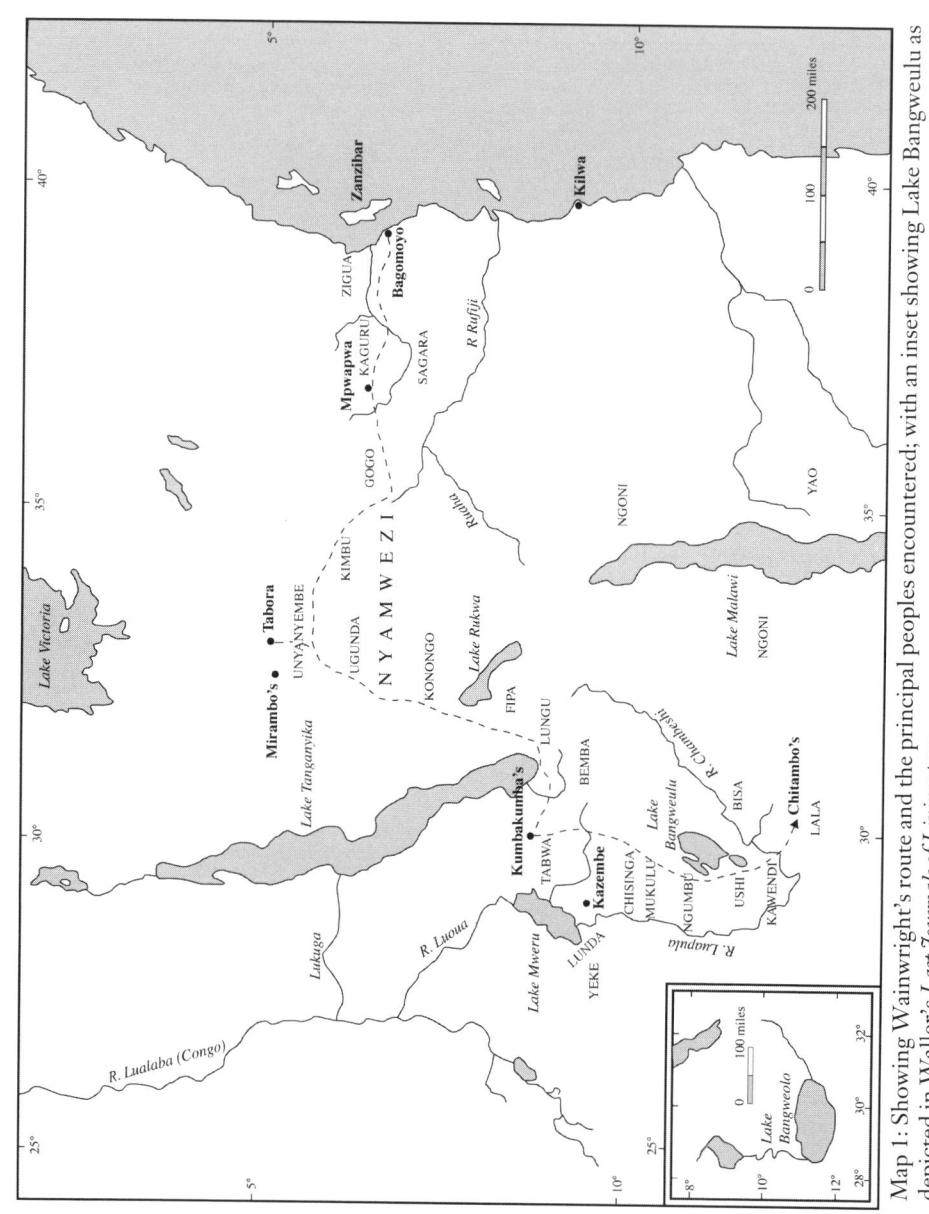

Map 1: Showing Wainwright's route and the principal peoples encountered; with an inset showing Lake Bangweulu as depicted in Waller's *Last Journals of Livingstone*.

local chief is Kitumbo,[1] he of the land of Muilala. After we had stayed there for a day, he died the following night, on 4th May 1873.[2]

On the following day, i.e., on the 5th May, we had no other means to save his body from decay than to salt him, and when his bowels were examined, nothing was found other than black blood, and his lungs were found to be consumed, as well.

On the next day we made a box or a coffin.[3] Normally we pitch our huts around a big shady tree. The name of the tree in this case is Mbura[4] into which we cut out the inscription and also the names of the three leaders, Yazuza, Mnyasere, and Ghopere.[5]

On the Northward Journey Home

When we saw that our master was dead, we had no need to go on but returned to Zanzibar, taking the dead body with us.[6] We had stayed in Kitumbo's town for twelve days, and on the next day we finally set out on our march. We came to a town whose mighty chief is Manawamzungu. Here we remained for several days on account of an illness: all the people in our caravan were ill, with the exception of five or six. On 25th June we reached the banks of the Luapula River,[7] which is the same river as the Chambezi.[8] This river is very wide[9] and important; it flows eastward or towards sunset.[10] The chief or sultan of this place was called Kasaramarama. When we crossed the river there was almost a fight and the natives did not want to give up their canoes. When we had gained the other side of the Luapula, this was the land of Kawenday.[11] They say that

---

[1] Better known as Chitambo. A headman among the Lala, who claimed himself to have travelled to the coast, he was deposed by Bemba raiders in 1884. Roberts, *Bemba*, p. 163; *Last Journals*, II, p. 314. The place, now marked by a memorial to Livingstone, is actually on the River Lube, a small tributary of the Lulimala at approximately 12°22′S, 30°16′E.

[2] On the question of the date, see Introduction, above, pp. 355–6.

[3] Wainwright does not mention the fact that he read the burial service from his Prayer Book as Livingstone's heart and viscera were interred. *Last Journals*, II, pp. 316–17. He later told the CMS, 'After the burial, we sat down and cried.' Stock, *History of CMS*, III, p. 78.

[4] Wainwright is using the correct Swahili word for *Parinari curatellifolia*; most sources have simply obtained the local Bemba name, Mupundu, without knowing what sort of tree it really was. See Fig. 7.

[5] Livingstone had divided his party into three sections of which these were the leaders. Yazuza is Susi: see above p. 355; Mnyasere is Manua Sera who served in many other expeditions and died in 1888; Chopere is Chowpereh, b. c. 1840. See Simpson, *Companions*, pp. 192, 195. The inscription is often transcribed with Chopere, shown as VCHOPERE as in Simpson p. 96; surely the 'v' must be a result of Wainwright's attempt to carve an ampersand.

[6] Wainwright later wrote: 'It must be remembered that it was entirely our own sence [*sic*] or consultation to preserve the body as [*sic*] to bring it to the coast.' NLS Acc12444, Wainwright to Moore, 23 May 1874.

[7] A note to the German text by Petermann says 'In the original text this name was written as Lwapla' '*Tagebuch*', p. 188n.

[8] This is correct; it is said that the continuous stream is discernible through lake and swamps.

[9] Four hundred yards wide near the Lake. Hughes, *Bangweulu*, p. 2.

[10] Presumably, Wainwright got east and west mixed up in his mind and should have written 'westward'.

[11] As the party crossed the Luapula they left the land of the Lala and were to move northwards through territories lying to the west of Lake Bangweulu occupied by the Ushi and the closely-related Kawendi, Ngumbu, Mukulu and Chishinga. There were also enclaves of Bisa and the lake-dwelling Unga and Twa. Whiteley, *Bemba and Related Peoples*, pp. 1–2, 12. All these peoples, too, are part of the 'Bemba cluster'. Murdock, *Africa*, p. 294.

Figure 7: The mupundu or mbura tree beneath which Livingstone's heart was buried. Photographed about twenty years later. Reproduced by permission of the Royal Geographical Society.

this country is full of wild animals like lions, tigers[1] and elephants, and that saying is quite true because the same night, at about 10 o'clock, when we were all asleep, two or a pair of lions came and killed our fine donkey which had served us well for our sick persons. That night, we did not sleep at all because of the roaring of the lions, but kept a watch and fired at them, whenever they tried to get near. We had some fear they might come back and attack human beings. The natives and all people said that people or men who know about black magic change into lions and roam around to kill people.[2] From that time onwards we were very careful when building our boma or fence.

As we did not find any village or houses on the next day, we went and took camp on an extensive hill.[3] On 30th June, we arrived at a town whose chief is called Mchisweesi;[4]

---

[1] Presumably leopards. Hughes, *Bangweulu*, p. 138, reports leopards in this area.
[2] Hughes also reports this idea which can be found among other peoples. Ibid., p. 72.
[3] In fact a large anthill. *Last Journals*, II, p.323.
[4] Susi and Chuma call the chief Kawinga. Ibid., p. 324.

the name of the town is Chitundwa. On the following day, 1st July, we came into a town whose chief or sultan is Makumba, but his brother's son rules over the town.[1] In that same city, it is said, the sultan's son does not succeed upon his father, but other relatives; at present the sultan ruling the town is named Inkoso. This chief had six cows. We intended to buy one for our soldiers but the cows of this country are different from the cows of India and not unlike the cows of England, they having no hump on their backs[2] and when they take to the meadows nobody will tend them. If the owner wants to have one of them killed, he slaughters them with a gun or spear, the reason being they are difficult to catch. When we wanted to slaughter the cow we had bought, we had to kill it with a bullet. On this occasion one of our bullets which had been shot carelessly by one of our soldiers, went and broke the thigh of a native.

Leaving this place we came to a town the chief of which is named Msanga. On 5th July we reached a town whose chief is called Insokoro; this native chief has a kind heart presenting us with half of his field of cassava, an edible root.[3] On 6th and 7th July we pitched camp in the bush, and on the next day, i.e. the 8th, the scant provisions we had carried with us were exhausted. On this day we had been travelling for longer than we usually went each day, in the hope of finding a village or town to buy some food so that we might regain strength. Fortunately we came upon a large town but the people are very bad and uncivil there and did not permit us to enter their town, and as we came from the wilderness, they believed us to be quite weak and unable to accomplish anything. One or two of our men who were at the front went and tried to get into the boma, but the people of the place shot their arrows and spears at them, and one spear went and hurt the arm[4] of one of our people; when we saw this we all dashed in, and within a minute they all turned their backs. During the night, however, they all came with their weapons to destroy us. But thanks to our watchfulness, they could not get near us. The people themselves had thought the town to be unconquerable.[5] On the third day they came on their own to ask for peace and brought us a slave; but we were in doubt as to whether this was sincere or not.[6] Consequently, to obtain more security, we sent them back to their chief telling them a number of conditions of peace, but we did not get an answer from them. The chief of this place had eight bomas some of which

---

[1] All the Bantu of the Bemba cluster are matrilineal; normally, brother succeeds brother and, after the latter, the sister's son. These societies are also characterized by 'positional succession' by which the inheritor takes on all the attributes of the person he succeeds, not least his title. Hence the confusion over names which occurs in accounts such as Wainwright's; a man may be known by his own name or his title or his praise-name. See Murdock, *Africa*, pp. 299–300; Whiteley, *Bemba and Related Peoples*, p. 27.

[2] *Bos taurus* are the cattle without humps but the evolution of breeds is a complicated story. Probably what Wainwright saw were 'sanga', a crossbreed of humped zebu and longhorns. Middleton, *Encyclopedia Africa*, I, pp. 45–50.

[3] Cassava or manioc, *Manihot aipi*, was introduced to Africa from the Americas at an unknown date and became an important secondary or even staple crop for many Africans. Hughes, *Bangweulu*, p. 69, says it is a biennial crop in this region.

[4] Waller's account of this episode says it was the man's thigh which was hurt. *Last Journals*, II, p.328.

[5] In Petermann's German text, 'Tagebuch', p. 189, the punctuation implies that this clause is linked to the previous one but I believe it relates to the one following and explains why the inhabitants now sued for peace. Bontinck also takes this view but considerably modifies the translation to make his point. Bontinck, 'Diaire', p. 416. Accordingly, 'had thought' seems a more appropriate tense for the verb.

[6] Waller's account says there was dissension among the inhabitants. *Last Journals*, II, p. 329.

Figure 8: The carving made on the mbura tree by Wainwright. Reproduced by permission of the Royal Geographical Society.

were burned.[1] On the 12th we proceeded,[2] leaving them everything. The name of the chief is Kiwenday, Kitondwa is the name of his city, Kapunda is a river in its vicinity, Tingatinga and Caneoswe are smaller streams in that area.[3]

---

[1] That is, by the Livingstone followers; they were now acting as predators in order to ensure they obtained supplies and passage through the area. Waller's account says six of the bomas were burned. Ibid., p. 328.

[2] Thus, after four days but Waller says the party remained there for a week. Ibid., p. 328.

[3] The information in this sentence makes it possible to correct that given by Waller in *Last Journals*, II, pp. 326–9. He assumed that the party had now reached the 'north shore' of the lake (p.326). Bontinck, 'Diaire', p. 417, n. 75, assumes that Waller misinterpreted what Susi and Chuma told him. Whilst this is part of the explanation, the basic problem lies in the fact that Livingstone made mistakes in his observations for longitude during his visit to the lake in 1868. Attempts to reconcile them with the observations of 1873 led to the elongated east–west shape given to the lake in the map in *Last Journals* as is explained by Debenham, *Way to Ilala*, pp. 251–2. In fact, the party were still at the south-western tip of the lake among the Ushi. The place Kitondwa or Kitondwe and the river Kapunda are identified by Bontinck by reference to the work of later travellers, especially Giraud. 'Diaire', p. 417n. Hughes, *Bangweulu*, shows 'Kawendi' on his map as a swampy area which may raise the possibility that Wainwright's name for the chief was actually an area or, more likely, the name of the tribal group (see Hughes, n. 19) who do actually occupy the area immediately to the south-west of the lake. The term 'Tingatinga' indicates a quicksand or swampy area which, again, suggests the location shown on Hughes' map.

On 12th and 13th July, [we were] in the bush. On the 14th we found our way among a maze of paths to Mumbu; we did not see anything along the way except destroyed towns and villages.[1] At last we reached a flooded plain, 900 feet wide;[2] when [we were] wading through it, the water made one of our small boys grow stiff, although it was not intensely cold. At 1 o'clock we camped on the banks of a river named Litandasi; on the other side the river surrounds a few houses on an islet. The name of the chief is Kusumwa.

Short Description of Lake Bemba
Bemba[3] is a lake formed by the Chambezi River.[4] It is of considerable size and surrounded by plains which together make an enormous flat expanse devoid of wood. It is totally unlike Lake Tanganyika, which is surrounded by green hills and mountains.[5] The water covering this enormous space[6] is fordable only in summer, while in the wet and winter times you can cross it only with canoes. There are numerous islands in it, and the inhabitants, who have uncountably many boats,[7] use these to carry on trade or communicate with one another. Many of the islands are found densely inhabited by people who leave the mainland for fear of enemy attacks.[8]

The natives are richly provided with excellent fish,[9] but although the occupants do not have a naturally beautiful countryside, they are very diligent in cultivating their fertile fields which offer pasture for their cattle.[10] When the observer sees the morning dew on the plains, they appear to him like a vast expanse of the sea. There are zebras,

---

[1] The Kawendi area had been heard of by Livingstone as having been overrun by Nyamwezi. *Last Journals*, II, p. 282. These must have been the 'Yeke', that is, the followers of the Nyamwezi warlord Msiri who was establishing his 'kingdom' further north in the Cazembe's territory and who was sending raiders south from there at this time. See Roberts, *Zambia*, p. 123 and also Introduction, above p. 370.

[2] Presumably the swamp area near the modern Kasoma.

[3] The now accepted name is 'Bangweulu' which signifies 'where the water meets the sky'. Hughes, *Bangweulu*, p. 1. Livingstone learned that the use of 'Bemba' properly related to the smaller (permanent?) part of the water and that 'Bangweolo', as he spelled it, was more correct. *Last Journals*, I, pp. 308–9. The lake was heard of by Lacerda in 1798 but not visited and put on maps until after Livingstone reached it in July 1868. The French traveller Giraud made a reasonably accurate survey in 1883.

[4] The Chambeshi flows through the swamp area to the south and east of the main water; it is not therefore strictly a feeder of the permanent part of the lake.

[5] No one had as yet realized that Tanganyika and many of the other great lakes of East Africa were rift valley features quite different from lakes like Bangweulu or Victoria. The distinction made here shows that Wainwright had at least something of a geographer's interest in the nature of the physical landscape he had encountered.

[6] The normal size is said to be 1,900 sq. miles but it is 4,500 in the wet season. 'Water, water, everywhere', said Livingstone. Ibid., II, p. 292.

[7] Long narrow dugout canoes according to Hughes, *Bangweulu*, p. 3.

[8] The lake area was something of a war zone in the mid-nineteenth century with Bemba–Bisa rivalry plus incursions from outside by the Yeke and the Mazitu (Ngoni). Livingstone refers to Bisa seeking refuge from the Mazitu by retreating to islands. *Last Journals*, II, p. 302. See also Roberts, *Bemba*, pp. 117, 140.

[9] Mostly various kinds of *Tilapia*. The Unga, Twa and the Bisa who had retreated to islands in the lake were renowned as fishermen and traded their catches.

[10] There are rich soils in certain areas among the swamps and intensive agriculture based on maize, cassava and cattle herding is carried on. See Roberts, *Zambia*, pp. 9, 11, 102.

antelopes,[1] and other deer. If anybody travels on these plains he sees only a tiring and sad monotony. The Chambezi River has a different name farther downstream (Luapula), and on its further course, the natives give it different names. It has been observed that Chambezi River has formed several lakes other than Lake Bemba.[2]

The March to Unyanyembe
On 16th July, after crossing the Ticali River, we came to the villages of Chief Chama,[3] and pitched our tent in the vicinity. On the next morning we arrived at the town itself on Nyinawambusi River (Goat's Mother), which empties into Lake Bemba. Here we spent a day because of the hardships we had had to endure. After crossing Cowloo and Rooda Rivers on 20th July, these rivers being slightly less important and flowing into the same lake, and after crossing several rivers and rivulets on 21st July, we finally reached the town of Sultan or Chief Mwege.[4] In the beginning we tried hard to find accommodation in the houses, but the great mass of the people appeared to oppose that when they had seen the corpse, and we obeyed when they told us we should have to go and build our huts at a distance of 300 feet from the town. After we had done this we watched them bringing their drums out and accompanying their dances with loud voices, and offering other foodstuffs for sale.

Here we left our two poor ill soldiers behind, and told the sultan to care for them until they would be able to accompany a caravan to Unyanyembe.[5]

On the 23rd after we had crossed the two rivers Onato and Mamiyanda[6] which both fall into Lake Bemba, we arrived at the town of Sultan Chiwaye. The wicked natives put up a strong opposition [to us]after having seen the corpse that we were carrying,[7] but after we had made it understood that we should not touch any of their property but spend the night in secluded places, they left us and went their ways. After one or two hours they caught up with us and brought food such as jowari or millet, batata,[8] flour, all kinds of beans, an edible root and other grain. In fact, these people busily till their fields year after year. In the afternoon the chief himself came into our camp to learn the true facts of the case. On being asked about his name he replied that his name was

---

[1] The sassaby is the commonest type according to Hughes, *Bangweulu*, p. 235.

[2] It is not clear whether Wainwright means several small lakes such as Chaya upstream or, when the river has become the Luapula, large lakes downstream, notably Mweru.

[3] A Bisa chiefly title. See Roberts, *Bemba*, pp. 108–9n.

[4] His ethnic designation is not clear but he was perhaps of the Ngumbu group. He is not mentioned in the account by Susi and Chuma but Waller writes of scattered huts of Ngumbu's; perhaps the misleading apostrophe was put in by accident. *Last Journals*, II, p. 330. Harry Johnston's map identifies the place in about 10°40′S, 29°35′W. Johnston, *Livingstone*, facing p. 317. This seems to be a little too far north and east. In any case if this were anywhere near correct, the party had now left the immediate lakeside but Waller says they were still in the inundated fringe. *Last Journals*, II, p. 329.

[5] This was not reported in *Last Journals*.

[6] Possibly an error of transcription by Brenner for Mwampanda which is in approximately 10°45′S, 29°25′E.

[7] Susi and Chuma reported that it was because of the flags the party were displaying. *Last Journals*, II, p. 330.

[8] Sweet potatoes, *Ipomoea batatas*.

Chiwaye, and Mtondo the name of his town.[1] We spent two days here, to barter for the food that seemed so cheap to us, and to prepare it for the regions that lay before us.

On 26th July, we slept by a stream near a village. On the 27th we reached the town of the chief whose name is Kanamamansata,[2] on Manika River. On 28th July we passed the town of Chilapanewa, who is a brother of Choongu's[3] wife, and we slept near a destroyed village.

At this time, and since our departure from chief Kitumbo,[4] we had been in the dark in respect of our route or the path we had taken before. When we sought information about the town of Chief Choongu,[5] we were told that it was situated at not so great a distance to the eastward from us. However, we were [now] in a position to locate our route or to guess it.[6] On 29th July, we came to Cumbru River, a tributary of another one named Lupupuse.[7] The current of this river is exceedingly strong even in summer; the banks are densely forested. This river is one of the biggest ones; crocodiles and other animals are hardly to be found in it. The chief named Mulalawampa is a subordinate of Choongu's. At the end of July we camped by a small river called Kasewa.

On 1st August, we reached the kingdom of Kapesa,[8] with whom we had previously stayed when we were with Dr Livingstone. The country is not very swampy, is of great extent and hills are scattered here and there. The people were assembled with their arms, believing a gang of robbers were approaching to attack their town.

The inhabitants themselves are much addicted to the intoxicating drink they prepare with their own hands, and consequently pay little attention to the tilling of fields. The river is the Chirima.

On 3rd August, we came to the border of Chief Chama's country[9] and slept in deserted houses. On the 5th [we were] at a place in the wilderness. On the 6th we took camp in a

---

[1] Mutondo is the area of the valley of the Luposashi River which enters Lake Bangweulu at its north-west corner.

[2] The uncle of Chiwaye according to *Last Journals*, II, pp. 330–31.

[3] See n. 5, below.

[4] That is, Chitambo's where Livingstone died.

[5] Chungu was a chief among the Mukulu, a group related to the Ushi and living to the north of the Lake. See Roberts, *Bemba*, pp. 110, 116. The modern settlement of Chungu lies on a tributary of the Luposashi at about 10°31'S, 29°37'E.

[6] Presumably because they had come around the western side of the Lake and were now to the north of it and nearing the route taken when they were coming south with Livingstone.

[7] Lopoposi according to Livingstone and flowing south towards the Chambeshi. *Last Journals*, II, p. 261.

[8] A chief whom Livingstone met in 1873 and called Katebé or Kapesha. *Last Journals*, II, pp. 261–2. More correctly Katele, he was a chief of the Tabwa people and the son of their paramount ruler Nsama III. Perhaps Katele welcomed Livingstone and now his followers as possible counterweights to Tippu Tip's men as Tippu extended his ivory and slaves empire in this region. For his defeat of Nsama, see his autobiography, *Maisha ya Hamed bin Muhammad*, paras. 21, 22.

[9] This was held by a group known as the Chisinga, closely related to the Ushi and noted for their ironworking skills. They were much affected by disorders consequent upon Tabwa–Bemba rivalry. Roberts, *Bemba*, pp. 157, 186.

little village in one of Chama's countries. On the 7th we crossed the river of Loofoomi, and came to a village, pitching our tent there; [this was] where we had spent Christmas Day of 1872 with Dr Livingstone.[1] The chief of the place is Mgosa. On 8th August [we reached] a small village with few houses, but the village was deserted for fear of the people or robbers.

On the 10th we crossed Kalangweese River,[2] a very important river. The width of this river amounts to about 300 yards (900 feet), or, if a stone is thrown from the opposite bank, it will not easily drop onto the other one. It is teeming with fish, alligators, hippopotamus and sea-horses.[3] On the following day, we slept in a little village near that same river, [and then] on 14th August at a small river of the name of Chikatwa, but the people of the place left their houses and everything, being afraid of the Waniamwezi who roam around to lay waste all the lands of Chama and Sama.[4] Within a short time it will be very difficult for travellers to traverse these distant regions, and it is unlikely that these people will settle down again at their former places. On 15th August [we camped] near a small village, but all the houses were burned down and deserted; we did not find anything more than elephants' tracks and those of other wild animals. In former times this village was convenient for travellers.

Before we arrived we met with a large snake which is called a Cobra.[5] It approached us at great speed and stood on the extreme tip of its tail. Three or four bullets were fired at it, but only a couple of them threw it to the ground; some of us rushed at it and cut into its head with a sabre. It was seven feet long.

On 16th August, we camped in the wilderness, on the 17th in an old town the home of Waniamwezi,[6] on the river of Mkowey in Sama-Land.[7] On 18th August, [we reached]

---

[1] *Last Journals*, II, pp. 258–9. The 'Loofoomi' must be the Lufubu, mentioned by Livingstone. Not to be confused with the river of the same name flowing into Lake Tanganyika, this flows north-west into the Chisaka which joins the Kalungwishi to feed Lake Mweru. The party had now, therefore, passed the watershed between Lakes Bangweulu and Mweru and must have been at roughly 10°S, 29°40'E.

[2] See previous note; the Kalungwishi is the major feeder of Lake Mweru apart from the Luapula.

[3] The 'alligators' are presumably crocodiles but it is not clear what is meant by 'sea-horses'; the German word given is 'Seepferden' (spelt thus rather than the usual modern rendering 'Seepferdchen'). Sea horses, *Hippocampina*, are entirely marine creatures. Perhaps Petermann misunderstood whatever Wainwright wrote. What seems most likely is that, in his usual style, Wainwright was giving two alternative words and actually wrote 'hippopotamus or sea-horse' assuming that sea-horse is a synonym for hippopotamus.

[4] In this context, 'Chama' refers to a Chisinga chief and 'Sama' (Nsama) to the leader of the Tabwa (Itawa). At this period, Msiri's Yeke were raiding into this area and so the reference to the Waniamwezi must mean them. See Roberts, *Zambia*, p. 123. Bontinck, 'Diaire', p. 421, n. 96, asserts that Wainwright must really mean Swahili followers of Arab traders but it is unlikely that he would not distinguish Swahili from Nyamwezi.

[5] The African cobra is *Naja haje*.

[6] Presumably the Yeke.

[7] That is, the land of the Itawa or Tabwa whose paramount chief was entitled Nsama. The third Nsama had been defeated in a battle of 1867 and died soon after. Nsama IV now reigned but real power in the area had passed to Tippu Tip. The 1867 battle was decisive because it meant that the Arab warlord and trader had gained control of the route around the southern end of Lake Tanganyika and so access to the rich ivory resources of the Upper Congo. More locally, the battle weakened the Tabwa in relation to the Bemba. *Maisha ya Hamed bin Muhammed*, paras 21–8; Roberts, *Bemba*, pp. 152–3, 195. On the growth of Arab/Swahili settlements and their impact on this region, see Wright and Lary, 'Swahili Settlements', pp. 555–6.

a village after a long laborious day's journey. One fell behind on the way and was lost in the bush, and what became of him is hard to say; he was one of our brothers named John Wainwright,[1] who commanded a considerable knowledge of the English language. He was searched for far and wide, but all in vain.

On 19th August, we arrived at the permanent town of Kumbakumba[2] who waited for the return of his brother Tipulipu, who has gone again into the far interior.[3] The number of his men is in fact large [and] he had stayed there almost four years in succession. He does not live here alone, but Pemba Moto[4] and Sayed-ben-Ali[5] are with him. The place where Kumbakumba lives appears desolate on account of the neighbouring stony hills, but in the valleys his numerous servants and slaves are occupied with the cultivation of his rice fields, from which a rich harvest is reaped year after year. Most of the soldiers live by robbery and when they have got the plundered spoils, they capture the natives and force them to carry the loot. Later they discharge the bearers or keep them tied up until their relatives come to ask for their freedom but [they have to] pay an amount.[6] The river is called Mtosi;[7] on the eastern side of the town there is a large pond, named Lamba, swarming with hippopotamus and crocodiles. The people themselves report, according to their superstitious ideas: 'an evil spirit [devil] lives in the pond.'

Early on the morning of the 27th August we set off, rather light-heartedly this time, because we had news of Unyanyembe and of the island of Zanzibar as well.[8] We were also informed that the march to Unyanyembe takes two full months. We had hoped to remain here several days but we went on because of the scarcity of our provisions.[9] [We] crossed the Mombasi River and stayed in Chief Chilenba's village. He said he was the son of King Sama. King Sama's country is called Eetawa (Itawa).[10]

---

[1] He was not, as far as can be ascertained, a brother in the biological sense but there is little information on him.

[2] This was Muhammad bin Masud el Wardi who was, strictly speaking, Tippu Tip's half-brother. He returned to Zanzibar in 1876. Bennett, *Arab versus European*, p. 245. Kumbakumba's town was at or near Chief Nsama's residence at approximately 8°55′S, 29°57′E.

[3] This is yet another version of Hamed bin Muhammed's nickname which he is said to have been onomatopoeia for the crackle of his guns in the 1867 battle. He remained in the Luapula valley building up his 'empire' before returning to the coast for a time in 1880 with a haul of 2,000 tusks. *Maisha ya Hamed bin Muhammed*, Historical Introduction, p. 15.

[4] A Swahili associate of Tippu Tip, the date of whose death is somewhat uncertain. Roberts, *Bemba*, p. 155n. This evidence shows that he was still alive in August 1873.

[5] Said bin Ali bin Mansur el Hinawi. Bontinck, 'Diaire' p. 422n.

[6] A succinct but revealing account showing how predation was now being substituted for trade after the Arab conquest and evidence, too, that in this region, unusually, slaves were forced to carry ivory

[7] Another sub-feeder of Lake Mweru.

[8] This was apparently news of the expedition led by Verney Lovett Cameron. These contacts with people from Unyanyembe and the coast meant also that news of Livingstone's death was carried ahead; by 26 January 1874, rumours had reached London. *Proc. RGS*, 18, pp. 131, 145.

[9] If Wainwright's dates are correct, they had remained eight days at Kumbakumba's but Susi and Chuma told Waller it was only five. *Last Journals*, II, pp. 332–3.

[10] The son was more correctly Chilembo. See also p. 370, n. 7 above.

On 28th August we reached Mkotwe River which runs into Lake Tanganyika. On 29th August we saw a great many columns of people at Ruvuga River where we slept.[1] These people were returning from the wars of Kumbakumba. When asked about the result of the war, they said it had been undecided because of the land being flooded and various other unfavourable circumstances.[2] They added that we had permission to cultivate our fields. The bed of this river seemed to be naturally paved. The stones are of various shapes, some like a triangle and others really the shape of a rowing boat. The banks are covered with very thick forest, the home of elephants and tigers.[3]

On the afternoon of 30th August, we camped in a small village of Chief Chimeata's, a subordinate of Choongu, on the river of Kalemba. On the 31st we spent the night in Choongu's town on the river of Tumba. It must be noted that Marungu land starts at Ruvuga River.[4]

The Character of the Wabisa Tribe

The Wabisa[5] are one of the tribes who use bows and arrows; some use spears without shields. They lack courage, cleanliness and honesty. In one part of the country people are found busy tilling the fields; the chief grains are maize or Indian corn, millet, ulazi (a grain of small round seeds).[6] There are also cucumbers, pumpkins [and] here and there bananas, batatas, and some edible roots.

In general, they never build permanent houses like other people, but leave their houses after two to four years to occupy another place. When they salute their chief, they kneel or roll on the ground clapping their hands. Their clothing is made of skins and barkcloth. All kinds of beads are useful in these regions, lead, too, and even paper can be used as a means of barter. The people are so ignorant that they [will] take tin bullets

---

[1] The Mkotwe seems to be a tributary of the large river, the Lufubu, which Wainwright renders as 'Ruvuga'and which, as he notes, drains into Lake Tanganyika. Significantly, the party had now crossed a major watershed, as no doubt Wainwright realized.

[2] Whether these were Kumbakumba's allies or his enemies is unclear but this is further evidence of the unsettled state of the area.

[3] Presumably, again, leopards.

[4] The party were now in the territory of the Lungu people ('Marungu'), specifically the more northerly of this people's two sections living nearer Lake Tanganyika. Chungu was the title of chiefs who claimed some sort of paramountcy; this one or his predecessor had lost power as a result of confrontations with the Bemba. Roberts, *Bemba*, pp. 148–9; Willis, *Fipa and Related Peoples*, pp. 39–46.

[5] The Bisa live mainly to the south and south-east of Lake Bangweulu and so it is not clear why Wainwright introduces his account of them at this point in his narrative of the journey. They had emerged to historical significance as traders who carried ivory and copper from the Kingdom of the Cazembe near Lake Mweru south-eastwards to the Yao in the Lake Malawi area and sometimes themselves went on to the coast at Kilwa. Yet Wainwright clearly knows nothing of this; his account is negative evidence of the decline of the Bisa and the diminished importance of the trade route which occurred both because of Ngoni attacks and the capture of the trade by Nyamwezi or Arabs who opened up more direct routes to the coast. St John, 'Kazembe and the Tanganyika–Nyasa Corridor', pp. 202, 212–3, 217. However, Gouldsbury and Sheane, *Great Plateau*, pp. 13, 15 reported that in 1911 there were still Bisa who had been to the coast and could speak Swahili and they generally give a much more favourable impression than Wainwright.

[6] The parentheses are Petermann's. 'Ulazi' or 'ulezi' or sometimes 'mawele' is 'bulrush millet', *Pennisetum malacochaete*.

for lead. When the king or sultan[1] dies, they never bury him on the same day but only after about fifteen days and in the meantime a large low-sounding drum is beaten at one-minute intervals until the appointed time comes when everyone is assembled to take part in the funeral ceremony. If an ordinary person dies, the corpse is made into a bundle after which it is buried in the grave.

***

Two months had gone by with our crossing the lands of Marungu and Fipa.[2] The rivers are more or less important, their banks covered with impenetrably dense forest in which the natives themselves keep all their provisions so useful to them in times of war. Some dig pits and keep all their grain in them.[3] The land of Fipa is very mountainous. Rivers and products are the same as in Marungu. Rice is unknown except at the depots of trading caravans from the coast. In the middle of the region[4] there are large fertile areas, immeasurable expanses almost totally treeless, tiring and sad for the human eye.

On 8th October we came to the country called Ukhonong[o] or Unyamwezi.[5] When we arrived at the town of the Sultan Mbowéa[6] and heard the interesting news of the arrival of Oswell Livingstone, the great explorer's son, we wrote a letter to him at Unyanyembe.[7] A few days later when the messengers returned, they told us he was not

---

[1] This must mean any leader; there was never an overall leader of the Bisa.

[2] The precise route across these two territories is difficult to establish. Waller's account in *Last Journals*, II, pp. 334–7 is rather general but by reference to this and his notes from Susi and Chuma in Bodleian Library MS Afr S 16/4, one can build up a succession of chiefs or villages after the visit to Chungu: Chitimbwa; Kasa Kalaisé; Nsombi; Kafufi (he was a powerful Fipa chief in 1860: Willis, 'Fipa' p. 90); across the plain to the north of Lake Rukwa to Konongo (an area and people part of Greater Unyamwezi to the east of the modern town of Mpanda and where the rivers drain into Lake Tanganyika via the Ugalla and its tributaries); Baula (Mbowra); Chitunda; Manueras; Machukuru; Ngundu (in the chiefdom of Ugunda); Kasekera (see p. 375, n. 5, below); Kanyonda; Unyanyembe. The route is almost certainly along the line of the trade route opened up in the 1850s by Arabs and Nyamwezi from Unyanyembe to the southern end of Lake Tanganyika.

[3] This practice may be a consequence of the Ngoni invasions of 1841–2 or the frequent incursions by Bemba raiders. Willis, 'Fipa', pp. 88–91.

[4] Bontinck, 'Diaire', p. 425, n. 118, refuses to translate 'der Mitte der Region' as 'au centre'. However, I believe Wainwright, who clearly had an eye for country, meant precisely this; the middle of (U)Fipa is different from the western margin which he had seen the year before. Willis, *Fipa and Related Peoples*, p. 17, echoes the succeeding comment by talking of 'an undulating and largely treeless plateau'.

[5] The Konongo are usually regarded as a distinct ethnic group within the larger Nyamwezi cluster. But precise definition was characteristic of colonial rule in the twentieth century; in the nineteenth, the situation over designations was more fluid.

[6] Mbowra would be a more accurate transcription. It is on the banks of a tributary of the Rungwa in about 6°40′S 32°E and in Konongo territory.

[7] The letter was written in a bold hand by Wainwright himself as follows:

Ukhonongo October 1873

Sir We have heared in the month of August that you have started from Zanzibar for Unyanyembe, and again and again lately we have heared your arrivel. Your father died by disease beyond the country of Bisa, but we have carried the corpse with us. 10 of our soldier are lost and some have died. Our hunger presses us to ask you some clothes to buy provision for our soldiers and we should have an answer that when we shall enter there shall be firing guns or not, and if you permit us to fire guns, then send some powder we have wrote these few words in the place of Sultan or king Mbowra.

The writer Jacob Wainwright Dr Livingstone Exped

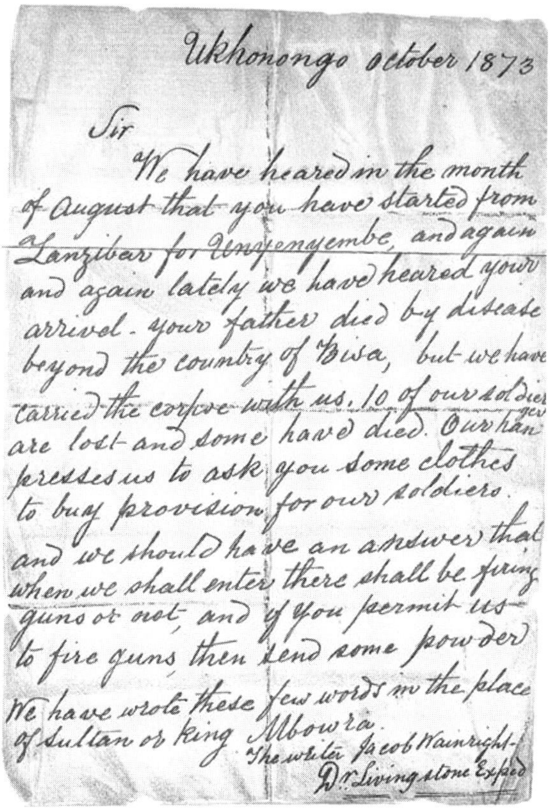

> Ukhonongo October 1873
>
> Sir
>
> We have heared in the month of August that you have started from Zanzibar for Unyenyembe, and again and again lately we have heared your arrival. Your father died by disease beyond the country of Bisa, but we have carried the corpse with us. 10 of our soldiers are lost and some have died. Our hunger presses us to ask you some clothes to buy provision for our soldiers. And we should have an answer that when we shall enter there shall be firing guns or not, and if you permit us to fire guns, then send some powder. We have wrote these few words in the place of Sultan or king Mbowra.
>
> The writer Jacob Wainright
> Dr. Livingstone Exped

Figure 9: Wainwright's letter to Captain Cameron. Reproduced from Cameron, *Across Africa*.

Livingstone's son[1,2] but [there were] several other gentlemen [including] Lieutenant Cameron. When we heard this our hearts sank or, rather, we were in despair. However, Lieutenant Cameron saved us by sending us a bale of cloth and two tins of gunpowder.

Cameron, *Across Africa*, I, facing p. 165 reproduces this letter in facsimile and it is reproduced again here as Fig. 9. 'Clothes' should be read as 'cloths', i.e. bales of cotton cloth used to barter for food. Whether to fire guns needed to be asked about as there could be misunderstandings about firing to announce one's arrival.

[1] The messengers were led by Chuma who arrived at Unyanyembe on 16 October. The confusion surrounding this date results from mistakes made by Cameron which he repeated in *Across Africa*, I, pp. 164–5 and which misled Waller in *Last Journals*, II, p. 338, and Simpson in *Companions*, p. 99. See Bontinck, 'Diaire', p. 426 n. 122. In fact, Cameron wrote a letter to the RGS on the day that Chuma arrived dated 16 October announcing the arrival. Murphy confirmed the date in a letter of his and also specifically referred to Wainwright's letter. These letters reached Zanzibar on 3 January 1874 and were under discussion at the RGS in London on 23 February. *Proc. RGS*, 18, pp. 177, 180.

[2] Oswell Livingstone had come to East Africa with the Dawson Relief Expedition in 1872 but when that expedition collapsed following the appearance of Stanley after his famous meeting with Livingstone, he decided, after much hesitation, not to go up country to meet his father. Bridges, 'Sponsorship', p. 100.

At the beginning of November, we reached Unyanyembe.[1] Two gentlemen had accompanied Lieutenant Cameron, Mr Moffic[2] and Dr Dillon. The latter died or killed himself with a gun in his tent as he was suffering from a painful disease and was buried on 24th November 1873. We had spent half a month or longer at Unyanyembe because of our business or with preparations for the caravan.[3]

The March from Unyanyembe to Zanzibar

On 18th November,[4] we left Unyanyembe with the two gentlemen who had come from Zanzibar to Unyanyembe with Lieutenant Cameron. On the same day, Lieutenant Cameron departed for Ujiji. On 20th November, we reached Kasekera, the town of Chief Mananywa;[5] in this same town Dr Dillon who had been very sick, killed himself with a gun in his own tent on 24th November, and was buried on the following day.[6]

10th December. It is raining a lot in Ugunda,[7] a big place about two days' journey from Kasekera. There is a great extent of field cultivation here.[8] 12th December, [we] set off and travelled for only three hours, and came into a village on the border of Ugunda where we could sleep in order to buy provisions, as we were about to cross an uninhabited region. 13th December, [we] slept in the wilderness. 14th December. When we were travelling in the same dreadful wilderness, a young girl was bitten on the way by a big snake. Many attempts were made to save her life and a large pair of fangs were pulled out of the wound in her thigh, but after 15 minutes, foam came from her mouth while the poison spread all over her body. Then, after [another?] fifteen minutes she took her last breath and she was immediately buried. 15th Dec., slept in the wilderness. 16th at 11 o'clock, [we] arrived at a town called Ngooru.[9] A famine rages in this town, and the natives themselves, women and children, enter the woods to

---

[1] There is confusion also about this date; it was probably 25 October. Bontinck, 'Diaire' p. 426 n. 123. In Unyanyembe was the Arab settlement of Tabora, approximately 5°03′S, 32°48′E.

[2] In fact, Lt Murphy.

[3] Unfortunately, Wainwright gives no details. Waller's account in *Last Journals*, II, pp. 338–40, is guarded but one can guess there were tensions; Waller clearly disapproved of Cameron's taking some of Livingstone's possessions for his own use and of the way the naval officer had tried to persuade the African party to give up Livingstone's body for immediate burial.

[4] Cameron says it was the 9th. *Across Africa*, I, p. 171.

[5] Kasekera or Kasegera is to the south of the Arab settlement of Tabora but still in the Nyamwezi chiefdom of Unyanyembe. Mananywa was the daughter of the chief of Unyanyembe and ally of the Arabs, Mkasiwa. Ibid., p. 178.

[6] Chuma and Susi described this episode in some detail to Waller and their testimony suggests that Murphy behaved rather casually or even callously: 'Murphy did not help or take any interest'. MS Afr S 16/4. See also *Last Journals*, II, p. 341, for further criticism of Murphy.

[7] The Nyamwezi chiefdom of Ugunda was in the 1840s on the main route from the coast to Lake Tanganyika but was now rather superseded by Unyanyembe a little to the north where the Arab settlement of Kaze, later known as Tabora, was established. Burton, *Lake Regions*, I, pp. 322–8; Roberts, 'Nyamwezi Trade', p. 50.

[8] Cameron echoes Wainwright's description saying that the name signifies 'a country of farms'. Cameron, *Across Africa*, II, p. 299.

[9] Ngulu was a chiefdom founded at about the end of the eighteenth century among the Kimbu. It was raided by Ngoni some time before 1845. Shorter, 'The Kimbu', pp. 104, 108. The Kimbu, usually regarded as part of the 'Nyamwezi cluster' claim to be an 'older' tribe than the Nyamwezi proper. Abrahams, *Political Organization*, p. 5.

gather wild fruit in order to make their bare living with them, and several also travel to Ugunda to buy grain for sowing and for household usage. The famine is supposed to have been caused by neglect of cultivation in the fields.[1] 17th Dec., departed, went and slept in a village belonging to chief Mahaloole.[2] 18th December in the wilderness. 19th in a village. 20th Kisaroosaroo.[3] 21st December, slept at a village called Kipira or Kipiri.[4] 22nd. Set out early in the morning, [and] arrived at a town of the name of Jiwe-Lasinga at 12 o'clock, so called after a stone upon which grass has grown like hair. The chief of the whole town is Masanja;[5] the inhabitants are the Wakimboo, lovers [sic] of cows and slaves.[6] In spite of their diligence with field cultivation, they are unfamiliar with rice fields; there is much jowari, bajree,[7] maize or Indian corn, and batatas. Many coastal inhabitants have made their homes here. The original king of this town is supposed to be Nyungoo, the son of the sister of Mkasiwa, king of Unyanyembe.[8]

Report on Unyamwezi

The Wanyamwezi are called Wakhonongo[9] by some, and are of a superior race in this part of Africa. They very much love fighting or war, are a courageous, industrious and strong race of men, and their music is the sweetest among all of Africa's nations that I have ever heard or seen.[10] Many of the wilderness areas are haunted by robbers. Only very few of the inhabitants use bows and arrows the main weapons being the gun and spear. Every man or soldier has a gun and a spear.[11]

The people of Unyanyembe are at present engaged in fighting the war with Mirambo.[12] This ambitious chief originally was the porter[13] of the Arabs, but through his riches and his power he gradually became the monarch or king of an extensive area.

---

[1] Possibly as a result of conflicts caused by the Kimbu warlord, Nyungu ya Mawe. Shorter, 'The Kimbu', p. 109.

[2] Muhalule was the chief of Ngulu. Nyungu ya Mawe made blood brotherhood with him. Shorter, *Chiefship*, p. 281.

[3] Kisalusala. Ibid., p. 206. The settlement is at approximately 6°05′S, 34°58′E.

[4] According to Cameron, *Across Africa*, I, p. 136, a place with fresh spring water.

[5] Jiwe La Nsinga was the most northerly of the chiefdoms controlled by Nyungu ya Mawe. Masanja was presumably one of his *vatwaale*, that is, appointed subordinate chiefs.

[6] On the Kimbu, see Abrahams, *Peoples*, p. ix. The standard historical account is Shorter, *Chiefship*.

[7] Jowari and bajree are both forms of millet, *Holcus sorghum* and *Panicum spicatum*.

[8] This is one of the few mentions of Nyungu ya Mawe in the literature of exploration; this important figure never met an European and so has tended not to appear in nineteenth-century travel works. Wainwright's testimony confirms oral traditions in affirming that he was drawn from the chiefly family of Unyanyembe. Nyungu, according to his modern biographer, was a small, one-eyed man who behaved in a 'cruel and violent' manner. Shorter, 'Nyungu', pp. 7–8, 26 and *passim*.

[9] See n. 373, n. 5, above.

[10] Abrahams, *People*, p. 37, writes of a variety of stringed instruments, rattles and of singing to accompany dances. Perhaps Nyamwezi music reminded Wainwright of Yao music heard when he was a child.

[11] The evidence of widespread possession of guns is notable; until about 1860, attempts were made at Zanzibar to curtail the sale of guns on the mainland. Smith, 'Southern Interior', pp. 276–7.

[12] The war went on from 1871 to 1875 and was principally a contest fought by the Arabs with their allies from the Unyanyembe chiefdom. It ended inconclusively partly because some Arabs, notably Tippu Tib, were prepared to do their own private deals with Mirambo to ensure that he allowed their trade caravans through. Bennett, *Mirambo*, pp. 51–68; *Maisha ya Hamed bin Muhammed*, para. 130.

[13] Petermann's text had 'poter' which was presumably a misreading by him or Brenner of Wainwright's writing.

Not long ago, he was the Arabs' friend, but some black people or servants of their own people have incurred his anger by going into his house and carrying on with his concubines.[1] Since then he has called for the assistance of the people of his own caste[2] to make war on the Arabs. Thus war began three years ago. Each year the war is supposed to end undecided for both sides. The difficulty of capturing or killing him lies in the countries of considerable power surrounding him, and the many [people who] have rallied to his call. If the people of the coast or the Arabs did not resist his growing power, all the lands of Unyamwezi would be under his sway.[3]

The land of Unyamwezi is well cultivated, and the wilderness is covered with [a] particular [kind of] tree, called Inyombo.[4] This tree is of great use to the inhabitants; it not only supplies them with wood for building their houses, but from the exceedingly durable bark which is peeled off it, enormously large baskets are manufactured for keeping the grain, these baskets fulfilling the purpose of barns. The bark can also be used for smaller round vessels and strong ropes, and may be made into clothing that is worn by the inhabitants; it can also be used for many other things as well.

Zebras, antelopes, buffaloes, giraffes, and many other animals are found here. The houses are low, with a flat roof covered with mud. There are only a few streams and rivers in this country.[5] The language is difficult to understand.[6]

Report on Ugogo

The country Ugogo[7] is an open, flat and extensive land, and the people are brave and very proud of their own country. They rarely travel from one country to another; they remain only in their own land. The tributes imposed in this country are hard for foreigners.[8] The land is mostly inhabited if not densely.[9] They very much love cattle, like the Brahmins in India who call the cows their mothers.[10]

---

[1] Various explanations have been given for the outbreak of the war and this is among the more unusual. The normal reason given is that noted by Stanley to the effect that Mirambo was 'levying blackmail' on trade caravans. *Stanley's Despatches*, pp. 22–3. This points to the basic issue of who was to control the trade route west from Tabora to the ivory-rich regions beyond Lake Tanganyika.

[2] The German is 'Kaste'; perhaps Wainwright with his Indian background did use the term but 'kin' or 'tribe' might be more appropriate.

[3] In fact, Mirambo's 'empire' did grow and he later received help from LMS missionaries hoping to make it the basis of a Christian society in East Africa. However, the empire was held together only by the terror tactics of Mirambo's 'ruga-ruga'; the empire fell apart on his death in 1884.

[4] The myombo or misombo consists of trees of the *Brachystegia* genus. Much of the area in the land of the Kimbu through which Wainwright passed is uninhabited. See Abrahams, *People*, p. 30.

[5] The streams tend to be seasonal becoming running water in the November–April period.

[6] Possibly because it is tonal and has seven vowels. Abrahams, *People*, pp. 28–9.

[7] There is no detailed modern historical literature on the Gogo although useful information can be gleaned from early travellers such as Burton, *Lake Regions*, I, pp. 294–312, and officials from the period of German occupation. The most important anthropological account in English is Rigby, *Cattle and Kinship*.

[8] The Gogo became notorious among nineteenth-century travellers because of their exactions. Burton referred to 'blackmail' but realized that it was, in effect, the 'customs-dues of government'. *Lake Regions*, I, p. 253. Earlier in the century, traders making for Unyamwezi had taken a more southerly route to avoid the Gogo but local wars and Ngoni incursions had rendered these alternatives hazardous. Roberts, 'Nyamwezi', p. 127.

[9] Petermann's text actually has '… (denoety)' presumably because either he or Brenner could not decipher precisely what Wainwright had written. 'Densely' seems the most likely word as Bontinck, 'Diaire', p. 429, agrees.

[10] The Gogo are what Koponen, *People and Production*, pp. 245–6, describes as 'agropastoralists'.

The soil is reddish sand and if more care is dedicated to cultivation, it yields rich harvests.[1] In each village that belongs to a different administration, the foreigner pays tribute or a toll.[2]

Concerning our march, we entered Ugogo first towards the end of the month of February [*sic*].[3] [At] the borders of Ugogo, we came to the country where the name of the local chief is Chiwiye; his town lies on Mdaboo River.[4] Here tribute had to be paid.

On 1st January, early in the morning, we departed for a town called Koko;[5] tribute was paid here. During the night while we were sleeping, a wild boar[6] attacked one of the dogs belonging to the people in our caravan or seized it by the head, and when the howling of the dog was heard, the boar was shot dead on the spot, but the dog got a very bad wound by the bite. 2nd January, to a town called Seke:[7] the third tribute or toll. Here we encountered a very large caravan from Zanzibar. 4th January, arrived at a country called Kanyenye in the centre of Ugogo; here the fourth tribute was paid. This king or sultan is supposed to be the greatest in Ugogo.[8] On this day, we had met with very heavy rain in the interior of Ugogo. It was about 6 o'clock p.m. The rain was accompanied by hail, as well, [and] some hailstones were as big as hens' eggs. The strong wind came from a north eastern direction, but it was much less than what is called a hurricane. 7th January, [we marched] to Magomba's son. 8th January, set out early in the morning, went and camped in the wilderness. On the 14th we had intended to travel up to the other border, but were intercepted by the people of an under chief and halted in order to pay some toll, which we were ready to do willingly, and on the following day we departed early in the morning and had marched for hardly an hour when we were halted again, for the same purpose of paying a tribute, which again we did. On the same day, we resumed our march to King Mawala,[9] [through] the most thorny part of the country. Some cloth and also two shotguns, powder, shot, flintstones, and bullets were sent to the king according to his demand. Normally in this country, as much property as the king needs is sent as tribute or toll. Sometimes,

---

[1] Wainwright implies that laziness accounts for agricultural production problems but the Gogo live in what a modern historian calls the 'arid core' of what is now Tanzania and points out that in the 1860s and '70s the Gogo expected to experience famine for at least one year in ten. Iliffe, *Modern History*, p. 70. Moreover, the Gogo leave land fallow after using it for three years or more and the sight of such areas may have misled Wainwright. Koponen, *People and Production*, p. 223.

[2] As Wainwright points out, lack of any political unity meant a series of demands. See also Koponen, *People and Production*, p. 204.

[3] A clear mistake for December as Petermann pointed out. 'Tagebuch', p. 192.

[4] Mdaburu is the settlement lying on the river of the same name which flows into the Kisiga and then the Ruaha. Stanley rendered 'Chiwiye' as 'Kiwyeh' when he met him two years before. *How I Found Livingstone*, p. 503.

[5] Khoko, according to Burton, had a chief named Maguru-Mafipi. *Lake Regions*, I, p. 259.

[6] This is my translation of the German text which has 'wilde Sau'. Bontinck, 'Diaire', p. 430, thinks the attack must have been by a hyena and translates accordingly.

[7] The modern town of Iseke in approximately 6°24′S, 35°E.

[8] Burton, *Lake Regions*, I, p. 259 identifies him as Magomba. This evidence and Cameron's indicate that he was still in power in 1873. Cameron, *Across Africa*, I, p. 109, thought he was a hundred years old. The settlement of Kanyenye was at approximately 6°40′S, 35°10′E.

[9] Bontinck identifies Mawala as the chief of Mvumi. 'Diaire', p. 431n.

however, it happens that the king refuses to take property as a tribute, and the caravan or the poor strangers must then till the fields, carry wood, or perform other work of this kind instead.[1]

18th January, slept in the wilderness;[2] in the dry periods of the year, definitely no water is found in this desert. 19th January, arrived at a small town called Chunyu,[3] but the inhabitants live in the mountains. 20th January, went on early in the morning, and reached the town called Mpwapwa[4] after about five hours of marching. We took quarters below the shady tree; here we met a European living in Mpwapwa. (This young man in Mpwapwa is a Swiss called Phillip Broyon who acquires ivory for us there – R. Brenner.)[5] He seems to be a good gentleman, for he accommodated Mr Moffic[6] well. On the next day, we sent several people to Zanzibar with a letter for the Consul, to give him news about our coming.[7] 22nd January, came upon a town named Toohugwe where there was a famine.[8] 23rd January, set out early, and crossed a flat plain where we found many animals like mules, zebras, and large herds of other animals. After a march of five hours, we made a stop for a time in order to have lunch; at 1 o'clock we resumed our march, and came to a halting place about 5 o'clock, where we slept; but the water was far off at a great distance. 24th January, departed early, and after walking over a very rough mountainous path, we reached a town called

---

[1] Or, possibly, these people were slaves; the Gogo were in the habit of selling ivory to obtain slaves to till their fields. Koponen, *People and Production*, pp. 98–9.

[2] The wilderness between Ugogo and Usagara known as Marenga Mkali, meaning 'bitter water'. Stanley, *How I Found Livingstone*, p. 139.

[3] The party were now at approximately 6°18'S, 36°20'E and into the country of the Kaguru.

[4] Mpwapwa was an important collection of settlements at approximately 6°21'S, 36°20'E where trading caravans regrouped and could recruit more porters before facing the rigours of Ugogo. It was and is high enough to be largely free of malaria. Koponen, *People and Production*, p. 154.

[5] Philippe Broyon (1844–84) was a Swiss who became a seaman in Marseilles ships trading to East Africa and then the representative there of the Roux de Frassinet firm. In 1872, he had decided to try to break the Arab monopoly on the ivory trade by himself going into the interior and he chose the strategic location of Mpwapwa as his base. Thus he was available to offer hospitality to Murphy. In 1875, he went further inland and formed a relationship with Mirambo who, for his own reasons, wished to break the Arab stranglehold on trade to the coast. Broyon even married a woman alleged to be the warlord's daughter. He helped the London Missionary Society set up stations in Unyamwezi and Ujiji, sent geographical information to the RGS and seemed set fair to become the first European trader to operate in the East African interior. However, the relationship with Mirambo soured, there was inevitable Arab hostility and so Broyon retreated to the coast where he tried his hand at tobacco growing before dying in mysterious circumstances in 1884. Broyon-Mirambo, 'Description of Unyamwezi', *Proc. RGS*, 22, pp. 28–35; Bennett, *Mirambo*, pp. 72–95.

[6] Viz., Lt Murphy.

[7] The practice of sending Chuma ahead bearing news had been used when Cameron was known to be in Unyanyembe and it was now repeated in relation to Zanzibar. Chuma reached Zanzibar on 3 February carrying a note from Murphy to the Acting British Consul, Prideaux, and another letter from Wainwright to him stating where and when Livingstone had died. This information Prideaux incorporated into a despatch sent to the Foreign Office on 10 February and a shorter telegram from Aden on 23 February. These communications are reprinted in Elton, *Lakes and Mountains*, pp. 107–8. The information confirmed what Prideaux had already learned from earlier messages sent by Cameron and Murphy but Prideaux, now knowing the imminence of the arrival of the body, could make arrangements for its reception.

[8] Tubugwé, according to Stanley, *Through Dark Continent*, I, p. 95. The famine may well have been the result of plundering by Dirigo people whom Cameron had heard about some six months before. *Across Africa*, I, pp. 83, 88. These appear to be a Masai or Masai-related pastoral people. Tubugwe is a dispersed settlement in about 6°21'S, 36°37'E.

Kitangay,[1] where we found a large trading caravan waiting for the arrival of others from Zanzibar. When we approached the town, they came to meet us with the usual drums and trumpets or horns. In the still of the night, when we were fast asleep, a boar[2] came into our huts and tried to seize one of our soldiers by his leg; but the man got away with only a slight injury. 25th January, departed early, but the path was very mountainous. We stopped on our way by one of the streams to drink water where we waited for about half an hour for some who had fallen behind. 30th January, in a village. 31st January, in a village.

1st February. Leaving Usagara, we arrived at the borders of Uzigua[3] on the river Mkata or Wame. At that time we found the river swollen or overflowing, and the former bridge had been washed away by the rapid current force, and so we had to build a new one across the river. So some of the soldiers got to the other side with much difficulty, but the bridge broke down again, and the rest had to stay to spend the night on the same bank and cross the river next morning. And so, with luck, we made it to the other side.[4]

Usagara

The kingdom of Usagara,[5] as far as I was able to gather news of it,[6] is nearly all mountainous. Many running rivers are seen in the valleys. The land is said to be remarkably fertile. The climate is milder than that of neighbouring countries and the products are jowari, maize or Indian corn, beans, bananas or paradise figs. The most numerous domestic animals are goats and sheep. In general, all native tribes from Ugogo to the coast of Zanzibar are circumcised.[7] Deceased persons are never buried in the ground, but they tie the body to a bed of strong sticks, and keep them in the hollow trunk of a big species of tree. This is also the case in Ugogo and Unyamwezi.[8]

---

[1] Stanley spells it Kitangeh. *Through the Dark Continent*, I, p. 91.

[2] The German text here reads: 'ein Bär (bear, vielleicht boar, Eber?)', 'Tagebuch' p. 193. Bontinck, 'Diaire', p. 432, omits the quoted words and without explanation simply inserts 'hyène'.

[3] They were on the southern border and hardly met many Zigula people. Beidelman, *Matrilineal Peoples*, pp. 66–72 provides basic information.

[4] Cameron had experienced some difficulty here in the previous year: *Across Africa*, I, pp. 63–4.

[5] Lying east of Ugogo, the country of the Sagara people must have seemed to Wainwright to be fertile and well-watered by comparison. In the event, he gives comparatively little information. The designation as a 'kingdom' presumably reflects the claims of the Saganza clan but there was no long tradition of paramountcy. Beidelman, *Matrilineal Peoples*, p. 51.

[6] This intriguing phrase may suggest that Wainwright made conscious efforts to seek informants who could give him data.

[7] Beidelman, *Matrilineal Peoples*, p. xiv, doubts that circumcision was general and implies that it is a custom resulting from relatively recent influence from the coast. However, this can be set against Wainwright's assertion for 1873–4.

[8] Modern anthropological data assembled by Beidelman for the peoples living in this region does not seem to bear out Wainwright's assertions about non-burial of those who have died except in the case of the Zigula people. *Matrilineal Peoples*, p. 71. On the other hand, practices could have changed because of Arab influences and, later, colonial regulations. Moreover, Wainwright's assertions about the Nyamwezi do seem to be echoed by other informants such as Grant, *A Walk Across Africa*, p. 84, who said that 'witches and slaves' at least were 'thrown into the jungle'. Burton implied that customs were changing among the Nyamwezi under Arab influence but claimed that the traditional way of disposing of a corpse had been to 'throw it into some jungle strip' and leave it to the hyenas. *Lake Regions*, II, p. 25.

On 3rd February, two of the people were seriously ill and we had to wait for another day to allow them to get better. 4th February, went on, reached a rock where there was water, and where we slept.[1] 5th February, slept in a town on Wame River[2] as we were now marching along with this river or parallel to it. 6th February, departed early, and after a six hours' march, came to the town named Kinmock, [where the] Chief [is] Moguba. 7th February, after eight hours of marching, we reached the town called Funi, [where the] Chief [is] Ngorido.[3] Here the water of the river is salty and drinking water is obtained from deep wells. 8 February, in Burahim's town.[4] 9th February, we went to the chief's house. Here we stayed for two days to rest and wait for the return of the men sent to Zanzibar; from them we learned all the news about Zanzibar.[5]

On 18th February, we arrived at Zanzibar.[6]

****

---

[1] It is difficult to identify this point precisely. Indeed, from here on, there are difficulties over names and designations. Bontinck, 'Diaire', p. 434, found himself forced to change or suppress all but one of his notes 163–8. Bontinck 'Encore sur le Diaire', pp. 603–4. Stanley's information about his routes in 1871 in *How I Found Livingstone*, and in 1874 in his *Through the Dark Continent*, seems to offer the best information on the routes and succession of stopping places in this region.

[2] Bontinck, 'Diaire', p. 434, n. 164 (which is not suppressed), identifies this town as Mfuteh. It seems to me more likely to be a settlement Stanley calls Changarikwa on his map or the Rubuti on his map and which he refers to in his text. Where Wainwright next stopped, what he (or Brenner) calls Kinmock, was really, I believe, Mfuteh. Stanley, *Through the Dark Continent*, I, pp. 89–90 and map.

[3] There is further difficulty here over the names. The chief's name must surely reflect rather the place Stanley designates Congorido, 'a populous village' but where the water was 'brackish'. *Through the Dark Continent*, 1, p. 89. The succession of settlements shown on Stanley's map from west to east, Rubuti, Changarikwa, Mfuteh, Congorido would accord with the interpretation offered here.

[4] Bontinck, 'Encore sur le Diaire', p. 604, identifies this as 'Brahim Ouakouéré', in other words, Brahim in Ukwere. This was east of Congorido and west of Rosako and possibly at or near the modern Msata which is in 6°20′S, 38°25′E.

[5] See n. 379, n. 7 [20] above. Unfortunately, Wainwright chooses not to say what the news was or what arrangements were now suggested to the party. The details are not altogether clear but it would seem that Chuma and his companions must have now returned to the main party, as this evidence implies, on 11 or 12 February. The whole group must then have gone on to Bagomoyo at the coast. Here a new double coffin of zinc and wood was made for Livingstone's body by the members of the French Holy Ghost Mission there who also held a service to honour Livingstone. Prideaux arrived from Zanzibar on HMS *Vulture* on the 14th and the coffin plus, presumably, Murphy was taken over to the island on the 16th, but, says Carus Farrar, 'leaving many of us behind.' In fact, it seems that the majority of the Party were ignored or shabbily treated by British officials and some missionaries in East Africa. Thomas, 'Carus Farrar's Narrative', p. 120; *Last Journals*, II, p. 345; Coupland, *Livingstone's Last Journey*, pp. 252–3.

[6] If Wainwright's dating is correct, it would seem that he, too, must have been left behind at Bagomoyo. Mathew Wellington, however, records the coffin going and 'we following in a dhow three hours later'. Whether he means just the Nasik boys or the whole party is unclear. 'Life of Mathew Wellington', p. 102. What subsequently happened to Wainwright and his 'Diary' is recorded in Sections 2 and 3 of the Introduction, above.

BIBLIOGRAPHY

Principal Sources Related Directly to Wainwright

Edinburgh, National Library of Scotland, Acc. 12444, Jacob Wainwright to the Rev. Joseph Moore, 23 May 1874, 10 July 1874, *carte de visite*, dated July 1874.
'Tagebuch von Jacob Wainwright über den Transport von Dr. Livingstone's Leiche, 4. Mai 1873–18. Februar 1874', *Mittheilungen aus Justus Perthes' Geographischer Anstalt*, 20, 1874, pp. 187–93.
Bontinck, François, 'Le Diaire de Jacob Wainwright (4 mai 1873–18 février 1874)', *Africa. Rivista trimestrale di studi e documentazioni dell'istituto Italo-Africano*, 32, 3, 1977, pp. 399–435.
Bontinck, François, 'Encore sur le Diaire de Jacob Wainwright. Corrigenda', *Africa. Rivista trimestrale di studi e documentazioni dell'istituto Italo-Africano*, 33, 4, 1978, pp. 603–4.

Other Works

Abrahams, R. G., *The Political Organization of Unyamwezi*, Cambridge, 1967.
Abrahams, R. G., *The Peoples of Greater Unyamwezi, Tanzania*, London, 1967.
Alpers, Edward A., *Ivory and Slaves in East-Central Africa*, London, 1975.
Beidelman, T. O., *The Matrilineal Peoples of Eastern Tanzania*, London, 1967.
Bennett, N. R., *Mirambo of Tanzania 1840?–1884*, New York and London, 1971.
Bennett, Norman Robert, *Arab versus European*, New York and London, 1986.
Bodleian Library, Oxford, Ms Afr S 16/4. Notes from Chuma and Susi concerning travels with Dr Livingstone, 1865–74, made by Horace Waller.
Bontinck, François, 'Voyageurs africains en Afrique Equatoriale; XVII, David Abdallah Susi', *Zaïre-Afrique*, 162 & 163, 1982, pp. 99–118, 169–84.
Bridges, R. C., 'The Sponsorship and Financing of Livingstone's Last Journey', [*International Journal of*] *African Historical Studies*, I, 1968, pp. 79–104.
Bridges, Roy C., 'Nineteenth-Century East African Travel Records', *Paideuma*, 33, 1987, pp. 179–96.
Bridges, Roy, 'Towards the Prelude to the Partition of East Africa', in Roy Bridges, ed., *Imperialism Decolonization and Africa*, Houndmills, 2000, pp. 65–113.
Broyon-Mirambo, Philippe, 'A Description of Unyamwezi', *Proc. RGS*, 22, 1877–8, pp. 28–35.
Burton, Richard Francis, *The Lake Regions of Central Africa*, 2 vols, London, 1860.
Cameron, Verney Lovett, *Across Africa*, 2 vols, London, 1877.
Clendennen, G. W. and Cunningham, I. C., *David Livingstone. A Catalogue of Documents*, Edinburgh, 1979.

Cooley, W. D., 'The Geography of Nyassi, or the Great Lake of Southern Africa ...', *Journal of the Royal Geographical Society*, 15, 1845, pp. 185–235.
Coupland, Reginald, *Livingstone's Last Journey*, London, 1947.
Elton, J. F., *Travels and Researches among the Lakes and Mountains of Eastern and Central Africa*, London, 1879.
Farar, Carras and Wellington, Matthew, 'David Livingstone: Two Accounts of his Death and Transportation of his Body to the Coast', comprising 'Carras Farar's Story of the Finding of Dr Livingstone in Central Africa' and 'Account of the Life of Matthew Wellington in his own Words, and of the Death of David Livingstone and the Journey to the Coast', *The Zambia [Northern Rhodesia] Journal*, 6, 1965, pp. 95–102.
Feierman, Steven, 'The Shambaa', in Andrew Roberts, ed., *Tanzania before 1900*, Nairobi, 1968, pp. 1–15.
Fraser, A. Z., *Livingstone and Newstead*, London, 1913.
Gamitto, A. C. P., *King Cazembe ... being the Diary of the Portuguese Expedition ... 1831 and 1832*, trans. Ian Cunnison, 2 vols, Lisbon, 1960.
Gouldsbury, Cullen and Sheane, Hubert, *The Great Plateau of Northern Rhodesia being some Impressions of the Tanganyika Plateau*, London, 1911.
Grant, James Augustus, *A Walk Across Africa*, Edinburgh and London, 1864.
[Hamed bin Muhammed], *Maisha ya Hamed bin Muhammed yaani Tippu Tip*, translation by W. H. Whiteley with a Historical Introduction by Alison Smith, Supplement to the *East African Swahili Committee Journals*, 28/2, 1958 and 29/1, 1959.
Helly, Dorothy, *Livingstone's Legacy. Horace Waller and Victorian Mythmaking*, Athens, Ohio, and London, 1987.
Hughes, J. E., *Eighteen Years on Lake Bangweulu*, London, n.d. [1933].
Ibn Baṭṭūta, *The Travels of Ibn Baṭṭūta AD 1325–1354*, II, ed. and trans. H. A. R. Gibb, Cambridge for the Hakluyt Society, 1962.
Iliffe, John, *A Modern History of Tanganyika*, Cambridge, 1979.
Johnston, H. H., *Livingstone and the Exploration of Central Africa*, London, 1891.
Koponen, Juhani, *People and Production in Late Precolonial Tanzania. History and Structures*, Helsinki, 1988.
Lacerda's Journey to Cazembe in 1798, trans. and annotated by R. F. Burton, London, 1873.
Middleton, John, ed., *Encyclopedia of Africa South of the Sahara*, 4 vols, New York, 1997.
Murdock, George Peter, *Africa. Its Peoples and their Culture History*, New York, 1959.
Omer-Cooper, J. D., *The Zulu Aftermath. A Nineteenth-Century Revolution in Bantu Africa*, London and Harlow, 1966.
Roberts, Andrew, 'The Nyamwezi', in Andrew Roberts, ed., *Tanzania before 1900*, Nairobi, 1968, pp. 117–50.
Roberts, Andrew, 'Nyamwezi Trade', in Richard Gray and David Birmingham, eds, *Pre-Colonial African Trade*, London, 1970, pp. 39–74.
Roberts, Andrew D., *A History of the Bemba. Political Growth and Change in North-eastern Zambia before 1900*, London, 1973.
Roberts, Andrew, *A History of Zambia*, London, 1976.
Royal Geographical Society Archives MSS File: Livingstone.
St John, Christopher, 'Kazembe and the Tanganyika-Nyasa Corridor', in Richard Gray and David Birmingham, eds, *Pre-Colonial African Trade*, London, 1970, pp. 202–27.

Sheriff, Abdul, *Slaves, Spices and Ivory in Zanzibar*, London, 1987.
Shorter, Aylward, 'The Kimbu', in Andrew Roberts, ed., *Tanzania before 1900*, Nairobi, 1968, pp. 96–116.
Shorter, Aylward, *Nyungu ya Mawe*, Nairobi, 1969.
Shorter, Aylward, *Chiefship in Western Tanzania. A Political History of the Kimbu*, Oxford, 1972.
Simpson, Donald, *Dark Companions*, London, 1975.
Smith, Alison, 'The Southern Section of the Interior 1840–1884', in Roland Oliver and Gervase Mathew, eds, *History of East Africa*, I, Oxford, 1963, pp. 253–96.
Stanley, Henry M., *Through the Dark Continent*, 2 vols, London, 1878.
Stanley, Henry M., *How I Found Livingstone in Central Africa*, London, [1872] 1895 [ed. with additional material, pp. ix–lxxix].
Stanley, Henry M., *Stanley's Despatches to the* New York Herald *1871–1872, 1874–1877*, ed. Norman R. Bennett, Boston, 1970.
Stock, Eugene, *The History of the Church Missionary Society*, 3 vols, London, 1899.
The Times Atlas of World Exploration, ed. Felipe Fernández-Armesto, London, 1991.
Thomas, H. B., 'The Death of Dr Livingstone: Carus Farrar's Narrative', *Uganda Journal*, 14, 1950, pp. 115–28.
Thomas, H. B., 'Jacob Wainwright in Uganda', *Uganda Journal*, 15, 1951, pp. 204–5.
Thomson, Joseph, *To the Central African Lakes and Back*, 2 vols, London, 1881.
The Times, 30 March 1874; 22 April 1874.
Waller, Horace, ed., *The Last Journals of David Livingstone in Central Africa*, 2 vols, London, 1874.
Whiteley, Wilfred, *Bemba and Related Peoples of Northern Rhodesia*, London, 1951.
Victoria, Queen, *Darling Child. Private Correspondence of Queen Victoria and the German Crown Princess*, ed. Roger Fulford, London, 1976.
Willis, Roy G., *The Fipa and Related Peoples of South-West Tanzania and North-East Zambia*, London, 1966.
Willis, Roy G., 'The Fipa', in Andrew Roberts, ed., *Tanzania before 1900*, Nairobi, 1968, pp. 82–95.
Wright, Marcia and Lary, Peter, 'Swahili Settlements in Northern Zambia and Malawi', *International Journal of African Historical Studies*, 4, 1971, pp. 547–73.
Wyse, Akintola, *The Krio of Sierra Leone. An Interpretive History*, Washington, D.C., 1991.

INDEX 1: BODEGA Y QUADRA

Academia de Guardias Marinas, 75, 78
Acapulco, Mexico, 27&n, 28, 30–31, 32n, 36–9, 43, 47, 52n, 67, 131
Afognak Island, Alaska, 72
Aguilar, Martín de, 40, 92
Aguirre, Fray Andrés, 35n
Aguirre, Martín, 32
Alava, José Manuel, 83
Aleutian Islands, Alaska, 41, 79, 118n
Alexander VI, pope 17
Alta California, 25n, 35–43, 50, 51n, 52, 54–6, 64, 66, 69
Alvarado, Pedro de, 24
Alzola, Tomás de, 36
Anson, George, 30n
Archipelago of San Lázaro, 117
Arellano, Alonso de, 27
Arriaga, Julián de, minister of the Indies, 55n, 56n, 61, 62n, 64, 65n, 66, 76
Arteaga, Ignacio de, 68, 70, 71, 72n, 77, 85
Avila, Pedro Arias de (Pedrárias Dávila), 17
Ayala, Juan Manuel de, 62–4, 66, 68, 76, 85, 86, 88

Bahia de la Asunción, Columbia River, 66
baidarkas, 70
Baja California, 22, 36, 39, 41, 43–4, 53, 55n, 62
Balboa, Vasco Nuñez de, 17–19, 27, 39
Barren Islands, Alaska, 72
Barrington, Daines, 16
Bautista de Aguirre, Juan, 69
Bellin, Jacques Nicolas, 89n, 93n, 94, 100n, 101n, 115
Bering, Vitus, 41, 70, 115
blue-footed booby (*Sula nebouxii*), 89n
Bodega y Quadra, Juan Francisco de la, 15, 62, 64, 67, 73–85, 130 *and passim*
Bolaños, Francisco de, 39
Bolton, Herbert E. 50
Bonin Islands, 28
Borneo, 27
Brazil, 17
brown booby (*Sula leucogaster*), 89n
Brunei, 27

Bucareli Bay, Alaska, 9n, 69–70, 116, 117n, 118
Bucareli y Ursúa, Antonio María de, Viceroy of New Spain (1771–79), 6, 7, 54–7, 59–62, 64–9, 72, 76–7, 85, 96
Buenos Aires, Argentina, 76

Caamaño, Jacinto, 79
Cabral, Pedro Alvares, 17–18
Cabrera Bueno, José González, 29–30, 47, 52
Cabrillo, *see* Rodriguez, Cabrillo
Cádiz, Spain, 56, 62, 76, 78–9
Camacho, Joseph, 67, 68n, 69
Campa Cos, Fray Miguel de la, 88n
Canary Islands, 18
Cañizares, José de, 43, 49, 55–6, 61–2, 66, 68n, 70–71
Cape Alava, Washington State, 33
Cape Arago, Oregon, 34–5
Cape Blanco, Oregon, 30, 40, 94, 122n
Cape Dezeado, Patagonia, 19
Cape Disappointment, Washington State, 59
Cape Elizabeth, Alaska, 71n, 72n
Cape Elizabeth, Washington State, 15, 102, 103n, 105, 128n
Cape Flattery, Washington State, 59
Cape Horn, Chile, 76
Cape of Good Hope, South Africa, 18, 32
Cape Mendocino, California, 30, 40, 59, 72, 94, 122n
Cape Saint Elias, Alaska, 70, 72
Cape San Lucas, Baja California, 30, 36, 39, 44–7, 48n, 49n, 126
Cape Suckling, Alaska, 70
Cape Verde Islands, 18
Cartagena, Spain, 75–6
Carvajal, Pedro, 83
Catalonian volunteers, 50
Cavendish, Thomas, 36–7
Cavite, Philippines, 28, 31
Cermeño, *see* Rodríguez, Cermeño
Cerro de Santa Rosalía, Mt Olympus, Washington State, 59
Chapultepec Castle, Mexico City, 61n
Charles III, king of Spain, 42, 68–9, 78n, 79, 85

385

Channel Islands, California, 35n, 47–8, 50, 51n, 126n
China, 27–8
Chirikov, Aleksei, 6, 41, 58, 115
Choquet, Diego, 62, 68, 76, 85–6
Chugach Islands, Alaska, 71
Chumash people, 51
Coast Miwok people, 123
Colchero, Alonzo Sánchez, 32
Colnett, James, 76, 80–81
Columbia River, 7, 16, 59, 66, 121n
Columbus, Christopher, 16–18, 23
Concepción, Chile 76
Concepción, ship, 19, 22, 32, 45, 81
Content, ship, 36–7
Cook, Captain James, 6, 23, 67, 70, 72
Cook Inlet, Alaska, 72
Cortés, Hernando, 24
Costansó, Miguel, 50, 52–3
Crespi, Fray Juan, 43, 50, 52–4
Croix, Carlos Francisco de, Viceroy of New Spain (1766–71), 42, 54–5

Delarov, Evstrat, 80
Desire, ship, 36
Destruction Island, *isla rasa*, Washington State, 101n, 102
Dixon Entrance, British Columbia, 58, 116, 119n, 120n
Drake, Sir Francis, 6, 31–5, 38, 67

El Camino Real, 54
Elcano, Juan Sebastián de, (Juan Sebastián del Cano), 22–4
Eliza, Francisco de, 79
Elizabeth Island, Alaska, 71–2
Ensenada de Nuestra Señora de la Regla, 71–2
Entrada de Bucareli, 'Bucareli's entrance', 117
Entrada de Fuca, 101n
Entrada de Hezeta, 'Hezeta's entrance', 66
Escuela de Guardias Marinas, Cádiz, 62, 75n
Espinosa, Gonzalo Gómez de, 22, 24
Espinosa, Rodrigo de la Isla, 26
Extremadura, Spain, 38

Fages, Pedro, 46, 50, 52n
Farallon Islands, California, 52, 59, 124n
Ferrelo, Bartolomé, 25
Fidalgo, Salvador, 79
Fletcher, Francis, 33
Fonte, Bartolomé de, 117
Friendly Cove, Nootka Sound, 80–83
frigatebird, magnificent (*Fregata magnificens*), 89

Fuca, Juan de (alias Apostolos Valerianos), 37, 100n

Gálvez, José de, 42, 44–5, 49, 54–6
Gerhard, Peter, 31
Gilmer Bay, Puerto de Guadalupe, Kruzov Island, Alaska, 111, 112n
Golden Hinde (formerly *Pelican*), ship, 31–5, 38
Gómez, Fray Francisco, 50, 52
Gonzáles, Juan, 65, 108–9
Graham Island, British Columbia, 58
Gray, Robert, 81
Grijalva, Hernando de, 89n, 91
Guadalajara, Mexico, 83
Guadalquiver River (Betis), Spain, 18, 23
Guadalupe Island, Mexico, 46, 112
Guam and Rota islands, 20, 24, 31
Guatulco, Mexico, 31–4
Güemes y Horcasitas, Francisco de, 11
Güemes Pacheco de Padilla, Juan Vicente de, Conde de Revillagigedo, Viceroy of New Spain (1789–94), 10, 79
Gulf of California, 38–9, 55n

Haida people, 58–9
Hakluyt, Richard, 33
Hawaiian Islands, 23, 29
Hesquiat Peninsula, Vancouver Island, British Columbia, 59
Hezeta, Bruno de, 6, 7, 10, 16, 62–6, 68
Hinchinbrook Island, Alaska, 70
Humboldt, Alexander von, 16

Ingraham, Joseph, 81
Iliamna Volcano, Alaska, 72
Isla de Cedros, Cedros or Ceders Island, 46
Islas de Ladrones, 20, 22
Islas Tres Marías, 88
Islas de Velas Latinas, 20

James Island, Washington State, 101n
Japan, 27–8
Jordán, padre Alejandro, 39

Kamchatka, Russia, 41
Kayak Island, Alaska, 70
kayaks, 70
Kenai Peninsula, Alaska, 71–2, 118
Knight in the Order of Santiago, 68, 77–8
Krenitsyn, Peter, 41
Kruzov Island, Alaska, 7, 66, 87, 110n, 111, 112n

La Paz, Baja California, 44–5

Levashev, Michael, 41
La Brea Tar Pits, 51
Langara Island, Queen Charlotte Islands, British Columbia, 58
Lima, Peru, 74–8, 93n, 117n
Lisianski Strait, 58
Loaysa, Francisco García Jofre de, 24–5
Lok the Elder, Michael, 37, 100n
London, England, 16, 74, 81
Long Beach, California, 47
López de Haro, Gonzalo, 79–80
López de Legaspi, Miguel, 26
Los Angeles, California, 47, 51

Mactan Island, Philippines, 21
Madrid, Spain, 16, 35, 37, 41, 55, 66–7, 76, 79–81, 82n, 97n
Magellan, Ferdinand, 18–25, 27, 31–2, 35
Manila, Philippines, 26–31, 37
Manila galleons, 26–7, 29–32, 32–42, 52n
Manrique, Miguel, 62, 64, 68, 85–6, 88
Martín, Lope, 27
Martínez, Esteban José, 56, 58, 79–82
Mártires, Punta de los, 15
Meares, John, 81
Menchaca, Antonio, 7, 74–5
Mendoza, Antonio de, Viceroy of New Spain (1535–50), 24, 25n, 30
Mendoza y Luna, Juan de, Marqués de Montesclaro, Viceroy of New Spain (1603–7), 41
Merganser, red-breasted (*Mergus serrator*), 110n
Mexico City, Mexico, 26, 35, 37–8, 41, 43, 54–5, 62, 76, 80, 83–5
Mission San Carlos Borromeo (Carmel), 54
Mission Santa María, Baja California, 44
Moluccas (East Indies or Spice Islands), 18, 22–6, 32, 35
Montague Island, Alaska, 71
Monterey, California, 41–2, 48–50, 52, 54–5, 57–8, 60–61, 64–5, 85–6, 90, 92–3, 104, 108, 121, 125
Monterey Bay, California, 7, 41–2, 44, 47–8, 50–54, 59, 66, 85n, 92–3, 104, 108, 121, 125n
Morga, Antonio de, 25, 28–30
Morison, Samuel Eliot, 17, 25
Morro Bay, California, 36
Mount Edgecumbe, Kruzov Island, Alaska, 69, 110n, 111, 112n, 118n
Mount St Elias, Alaska, 70
Mourelle, Francisco Antonio, 10, 16, 27, 62–3, 65–6, 69, 77–8, 91, 93, 98–9, 101n, 103, 108, 110, 112n, 127–8, 132

Moya de Contreras, Pedro, Archbishop and Visitador, Viceroy of New Spain (1584–85), 35

Navidad, Mexico, 24–7
New Guinea, 26
New Mexico, 35
New Spain (Mexico), 24, 26, 28, 34–5, 37–40, 42–3, 54n, 55n, 62, 68, 76, 79, 85
Nootka Sound, Vancouver Island, British Columbia, 58–9, 73, 80–83
Nova Albion, 34
Nuestra Señora de Cabadonga, ship, 30n
Nuestra Señora de Esperanza, ship, 36
Nuestra Señora de Guadalupe, alias *Sonora*, ship, 62, 86, 131
Nuestra Señora de la Concepción, ship, 32
Nuestra Señora de los Remedios, alias *Favorita*, ship, 68–9, 77
Nuestra Señora del Rosario, ship, *see Princesa*
Nuttall, Zelia, 31

Palóu, Fray Francisco, 42n, 44n, 45, 48, 54n
Panama, 2, 32, 78
  Isthmus of, 17
Pantoja y Arriaga, Juan, 69, 71
Patagonia, 19
Pérez, Juan, 6, 45, 47–8, 50, 53–62, 85n, 107–9
Perrón, Fray Fernando, 52
Philip II, king of Spain, 21, 25, 35, 37–8, 63n
Philippine Islands, 20n, 21–32, 36–7, 47n, 71n
Pigafetta, Antonio, 18–22, 27–8
Pimería Alta, Arizona, 55
Point Grenville, Washington State, 104n, 105
Point Reyes, California 34, 52, 54, 124n, *see also* Punto de los Reyes
Portolá, Gaspar de, 44, 49–55, 59–60, 85n
Prince William Sound, Alaska, 70
Princesa, alias *Nuestra Señora del Rosario*, ship, 67–72, 75, 77, 79–81
promyshlenniki, 41
Puerto de la Trinidad, *see* Trinidad Harbour
Puerto de Santiago, Port Etches, Alaska, 70–72
puffins: tufted (*Lunda cirrhata*); horned (*Fratercula corniculata*), 110n
Punta de Arena, Sand Point, Bodega Bay, California, 122
Punta de Santa Margarita, Langara Point, Queen Charlotte, British Columbia, 58
Punta del Cordón, Bodega Bay, California, 122, 124
Punto de los Reyes, California, 40, *see also* Point Reyes

Queen Charlotte Islands, British Columbia, 58, 60n, 121n
Quimper, Manuel, 79, 83
Quinault people, 7, 15, 103n
Quinault River, 15, 128n
Quiros y Miranda, Fernando, 62, 68, 85

Rada de Bucareli, 108
Redoubt Volcano, Alaska, 72
Revilla, Cristóbal, 62, 65–6, 107, 129–30
Rio de Martín de Aquilar, 40, 122
Rivera y Moncada, Fernando de, 43–4, 49–50, 52–3
Rodríguez Cabrillo, Juan, 24–5, 39
Rodríguez Cermeño, Sebastián, 38–9
Rodríguez, Esteban, 26

Saavedra Cerón, Alvaro de, 24, 27n
Saavedra, Ramón, 79
Salcedo, Philip de, 26
San Agustín, ship, 37–40
San Antonio alias *El Príncipe*, ship, 19, 44–5, 47–56, 62, 85
Santana/Santa Ana, Pedro, 106, 128n
San Bernabé, Baja California, 45
San Bernardino Strait, Philippines, 28, 30n, 31
San Blas, Mexico, 9, 10, 15, 43–5, 48–50, 55–6, 57n, 59–68, 76–81, 91, 99, 104, 107–8, 112, 125, 131
San Carlos alias *El Toisón de Oro*, ship, 44–50, 52–7, 62–4, 74–80, 85–6, 88, 125
San Carlos Island/Forester Island, 118–19
San Diego, California, 42, 44–54, 57, 60, 85, 90
San Diego, ship, 39–40
San Diego Bay, California, 44–5, 47, 49, 53
San Diego River, 48n
San Francisco, Mount Edgecumbe, Alaska, 112
San Francisco Bay, California, 34, 38, 40, 51–2, 59–60, 64, 66, 85n
San José alias *El Descubridor*, ship, 49–50, 52, 55
San Lucas, ship, 27
Sanlúcar, Spain, 18
San Miguel Island, California, 47
San Nicolas Island, Baja California, 47
San Pedro, ship, 26–7
San Pedro Bay, California, 47
San Pedro Channel, California, 47
Santa Ana, Manila galleon, 36–8
Santa Barbara Channel, California, 47
Santa Catalina Island, California, 47
Santa Cruz Island, California, 47
Santa María de la Antigua de Darien, Panama, 17
Santa Rosa Island, California, 47

Santiago (*Nueva Galicia*), ship, 6, 7, 56–7, 59–62, 64–7, 76, 78, 85–8, 91–2, 101, 103n, 110n, 128n, 130
Santo Tomás (island), 89n, 91
Santo Tomás, ship, 39–40
Sarmiento, Pedro, 37
Schurz, William Lytle, 26
scurvy, 20, 24, 29, 46, 48–9, 52, 60, 65–6, 69n, 70, 72, 106, 110n 120–21, 128
Sea Lion Cove, Puerto de los Remedios, Kruzov Island, Alaska, 66, 87, 111, 112n, 113–14
Sea of Cortez, 44
Serra, Fray Junípero, 43–4, 50, 52–6
Seville, Spain, 18, 22–3, 56
Shelikov Bay, ensenada de Guadalupe, Kruzov Island, Alaska, 112n
Sitka Sound, ensenada de Susto, Bay of Fright, Alaska, 69, 110n, 112n
Sonora, Mexico, 55, 62, 131
Sonora (*Nuestra Señora de Guadalupe*), ship, 6, 9–10, 16, 27, 62–3, 76, 78, 81, 85–7, 89, 91–2, 101n, 103n, 112n, 115n, 120n, 121n, 123n, 127, 131–2
Sonora Reef, 103n
South Cove, Cape Arago, Oregon, 34
Strait of Anian, 37
Strait of Gibralter, 76
Strait of Juan de Fuca, 59, 82n
Strait of Magellan, 19, 31–2, 35
Surgidero de San Lorenzo, St Lawrence's Roadstead, Nootka Sound, 58

Tehuantepec, Mexico, 91
Tello, Rodrigo, 32, 34–5
Tepic, Mexico, 62–3, 76–7, 83, 86
Ternate, Moluccas, 22, 35
Tierra del Fuego, 19
Tlingit people, 69, 112n
Tordesillas, Spain, 16
Tordesillas, Treaty of, 17–18, 23, 25–6
Torre del Oro, Seville, 18
Tres Reyes, ship, 39–40, 122n
Trinidad, California, 101, 103, 114, 122
Trinidad, ship, 19, 22
Trinidad Harbour, California 96n, 97n, 98n, 99, 103, 106, 128
Trinidad Head, California, 38, 66, 95n, 99n
tropicbird, red-billed (*Phaeton aethereus*), 89
tsunami, 74

Unamuno, Pedro de, 36–7
Urdaneta, Andrés, 24–8
US National Archives, Washington DC, 61

Valdés y Bazán, Antonio, minister of the Navy, 79
Valdés, Salvador Menéndez, 79
Valladolid, Spain, 16
Valparaiso, Chile, 31
Vancouver, George, 6, 81–3
Vancouver Island, 58, 66, 73, 121n
Velicatá, Mexico, 42n, 44
Venice, Italy, 37
Victoria, ship, 19, 22–3
Viana, Francisco José, 81
Vila, Vicente, 44–50, 55, 57
Villalobos, Ruy López de, 24–5

Vizcaíno, Fray Juan, 52
Vizcaíno, Sebastián, 35n, 38–42, 45, 47, 51, 122n
Volcán de Miranda, Alaska, 72
Volcano Islands, 28

Wagner, Henry Raup, 25–6, 31, 83n, 93n, 103n

Yaquina Head, Newport, Oregon 59

Zaikov, Potap, 80
Zúñiga, Gaspar de, Conde de Monterey, Viceroy of New Spain (1595–1603), 38

INDEX 2: STOKES

This index lists proper names and subjects in alphabetical order. Names of people are given with the family name first followed by the given name (ranks are not included), places with the toponym followed by the generic term. The definite article, if part of a name, precedes the proper name, e.g. El Morrión. In general toponyms are indexed under the form used by the state having sovereignty over the position, as given on modern maps, with names, which, where they differ in the text of the journal, are cross referenced. Those that differ markedly from the modern form are also given in brackets after the modern form.

Ships are all listed together under a heading of Ships.

The page numbers of illustrations are given in bold type.

Adams, William, pilot with Jacob Mahu, 162n
Africa, west coast, 153, 216
Agua Dulce, Caleta (Muscle Cove), 159n, 247
Agua Fresca, Bahía, 181&n
Amsterdam, 163–4
Anson, George, 169, 195n
Apóstoles, Rocas (Apostles), 227
Arauco, Golfo, 161n
Argote, Roldan de, 156, 247
Armiger, Thomas, Lieutenant, *Sweepstakes*, 166, 186n
artificial horizon, 187, 225
Asses Ears, 195n
Atacame, 161
Atrill, J., purser, *Beagle*, 154
azimuths, 187
  celestial, 222, 224

Bachelor, River, *see* Batchelor, Río
Bahia, 158
Baptiste, cook, *Beagle*, 209n
Bárbara, Canal, 169, 175n, 230&n, 231n
Barbosa, Odoardo, 155
barometer, 200
Batavia, *see* Jacatra
Batchelor, Río, 191n
Beagle, Canal (Channel), 157
Beans, Robert, Lieutenant, *Wager*, 195n
Bear Haven, 161
Beaubasin, Caleta, 172, 183n
Beauchêne Island, 169
Beauchesne, Jacques Gouin de, 167
Beaufort, Sir Francis, Hydrographer 1829–55, 151

Beggarly Bay, *see* Guesen Bay
Bell, Bahía (Campana, Bahía), 156, 247
Belmonte, Alexandro, 174n
Berlenga, Ilha da, 223n
Bermejo, Puerto, 211n
Bordeaux, 158
Borja, Bahía (Bay), 150, **192**
Borja, Don Francisco de, Prince of Esquilache, Viceroy of Peru, 159n
Bougainville, Bahía, 172, 175n, 181&n
Bougainville, Louis-Antoine de, 171–2, 176, 177n, 183&n, 190&n, 191, 204, 216, 230n
  chart, **172**, 176, 183n
  lunar distances, 172n
Bowen, Evan, surgeon, *Beagle*, 152n, 154, 190n
Brisbane, Mathew, master, *Prince of Saxe-Cobourg*, 222n, 230, 232n
Buckley's Sound, *see* Xaultegua Golfo
Buckly, Cape, 194, 195&n
Bukley, Punta de, *see* Havannah, Punta
Bulkeley, John, gunner, *Wager*, 169, 195&n
Bulkeley's Island, *see* Santa Ana, Islas
Burney, James, 162, 176
Burns, Robert, 209n
Butler, Thomas, Earl of Ossory, 165n
Bynoe, Benjamin, assistant surgeon, *Beagle*, 152, 154
Byron, Hon. John, in charge *Dolphin* and *Tamar* expedition, 169, 171, 176, 180–81&n, 195, 203n, 216, 229n

Cádiz, 158
Callao, 157
Campana, Bahía, *see* Bell, Bahía

INDEX 2: STOKES

Candish, Thomas, *see* Cavendish, Thomas
cap topsails, 182&n, 207
Carlos, Don, 166, 186n
Carlos III, Isla (Louis le Grand, Isle of), 191n
Camargo, Alonso de, 157
Carteret, Philip, commanding officer, *Swallow*, 169n, 171, 176, 195&n, 198n, 216, 217n
Cascada, Caleta, 183n
Castle of Otranto, The, 194n
Castro, Don Beltran de, 161
Cavendish (Candish), Thomas, 159–61, 175, 176n, 179n, 181n, 186n, 196n, 248
    charts, 160n, 175n
Cevallos, Ciriaco, 174n
Charles II, King of England, 165
Charles V, Holy Roman Emperor (Charles I, King of Spain), 167
Cheap, David, commanding officer, *Wager*, 169, 195n
Cherbourg, 160
Chiloé, 149, 177, 195n
chronometers, 149–50, 174&n, 179, 187&n, 194, 225, 243-4
chronometric difference of longitude, 198, 213, 222, 224, 227, 243
chronometric journals, 151
Churruca, Cosme de, 174n
Clusius, Carolus, 185n
Coen, Jan Pietersz, VOC Governor General, 164n
Cook, James, 176
Cordes, Bahía (Bay de), 162–3, 171n, 175, 186&n; *see also* Gallant, Caleta
Cordes, Simon de, 162, 186n
Córdoba (Cordova), Antonio de, 172, 177, 182, 189n, 190&n, 198n, 214n, 221n, 245
    canoes 245–6
    charts, **173**, **174**, 176, 177, 189n, 195n
    chronometers, 174&n
Cortado, Cabo (Cape), **215**, 216, 227n, 228
Crozet, Isles (Islands), 196n
Cruz, Cerro de la, 190n
Cumming, James, commanding officer, *Tamar*, 169n
Cummins, John, carpenter, *Wager*, 169, 195n
Curran, J. B. H., commanding officer, *Tyne*, 152

Davis, John, master, *Desire*, 160, 196n
Davis's Southern Land, *see* Falkland Islands
del Cano, Juan Sebastian, 155–6
Deseado, Cabo, 156, 159, 160n, 164, 200n, 214n, 223n

Deseado, Puerto, 160
Desire, Cape, *see* Deseado, Cabo
Desire, Port, *see* Deseado, Puerto
Desseada, Cape, *see* Deseado, Cabo
Dickson, Sir Archibald, commanding officer, *Orion*, 151
Diego Ramírez, Islas, 164
Diseada, Cape, *see* Deseado, Cabo
Dover, Strait of, 198
Drake, Sir Francis, 157, 158, 160n, 176n, 181n, 185n
ducks, 185, 210, 218, 221n
    logger-headed, 218n, 228&n

Echeñique, Punta, 246
Egmont, Port, 171
El Morrión, Cerro (Saint David's Head), 194&n, **193**
Eleven Thousand Virgins, Cape of, *see* Vírgenes, Cabo
Elizabeth I, Queen of England, 158
Elizabeth Bay, *see* Isabel, Bahía
English Bay, *see* Wood, Bahía
Enriquez, Don Martin, Viceroy of New Spain, 157
Estrecho, Cape del *see* Froward, Cabo
Evangelistas, Islotes (Direction, Islands of), 179&n, 196, 222, 223n, 225, **226**, 227&n
Eyre, Sir George, Commander in Chief, 153

Falkland Islands, 160, 169, 171, 181&n, 196n, 230n
Famine, Port, *see* San Juan de la Posesión, Puerto
Farquhar, Arthur, commanding officer, *Desiree*, 151
Fendu, Cap, *see* Notch, Cabo
FitzRoy, Robert, subsequent commanding officer, *Beagle*, 150–51, 153, 227n, 242-3
Fleming, Humphrey, commanding officer, *Bachelour*, 165n
Flinn, Samuel S., master, *Beagle*, 154, 200, 203, 210–11, 212n, 231
Forte Escudo, Bahia de, *see* Fortescue, Bahía
Fortescue, Bahía, 171n, 175n, 186&n, 187n, 189&n
Fortescue, John, volunteer, *Sweepstakes*, 166, 186n
Forward, Capo, *see* Froward, Cabo
French Bay, *see* San Nicolás, Bahía
Fresh Water Bay, *see* Agua Fresca, Bahía
Frézier, Amédée, François, 169, 230n
    chart, **168**, 176
Froger, François, 167, 176

391

Froward, Cabo, 156n, 157, 159–62, 166, 172, 179, 181–3, **184**, 196, 200n, 216, 229, 244, 248
Fuegian natives, 147, 156, 210, 217–22, 231&n
  appearance and clothing, 217
  arms, 221
  barter, 222
  canoes, 220–21, 245–6
  children, 219–20
  dwellings, 220
  fire lighting, 156, 194, 210, 220, 246
  food, 218–19
  intercourse of the sexes, 219
  language, 221–2
  man, **208**
  religion, 222
  toilette, 218
  woman and child, **209**
Fuller, Thomas, master, *Desire*, 181n
  sailing directions, 159&n, 160n
Furia, Bahía, 230&n, 231
Fury Harbour, *see* Furia, Bahía

Galiano, Dionisio Alcalá, 174n
Gallant, Cabo (Cape), 186–7, 189–91, 231, 242–4
Gallant, Caleta (Puerto) (Port) (Great Bay) (San Josef, El Puerto de), 150, 159, 162, 164–5, 171n, 175, 186, **188**, 189–90, 230–31
  geese, 181, 185, 210, 218, 228–9
Gennes, Jean-Baptiste de, 166, 180n
  chart, **167**, 176
Gennes, Río de, 166, 167n
Geographical Position Book, extract, 150, 190n, 224n, 242–3
Good Hope, Cape of, 159, 162n, 163n, 196n
Grande, La Isla, 166
Great Bay, *see* Gallant, Caleta
Grenville, Sir Richard, 158, 196n
Guatulco, 157
Guesen (Beggarly) Bay, 162

Hakluyt, Richard, 159&n, 160n, 175n, 196n
Hambre, Puerto del, *see* San Juan de la Posesión, Puerto
Harrison, John, purser with Wallis, 171n
Havannah, Punta (Bukley, Punta de), 195n
Hawkesworth, John, 171&n, 180, 198n
  chart (showing the work of Byron, Wallis and Carteret), **170**, 171, 176, 229&n
  plans, 171n, 195&n, 229n
Hawkins, Sir Richard, 161, 185n, 196n, 245
  canoes, 221n, 245

Hawkins's Maiden Land, *see* Falkland Islands
Heemskercke, Jacob van, 163n
Hernandez, Tomé, survivor from Sarmiento's settlements, 159
Hewett, William, commanding officer, *Protector*, 152
Holland, Cabo (Capo), 162, 176, 182&n, 183, **184**, 185–6, 189
Hood, Sir Samuel, commanding in *Minden*, 152
Horn, Cape, *see* Hornos, Cabo de
Hornos, Cabo de (San Ildefonso, Cabo de), 147, 164, 169, 235–6
humming bird, 189
  Green-backed Firecrown, *Sephanoides sephaniodes*, 190n
Hurtado, Don Garcia de, Governor of Chile, 157

instruments, surveying, 185, 187, 191, 214, 222, 224–5, 227; *see also* artificial horizon; chronometers; sextant
Isabel, Bahía, 159
Isles of Direction, *see* Evangelistas, Islotes

Jacatra, 164
Jack, Puerto, *see* Bougainville, Bahía
Jane, John, 161n
Japan, 161–2
Jerom's Channel, *see* Jerónimo, Canal
Jerónimo, Canal, (Saint Jerome, Channel of) 159n, 160n, 191&n, 194
Jones, Thomas, instrument maker, 187&n, 242
Jones, W., volunteer, *Beagle*, 154, 190n, 210n, 212n
Joy, Cape, *see* Vírgenes, Cabo

kelp, *see* seaweed
King, Cabo (Cape), 213&n
King, Phillip Parker, commanding officer, *Adventure*, 147&n, 149–52, 177, 179–80, 189n, 190&n, 210, 213, 214, 225, 229–30, 243–4, 247
  chart, **178**
  orders, 149, 177, 179
  sailing directions, 151, 180n, 187n, 194n, 198n, 200n, 203n, 206n, 213n, 214n, 217n, 222n, 223n, 227n, 228n, 229n
Kirke, J., mate, *Beagle*, 154, 190n, 231n

Lacaille, Abbé Nicolas-Louis, 172n
La Coruña, 156
Ladrilleros, Juan de, 157

Langford, George, commanding officer, *Alpheus*, 152
l'Ecluse, Charles de, *see* Clusius, Carolus
Le Maire, Estrecho de (Strait of), 157, 164&n
Le Maire, Jacob, merchant, *Unitie*, 164&n
Loaysa, Garcia Jofre de, 156, 223n
Long-Reach, Long-Lane, 191n, **192**
Louis XIV, King of France, 166
Louis XV, King of France, 171
Louis, Port (Saint Louis, Port), 171, 191n, 230n
Louis le Grand, Isle of, *see* Carlos III, Isla
Low Countries, 161n
lunar distances, 171&n, 172&n, 174, 176
Lunie, R. F., volunteer, *Beagle*, 154, 190n

Macdouall, John, clerk, *Beagle*, 151, 154, 190n
    finding of papers on Cerro de la Cruz, 190n
    narrative, 151, 190n, 191n, 209n, 210n, 212n, 217n, 221n, 231n
Madre de Dios, Estrecho de la, *see* Magallanes, Estrecho de
Madrid, 171
Magalhaens, Straits of, *see* Magallanes, Estrecho de
Magallanes, Estrecho de, 147, 149–51, 155–237 *passim*, 242, 247
    charts, 150–51, 158, 160&n, 161, 163, 164, 165&n, 166–8, 169, 170, 171, 172–5, 176–7, 178, 179, 233
    current and tides, 182&n, 186, 195, 198–9, 200, 212, 227, 235
    flora and *fauna*, 181, 185, 189–90, 210, 213, 218, 228–9
    plans, 150, 167n, 171&n, 172, 175&n, 179, 182, 185–7, 188, 189&n, 192, 195, 198, 200, 201–2, 204–5, 206, 209, 213–14, 227, 229
    sailing directions, 151, 157, 159&n, 160n, 165&n, 171, 198, 206, 213–14
    tide pole erected, 212
    tide ripple, 191
    warning of approaching squall, 208
    winds and squalls, 182–3, 189, 207–8, 211, 228
Magellan, Ferdinand, 155–6, 159, 176n, 181n
Magellan, Strait of, *see* Magallanes, Estrecho de
Mahu, Jacob, 161–2
Malouines, The, *see* Falkland Islands
Marian Cove, *see* Marión, Caleta
Marión, Caleta, 200&n, **201**, **202**, 204n, 229n
Maskelyne, Nevil, 171n, 172n
Massacre, river du, *see* Batchelor, Río
Mauritius Bay, 162
Maurits, Prince, Stadtholder, 162n
Maxwell, Murray, commanding officer, *Daedalus*, 151
May, John Cornelitz, shipmaster with van Speilbergen, 176
Mendoza, Don Antonio de, Viceroy of Peru, 157
Mends, Sir Robert, commanding officer, *Iphigenia*, 152
Mercator, chart, **175**
Mercy, Harbour of, *see* Misericordia, Puerto
Merick, Andrew, master, *Delight*, 160, 196n
meteorological journal of HMS *Beagle*, 237–41
Misericordia, Puerto (Mercy, Harbour of), 150, 164, 210&n, 211n, **215**, 216&n, 217n, 219, 222&n, 227, 242–4, 247
Moluccas, 156–7, 161–2
Monday, Cabo (Cape), 203n
Montevideo, 172, 231n
Monville de Hague, 160
Mouat, Patrick, commanding officer, *Tamar*, 169n
Muscle Cove, *see* Agua Dulce, Caleta

Narborough, Grupo, 223n
Narbrough (Narborough), John, commanding officer, *Sweepstakes*, 165&n, 166, 176, 177n, 179n, 181n, 186n, 191&n, 194, 200&n, 203&n, 220, 223n, 227n, 228
    chart, **166**, 176, 181n, 200n, 203n
natives, *see* Fuegean natives
Needles, The (Isle of Wight), 214&n
Negro Río, 230
Negro, Cabo (Cape), 217
Nodal Gonzalo de, 164, 172, 180n, 225n
Nodal, Bartolomè Garcia de, 164, 172, 180n, 225n
Nodales, Pico (Peak), 180n
Nombre de Jesus, Ciudad del, 158, 196n
Noort, Olivier van, 162–3, 175–6
    chart, **163**
Notch, Cabo (Cape), (Fendu, Cap) (Tajado, Cabo), 150, 194&n, **197**, 198&n, 199, 200n, 201, 202, 229n
Nouvelle, Isles, *see* Falkland Islands

Observation Islets (Puerto Misericordia), 222n, 224n, 244
Observation Mount, 223n, 224&n, 242, 244
Olivier, Bay d', 162
Otranto, Castle of, 194
Otway, Sir Robert, Commander in Chief, Rio de Janeiro, 153

Pacific Ocean (South Sea), 157, 159–60, 162, 164, 167, 169, 171, 211n, 214, 225, 235

Pan de Azúcar, Islote (Sugar Loaf), 224n, 225&n, 227n
papers from previous expeditions found, 190&n
Paris, 158
Parker, Cabo (Cape), 213, 228, 242–3
parrots, 189
    Austral parakeet, *Enicognathus ferrugineus*, 189n
Parry, Sir William. E, Hydrographer 1823–29, 150, 187n
Pathagonico, Strait of, *see* Magallenes, Estrecho de
penguins, 228–9
Pepys's Island, 169
Peru, 162
Philip II, King of Spain, 176n, 179n, 157
Philippines, 161
Phillip, Cabo (Cape), 209n, 213&n, 242–4
Phillip, Charles, patent capstan, 185&n
Pigafetta, Antonio, 155
Pilar, Cabo (Pillar, Cape), 151, 157, 159n, 164, 171n, 172, 174, 179, 190n, 209n, 210, 214&n, **215**, 216, 217n, 222, 227–8, 244
    table of positions, 177n
Pillar, Cape, *see* Pilar, Cabo
Pillolco, Isla, 195n
Plasencia, Bishop of, 157
Plata, Río de la, 149, 165, 177, 196n, 236
Plate, River, *see* Plata, Río de la
Plymouth, 149, 158–61
Pope, Alexander, 211&n
Providencia, Cabo (Providence, Cape), 171n, 203&n, 206, 213

Quad, Cape, *see* Quod, Cabo
Quod, Cabo, 191n, **193**, 194–5, 200n

Rathery, T. H., commanding officer, *Snapper*, 152
Redond, Cap, *see* San Isidro, Cabo
rescue of crew of *Prince of Sax Cobourg*, 222n, 229–31
Rey Don Felipe, Cuidad del, 158–9, 179n, 196n
Ribeiro, Diogo, cosmographer, 155–6
    planispheres, 155, 156&n
Rio de Janeiro, 149, 152–3, 158, 164, 172, 195n, 236
Robertson, George, master, *Swallow*, 190n
Roldan, Campana de, 156, 247
Roldan's Bell, *see* Roldan Campana de
Rosa, Bahía, 182n
Rotterdam, 161–3

Saint David's Head, *see* El Morrión, Cerro

Saint Jerome, Channel of, *see* Jerónimo, Canal
Saint Louis, Port, *see* Louis, Port
Saint Ursula, 155n, 181n
Saint Valentine, 214n
Salano, Puerto, *see* Wood, Bahía
San Alfonso, Cape, *see* Victoria, Cabo
San Antonio, Cabo (Cape), 149, 177
San Ildefonso, Cabo de, *see* Hornos, Cabo de
San Isidro (Isidor), Cabo (Redond, Cap) (Shut up, Cape), 180&n, 248
San Josef, El Puerto de, *see* Gallant, Caleta
San Joseph, Bay, 164
San Juan de la Posesión, Puerto (Famine, Port) (Hambre, Puerto del), 149, 159–60, 166, 175n, 179&n, 181, 183n, 189, 190n, 216, 224n, 225, 229, 232, 243–4, 248
San Julián, Puerto (Port), 153, 165
San Lúcar de Barrameda, 155, 165
San Miguel, Puerto (Bahía), 175n, 186&n
San Nicolás, Bahía, 165–6, 175n, 180&n, 181
Santa Ana, Islas (Bulkeley's Island), 195n
Santa Ana, Punta (Saint Anna, Point), 158, 179n, 248
Santa Aguada, Morro, 183&n, 248
Santa Brigida y Santa Aqueda, Bahía, *see* San Nicolás, Bahía
Santa Maria, Isla (Isle), 161
São Tiago (Ilhas do Cabo Verde), 157
Sarmiento de Gambóa, Pedro, 157–60, 179n, 180n, 183n, 196n, 211&n, 230n
    table of names, 247–8
Sarmiento, Monte (Mount), 156n
Sayer, George, commanding officer, *Leda*, 152
Schouten, William Cornelison, master, *Unitie*, 164
Scilly Islands (English Channel), 223n; *see also* Westminster, Grupo
seal, 217–18, 220–23, 225, 227, 230, 246
seaweed, 195, 196&n, 198, 206n, 214
    Ecklonia buccinalis, 196n
    Durvillea, 196n
    Lessonia, 196n
    Macrocystis pyrifera, 196n, 218n
Sebaldines, The, *see* Falkland Islands
Separation Harbour, *see* Misericordia, Puerto
sextant, 187, 225, 242
shells used by natives
    as cups, 220; *Concholepas concholepas*, 220n
    as necklaces, 218; *Diloma nigerrima*, 218n; *Margarella violacea*, 218n
ships
    Adelaide, 229n

Adventure, 147, 149–53, 177, 179, 181n, 190, 227n
    medals, 210
Aid, 152
Alecto, 228n
Alpheus, 152
Ariadne, 151
Æolus, 163
Ætna, 152
Bachelour, 165
Beagle, 147, 149–54, 177, 179–237 *passim*
    grounding, 206n, 209&n
    medals, 210
Beaufoy, 230n, 232n
Capitana, 211n
Content, 159
Daedalus, 151–2
Daintie, 161
Delight, 160
Desire, 160, 181n
Desiree, 151
Dolphin, 169, 171, 203n
Elizabeth, 185n
Galleon Leicester, 160
Groote Mane, 163
Groote Sonne, 163
Hendrick Fredrick, 162
Hugh Gallant, 159, 186n
Iphigenia, 152, 216n
Jager, 163
Jane, 230n, 232n
Leda, 152
Lightening, 228n
Maria, 158
Maurepas, 167
Mauritius, 162, 176
Meeuwe, 163
Minden, 152
Morgenstern, 163
Nuestra Señora de Atocha, 164
Nuestra Señora del Buen Soceso, 164
Orion, 151
Owen Glendower, 152
Phelippeaux, 167
Prince of Saxe-Coburg, 222n, 230–31
Protector, 152
Rattler, 228n
Roebuck, 160
Saint Michael, 165 n
Santa Barbara, 169
Santa Casilda, 172, 174n, 190n
Santa Eulalia, 172, 174n, 190n
Santa Maria de la Cabeza, 172, 174n, 176, 182, 190n

Snapper, 152, 216n
Speedwell, 195n
Swallow, 171, 190n, 198n
Sweepstakes, 165
Tamar, 169, 203n
Trinidad, 158
Tyne, 152
Victoria, 155
Wager, 169, 195n
Sholl('s), Bahía (Bay), 213n, 242, 244
Sholl, Robert H., Lieutenant, *Beagle*, 154, 183, 230n
    biography, 153,
Shut up, Cape, *see* San Isidro, Cabo
Silva, Nuño da, pilot with Sir Francis Drake, 157, 181n
Skyring, William George, Senior Lieutenant and Assistant Surveyor, *Beagle*, 152, 154, 185, 187, 191, 200, 204, 206n, 209–10, 216n, 222n, 224, 231
    biography, 152–3
Sloane, Sir Hans, 218n
Solano, Juan, Lieutenant Governor of Costa Rica, 157n
Sorlinges, 223n
South Desolation, 191
South Sea *see* Pacific Ocean
Speilbergen, Joris van, 163–4, 176, 223n, 230n
    chart, **164**, 176, 223n
steamer, *see* ducks, logger-headed
Stokes, John Lort, Midshipman, *Beagle*, 154, 190n
Stokes, Pringle, commanding officer, *Beagle*, 147–54, 177, 179–232 *passim*, 237, 241–2
    biography, 151–2
    chart, 150, 233, 227
    journal, 150–51, 169, 179–232
    plans, 150, 179, 185–7, 188, 189, 192, 198, 200&n, 201–2, 204–5, 206, 209, 213–14, 227, 229
Sugar Loaf, *see* Pan de Azúcar, Islote
Swallow, Bahía (Harbour), 171n, 175n, 195, 198&n,

Tajado, Cabo, *see* Notch, Cabo
Tam O'Shanter, 209n
Tamar, Cabo (Cape), 171n, 204, 206&n, 209, 213, 242–4
Tamar, Isla (Isle), 150, **205**
Tamar, Península, 203n
Tamar, Puerto, (Harbour, Bay), 150, 203, 203, **204**, **205**, 206n, 209, 211, 212
Tarn, John, surgeon, *Adventure*, 152

Texel, 163
theodolite, 187, 191, 214, 243
Tidore, 156
Tierra de los Fuegos, *see* Tierra del Fuego
Tierra del Fuego, 149, 150n, 152, 156–7, 201n, 210n, 213, 217, 230–31
Tofiño, Don Vicente, 174
Toledo, Don Francisco de, Viceroy of Peru, 158
Transylvanus, Maximillianus, 156
Tuesday, Bahía (Bay), 227, 228&n, 242
Turner, Dr Peter, 185n

Upright, Bahía (Bay), 229n
Upright, Cabo (Cape), 171n, 194–5, **197**, 201, 203, 229, 243, 248
Urdaneta, Friar, 158
Uriarte, Martin de, pilot on Loaysa expedition, sailing directions, 157, 223n

Valdes, Don Diego Flores de, 158
Valdivia, 157, 161, 165, 186n
Valentina, Bahía (Valentine's Bay, Harbour), **202**, 214&n
Valentina, Cabo (Cape Valentine), 150
Valparaíso, 162, 195n, 236
Vargas y Ponce, Josef de, 175
Verenigde Oostindische Compagnie (VOC), 162n, 163n, 164n
Véron, Pierre-Antoine, astronomer with Bougainville, 172
Victoria, Cabo, 174, 190n, 200n, 217, 223n, 224–5, 227n, 242–4
Victoria, Strait of, *see* Magallanes, Estrecho de
Victory, Cape, *see* Victoria, Cabo

Villalobos, Juan de, 158
Virgen María, Cabo, *see* Vírgenes, Cabo
Vírgenes, Cabo (Joy, Cape), 149, 156–7, 159, 160n, 162, 164–6, 171n, 172n, 181n, 190, 206n
  table of positions, 177n
Virgin Mary, Cape, *see* Vírgenes, Cabo

Wallis, Samuel, commanding officer, *Dolphin*, 171, 176, 190n, 195&n, 198n, 216
Walpole, Horace, 194
Walter, Richard, 169
Weddell, James, 230n, 232n
Weert, Sebald de, 162, 176
weighing anchor, 199n
Westminster, Grupo (Scilly Islands), 223&n, 224
Westminster Hall, Isla, 179&n, 222, 223&n, 228n, 244
whale, 218, 225, 245
Wight, Isle of, 214&n
William, Duke of Clarence, Lord High Admiral, later William IV, king of England, 150&n
Winter John, commanding officer, *Elizabeth*, 181n, 185n
Winter's bark tree, 185&n
Wood, Bahía (English Bay), 161&n, 171n, 186n
Wood, John, mate, *Sweepstakes*, 165n, 186n, 214n
  sailing directions, 165n

Xaultegua Golfo (Buckley's Sound), 195n, 248

York, Punta (Point), 171n, 191&n, 242, 244

INDEX 3: KAY

Note: The unfamiliar flora and fauna of this region was not well understood at this time. Kay simply recorded the names he was given, which are indexed here under three heads: 'birds', 'mammals' and 'plants'.

Abrolhos Shoals, 287
Adventure, HMS, 265, 279, 301, 322–4
Albatross, HMS, 284
Annawan, US brig, 263
Antarctic Peninsula, 262–3
Ascension Island, 275, 280, 287
Astrea, HMS, 298–9
Austin, Horatio Thomas (d. 1865), 265, 269, 271, 279–80, 292, 305
Azores, 265

Baily, Francis (1774–1844), 262, 272, 274–5
Beechey, Frederick William (1796–1856), 290
birds, 258, 264, 285, 292, 298
    albatross (*Diomedea*), 265, 311
    booby, 286
    eglet [*sic*], eaglet, 311, 313
    goose, 'brent' (*Chloephaga rubidiceps*), 298
    goose, kelp, 298, 322
    grey (nelly), 311, 313, 327, 329
    loggerhead duck (steamer duck, *Tachyeres cinereus, T. cinereus*) 298
    Mother Carey's chicken, 294
    partridge, 292
    penguin, 264–5, 268, 295–7, 304–6, 310–11, 313, 315–16, 364
    peroquites [*sic*], 292
    petrel *spp*, 294n, 311n
    silvery gull (*Larus argentea*), 311, 313
    teal, 292
    tern, 313
Blonde, HMS, 281–2
Blossom, HMS, 290
Boston, US corvette, 291
Boteler (Captain), 284
Brown, William (1776–1857), 294
Browne, Henry, 274
Buenos Aires, 293–4

Cadmus, HMS, 288

Cape Foster, James Ross Island, 267
Cape Frio, 287–8
Cape Horn, 259, 261–3, 265, 267–77, 276, 279–80, 287, 292, 294–7, 301–3, 306, 312, 317–20, 322, 324
    Wigwam Cove (St Martin's Cove), 301–2, 319, 323
Cape of Good Hope, 274, 279
    Cape Observatory, 265, 274
Cape Possession, *see* Hoseason Island
Cape Spencer, *see* Hermite Islands
Cape [de] Verde Islands, S. Antonio, S. Paul, 263, 270, 284–5
Carrick Roads, 282
Carrigan, William Perceval, 279–80
Caught, John, 279
Chagres, 265
    Chagres River, 265, 281, 283–4
Chanticleer, HMS ('Chantie'), 259, 261–3, 265, 267–77, 279–80, 287, 292, 294–7, 307, 312, 317–20, 322, 324
Christie, Samuel Hunter (1784–1865), 277
clams, 305
chronometers, clocks, *see* scientific apparatus
Clavering, Douglas (d. 1827), 261, 272
Collett, Charles Frederick, 280, 316
Collinson, Richard (1811–83), 270–71, 277, 280, 315
Conolan, Peter, 270, 279
Conway, HMS, 261, 272
Corvocado, *see* Rio de Janeiro
Crozier, Francis Rawdon Moira (1796–1848), 268
Cural, *see* Madeira Island

Daniell, John Frederick (1790–1845), 284
Deadman's Island, *see* Staten Island
Deception Island, *see* South Shetland Islands
Donkin's preserved meat, 265, 281
Dundas, James Whitley Deans (1785–1862), 291

East Greenland, 261
Eights, James, 263
Erebus, HMS, 268

Fallows, Fearon (1789–1831), 265, 274
Falmouth, Cornwall, 282
Fernando (de) Noronha Island, 265, 275, 286
Flores island, *see* Monte Video
Foster, Henry (1796–1831) ('the Captain'), 259, 261–3, 265–70, 272, 274–7, 279–80, 290, 295, 297, 300–301, 307, 312
Franklin, Sir John (1786–1847), 263, 267–71, 290, 315, 319
Fraser's stove, 281, 303
Funchal, *see* Madeira Island

Ganges, HMS, 290
Grenville [*sic*], Grenfell, John Pascoe (1800–1869), 288
Griper, HMS, 261, 270, 272

Hall, Basil (1788–1844), 261–2, 272
Hall, John, 281n
Hecla, HMS, 270, 272, 279–80, 284
Hermite Islands, 319, 323
  Cape Spencer, 323
  Maxwells Harbour, 323
  St Francis Bay, 302, 319, 323
Hobart (Tasmania), 368–9
  Rossbank Magnetic Observatory, 368
Hodgkin, George, 279–80
Hodgskin, James Archibald, 277, 280
Horsburgh, James (1762–1836), 262
Hoseason Island, 263, 305
  Cape Possession, 263, 305

Inglis, William James, 290n

Jeffery (Mr), 280

Kater, Henry (1777–1835), 272–4, 276
Kay, Joseph Henry (1815–1875), 261, 267–70, 277–80, 301
Kendall, Edward Nicholas (1800–1845), 263–5, 267, 270, 273, 279–80, 299, 301, 305, 310–17, 320
King, Philip Parker (1793–1856), 265, 279, 301, 322–3

Le Maire Strait, 297, 300
limpets, 298
Livingston Island, *see* South Shetland Islands
Lobos Island, *see* Monte Video

Madeira Island, 263, 283–4
  Funchal, Cural [*sic*] Caral, 283
  Porto Santo, 283
mammals
  otter, 298
  porpoise, 283
  sea leopard, 270, 311–13, 364
  seal, 291, 301, 306, 317, 321
  'tiger', 292
  whale, 304
Maranham (San Luis de Maranhão), 275
Maria Isabela, Brazilian corvette, 288
Maxwells Harbour, *see* Hermite Islands
Meredith, Joseph, 279–80
Miers, Henry, 280
Monte Video (Montevideo), 263, 265, 273, 291–4, 306
  Flores Island, 291
  Lobos Island, 291
  Point Braba, 293
  Rat Island, 292
  River Plate, 290n, 291–2
Mount Foster, *see* South Shetland Islands

Nelson, Horatio (1758–1805), 269, 284
Neptune, 285
Neptune's Bellows, 263, 307

Onyx, HMS, 283
Otway, Sir Robert (1770–1846), 290

Pacific, US whaler, 302
Palmer, Alexander S., 263, 295, 297
Palmer, Nathaniel B., 263, 295n
pampero wind, 292
Panama, 265
Parry, Sir William Edward (1790–1855), 261–2, 269, 272, 276, 279, 319
patent illuminators (bullseyes), 281–2
Pendleton, Benjamin, 263
Pendulum Cove, *see* South Shetland Islands
Penguin, US schooner, 263, 297
Phipps, Constantine John (later Lord Mulgrave) (1744–1792), 272
plants
  Arbutus aculeata, 296, 298, 321, 323
  Asclepius venosa, 285
  Fagus antarctica, 298–9, 323
  land cress, 298
  wild celery, 321, 296, 298, 300
  Wintera aromatica, 298
Point Braba, *see* Monte Video
Pond, John (1767–1836), 316n

Port Bowen, 262
Port Cook, *see* Monte Video
Port Foster, *see* South Shetland Islands
Port Vancouver, *see* Monte Video
Porto Santo, *see* Madeira Island
Portsmouth, 270, 281
Preston, Lancashire, 265
Prothee, RN hulk, 281

Rainbow, brig, 268
Rat Island, *see* Monte Video
Rio de Janeiro, 263, 268, 288–9, 293
    Corcovado, 289
    Sugar Loaf, 285, 289
    Villegagnon ('Ville Gagnon') Island, 289
River Plate, *see* Monte Video
Robinson, Frederick, 280
Ross, James Clark (1800–1862), 261, 268, 277
Rossbank Magnetic Observatory, *see* Hobart
Royal Astronomical Society, 267
Royal Society of London, 261–3, 268, 272, 274, 277
    Copley medal, 262, 266, 272
Royal Society of Tasmania, 268

Sabine, Sir Edward (1788–1883), 272, 277, 284n
St Catherine's Island, 306–7
St Francis Bay, *see* Hermite Islands
St George's Mount, *see* South Shetland Islands
St Helena Island, 281, 291, 303
Samwick (Mr), 279
Sandercombe, James B., 279–80
Sapphire, HMS, 291, 293
scientific apparatus, 268, 271–7, 280, 292, 299–30, 305, 308–11, 315–16, 321–4
    chronometers, clocks, 271–2, 274, 276, 284, 315
Seraph, US brig, 263
shark, 284, 286
Smiley, William H., 264
Smith, William (1790–1847), 263, 306
Smith Island, *see* South Shetland Islands
South Shetland Islands, 262–4, 275, 295n, 301, 306–10, 312, 314
    Deception Island, 261, 263–5, 267–8, 280, 305–6, 308, 312, 314, 316–17

Livingston Island, 313
Mount Foster, 267
Pendulum Cove, 263, 307, 310
Port Foster, 263, 267, 307–9
Smith Island, 267, 304, 318
St George's Mount, 316
Spithead, 297
Spitsbergen, 277
Staten Island, 263–4, 278–80, 286, 294–5, 297–9, 303–9, 311, 313–15, 322, 325, 338
    Cape St Bartholomew, 297, 314
    Cape St John, 295, 297, 311, 314, 317
    Deadmans Island, 295, 297, 301, 311–12, 314
    Hatchetts harbour, 311, 313–14
    New Years Harbour 295, 297
    Port Cook, 298, 313, 315, 317
    Port Vancouver, 313, 317
Stopford, Sir Robert (1768–1847), 281
Sugar Loaf, *see* Rio de Janeiro

Tasmania (Van Diemen's Land), 269–9
Teneriffe [*sic*] (Canary Islands), 384
Terror, HMS, 268
Tierra del Fuego, 300, 320n, 322
    Fuegian Indians, 319–21, 323
Thompson, Sir T. B. [*sic*], Thompson, Sir Thomas Raikes Trigge (1804–65), 288
Trinidad, 265
Troubridge, Edward Thomas (c.1787–1852), 284

Valparaiso, Chile, 306, 324
Villegagnon ('Ville Gagnon') Island, *see* Rio de Janeiro
Vitch [*sic*] Veitch, H., 283

Webster, William Henry Bayley (1793–1875), 261, 263–5, 267–70, 279, 296n, 306n, 320n, 323
Williams, British merchant brig, 263
Wigwam Cove, *see* Cape Horn
Williams, George, 279–80 (C. Williams, 277, possibly same man)
Woodplumpton, Lancashire, 265

Young, Thomas (1773–1829), 272

INDEX 4: WAINWRIGHT

Africa, East, 348–52
  African traders in, 349–50
  coast and trade, 349
  political authority in, 351
  trade routes in, 355
Albert, lake, 353
Alexandria, 341
alligators, *see* crocodiles
Anglican Church, divisions in, 343n
antelopes, 368, 377
Arab traders, methods, 371
  settlements, 371
Arabs in East Africa, 350, 353

Bagomoyo, 337, 338, 352, 381n
bajree, *see* millet
Baker, Samuel, 353
bananas, 372, 380
Bangweolo, lake, *see* Bangweulu
Bangweulu, (Bemba) lake, 338, 349, 353, 359, 361n, 362, 363n, 367n, 367–8, 369n, 370n, 372n
Bantu peoples, 349, 365n
  languages, 361
batata, *see* sweet potatoes
beans, 368, 380
Bemba, 'cluster', 361n, 363n, 365n
  matriliny, 365n
  people, 334
  rivalry
    with Bisa, 367n
    with Lungu, 372n
    with Tabwa, 369n, 370n
Bemba, lake, *see* Bangweulu
Bennett, Norman R., 334
Berlin, 359
Bisa, people and area, 338, 350, 361n, 362, 363n, 367n, 368n, 370n, 372n
  funerary practices, 373
  Wainwright's description and verdict, 372–3
Black magic, 364&n
boars, attacks by, 378, 380
Bombay (Mumbai), 338, 346
Bontinck, François, 333–4, 339, 357, 365n, 366n
  and map, 359
Brahim (Burahim), village, 381&n
Brenner, Richard, 345, 346n, 356, 361, 368n, 379
  transcription of text, 339–40, 356–7
Broyon, Philippe, 379&n
buffaloes, 377
Buganda, kingdom, 348
bulrush millet, 372&n
Burton, Richard F., 353, 377n

Cameron, Verney Lovett, 338&n, 339, 353, 371n, 374–5
  receives Wainwright's letter, 374&n
  relations with Wainwright's party, 375&n
canoes, 363, 367&n
caravans, trading, 350, 351, 373, 376n, 377n, 378, 379n, 380
cassava, 365&n, 367n
Catholic fathers, *see* Holy Ghost Fathers
cattle, 365&n, 367&n, 377
Cazembe's Empire, 349, 350, 352, 372n
Chama, chief, 368&n, 370&n
Chambeshi (Chambezi), river, 361&n, 367&n, 369n
chiefs and chiefship, 351&n, 356, 372n, 376n
Chilapanewa, chief, 369
Chilembo (Chilenba), chief, 371&n
Chimeata, chief, 372
Chirima, river, 369
Chisinga, people, 369n, 370n
Chitambo, *see* Kitumbo
Chitundwa, village, 365
Chiwaye, chief, 368, 369&n
Chiwiye, chief, 378&n
Chowpereh (Chopere; Ghopere), porter, 363&n
Christie, James, 339
Chuma, James, 334, 335&n, 337–8, 340–43, 346–7, 352, 355–6, 361n, 366n, 368n, 371n, 373n, 374n, 375n, 379n, 381n
  as a leader, 352
  as an emissary, 379n
Chungu (Choongu), chief, 369&n, 372&n
Chunyu, village, 379

Church Missionary Society (CMS), 337, 341, 342, 343, 346–8, 353
  and Buganda mission, 348
circumcision, 380&n
climate and weather, 378, 379, 380
cloves, 350
cobra, 370&n, *see also* snakes
Comoro Islands, 335n
conflict at Kitondwa, 365–6
Congorido, settlement, 381&n
Consulate, British at Zanzibar, 339, 379
Cooley, William D., 350n
crocodiles, 369, 370n, 371
cucumbers, 372

Daily Telegraph, 344
Dawson, Llewellyn S., 346, 374n
Decken, Claus von der, 340
Diary of Jacob Wainwright, 333, 335, 337–43, 346, 348, 356, 357n
  Bontinck and, 333–4, 339&n
  composition, 337–9, 348
  English version, 356–8
  French translation, 339, 357–8
  German translation, 335, 339, 341, 355, 356–7&n, 358n
  history of text, 339–343
  punctuation, 358
  Waller and, 342–3
Dillon, W. E. and death, 375

East Africa, *see* Africa, East
Egypt, 354n
elephants, 364, 370, 372
Ethiopia, 352
European position in East Africa, 352–5
expedition, losses, 368
explorers, European, 351, 353, 355
  scientific, 352, 354–5

famine conditions, 375–6, 379&n
figs, 380
Fipa, people and area, 338, 358, 359, 373&n
firearms, 351
  numbers among Nyamwezi, 376&n
fish, 367&n, 370
Fort Jesus, Mombasa, 350
France, 349, 354n
Fraser, Mrs A. Z., 343–4, 346
Frere, Sir E. Bartle, 340–41
Frere Town, 348
Fundikira, ruler of Unyanyembe, 350, 351
Funi, settlement, *see* Congorido

Gamitto, Antonio, 352&n
Germany, 354n
Ghopere, *see* Chowpereh
giraffes, 377
Giraud, Victor, 367n
Gogo, area and people, 339, 377&n, 379n
  and cattle, 377
  and circumcision, 380
  and 'hongo', 377&n
  report on by Wainwright, 377–9
Grant, James A., 353
Great Britain, government, 354–5

hailstorm, 378
Hakluyt Society, 339n, 350n
Hamed bin Muhammed, *see* Tippu Tib
Helly, Dorothy, 339
Herodotus, 353
hippopotamus, 370&n, 371
Holy Ghost Fathers (Congregation of the Holy Spirit) mission, 337&n, 338, 381n
'hongo', 351, 377&n, 379
  and labour services, 379&n
Hutchinson, Edward, 340–41

Ibn Battuta, 349
Ilala, 335, 346, 361n
India, administration, 354–5
inscription on tree at Ilala, 346, 363n, 366
Insokoro, chief, 365
inyombo, *see* myombo
Iseke (Seke), town, 378&n
Itawa, *see* Tabwa
ivory trade, 349–50, 369n, 370n, 371n, 372n, 377n, 379

Johnston, Harry, 359
jowari, *see* millet

Kaguru, people and area, 379&n
Kalangwishi (Kalangweese), river, 370
Kalemba, river, 372
Kalongandyofu, village, 361
Kanamamansata, chief, 369&n
Kanyenye, settlement, 378
Kapesha (Kapesa), *see* Katele
Kapunda, river, 366
Kasaramarama, chief, 363
Kasegera (Kasekera), settlement, 375
Katele, chief, 369&n
Katenkera, chief, 361
Kawendi, people, 363n, 366n, 367&n
Kaze, *see* Tabora

Kenya, mountain, 353
Kessingland, Norfolk, 347
Khoko (Koko), village, 378&n
Kilimanjaro, mountain, 353
Kilwa, city, 345–6, 350
Kimbu, people, characterized by Wainwright, 375–6&n
Kinmock, settlement, *see* Mfuteh
Kipira (Kipiri), village, 376&n
Kisalusala (Kisaroosaroo), village, 376&n
Kisuludini, CMS station, 348
Kitangeh (Kitangay), 380
Kitondwa, village, 366
Kitumbo, chief, 363&n, 369
Kiwenday, chief, 366
Konongo, people and area, 373&n
Kopa, chief, 361
Krapf, Johann L., 353
Kumbakumba, *see* Muhammad bin Masud 350, 371&n, 372&n
Kusumwa, chief, 367

Lacerda, Dr Francisco, 350, 352, 367n
Laing, Arthur, 340
 opinion of Wainwright, 347, 356
Lala, people and area, 361–3&n
Last Journals, 335, 337n, 339, 341n, 342, 357, 359
Leo Africanus, 339n
leopards, 364n, 372&n,
Leopold II, 354n
lions, 364
Litandasi, river, 367
Livingstone, David
 as an explorer, 352
 burial of internal organs, 346, 363n
 death, 335, 337, 341–2, 354n, 355, 363&n
 inscription at place of death, 346, 363&n
 funeral at Westminster, 335, 343, 344, 347
 embalming of, 335, 363
 and Makololo, 352
 meeting with Stanley, 346, 353, 374n,
 Narrative of an Expedition to the Zambesi, 341n, 343
 and Nile sources, 353, 355
 reception of body at Zanzibar, 379n, 381&n
 search expedition, 346
 Zambesi book, *see* Livingstone, David, *Narrative of an Expedition*
 see also Last Journals
Livingstone, family of, 341
Livingstone, Oswell, 373–4&n
 Wainwright's letter to, 373&n

Livingstone, Tom, 341&n, 342
London Missionary Society
 and Urambo mission, 348, 377n
Loofoomi, *see* Lufubu
Lookulu, river, 361
Lowestoft, 347
Lualaba, river (Congo), 350, 353
Luapula, river, 351, 363, 370n
Lufubu, river, 338, 370&n, 372&n
Lulimala (Molilamo), river, 361&n, 363n
Lunda, people and area, 349
Lungu, people and area, 338, 358, 372&n, 373
Lupupuse (Lopoposi), river, 369&n

Magomba, chief, 378&n
maize, 367n, 372, 376, 380
Majwara, 337&n
Makololo, people, 352
Makumba, chief, 365
Malawi, lake, 346, 350, 352, 361n
Malwa, SS, 340, 341, 344
Mananywa, chieftainess, 375&n
Manawamzungu, chief, 363
Manika, river, 369
maps, *see* route
Marenga Mkali, desert area, 379&n
Marseilles, 340, 379n
Marungu, *see* Lungu
Masanja, chief, 376
matriliny, 351, 365&n
Mawala, chief, 378&n
Mazawamba, village, 361&n
Mazitu, *see* Ngoni
Mbowra (Mbowea), chief and settlement, 373&n
Mbura tree, 363&n, Figs 7 and 8
Mchisweesi, chief, 364&n
Mdaburu (Mdaboo), settlement and river, 378&n
Mfuteh, settlement, 381&n
Mgosa, chief, 370
millet, 368, 372&n, 376n
Mirambo, Nyamwezi 'warlord', 349, 351, 379n
 'empire', 377n
 war with Unyanyembe and Arabs, 376–7&n
Mitikira, river, 361&n
Mkasiwa, chief of Unyanyembe, 376
Mkata, river, *see* Wame
Mkotwe, river, 372
Mnwa Sele (Manua Sera), chief, 351
Mnyasere (Manua Sera), porter, 363&n
Moffat, Robert, the younger, 337n
Moffic, *see* Murphy
Moguba, chief, 381

Molilamo, *see* Lulimala
Mombasa, 347–8, 350
Mombasi, river, 371
Monteiro, José, 352
Moore, Joseph, 335n, 356–8
Moravian Mission, 348
Mozambique, 346, 350
Mperembe, Ngoni leader, 349
Mpwapwa, town, 379&n
Msanga, chief, 365
Msiri, warlord, 351, 370n
Mtosi, river, 371&n
Muhalule (Mahaloole), chief, 376&n
Muhammad bin Masud (Kumbakumba), Swahili trader, 350, 371&n, 372&n
Mulalawampa, chief, 369
mules, 379
Mumbai, *see* Bombay
Munikazi (Manikazi), river, 361&n
Murphy, Cecil, ('Moffic') 337n, 374n, 375&n, 379&n, 381n
Murray, John, 341–2
music, 348, 376&n
Muslims in expedition, 337&n
Mutesa, *kabaka* of Buganda, 348
Mutondo (Mtondo), area, 369&n
Mwampanda (Mamiyanda) river, 368&n
Mwege, chief, 368&n
Mweru, lake, 351, 352, 353, 370n
myombo tree and its bark, 377&n

Nasik School, 337–8, 340–41, 344–5, 348, 355–6, 381n
New York Herald, 354n, 361n
Newstead Abbey, 341&n, 342–3&n, 348
Ngami, lake, 352
Ngoni, peoples and migrations, 349, 351, 367n, 372n, 373n, 375n, 377n
Ngorido, 'chief', *see* Congorido
Ngulu (Ngooru), chiefdom, 375&n, 376n
Nile, river sources, 353, 355
Nsama, chief, 369n, 370n, 371n
Nyamwezi, area and people, 339, 348, 349, 350–51, 367n, 372n, 373, 375n, 380&n
  language, 370n
  report on by Wainwright, 376–7
Nyasa, lake, *see* Malawi
nyika, 358
Nyinawambusi, river, 368
Nyungu ya Mawe, Kimbu warlord, 376&n

O'Flaherty, Philip, 348
Oman, 346n, 350

Onato, river, 368

patriliny, 351
Pemba Moto, trader, 371&n
Petermann, August, 340, 341, 356–8, 361
  and maps, 358n
  see also Diary
pombeiros, 352
porters, 'caravan', 355
Portuguese in East Africa, 350, 352
Price, W. S., 338, 340, 343, 344–5, 346–8, 356n
Prideaux, W. F., 340, 379&n, 381n
pumpkins, 372

Rebmann, Johann, 353
Reimer, Dietrich, 359
rice growing, 371, 373, 376
Roberts, Andrew, 335
route of the expedition, 338, 341, 358–9, 363&n, 366n, 369, 381n
  through Lungu and Fipa areas, 373n
  and maps, 358–9&n
Royal Geographical Society (RGS), 333, 340, 346, 353–4
Royal Navy, 346&n
ruga-ruga, 349, 377n
Ruvuga *see* Lufubu

Sagara, area and people, 338, 339, 379n, 380n
  funerary customs, 380&n
Sama, *see* Nsama
Sayed ben Ali, trader, 371&n
sea-horse, 370&n
Seke, *see* Iseke
Seyyid Said, ruler of Muscat and Zanzibar, 350
Sharanpur School, *see* Nasik
Shire, river, 335n, 352–3
Simpson, Donald, 334
slave trade, 346, 348, 350, 354–5
  and treaty with Zanzibar, 355
sleeping sickness, 350
snakes, encounters with, 370, 375
Speke, John H., 353
Stanley, Henry Morton, 341, 346, 353
steamboats, 352, 354
Suez, 340, 356, 361n
Susi, David Abdallah, 334, 335&n, 337, 340–43, 346–7, 352, 355–6, 361n, 363n, 373n 375n
  as a leader, 352, 363&n
  skills, 355
Swahili, people, 350, 370n
  as settlers in the interior, 376
  as traders, 350, 353

language, 350, 355, 372n
sweet potatoes, 368&n, 372, 376

Tabora, Arab station, 350, 375n, 377n
Tabwa, people, 369n, 370n
Tanganyika, lake, 338, 350, 353, 359, 361n, 367&n, 370n, 372, 373n, 375n, 377n
Tanzania, 349, 359, 378n
Thomson, Joseph, 348
Ticali, river, 368
tigers, *see* leopards
tilapia, 367n
Times, The, 347
Tippu Tib (Hamed bin Muhammed), 350, 369n, 370n, 371n, 376n
Tipulipu, *see* Tippu Tib
towns, 358
travel literature, 354
tribute, *see* 'hongo'
Tubugwe (Toohugwe), village, 379&n

Ugogo, *see* Gogo
Ugunda, settlement and chiefdom, 375&n, 376
Ujiji, town, 353, 359, 375, 379n
Ukhonongo, *see* Konongo
Universities' Mission to Central Africa, (UMCA), 335n, 342, 343, 348n
'unofficial empire', 355&n
Unyamwezi, *see* Nyamwezi
Unyanyembe, settlement and chiefdom, 337, 338, 346, 350–51, 368, 371, 375, 379n
Usagara, *see* Sagara
Ushi, people, 363n, 366n
Uzigua, *see* Zigua

Victoria, lake, 349, 353, 367n
Victoria, Queen, 347&n, 348
villages, 358

Wabisa, *see* Bisa
Wainwright, Jacob, 334, 335, 337
achievements, 348, 353, 355, 356
age, 344–5
and Tom Livingstone, 341
as an *evolué*, 346&n
attitudes and outlook, 348
birth and origins, 333, 344–5, 350
character and career, 343–8
death of, 348
early career, 345–6
knowledge of English, 355–6
later career, 347–8
methods, 356
name, 346n
pall bearer at Westminster Abbey, 335, 347
Wainwright, John, 371&n
Waller, Horace, 335, 337, 341–3
and Livingstone, 341–3
and Wainwright's Diary, 342–3
attitude to Wainwright, 341–2
Wame, river, 380–81
weapons among Nyamwezi, 376&n
Webb, W. F., and family, 343–4
Wellington, Mathew 337&n
wilderness, *see* nyika

Yao, people, 346, 350, 351, 376n
language, 355
Yeke, people, 351, 367n, 370&n
Young, James, 335, 341

Zambesi, expedition, 342, 352–3
river, 350, 352
Zanzibar, 337n, 338–40, 342, 346, 348, 352, 353, 356, 363, 371&n, 375, 378, 379, 380, 381
and 1873 treaty, 355
as an Omani polity and trading empire, 350, 355
Zawamba, chief, 361
zebra, 367, 377
Zigua, area and people, 380&n
Zwangendaba, Ngoni leader, 349